Multi-Phase Flow and Heat Transfer III

PART B — APPLICATIONS

Process Technology Proceedings, 1

Multi-Phase Flow and Heat Transfer III

Part B: Applications

Proceedings of the Third Multi-Phase Flow and Heat Transfer Symposium —
Workshop, Miami Beach, Florida, U.S.A., April 18–20, 1983

Edited by

T. Nejat Veziroğlu

University of Miami, Coral Gables, Florida, U.S.A.

and

Arthur E. Bergles

Iowa State University, Ames, Iowa, U.S.A.

ELSEVIER — Amsterdam — Oxford — New York — Tokyo 1984

ELSEVIER SCIENCE PUBLISHERS B.V.
Molenwerf 1
P.O. Box 211, 1000 AE Amsterdam, The Netherlands

Distributors for the United States and Canada:

ELSEVIER SCIENCE PUBLISHING COMPANY INC.
52, Vanderbilt Avenue
New York, NY 10017

Library of Congress Cataloging in Publication Data

Multi-phase Flow and Heat Transfer Symposium Workshop
    (3rd : 1983 : Miami Beach, Fla.)
    Multi-phase flow and heat transfer, III.

    (Process technology proceedings ; 1)
    Contents: v. 1. Fundamentals -- v. 2. Applications.
    Includes bibliographies and indexes.
    1. Multiphase flow--Congresses. 2. Heat--
Transmission--Congresses. I. Veziroğlu, T. Nejat,
1924-   . II. Bergles, A. E., 1935-
III. Title. IV. Series.
TA357.M83 1983   620.1'064   84-10237
ISBN 0-444-42379-6 (U.S. : v. 1)
ISBN 0-444-42380-X (U.S. : v. 2)

ISBN 0-444-42381-8 (U.S. : set)

ISBN 0-444-42380-X (Part B)
ISBN 0-444-42381-8 (Set)
ISBN 0-444-42382-6 (Series)

Printed in The Netherlands

MULTI-PHASE FLOW & HEAT TRANSFER III

PART A   FUNDAMENTALS

PART B   APPLICATIONS

Proceedings of the Third Multi-Phase Flow and Heat Transfer Symposium-Workshop, held in Miami Beach, Florida, U.S.A., on 18-20 April 1983, and presented by the Clean Energy Research Institute, College of Engineering, University of Miami, Coral Gables, Florida, U.S.A.; sponsored by the National Science Foundation, Washington, D.C., U.S.A.; in cooperation with the International Association for Hydrogen Energy, International Journal of Heat and Mass Transfer, and the Department of Mechanical Engineering, University of Miami, Coral Gables, Florida, U.S.A.

EDITORS

T. Nejat Veziroğlu
Clean Energy Research Institute
University of Miami
Coral Gables, Florida, U.S.A.

Arthur E. Bergles
Department of Mechanical Engineering
Iowa State University
Ames, Iowa, U.S.A.

MANUSCRIPT EDITOR

Sheila M. Puryear
Clean Energy Research Institute
University of Miami
Coral Gables, Florida, U.S.A.

EDITORIAL ASSISTANTS

İlker Gürkan
University of Miami
Coral Gables, Florida, U.S.A.

Aykut Menteş
University of Miami
Coral Gables, Florida, U.S.A.

ACKNOWLEDGEMENTS

The Symposium Organizing Committee gratefully acknowledges the sponsorship of the National Science Foundation and the assistance and cooperation of the International Association for Hydrogen Energy, International Journal of Heat and Mass Transfer, and the Mechanical Engineering Department of the University of Miami.

We wish to extend our sincere appreciation to the Invited Speakers: Dr. Win Aung, Director, Heat Transfer Program, National Science Foundation; Dr. Adrian Bejan, Associate Professor, Department of Mechanical Engineering, University of Colorado, Boulder, Colorado; and Dr. Jack T. Sanderson, Assistant Director for Engineering, National Science Foundation. We also thank Dr. Arthur E. Bergles, Distinguished Professor, Department of Mechanical Engineering, Iowa State University, Ames, Iowa, for his efforts in organizing and directing the workshop sessions.

Special thanks are due to our authors and lecturers, who have provided the substance of the symposium as published in the present proceedings.

And last, but not least, our debt of gratitude is owed to the Session Chairpersons and Co-Chairpersons, and to the Workshop Group Leaders, for the organization and execution of the technical and workshop sessions.

The Organizing Committee
Third Multi-Phase Flow and Heat
Transfer Symposium-Workshop

FOREWORD

The Third Multi-Phase Flow and Heat Transfer Symposium-Workshop continued the tradition established by its predecessors, the Two-Phase Flow and Heat Transfer Symposium-Workshop of October 1976 and the Second Multi-Phase Flow and Heat Transfer Symposium-Workshop of April 1979. It provided an event worthy of this important topic, including the latest information on the status of multi-phase flow and heat transfer research, development and applications. It also sought to identify important areas of multi-phase flow and heat transfer in urgent need of further research and development. In keeping with this latter objective, workshop discussions were held among the 188 attendees who represented universities, research establishments and industrial organizations. The Workshop Summaries are presented in a separate report.

It is particularly significant that this Third Symposium-Workshop included participants from 28 countries. This assured the international character of the information transfer, in keeping with the widespread need for better understanding of the behavior of multi-phase flows.

This two-volume set of proceedings includes 119 papers from the Symposium program. The great majority of the papers addresses gas-liquid flow; however, studies of gas-solid and liquid-solid flows are also included. While the heat transfer papers stress boiling (evaporation) or condensation, several papers consider phase change via freezing or melting. The papers have been divided into those that relate to fundamentals (Part A) and those that emphasize applications (Part B).

Included in Part A are papers dealing with the formulation of the multi-phase flow equations and solutions of these equations, flow regime transitions for gas-liquid flows, flow structure details and pressure drop of gas-liquid flows, steam separation and distribution, pressure wave propagation in gas-liquid and gas-solid flows, flashing and critical flow, heat transfer in pool boiling and forced convection boiling, heat transfer in dispersed gas-liquid flows, heat transfer in condensing, and thermal hydraulic instabilities.

Part B includes a large collection of papers addressing safety issues in nuclear reactor technology: thermal hydraulic code development, experiments to assess performance and validate codes and modelling of selected accident phenomena. Other papers in this part consider gas-solid flows, fluidized beds, liquid-solid flows, freezing and melting, mass transfer and chemical reactions, porous media, measurement techniques, and miscellaneous industrial applications.

The papers included in these two volumes are of both current and archival interest. The volumes should, therefore, serve as both a proceedings of the Symposium-Workshop and as a reference covering important topics in the field of multi-phase flow and heat transfer.

T. Nejat Veziroğlu
Arthur E. Bergles

PART B — APPLICATIONS

CONTENTS

12. THERMAL HYDRAULIC CODE DEVELOPMENT

13. THERMAL HYDRAULIC EXPERIMENTS

21. MEASUREMENT TECHNIQUES

22. MISCELLANEOUS INDUSTRIAL APPLICATIONS

SYMPOSIUM COMMITTEE AND STAFF

SYMPOSIUM COMMITTEE

Arthur E. Bergles
Iowa State University, U.S.A.

Xuejun Chen
Xian Jiaotong University, China

Dimitri Gidaspow
Illinois Institute of Technology, U.S.A.

Sadik Kakaç
University of Miami, U.S.A.

R. T. Lahey, Jr.
Rensselaer Polytechnic Institute, U.S.A.

Samuel S. Lee
University of Miami, U.S.A.

Robert Lyczkowski
Argonne National Laboratory, U.S.A.

Samuel Sideman
Technion-Israel Institute of Technology, Israel

K. Stephan
Universität Stuttgart, F.R.G.

T. Nejat Veziroḡlu (Chairperson)
University of Miami, U.S.A.

STAFF

Symposium Coordinators:	Sheila M. Puryear
	Lucille J. Walter
Assistant Coordinators:	Barbara Berman
	Marina M. Blanco
	Ann Raffle
Manuscript Editor:	Sheila M. Puryear
Graduate Assistants:	O. Eser Ateşoglu
	İ. Gürkan
	A. Menteş
	O. T. Yıldırım
Undergraduate Assistants:	Mark Drews
	Clarence Mackey
	Cristina Robu

SESSION OFFICIALS

PLENARY SESSION | SYMPOSIUM OPENING

Chairperson: A. E. Bergles, Iowa State University, Ames, Iowa, U.S.A.

Co-Chairperson: S. S. Lee, University of Miami, Coral Gables, Florida, U.S.A.

SESSION 1A | MULTI-PHASE FLOW FUNDAMENTALS

Chairperson: R. W. Lyczkowski, Argonne National Laboratory, Argonne, Illinois, U.S.A.

Co-Chairperson: G. Kvajic, U-MAXS, Electromag Inc., Miami, Florida, U.S.A.

SESSION 1B | TWO-PHASE FLOW INSTABILITIES

Chairperson: S. Kakaç, University of Miami, Coral Gables, Florida, U.S.A.

Co-Chairperson: A. Awad, University of Miami, Coral Gables, Florida, U.S.A.

SESSION 1C | FLUIDIZED BEDS

Chairperson: H. K. Fauske, Fauske and Associates, Inc., Burr Ridge, Illinois, U.S.A.

Co-Chairperson: S. Sideman, Technion-Israel Institute of Technology, Technion City – Haifa, Israel

SESSION 2A | MULTI-PHASE FLOW REGIMES

Chairperson: M. E. Salcudean, University of Ottawa, Ottawa, Ontario, Canada

Co-Chairperson: E. E. Michaelides, University of Delaware, Newark, Delaware, U.S.A.

SESSION 2B | REACTOR SAFETY

Chairperson: R. W. Lyczkowski, Argonne National Laboratory, Argonne, Illinois, U.S.A.

Co-Chairperson: J. S. Chang, McMaster University, Hamilton, Ontario, Canada

SESSION 2C | MEASUREMENTS / SUSPENSIONS

Chairperson: S. C. Kranc, University of South Florida, Tampa, Florida, U.S.A.

Co-Chairperson: S. M. Sami, Ecole Polytechnique de Montreal, Montreal, Quebec, Canada

SESSION 3A   MULTI-PHASE FLOW PRESSURE DROP AND HEAT
                     TRANSFER

Chairperson: S. G. Bankoff, Northwestern University, Evanston,
                     Illinois, U.S.A.

Co-Chairperson: J. P. Adams, EG & G Idaho, Inc., Idaho Falls,
                     Idaho, U.S.A.

SESSION 3B   MULTI-PHASE FLOW MODELLING

Chairperson: D. Gidaspow, Illinois Institute of Technology,
                     Chicago, Illinois, U.S.A.

Co-Chairperson: T. S. Andreychek, Westinghouse Electric Corpora-
                     tion, Pittsburgh, Pennsylvania, U.S.A.

SESSION 3C   MULTI-PHASE FLOW APPLICATIONS

Chairperson: T. M. Romberg, CSIRO Division of Mineral Physics
                     Sutherland, NSW, Australia

Co-Chairperson: A. M. Tentner, Argonne National Laboratory,
                     Argonne, Illinois, U.S.A.

SESSION 4A   MASS TRANSFER AND PHASE CHANGE

Chairperson: K. V. Wong, University of Miami, Coral Gables,
                     Florida, U.S.A.

Co-Chairperson: J. J. J. Chen, University of Hong Kong,
                     Hong Kong

SESSION 4B   STEAM GENERATION AND DISTRIBUTION

Chairperson: J. H. Kim, Electric Power Research Institute,
                     Palo Alto, California, U.S.A.

Co-Chairperson: F. Dobran, Stevens Institute of Technology,
                     Hoboken, New Jersey, U.S.A.

SESSION 4C   DROPLET DYNAMICS

Chairperson: K. Johannsen, Technische Universität Berlin,
                     Berlin, Federal Republic of German

Co-Chairperson: S. C. Yao, Carnegie-Mellon University,
                     Pittsburgh, Pennsylvania, U.S.A.

WORKSHOP
SESSION 5A   BOILING AND CONDENSATION

Workshop Leader: J. Lienhard, University of Houston, Houston,
                     Texas, U.S.A.

WORKSHOP
SESSION 5B   TWO-PHASE GAS-LIQUID FLOW REGIMES AND PRESSURE
                     DROP

Workshop Leader: J. Weisman, University of Cincinnati, Cincinnati
                     Ohio, U.S.A.

WORKSHOP
SESSION 5C   TWO-PHASE GAS-LIQUID FLOW ISSUES IN NUCLEAR
                      REACTOR AND CHEMICAL PLANT SAFETY

Workshop Leaders: S. G. Bankoff, Northwestern University, Evanston,
                      Illinois, U.S.A.
                      M. E. Salcudean, University of Ottawa, Ottawa,
                      Ontario, Canada

WORKSHOP
SESSION 5D   GAS-SOLID AND SOLID-LIQUID FLOW AND HEAT
                      TRANSFER

Workshop Leader: R. Lyczkowski, Argonne National Laboratory,
                      Argonne, Illinois, U.S.A.

WORKSHOP
SESSION 5E   TWO-PHASE EQUIPMENT FOR THE POWER AND PROCESS
                      INDUSTRIES

Workshop Leader: G. Breber, Heat Transfer Research, Inc.,
                      Alhambra, California, U.S.A.

WORKSHOP
SESSION 5F   OPEN FORUM (Subjects Not Covered in Other
                      Sessions)

Workshop Leader: W. M. Rohsenow, Massachusetts Institute of
                      Technology, Cambridge, Massachusetts, U.S.A.

SESSION 6A   BOILING AND CONDENSATION

Chairperson: P. Saha, Brookhaven National Laboratory,
                      Upton, New York, U.S.A.

Co-Chairperson: H. Auracher, Institute fur Techn. Thermodynamik,
                      Stuttgart, Federal Republic of Germany

SESSION 6B   TRANSIENTS AND WAVE PROPAGATION

Chairperson: T. M. Romberg, CSIRO Division of Mineral
                      Physics, Sutherland, NSW, Australia

Co-Chairperson: R. Taleyarkhan, Rensselaer Polytechnic Insti-
                      tute, Troy, New York, U.S.A.

SESSION 6C   HEAT TRANSFER / ENERGY CONVERSION

Chairperson: T. W. Fogwell, Texas A & M University, College
                      Station, Texas, U.S.A.

FINAL
PLENARY SESSION   WORKSHOP REPORTS

Chairperson: A. E. Bergles, Iowa State University, Ames,
                      Iowa, U.S.A.

Multi-Phase Flow and Heat Transfer III. Part B: Applications
edited by T.N. Veziroğlu and A.E. Bergles
Elsevier Science Publishers B.V., Amsterdam, 1984 — Printed in The Netherlands

A COUNTER CURRENT FLOW MODEL FOR RELAP5

E.J. Kee and J.W. Spore
EG&G Services, Inc.
Idaho Falls, Idaho 83401, U.S.A.

ABSTRACT

Calculations of a German upper head injection pressurized water reactor indicate that RELAP5 does not calculate counter current flow properly. Investigation shows that the RELAP5 flow regime map is not appropriate for gravity dominated flow. The RELAP5 flow regime map is modified to be a function not only of void fraction but entrainment based on Ishii's entrainment model. To prevent the code from exceeding an upper limit on liquid downfall, a counter current flow limiting (CCFL) model is incorporated. The code using the modified flow regime map and CCFL model is greatly improved over the unmodified code when compared with GE single bundle upper tie plate CCFL data.

1.   INTRODUCTION

The RELAP5 computer code [1] is being developed as an analysis tool for pressurized water reactor (PWR) and boiling water (BWR) transients. The initial development efforts have been directed primarily at PWR analysis with BWR analysis capability as a later effort. As a result of this development history, attention has not focused on the code's ability to simulate accurately counter current flow or counter current flow limiting (CCFL) even though both liquid and gas field equations are used. In German vendored PWR's however, the ability to simulate counter current flow becomes very important because unlike U.S. designs, the German plants use injection of emergency core coolant in the upper head. If counter current flow is not simulated accurately, unrealistic clad surface temperatures are calculated. In fact, the RELAP5 code has been used to analyze a small break in a German vendored PWR and the results indicated that the core would completely dry out and the upper head completely fill with water before any ECC penetration would occur. Analysis of these results indicated that counter current flow should have occurred during the period the core was drying out based on the volumetric gas flux at the upper tie plate using a Kutateladze [2] flooding correlation.

In this paper, the RELAP5 code is modified so that the flow regime map has an annular—mist regime and for flooding, a Kutateladze correlation limits the downward liquid velocity. The new code is compared with General Electric 8x8 upper tie plate flooding data over a typical range of volumetric gas fluxes ($1.72 > Kg^{1/2} > 1.05$).

2. ANALYSIS

Recent application of the RELAP5 code to small break analyses in a German vendored reactor system indicated the need for improvements in the way the code calculates liquid downfall in gravity dominated flow situations. For high void fraction flows, interphase friction was numerically set to infinity (1×10^{20}) resulting in cocurrent flow whereas data in the literature [2,3] show relatively high slip that produces counterflow. In order to achieve more realistic results and better comparison to the data, several modifications were made to the RELAP5 code.

1. The annular and mist flow regimes for vertical flow were replaced with an annular-mist flow regime.

2. The amount of mist present was calculated from Ishii's entrainment model.

3. The averaging of interphase friction was eliminated.

4. Temporal averaging of interphase friction was modified to use one-tenth new time and nine-tenths old time interphase friction.

5. The flow regime map for horizontal or vertical flow was changed from being donored to being a vertical map for junctions connecting any vertical volume or horizontal when the junction connects two horizontal volumes.

6. A Counter Current Flow Limiting (CCFL) model was implemented.

The RELAP5 flow regime map assumed that mist flow was a function of void fraction only, so that at high void fractions ($\alpha_g > 0.90$), the relatively large interphase friction associated with mist flow was used. This restriction was removed by including an annular-mist regime where the interphase friction was computed using an interpolation between the annular and mist friction correlations based on Ishii's [4] entrainment correlation:

$$F_{A-M} = E \, F_M + (1. - E) \, F_A \qquad (1)$$

where

$$E = \tanh (7.25 \times 10^{-7} \cdot We^{1.25} \cdot Re_f^{0.25}) \qquad (2)$$

$$We = \rho_g \, j_g^2 \, D/\sigma \, (\Delta\rho/\rho_g)^{1/3} \qquad (3)$$

$$Re_f = \rho_f \, j_f \, D/\mu_f \qquad (4)$$

The new vertical flow regime map configuration is shown in Figure 1.

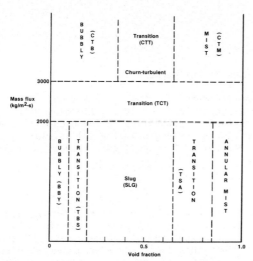

FIGURE 1. Flow Regime Map with Annular–Mist Regime

The test section used by Jones [5] to obtain flooding on a simulated General Electric 8x8 fuel bundle was modeled in RELAP5 by an input description shown schematically in Figure 2. In this noding, the upper tie plate is represented by junction 004. The bundle (volume 005) drains to an active volume (007) through junction 006. Liquid is supplied to the simulated upper plenum (003) through a time dependent junction (101). The test section overpressure was specified by a time dependent volume (001). Different flooding data were generated by varying the steam supply to the bottom of the bundle through a time dependent junction (201) and allowing the problem to run to a steady state.

The data generated by these calculations were compared to a "best fit" [2] of Jones' data for a capped sparger as shown in Figure 3. The equation for the "best fit" is given by:

$$K_g^{1/2} + K_f^{1/2} = (3.57)^{1/2} \qquad (5)$$

This comparison indicated the need to limit liquid downward velocity and is indicative that the entrainment correlation is not appropriate for counter current flow. However, when the code was used to analyze the same data without the annular–mist regime included, only cocurrent flow at the upper tie plate was calculated.

The code was further modified to include a CCFL model which prevents the relative velocities from exceeding those in experimental

4

data. The Wallis type flooding correlation [6] has been widely accepted and was used in this work to limit the liquid downward velocity:

$$K_g^{1/2} + mK_f^{1/2} = K^{1/2} \tag{6}$$

where

$$K_{g,f} = j_{g,f} \sqrt{\rho_{g,f}} \, / \, [g\sigma \, \Delta\rho]^{1/4} \tag{7}$$

m and K are constants related to the geometry.

Use of the Kutateladze parameter for the gas and liquid fluxes has become more common in recent years for correlating flooding data and includes the effect of interfacial instability in the hydraulic diameter commonly used for the characteristic length in the Wallis flux parameters. Equation (5) was solved for the liquid velocity:

$$V_{f,max} = - \frac{[g\sigma \, \Delta\rho]^{1/4}}{\alpha_f \sqrt{\rho_f}} \, (1.89 - K_g^{1/2})^2 \tag{8}$$

This velocity was used in the code as a maximum so that if the liquid velocity calculated by the momentum equation exceeds the velocity computed by Equation (8), it was reset to this value. With the code additionally modified to include this limit, data were generated using the approach outlined for the first cases and compared with the "best fit" to the General Electric flooding data (Equation (5)) as shown in Figure 4. Use of this flooding correlation in conjunction with the annular-mist regime produces very good agreement with the General Electric upper tie plate flooding data "best fit".

Because the slope and intercept constants (m and K in Equation (6)) are dependent on the geometry of the problem, the code was additionally modified to allow the user to input these parameters in the normal junction input data for the RELAP5 components single junctions, branches, valves, pumps and pipes.

3.   CONCLUSIONS

The RELAP5 code can be modified so that counter current flow is produced from the two-equation momentum balance at high void fractions by using a flow regime map that includes an annular-mist regime. Excessive slip is calculated by the code when Ishii's entrainment correlation is used to interpolate between the annular and mist flow regimes during flooding. When a Kutateladze flooding correlation is included, the code compares well with a best fit to General Electric 8x8 bundle upper tie plate flooding data.

ACKNOWLEDGEMENTS

The work reported here was sponsored by Control Data Corporation and the code improvements have been incorporated in their version of RELAP5 which resides on the Cybernet time sharing system. The authors are grateful to Dr. Wolfgang Mayr and Dr. Kenneth L. Kin of Control Data Germany and Control Data Japan, respectively, for their help in providing computer time for model development and assessment.

NOMENCLATURE

E   Fraction of liquid entrained as mist.
D   Hydraulic diameter (m).
g   Acceleration due to gravity (9.8 m/s^2).
j   Superficial velocity (m/s).
α Volumetric fraction.

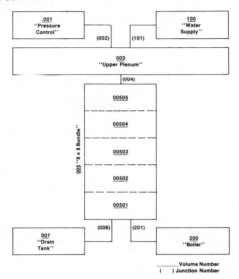

FIGURE 2.   Schematic Representation of RELAP5 Nodalization of Jones' Experimental Apparatus.

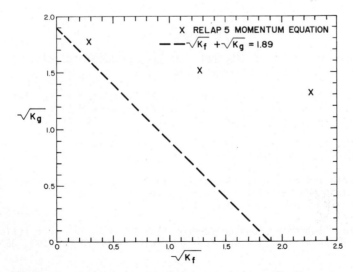

FIGURE 3.   Comparison of RELAP5 Calculated Flooding Curve With Sun's Best Fit to Jones' Data

6

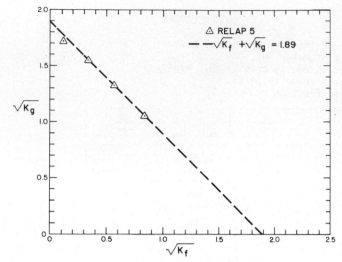

FIGURE 4. Comparison of RELAP5 Calculated Flooding Curve Using
CCFL Model with Sun's Best Fit To Jones' Data

ρ Density (kg/m^3)
$\Delta\rho$ Difference between liquid density and gas density (kg/m^3).
σ Surface tension (N/m).
μ Dynamic viscosity (N–S/m^2).

Subscripts

g   Gas
f   Liquid

REFERENCES

1.   V. H. Ransom, et al,"RELAP5/MOD1 Code Manual", NUREG/CR–1826,
     EGG–2070 DRAFT REV 2, September, 1981.

2.   K. H. Sun, "Flooding Correlations for BWR Bundle Upper Tie Plates
     and Bottom Side Entry Orifices," Proceedings, Second Multi-Phase
     Flow and Heat Transfer Symposium-Workshop, Miami Beach, FL, U.S.A.

3.   H. J. Richter, et al., "Effect of Scale on Two–Phase Countercurrent
     Flow Flooding", NUREG/CR–0323 R2, June, 1979.

4.   I. Kataoka and M. Ishii, "Mechanism and Correlation of Droplet
     Entrainment and Deposition in Annular Two-Phase Flow,"
     NUREG/CR-2885, ANL-82-44, July, 1982.

5.   D. Jones, "Subcooled Counter-Current Flow Limiting Characteristics
     of the Upper Region of a BWR Fuel Bundle," NEDG-NUREG-23549, July,
     1977.

6.   G. B. Wallis, Flooding Velocities for Air and Water in Vertical
     Tubes, UKEA Report AEEW-R123, 1961.

Multi-Phase Flow and Heat Transfer III. Part B: Applications
edited by T.N. Veziroğlu and A.E. Bergles
Elsevier Science Publishers B.V., Amsterdam, 1984 — Printed in The Netherlands

A LOCAL VOID AND SLIP MODEL USED IN BODYFIT-2PE

B. C-J. Chen and T. H. Chien
Argonne National Laboratory
Argonne, Illinois 60439, USA

J. H. Kim and G. S. Lellouche
Electric Power Research Institute
Palo Alto, California 94303, USA

ABSTRACT

A local void and slip model has been proposed for a two-phase flow without the need of fitting any empirical parameters. This model is based on the assumption that all bubbles have reached their terminal rise velocities in the two-phase region. This simple model seems to provide reasonable calculational results when compared with the experimental data and other void and slip models. It provides a means to account for the void and slip of a two-phase flow on a local basis. This is particularly suitable for a fine mesh thermal-hydraulic computer program such as BODYFIT-2PE.

1.0   INTRODUCTION

In most of the computer codes[1-3] for subchannel, two-phase flow problems, empirical correlations are often used to calculate the sub-channel void fractions and slip ratios. These correlations are used to account for the phase slip and the non-homogeneities in two-phase flows. Some of the commonly used ones are the Levy correlation[4], the Lellouche and Zolotar correlation[5], the Zuber-Findlay drift-flux model [6], and the Smith correlation.[7]   All these correlations are applied on a global basis on the subchannel of the rod bundle.   Therefore, the detailed distribution of the void within the bundle is lost.

2.0   OTHER GLOBAL MODELS

The Levy correlation and the Lellouche and Zolotar correlation are similar in their approaches. They calculate the void fraction by first determining the bubble detachment point and then determining the relationship between the true flow quality and the thermal-equilibrium quality.   In the Levy correlation, the relationship among the actual quality x, the equilibrium quality x_e, and the detachment quality x_d is given as

$$x = x_e - x_d \exp\left(\frac{x_e}{x_d} - 1\right).$$ (1)

This form satisfies the curves of x versus x_e derived from experiment. The detachment quality x_d is then expressed in terms of the specific heat of saturated liquid, c_p, the latent heat of vaporization, h_{fg}, and the bulk subcooling of the fluid at the bubble departure point, ΔT_d, as

$$x_d = - \frac{c_p \, \Delta T_d}{h_{fg}}. \tag{2}$$

The bulk subcooling is then expressed as a complicated function of the Prandtl number, the surface heat flux, the shear at the wall, the surface tension, and the buoyancy force on the bubble. The actual quality is then substituted into the homogeneous equation for void fraction,

$$\alpha = \frac{x}{x + \dfrac{\rho_g}{\rho_f}(1-x)}. \tag{3}$$

In the Lellouche and Zolotar model, a profit fit of the void fraction curve from single-phase to fully-developed forced convection gives

$$x = \frac{x_e - x_d \left[1 - \tanh\left(1 - \dfrac{x_e}{x_d}\right)\right]}{1 - x_d \left[1 - \tanh\left(1 - \dfrac{x_e}{x_d}\right)\right]} \tag{4}$$

instead of Eq. (1). The detachment quality x_d is similar to Eq. (2) as

$$x_d = - \frac{c_p Z}{h_{fg}}, \tag{5}$$

where Z is expressed in terms of heat flux, heat transfer coefficients, and pressure at the detachment point. Once x is known from Eq. (4), void fraction α is found from

$$\alpha = \frac{x}{C_o \left[x + \dfrac{\rho_g}{\rho_f}(1-x)\right] + \dfrac{\rho_g \, V_{gj}}{G}} \tag{6}$$

where C_o is the distribution factor, V_{gj} is the drift flux velocity, and G is the mass flux. Note that if $C_o = 1$ and $V_{gj} = 0$, Eq. (6) reduces to Eq. (3) for the homogeneous equilibrium model. The C_o is an empirical constant and is determined on a global basis across the channel. In order to avoid this empiricism in the above models, the following simple model has been proposed by EPRI.

3.0 THE PRESENT LOCAL MODEL

In this model, one assumes the bubbles have reached their terminal rise velocities locally as

$$V_g - V_\ell = 1.41 \left[\frac{\sigma g \, (\rho_\ell - \rho_g)}{\rho_\ell^2}\right]^{1/4}, \tag{7}$$

where V_g and V_ℓ are the axial velocities of the vapor and the liquid respectively. The lateral slip between the two phases are assumed to be zero.

The exact relationship between the quality and the void fraction is given by

$$\alpha = \frac{x}{x + \dfrac{\rho_g}{\rho_\ell}(1 - x)S}, \qquad (8)$$

where S is the slip ratio of V_g/V_ℓ .

To express S in terms of $V_g - V_\ell$, the denominator can be rearranged to yield

$$x + \frac{\rho_g}{\rho_\ell}(1 - x)\left(1 + \frac{V_g}{V_\ell} - 1\right)$$

$$= x + \frac{\rho_g}{\rho_\ell}(1 - x) + \frac{\rho_g(1 - x)}{\rho_\ell V_\ell}(V_g - V_\ell)$$

$$= x + \frac{\rho_f}{\rho_\ell}(1 - x) + \frac{\rho_g(1 - \alpha)}{G}(V_g - V_\ell) .$$

Therefore, α can be written to give

$$\alpha = x \bigg/ \left[x + \frac{\rho_g}{\rho_\ell}(1 - x) + \frac{\rho_g(1 - \alpha)}{G}(V_g - V_\ell)\right] \qquad (9)$$

This equation can be solved for α as

$$\alpha = \frac{B' - \sqrt{B'^2 - 4A'x}}{2A} \qquad (10)$$

where

$$A' = \frac{\rho_g}{G}(V_g - V_\ell),$$

$$B' = x + \frac{\rho_g}{\rho_\ell}(1-x) + \frac{\rho_g}{G}(V_g - V_\ell),$$

and $V_g - V_\ell$ is given by Eq. (7). In order to insure that $\alpha = 0$ when $x = 0$, the + solution should be discarded in solving the quadratic equation. Also, when $x = 1$, $\alpha = 1$. Once α is known, S can easily be obtained through Eq. (8).

The advantage of this model is that it is simple and yet it provides direct information for the local void fraction and slip ratio without involving fitting of parameters. This model is programmed in the BODYFIT-2PE[8] code. BODYFIT-2PE is a steady-state/transient, three-dimensional computer program for thermal-hydraulic analyses of two-phase flow problems in light-water-reactor fuel assemblies. It uses the technique of the boundary-fitted coordinate system where all the physical boundaries are transformed to be coincident with constant coordinate lines in the transformed space. Therefore, the boundary conditions can be accurately represented without interpolation. The code uses the parabolic approximation for saving computer running time and storage. The physical models and the numerics are described in Ref. 8.

4.0 COMPARISON WITH OTHER MODELS AND EXPERIMENTAL DATA

This local void and slip model is used to calculate the hot and cold channel mass fluxes, qualities, and void fractions for a 4x4 rod bundle. Results of the calculation were compared with the test data measured by Columbia[9] for the 4x4 rod bundle. In the experiment, isokinetic sampling measurements of mass flow rate and enthalpy were carried out on the rod bundle array typical of a Boiling Water Reactor geometry. The physical dimension of the rod bundle as well as the computational meshes used in the BODYFIT calculation are given in Fig. 1. The mass flow rate and enthalpy of two selected subchannels, a hot and a cold channel at centrally symmetrical locations were measured by isokinetically extracting the entire flows and sending them through two calorimetry loops. These hot and cold subchannels are also outlined and marked H and C, respectively, in Fig. 1. The rod bundles were electron beam welded to the grid plate. The rod to rod spacing was maintained by placing specially machined grids at various axial locations. The grids were designed to have enough strength while producing minimum mixing. They are modeled in the BODYFIT calculation by head loss terms for different subchannels.

The 16 rods were electrically heated with a power distribution of two to one; two on the left half and one on the right half. After the steady state condition was attained, heat loss calibrations were performed. Enthalpy and flow data were taken at selected powers and flow rates for a series of inlet temperatures while the pressure was held constant at 1000 psia.

Six sets of experimental data with diversified test conditions were chosen to compare with the BODYFIT calculations. The test conditions were given in Table 1 along with the comparison between the experimental measurements (A) and BODYFIT-2PE calculations for various void and slip models - Homogeneous Equilibrium Model (B), Lellouche and Zolotar model (C), and the present local model (D). Figure 2(a) and 2(b) show the cold and the hot channel mass flow rates at the exit of the rod bundle. Figure 3(a) and 3(b) show the cold and the hot channel quality at exit. Comparisons were made between the experimental measurements and homogeneous equilibrium model (HEM) - model 1, the Lellouche and Zolotar model - model 2, and the present local model -model 3. In general, all three models agree well with the experimental measurements. The calculations based on the present local model were very close to those based on the HEM. This result seems expected from the fact that the relative phase velocity, $V_g - V_\ell$, is small compared to the flow rate G in the cases we have studied. This is evident in examining Eq. 9. Further comparison with cases where G is small will be a crucial test of the model. It should also be noted from Fig. 2 that there are significant amounts of cross flow going from the hot channels to the cold channels.

5.0   CONCLUSION

In summary, a simple void and slip model has been proposed for two-
phase flow without the need for fitting any empirical parameters.   This
model is based on the assumption that all bubbles have reached their
terminal rise velocities in the two-phase region.   This assumption seems
reasonable when the calculational results are compared with the
experimental data and other void and slip models.

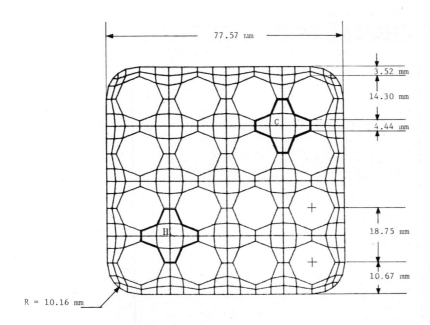

Hydraulic Diameter = 13.28 mm

Flow Area   = 3.36 x 10^{-3} m^2

Heated Length   = 3.658 m

Power Ratio   = 2 to 1

Fig. 1.   Mesh structure and geometrical dimensions
          for the Columbia University 4x4 rod bundle.

This does not mean the current model is perfect. It is a first attempt to use a void and slip model on a local basis suitable for a fine mesh computer program like BODYFIT. Further testing and improvement on the model is currently under way to study the effect of different flow regimes on the terminal rise velocities and low flow rates on the validity of the model.

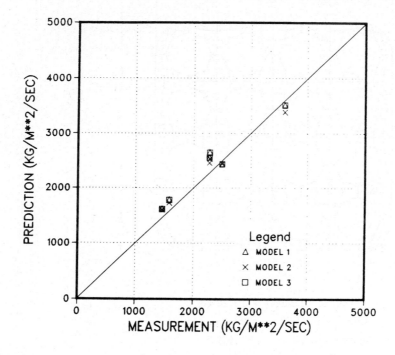

Fig. 2(a). Comparison of cold channel mass flow rate at exit between the experimental measurements and the BODYFIT-2PE calculations for various void and slip models.

Table 1   Comparison between the experimental measurements and BODYFIT-2PE Calculations

Run No.	BUNDLE AVERAGE						COLD CHANNEL			HOT CHANNEL		
	G	H_{in}	H_{out}	X_{out}	HFLUX		G	H	X	G	H	X
219	1949	1.088	1.299	2.4	0.524	(A)	2274	1.258	− 0.3	1278	1.415	10.1
						(B)	2658	1.244	− 0.4	1590	1.399	9.7
						(C)	2535	1.238	− 0.7	1683	1.396	9.5
						(D)	2647	1.244	− 0.4	1596	1.398	9.6
224	1345	0.875	1.189	−4.8	0.539	(A)	1457	1.114	− 9.8	990	1.310	3.2
						(B)	1619	1.102	− 9.9	1253	1.297	2.8
						(C)	1603	1.098	−10.1	1314	1.300	3.0
						(D)	1612	1.101	− 9.9	1265	1.297	2.8
225	1345	0.875	1.236	−1.7	0.621	(A)	1581	1.158	− 6.9	902	1.392	8.6
						(B)	1789	1.145	− 7.0	1092	1.384	8.5
						(C)	1726	1.136	− 7.6	1181	1.381	8.3
						(D)	1780	1.145	− 7.1	1100	1.383	8.4
232	2717	1.072	1.276	0.9	0.708	(A)	3594	1.226	− 2.4	1785	1.390	8.4
						(B)	3523	1.227	− 1.1	2197	1.366	7.8
						(C)	3390	1.221	− 1.6	2305	1.365	7.8
						(D)	3513	1.227	− 1.1	2205	1.365	7.8
242	2021	1.201	1.384	8.0	0.470	(A)	2269	1.354	6.1	1538	1.493	15.3
						(B)	2561	1.321	4.9	1739	1.487	15.6
						(C)	2457	1.322	4.9	1825	1.481	15.2
						(D)	2559	1.322	4.9	1739	1.486	15.5
245	2033	0.755	1.157	−6.9	1.045	(A)	2497	1.065	−13.0	1373	1.340	5.1
						(B)	2440	1.048	−13.1	1760	1.303	3.4
						(C)	2441	1.041	−13.5	1847	1.310	3.9
						(D)	2433	1.048	−13.1	1774	1.303	3.4

NOMENCLATURE: G – Mass Flux (Kg/m^2/sec.), H – Enthalpy (MJ/Kg), HFLUX – heat flux (MJ/m^2/sec), X – Percent (%)
(A) – Experimental Measurements, (B) – HEM Model, (C) – Lellouche & Zolotar Model, (D) – Local Model

14

References

1. C. W. Stewart, et al., "VIPRE Code Manual: Volume 1, Models and Constitutive Relations," to be published under EPRI project RP-1584.

2. "RETRAN – A Program for One-Dimensional Transient Thermal-Hydraulic Analysis of Complex Fluid Flow Systems, Vol. 1: Equations and Numerics," EPRI-CCM-5, (Dec. 1978).

3. "TRAC-PIA, An Advanced Best-Estimate Computer Program for PWR LOCA Analysis," NUREG/CR-0665, LA-7777-MS, (May 1979).

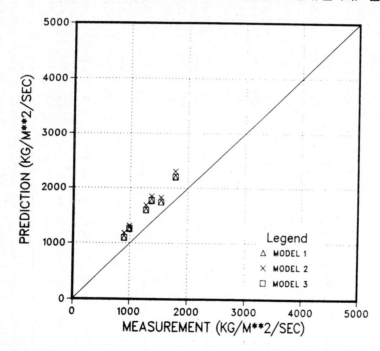

HOT CHANNEL MASS FLOW RATE AT EXIT

Fig. 2(b). Comparison of hot channel mass flow rate at exit between the experimental measurements and the BODYFIT-2PE calculations for various void and slip models.

4. S. Levy, "Forced Convection Subcooled Boiling--Prediction of Vapor Volumetric Fraction." Int. J. of Heat & Mass Transfer, Vol. 10, pp. 951-965 (1967).

5. G. S. Lellouche and B. A. Zolotar, "Mechanistic Model for Predicting Two-Phase Void Fraction for Water in Vertical Tubes, Channels, and Rod Bundles, EPRI NP-2246-SR (Feb. 1982).

6. N. Zuber and J. A. Findlay, "Average Volumetric Concentration in Two-Phase Flow Systems, J. Heat Transfer, Vol. 87C, pp. 453-468 (1965).

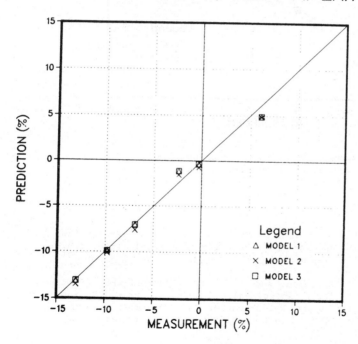

Fig. 3(a).   Comparison of cold channel qualtiy at exit between the experimental measurements and the BODYFIT-2PE calculations for various void and slip models.

HOT CHANNEL QUALITY AT EXIT

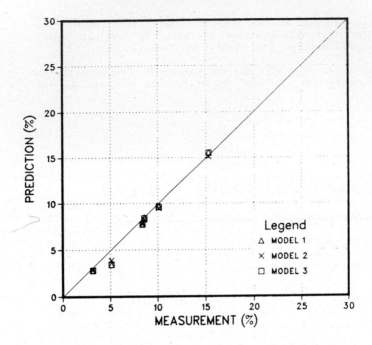

Fig. 3(b). Comparison of hot channel quality at exit between the experimental measurements and the BODYFIT-2PE calculations for various void and slip models.

7. S. L. Smith and J. A. Findlay, "Void Fractions in Two-Phase Flow; A Correlation Based upon an Equal Velocity Head Model, _Proc. I.M.E._, Vol. 1, Pt. 1, No. 38, p. 647 (1969-70).

8. B. C-J. Chen, T. H. Chien, and W. T. Sha, "BODYFIT-2PE: Steady-State/Transient, Three-Dimensional, Two-Phase Thermal-Hydraulic Computer Code for Light Water-Reactor Fuel Assemblies; Vol. 1, Theory and Formulation," to be published as EPRI report under RP-1383-1.

9. Sastry Sreepada, et al., "Outlet Sampling Measurement of Mass Flux, Enthalpy and Void Fraction in Rod Bundles," to be published as an EPRI report under research project RP-345-1.

Multi-Phase Flow and Heat Transfer III. Part B: Applications
edited by T.N. Veziroğlu and A.E. Bergles
Elsevier Science Publishers B.V., Amsterdam, 1984 — Printed in The Netherlands

DEVELOPMENT AND ASSESSMENT OF A CRITICAL HEAT FLUX MODEL FOR LIGHT–WATER REACTOR SYSTEM ANALYSIS

James N. Loomis
Yankee Atomic Electric Company
1671 Worcester Road
Framingham, Massachusetts   01701, U.S.A.

ABSTRACT

A new critical heat flux (CHF) model was developed to provide accurate CHF predictions in the RELAP5YA light–water reactor system analysis code [1].   Accurate CHF predictions are needed to obtain reliable temperature calculations in Loss-of-Coolant Accident (LOCA) and other transient analyses.   The model uses a modified form of the Biasi [2] and Griffith-Zuber [3] correlations to cover a broad range of conditions.   The modification of the Biasi correlation accounts for bundle geometry.   The new CHF model was assessed against twelve steady–state tests from four electrically heated rod bundle experiments.   The experiments included non–uniform and uniform axial and radial power distributions.   Agreement between RELAP5YA predictions and the experiments is generally good considering the broad range of conditions represented by these twelve tests.   Further assessment of the model has been performed at Yankee Atomic Electric Company (YAEC) for tests conducted in the Thermal-Hydraulic Heat Transfer Facility (THTF) at Oak Ridge National Laboratory (ORNL), the Two Loop Test Apparatus (TLTA) at General Electric, and the LOFT Facility at Idaho National Engineering Laboratory.   Those results, presented in Reference 1, provide added confidence in the use of this model.

1.   INTRODUCTION

Accurate CHF predictions are needed to obtain reliable temperature calculations in LOCA and other reactor system transient analyses.   CHF predictions are used to determine convective heat transfer modes.   Heat transfer is sharply degraded when the heat transfer mode switches from pre–CHF to post–CHF.   A new CHF model was incorporated in the RELAP5YA reactor system thermal-hydraulic analysis code during its development from the RELAP5 MOD1 code.   The original CHF model from RELAP5 MOD1 gave inaccurate temperature predictions when it was used for steady–state experiments.   Since reliable fuel rod temperature prediction is a major goal of reactor transient analysis, a more accurate CHF model is needed.

2.   DEVELOPMENT OF THE CRITICAL HEAT FLUX (CHF) MODEL

A new critical heat flux (CHF) model was developed to provide accurate CHF predictions in the RELAP5YA code.   Development of the model was based, in part, upon recommendations from the transient CHF and blowdown heat transfer studies by Leung [5] and by Hsu [6].

The CHF model uses several correlations to cover a broad range of conditions. At high mass flux values, the Biasi correlation developed from tube data [2] is used in a modified form, as shown in Table 1. The modification, recommended by Phillips and Shumway [7], accounts for bundle geometry. At low mass flux values, the Griffith-Zuber correlation [3] is used. This correlation is defined in Table 2. At intermediate mass flux values, the CHF value is obtained by linear interpolation between the Griffith-Zuber and the modified Biasi values. In addition, a critical void fraction criterion is used, beyond which wall dryout is assumed to occur. The critical void fraction is a function of mass flux. The range for the CHF correlations are shown in Figure 1 as a function of mass flux and void fraction.

3. ASSESSMENT OF THE CRITICAL HEAT FLUX (CHF) MODEL

3.1. Introduction

The new CHF model was assessed against twelve steady-state tests from four electrically heated rod bundle experiments. The experiments included non-uniform (chopped cosine and top skewed) and uniform axial and radial power profiles. The tests also covered a broad range of pressures, mass fluxes, heat fluxes, heat fluxes and inlet qualities for several rod geometries. The steady-state tests are as follows:

a. Five tests in the Medium Pressure Heat Transfer Flow Loop at the Chemical Engineering Research Laboratory of Columbia University [8 and 9].

b. Four tests in the Nine Rod Test Section at General Electric [10].

c. Three tests in the Thermal-Hydraulic Test Facility (THTF) at ORNL [11].

Each facility used a vertical array of electrically heated rods to simulate fuel rod clusters. The five Columbia tests were selected primarily because their test sections had representative, non-uniform power distribution and full-length, prototypical fuel rod simulators. The four GE nine rod tests were selected because they had larger rod diameters and a fairly broad mass flux range. The three ORNL THTF tests were selected because they involve a fairly large rod cluster, a fairly broad pressure range and intermediate to high qualities at CHF. Table 3 summarizes the range of thermal-hydraulic conditions and geometry for these tests.

3.2. Columbia CHF Tests

As stated above, five Columbia tests were selected for CHF assessment. These tests include data from two separate test sections (numbers 68 and 71). Both test sections simulated a region within Combustion Engineering's 16x16 fuel assemblies, including 21 heated rods and a control rod guide tube. The 21 heated rods in the Columbia test sections were represented by one average rod in RELAP5YA analyses. Test Section 68 had a chopped cosine axial power profile with a normalized axial peak of 1.472. The non-uniform radial power distribution of the rods ranges from 0.963 to 1.199. Test Section 71 had a top skew axial power profile, with a normalized axial peak of 1.480. The non-uniform radial power distribution of rods ranges from 0.980 to 1.073. The thermal-hydraulic test conditions for these five experiments are presented in Table 4.

Table 1

Modified Biasi Critical Heat Flux Correlation

For $|G| \geq 300$ kg/m^2 second:

$\qquad q_{B-CHF} = $ Maximum $[q_1, q_2, 1000$ W/m$^2]$

For $|G| < 300$ kg/m^2 second:

$\qquad q_{B-CHF} = $ Maximum $[q_2, 1000$ W/m$^2]$

where:

$$q_1 = 1.883 \times 10^7 \ [f_1/(0.1|G|)^{1/6} - X_B][(100 \ D_{HY})^n \cdot (0.1|G|)^{1/6}]^{-1}$$

$$q_2 = 3.78 \times 10^7 \ [f_2(1-X_B)][(100 \ D_{HY})^n \cdot (0.1|G|)^{3/5}]^{-1}$$

$$f_1 = 0.7249 + 0.099 \ P_B \ \exp \ (-0.032 \ P_B)$$

$$f_2 = -1.159 + 0.149 \ P_B \ \exp \ (-0.019 \ P_B) + 8.99 \ P_B/(10 + P_B^2)$$

$$n = 0.4 \quad \text{for} \quad D_{HY} > 0.01 \text{ m}$$

$$n = 0.6 \quad \text{for} \quad D_{HY} \leq 0.01 \text{ m}$$

$$X_B = X_D \ [D_{HT}/D_{HY}]$$

$$X_D = \frac{\alpha_g \ \rho_g \ |V_g|}{\alpha_g \ \rho_g |V_g| + \alpha_f \ \rho_f \ |V_f|} \qquad \text{for} \quad (V_g - V_f)V_f > 0.0$$

$$X_D = X_E \quad \text{for} \ (V_g - V_f)V_f \leq 0.0$$

Definitions:

A	=	flow area, m^2
D_{HT}	=	$(4A/P_{HT})$, heated diameter, m
D_{HY}	=	$(4A/P_W)$, hydraulic diameter, m
G	=	mass flux, kg/m^2 second
P_B	=	absolute pressure, bars
P_{HT}	=	heated perimeter, m
P_W	=	wetted perimeter, m
q_{B-CHF}	=	modified Biasi critical heat flux, W/m^2
V	=	phasic velocity
X_B	=	dynamic quality corrected for bundle geometry
X_D	=	dynamic quality for single tube

Table 1
(Continued)

Modified Biasi Critical Heat Flux Correlation

X_E =   equilibrium static quality

α =   phasic volumetric fraction

ρ =   phasic density

Table 2

Griffith-Zuber Critical Heat Flux Correlation

q_{GZ-CHF} = Maximum [q_3, 1000 W/m^2]

where:

q_3 = $1.31(0.96 - \alpha_g)\ \rho_g\ h_{fg}\ [\sigma\ g(\rho_f - \rho_g)/\rho_g^2]^{1/4}$

g =   gravitational constant

h_{fg} =   latent heat of vaporization

α_g =   void fraction

ρ =   phasic density

σ =   surface tension

Table 3
Steady State CHF Test Conditions

Experiment	Pressure (psi)	Average Mass Flux (Mlb/hr-ft^2)	Average Heat Flux (MBtu/hr-ft^2)	Heated Rod Diameter (in.)	Heated Length (in.)
Columbia	1500 - 2005	1.968 - 2.008	0.282 - 0.436	0.382	150.0
GE Nine Rod	997 - 1005	0.249 - 1.248	0.289 - 0.522	0.570	72.0
ORNL THTF	635 - 1849	0.166 - 0.525	0.14 - 0.29	0.374	144.0

The RELAP5YA calculations using the new CHF option were in good agreement with the Columbia CHF measurements. Calculations always showed CHF in the vicinity of axial locations where CHF conditions were measured. The calculations did well considering that the 21 heated rods were represented by an average rod in the calculation, and considering that often CHF was experimentally detected on rods with the highest normalized radial power factor.

Table 5 compares minimum dynamic quality and axial CHF location from the test data and RELAP5YA calculations for the five Columbia cases. The minimum quality at CHF in the test data is the bundle average quality corresponding to the minimum axial elevation where CHF was detected. Minimum experimental critical qualities range from 12 percent to 19 percent. The minimum calculated critical qualities ranged from 14 percent to 17 percent, indicating good agreement for both chopped cosine and top skew power profiles.

3.3. General Electric CHF Tests

The General Electric CHF tests were performed in the Nine Rod Test section using a 3x3 bundle geometry. The rods were electrically heated over

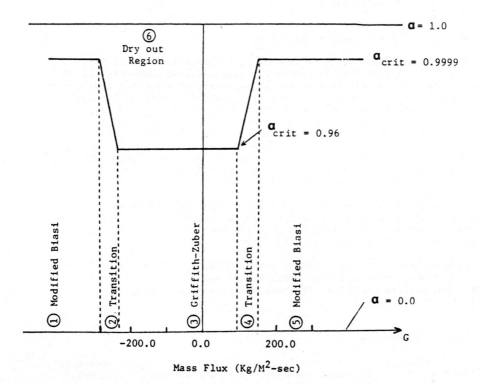

Figure 1: Map For the Modified Biasi and Griffith-Zuber Critical Heat Flux Option

a 72-inch axial height. The bundle had uniform axial and radial power profiles. The procedure used for each test was to set the inlet hydraulic conditions, then increase power until CHF was detected at a 71.5-inch axial height by one or more rods. Four tests were analyzed using the new CHF option in RELAP5YA. The test conditions are listed in Table 6. These tests were selected primarily because they had large rod diameters and a fairly broad mass flux range.

The nine heated rods were represented as an average rod in RELAP5YA analysis. The fluid in the test bundle was represented by a single channel divided into eight axial volumes. Axial portions of the nine rods were represented by a single heat structure for each hydrodynamic volume. The analysis bundle was extended 14 inches longer than the bundle used in the experiment in order to find CHF fluid conditions for cases where CHF was not calculated to occur within the experiment's 72-inch bundle length. The extended portion of the bundle used the same value of heat flux as was used for the bottom 72-inch portion.

Calculated results are compared to experimental data in Table 7. The measured CHF location was always at 71.5 inches, and was detected to occur mostly on the corner rods. Three of the four calculated critical qualities are within five percent of the qualities corresponding to measured CHF locations. Run 296 showed the maximum difference (20 percent) between calculation and experimental data.

3.4 Oak Ridge National Laboratory Tests

Three steady-state tests performed in the Thermal-Hydraulic Test Facility at Oak Ridge National Laboratory were analyzed using the new CHF option. These tests were part of the steady-state film boiling test series 3.07.9 and were conducted with water flowing upward through an 8x8 rod bundle. The bundle cross sections contained 60 heated rods and four unheated rods. The heated rod diameter and pitch are typical of later generation PWRs with 17x17 fuel assemblies. The rod bundle had flat axial and radial power profiles. The rods were 12 feet long and 0.374 inch in diameter. The test conditions are summarized in Table 8. These tests were selected primarily because they involve a fairly large rod cluster, a fairly broad pressure range, and intermediate to high qualities at CHF.

For RELAP5YA modeling, the bundle was divided into 19 axial volumes, chosen for ease of comparing calculated and measured results. The heated rods were represented by a single heat structure divided into 19 axial nodes to correspond to the 19 fluid volumes.

Calculated results are compared to experimental data in Table 9. CHF is calculated to occur below the measured CHF axial locations for all cases. The maximum difference in experimental and calculated critical qualities is 19 percent.

3.5 Summary of Results

The RELAP5YA results for the twelve steady-state tests are summarized in Figure 2. The figure shows a comparison of the calculated critical quality and the experimental critical quality for all twelve tests. The agreement is generally good considering the broad range of conditions represented by these twelve tests. The maximum difference between the

Table 4

Columbia Thermal-Hydraulic Test Conditions

Test Section Number	Run Number	Inlet Temperature (^{o}F)	Pressure (psia)	Average Mass Flux (Mlb/hr-ft^2)	Average Heat Flux (MBtu/hr-ft^2)
68	18	581.7	1755.	1.980	0.324
68	22	553.4	1755.	2.008	0.379
68	27	520.5	1500.	1.995	0.436
71	20	604.8	2005.	1.995	0.282
71	39	547.4	1750.	1.968	0.343

Table 5

Columbia CHF Results

Test Section Number	Run Number	z_{CHF} (in.)		$\bar{x}_{CHF}$	
		Experiment	Calculation	Experiment	Calculation
68	18	101 in.	101 in.	0.1509	0.1509
	22	101	115	0.1185	0.1535
	27	101	115	0.1328	0.1733
71	20	129	128	0.1595	0.1548
	39	143	136	0.1911	0.1381

predicted and experimental critical quality is 20 percent. Further assessment of the new CHF model has been performed at YAEC for tests conducted in the Thermal-Hydraulic Heat Transfer Facility at Oak Ridge National Laboratory, the Two Loop Test Apparatus at General Electric, and the LOFT Facility at Idaho National Engineering Laboratory. Those results, presented in Reference 1, provide added confidence in the use of this model.

4. SUMMARY

A new CHF model was developed to provide accurate CHF predictions in the RELAP5YA Light-Water Reactor System analysis code. The Biasi CHF correlation included in the model has been modified by using the ratio of heated to wetted perimeters. This modification extends the use of the correlation to bundle geometries. The new CHF option has been assessed

Table 6

General Electric CHF Test Conditions

Run Number	Inlet Temperature ($^\circ$F)	Pressure (psia)	Average Mass Flux (Mlb/hr-ft^2)	Average Heat Flux (MBtu/hr-ft^2)
266	539.0	1005.0	1.008	0.510
279	486.4	1000.0	0.500	0.474
286	510.7	997.0	0.249	0.289
296	534.5	1000.0	1.248	0.522

Table 7

General Electric CHF Results

Run Number	z_{CHF} (in.)		$\bar{x}_{CHF}$	
	Experiment	Calculation	Experiment	Calculation
266	71.5 in.	80 in.	0.2957	0.3048
279	71.5	72	0.4885	0.4869
286	71.5	55	0.6334	0.6016
296	71.5	84	0.2338	0.2809

against twelve steady-state tests from four electrically heated rod bundle experiments. The experiments included non-uniform and uniform axial and radial power distributions. The tests also covered a broad range of pressures, mass fluxes, heat fluxes and inlet qualities for several rod geometries. RELAP5YA predictions were in generally good agreement with the test measurements. The maximum difference between the predicted and experimental critical qualities is 20 percent. Results from system tests, such as LOFT, THTF and TLTA, described in Reference 1 provide added confidence in the use of this model.

Table 8

ORNL THTF CHF Test Conditions

Test Number	Inlet Temperature (°F)	Pressure (psia)	Average Mass Flux (Mlb/hr-ft^2)	Average Heat Flux (MBtu/hr-ft^2)
3.07.9B	590.2	1849	0.525	0.29
3.07.9K	410.0	635	0.166	0.14
3.07.9X	513.9	872	0.250	0.19

Table 9

ORNL THTF CHF Results

Run Number	Z_{CHF} (in.)		$\overline{X}_{CHF}$	
	Experiment	Calculation	Experiment	Calculation
3.07.9B	59.7	56.	0.368	0.34
3.07.9K	118.9	99.	0.887	0.72
3.07.9X	108.3	90.	0.84	0.71

REFERENCES

1. Fernandez, R. T., et al., "RELAP5YA, A Computer Program for Light-Water Reactor System Thermal-Hydraulic Analysis", Yankee Atomic Electric Company, October 1982.

2. Biasi, L., et al., "Studies on Burnout, Part 3", Energia Nucleare, 14, 530 (1967).

3. Griffith, P., et al., "Critical Heat Flux During a Loss-of-Coolant Accident", Nuclear Safety, 18, May-June 1977.

4. Ransom, V. H., et al., "RELAP5/MOD1 Code Manual", EGG-2070 DRAFT, EG and G Idaho, Inc., Idaho Falls, ID, November 1980.

5. Leung, J. C., "Transient Critical Heat Flux and Blowdown Heat Transfer Studies", ANL/RAS/LWR-80-2, Argonne National Laboratory, April 1980.

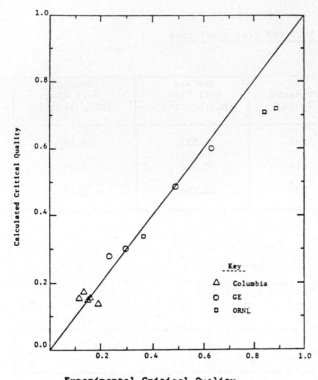

Figure 2   Comparison of Calculated and Measured
Critical Qualities

6.   Hsu, Y. Y., and H. Sullivan, "Updating of 'Best-Estimate' Heat Transfer
     Recommendation for Transient CHF and Post-CHF Heat Transfer Modes",
     CSNI Two-Phase Specialists' Meeting, Pasadena, California, 1982.

7.   Phillips, R. E., R. W. Shumway and K. H. Chu, "Improvements to the
     Prediction of Boiling Transition During Boiling Water Reactor
     Transients", 20th ASME/AICHE National Heat Transfer Conference,
     Milwaukee, Wisconsin, August 1981.

8.   "Critical Heat Flux Correlation for CE Fuel Assemblies with Standard
     Spacer Grids, Parts 1, 2, Non-Uniform Axial Power Distribution",
     Combustion Engineering Topical Report CENPD-207, June 1976.

9.   Electrical Power Research Institute Report, EPRI-RP-813-1, to be
     published.

10. Janssen, E., "Two-Phase Flow and Heat Transfer in Multirod Geometries,
     Final Report", General Electric Company Report GEAP-13347, March 1971.

11. Yoder, G. L., et al., "Dispersed Flow Film Boiling in Rod Bundle
     Geometry - Steady-State Heat Transfer Data and Correlation
     Comparisons", ORNL/5822, to be published.

Multi-Phase Flow and Heat Transfer III. Part B: Applications
edited by T.N. Veziroğlu and A.E. Bergles
Elsevier Science Publishers B.V., Amsterdam, 1984 — Printed in The Netherlands

A MECHANISTIC JET PUMP MODEL FOR BWR TRANSIENTS AND ACCIDENTS

Jamal Ghaus and R. T. Fernandez
Yankee Atomic Electric Company
Framingham, Massachusetts 01701, U.S.A.

ABSTRACT

A mechanistic jet pump model has been developed for Boiling Water
Reactor (BWR) safety analysis. This model is intended to predict the jet
pump behavior during normal operation and abnormal BWR conditions, including
Loss-of-Coolant Accident (LOCA) events. The momentum mixing phenomena that
occurs in the jet pump throat region during normal operation was accounted
for by applying integral mass and momentum equations. The flow dependent
mechanical energy losses that dominate during abnormal conditions were
developed by utilizing integral mass and mechanical energy conservation
equations and empirical data.

The generalized jet pump model was implemented into the RELAP5YA [1]
computer code. This code is a modified version of the RELAP5 MOD1 [2].

The generalized jet pump model was benchmarked against 1/6 scale jet
pump data [3]. These test data include single-phase, steady-state
conditions and two-phase transient conditions. The comparison of model
predictions to data indicate that the jet pump model can predict the
relevant hydrodynamic phenomena that occur in BWR jet pumps during normal
and off-normal operating conditions.

1. INTRODUCTION

Jet pumps are an integral part of modern Boiling Water Reactor (BWR)
Recirculation Systems [4]. The flow rate through a BWR core is strongly
influenced by the behavior of the jet pumps. This necessitates accurate
modeling of the jet pumps to predict the behavior of a BWR System during
normal operation, transients and accidents. During normal jet pump
operation, a high velocity jet exiting from the drive nozzle imparts
momentum to the suction stream causing it to flow in the same direction.
The momentum transfer from the drive to suction stream causes a flattening
of the radial velocity profile and an asymptotic pressure rise in the throat
as shown in Figure 1. The diffuser, located downstream of the throat,
decelerates the flow and further increases the static pressure. Abnormal
operating conditions encountered during BWR transients and accidents can
lead to the reversal of any one or all three of the drive, suction and
throat discharge streams as indicated by the schematics in Table 2.

The publicly released version of RELAP5 MOD1 [2] does not simulate the momentum mixing phenomenon that occurs in the throat region during normal operation of a jet pump [5]. Additionally, for off-normal flow configurations, the mechanical energy loss coefficients for the jet pump throat region are flow dependent, but RELAP5 MOD1 does not provide for this. To overcome these deficiencies, a new jet pump model was added in the RELAP5YA computer code. This paper discusses the development of a mechanistic model for predicting jet pump behavior, its implementation into the RELAP5YA code and the model assessment results.

2. MODEL DEVELOPMENT

The main focus of the development effort was to model the hydrodynamic phenomena in the throat region of a jet pump. The diffuser and tail pipe regions can be modeled by the one-dimensional, two-fluid field equations in

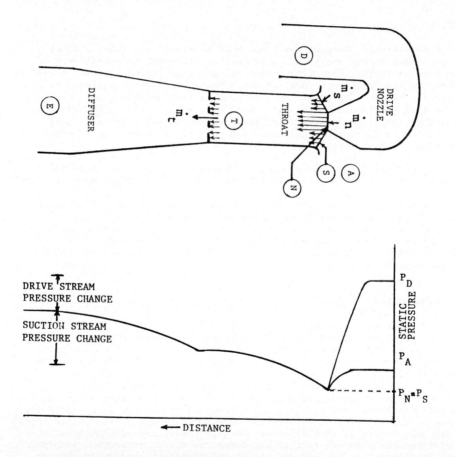

Figure 1
JET PUMP

RELAP5 [2]. The strategy for developing a generalized jet pump model was to first develop a model for single-phase, steady-state conditions. This model was subsequently extended to two-phase conditions and implemented into the RELAP5YA code.

To obtain the single-phase, steady-state model for the throat region, a control volume (see Figure 1) bounded by planes S, N, T and the solid boundary was defined. For single-phase, steady-state flow, the following assumptions were made:

a. At the inlet of the throat, the static pressure in the drive stream was assumed to be equal to that in the suction stream.

$$P_N = P_S = P_i \tag{1}$$

Table 1

Jet Pump Nomenclature

SYMBOL	DEFINITION
P_i	Static pressure at location i
$\overline{P}_i$	$(P_i + \frac{1}{2} \rho_i V_i^2)$ Total pressure at location i.
$\dot{m}_i$	Mass flow rate at i.
$\dot{Q}_i$	Volumetric flow rate at i
H_i	$(\frac{1}{2} \rho_i V_i^2)$ Dynamic head at i.
EL_{ij}	Total mechanical energy losses from station i to station j.
ELF_{ij}	Specific mechanical energy losses from station i to station j.
K_{ij}	Form loss coefficient for mechanical energy loss from station i to station j.
K_{mxr}	Frictional energy loss coefficient for throat region.
ρ_j	Density at location j.
α_g	Gases void fraction.
α_f	$(1-\alpha_g)$ liquid void fraction.
G_j	Mass flux at location j.

Table 2

Jet Pump Conservation Equations

	$-\infty < M < -1$	$-1 \leq M \leq 0$	$0 < M$
	POSITIVE DRIVE	POSITIVE DRIVE	POSITIVE DRIVE
CONTINUITY	$\dot{m}_N - \dot{m}_S = -\dot{m}_T$ (c-1)	$\dot{m}_N - \dot{m}_S = \dot{m}_T$ (b-1)	$\dot{m}_N + \dot{m}_S = \dot{m}_T$ (a-1)
MECH. ENERGY	$\dot{Q}_N\bar{P}_D - \dot{Q}_S\bar{P}_A - \dot{Q}_T\bar{P}_T = EL_{TH} + EL_{DN} + EL_{SA}$ (c-2)	$\dot{Q}_N\bar{P}_D - \dot{Q}_S\bar{P}_A - \dot{Q}_T\bar{P}_T = EL_{TH} + EL_{DN} + EL_{SA}$ (b-2)	$\dot{Q}_N\bar{P}_D + \dot{Q}_S\bar{P}_A - \dot{Q}_T\bar{P}_T = EL_{TH} + EL_{DN} + EL_{AS}$ (a-2)
	NEGATIVE DRIVE	NEGATIVE DRIVE	NEGATIVE DRIVE
CONTINUITY	$-\dot{m}_N + \dot{m}_S = \dot{m}_T$ (f-1)	$-\dot{m}_N + \dot{m}_S = -\dot{m}_T$ (e-1)	$-\dot{m}_N - \dot{m}_S = -\dot{m}_T$ (d-1)
MECH. ENERGY	$-\dot{Q}_N\bar{P}_D + \dot{Q}_S\bar{P}_A - \dot{Q}_T\bar{P}_T = EL_{TH} + EL_{TA}$ (f-2)	$-\dot{Q}_N\bar{P}_D + \dot{Q}_S\bar{P}_A + \dot{Q}_T\bar{P}_T = EL_{TH} + EL_{TA}$ (e-2)	$-\dot{Q}_N\bar{P}_D - \dot{Q}_S\bar{P}_A + \dot{Q}_T\bar{P}_T = EL_{TH} + EL_{TA}$ (d-2)

where P_N and P_S are the static pressures at plane N and plane S shown in Figure 1. These two planes form the inlet surface for the throat region, defined as surface i.

b. A flat velocity profile was assumed at the throat exit plane. A step function velocity profile was assumed at the inlet surface i. This implies that at the inlet plane, the total pressure $(\overline{P})$ in the drive stream is not equal to that in the suction stream.

c. Incompressible, adiabatic flow is assumed; this assumption is relaxed for the extension of the model to two-phase, transient conditions.

With these assumptions, the objective of the steady-state model development effort is to obtain correlations for $(\overline{P}_D - \overline{P}_T)$ and $(\overline{P}_A - \overline{P}_T)$ for each of the flow configurations shown in Table 2. The correlations will be expressed in terms of density, velocity, jet pump geometry and empirical constants:

$$\overline{P}_D - \overline{P}_T = f\ (\rho, V, \mathrm{Geom}, K) \tag{2}$$

$$\overline{P}_A - \overline{P}_T = f\ (\rho, V, \mathrm{Geom}, K). \tag{3}$$

To obtain these correlations, conservation principles were utilized.

For normal operation (Case a), conservation of mass for the throat control volume is given by:

$$\dot{m}_N + \dot{m}_S - \dot{m}_T = 0. \tag{4}$$

The conservation of mechanical energy equation, which accounts for losses in the drive line, the suction inlet and the throat region, is given by:

$$\dot{Q}_N \overline{P}_D + \dot{Q}_S \overline{P}_A - \dot{Q}_T \overline{P}_T = EL_{DN} + EL_{AS} + EL_{TH}. \tag{5}$$

where EL_{DN} is the mechanical energy loss between station D and plane N, and EL_{AS} represents losses between station A and plane S. The term EL_{TH} represents the sum of all the losses in the throat region.

The mass and mechanical energy conservation equations for the other flow configurations are given in Table 2, where in-flow quantities are defined as positive and out-flow quantities as negative. The flow configuration in a jet pump is defined by specifying the direction of the drive flow and the parameter, M, defined as:

$$M = \frac{\dot{m}_S}{\dot{m}_N} \tag{6}$$

To derive correlations of the form given by Equation (2) and Equation (3), the equations of Table 2 were partitioned. The partitioning process consisted of substituting the appropriate continuity equation into the associated mechanical energy equation given in Table 2. The resulting equation was partitioned into two mechanical energy equations, one for each stream through the jet pump. The mechanical energy equations obtained by the partitioning procedure were normalized by the associated flow rate to obtain the specific mechanical energy equations (Bernoulli form) given in Table 3.

In the equations of Table 3, the terms ELF_{DN}, ELF_{AS}, ELF_{ND} and ELF_{SA} represent the flow independent mechanical energy losses in the drive and suction nozzles adjacent to the throat region. These losses can be represented in terms of loss coefficients derived from standard handbooks [6] for a particular jet pump geometry and the associated dynamic head. Additional mechanical energy loss terms ELF_{ij} represent the flow dependent losses. These terms arise from partitioning the term EL_{TH} of the equations of Table 2. These terms represent the mechanical energy changes in the throat region and are discussed in the following sections.

2.1 Normal Flow Configuration

This flow configuration is representative of the normal operation of a jet pump and is shown in Table 2 as Case a. The term ELF_{NT} which represents the mechanical energy change in the drive flow between planes N and T can be written as:

$$ELF_{NT} = f(\rho, V, Geom) \tag{7}$$

or
$$ELF_{NT} = \bar{P}_N - \bar{P}_T \tag{8}$$

$$ELF_{NT} = (P_N - P_T) + (H_N - H_T) \tag{9}$$

Table 3
Jet Pump Mechanical Energy Equations

Case	Drive Flow Direction	Range of M	Partitioned Mechanical Energy Equations	
a	P O S I T I V E	$M \geq 0$	$\bar{P}_D - \bar{P}_T = ELF_{NT} + K_{DN}H_N$ (a-3)	$\bar{P}_A - \bar{P}_T = ELF_{ST} + K_{AS}H_S$ (a-4)
b		$-1 \leq M \quad 0$	$\bar{P}_D - \bar{P}_A = ELF_{NS} + K_{DN}H_N + K_{SA}H_S$ (b-3)	$\bar{P}_D - \bar{P}_T = K_{DN}H_N + ELF_{NT}$ (b-4)
c		$M < -1$	$\bar{P}_D - \bar{P}_A = ELF_{NS} + K_{DN}H_N + K_{SA}H_S$ (c-3)	$\bar{P}_T - \bar{P}_A = ELF_{TS} + K_{SA}H_S$ (c-4)
d	N E G A T I V E	$M \geq 0$	$\bar{P}_T - \bar{P}_D = ELF_{TN} + ELF_{ND}$ (d-3)	$\bar{P}_T - \bar{P}_A = ELF_{TA}$ (d-4)
e		$-1 \leq M \leq 0$	$\bar{P}_T - \bar{P}_D = ELF_{TN} + ELF_{ND}$ (e-3)	$\bar{P}_A - \bar{P}_T = ELF_{AT}$ (e-4)
f		$M \leq -1$	$\bar{P}_A - \bar{P}_D = ELF_{ND} + ELF_{AN}$ (f-3)	$\bar{P}_A - \bar{P}_T = ELF_{AT}$ (f-4)

The term $(P_N - P_T)$ is evaluated by applying the axial component of the integral momentum equation to the throat control volume. Considering previously stated assumptions, the momentum equation for the control volume is:

$$P_T A_T - P_N A_N - P_S A_S Cos\emptyset + \int\limits_{A_T}^{A_i} P(Z)dA = \rho V_N^2 A_N$$

(10)

$$+ \rho V_S^2 A_S Cos\emptyset - \rho V_T^2 (1 + \frac{K_{mxr}}{2}) A_T$$

where

$$P_T A_T, \ P_N A_N, \ P_S A_S Cos\emptyset = \text{Vertical forces on the fluid control volume.}$$

$$\int\limits_{A_T}^{A_i} P(Z)dA_H = \text{Vertical force due to elliptic contour of the suction nozzle.}$$

$$dA_H = \text{Horizontal component of differential area in suction nozzle.}$$

$$A_i = A_N + A_S Cos\emptyset$$

(11)

$$\rho V_N^2 A_N, \ \rho V_S^2 A_S Cos\emptyset = \text{Momentum into the control volume.}$$

$$\rho V_T^2 (1 + K_{mxr}/2) A_T = \text{Momentum out of the control volume plus the force due to wall friction in the control volume.}$$

The quadrature in Equation (10) was approximated by assuming that the static pressure in the short suction nozzle remains constant; therefore:

$$\int\limits_{A_T}^{A_i} P(Z)dA_H = P_i(A_i - A_T)$$

(12)

Substituting Equations (1), (11) and (12) into Equation (10) yields:

$$P_T A_T - P_i A_i + P_i A_i - P_i A_T = \rho V_N^2 A_N + \rho V_S^2 A_S Cos\emptyset$$

$$- \rho V_T^2 (1 + K_{mxr}/2)A_T$$

Dividing the above equation by A_T and simplifying, yields:

$$P_T - P_i = \rho V_N^2 \frac{A_N}{A_T} + \rho V_S^2 \frac{A_S Cos\emptyset}{A_T} - \rho V_T^2 (1 + K_{mxr}/2)$$

(13)

Utilizing Equation (1) in Equation (13) and substituting the resulting equation into Equation (9) yields:

$$ELF_{NT} = -\left\{\left[\frac{A_N}{A_T} - \frac{1}{2}\right]\rho V_N^2 + \frac{A_S Cos\emptyset}{A_T}\rho V_S^2 - \frac{1}{2}(1 + K_{mxr})\rho V_T^2\right\} \qquad (14)$$

The second term on the right-hand side of Equation (a-3) of Table 3 represents the mechanical energy losses in the drive line and is given by:

$$ELF_{DN} = K_{DN}\rho\frac{V_N^2}{2} \qquad (15)$$

Substituting Equation (15) and Equation (14) into Equation (a-3) of Table 3 and simplifying, yields:

$$\overline{P}_D - \overline{P}_T = -\left\{\left[\frac{A_N}{A_T} - \frac{1}{2}\right]\rho V_N^2 + \frac{A_S Cos\emptyset}{A_T}\rho V_S^2 - \frac{1}{2}(1 + K_{mxr})\rho V_T^2\right\}$$
$$+ K_{DN}\rho\frac{V_N^2}{2} \qquad (16)$$

An expression for the term ELF_{ST} in Equation (a-4) of Table 3 was developed in a similar way as that for ELF_{NT}, where:

$$ELF_{ST} = \overline{P}_S - \overline{P}_T \qquad (17)$$

$$ELF_{ST} = (P_S - P_T) + H_S - H_T \qquad (18)$$

From Equation (1) ($P_S = P_i$), Equation (13) can be utilized in Equation (18) to yield:

$$ELF_{ST} = -\left\{\rho V_N^2\frac{A_N}{A_T} + \frac{A_S Cos\emptyset}{A_T} - \frac{1}{2}\quad\rho V_S^2 - \frac{1}{2}(1 + K_{mxr})\rho V_T^2\right\} \qquad (19)$$

The term ELF_{AS} in Equation (a-4) is formulated as:

$$ELF_{AS} = K_{AS}\rho\frac{V_S^2}{2} \qquad (20)$$

Substituting Equation (19) and Equation (20) into Equation (a-4) and simplifying, yields:

$$\overline{P}_A - \overline{P}_T = -\left\{\frac{A_N}{A_T}\rho V_N^2 + \left[\frac{A_S Cos\emptyset}{A_T} - \frac{1}{2}\right]\rho V_S^2 - \frac{1}{2}(1 + K_{mxr})\rho V_T^2\right\}$$
$$+ K_{mxr}\rho\frac{V_S^2}{2} \qquad (21)$$

2.2 Off-Normal Flow Configurations

For the configurations associated with Case b through Case f of Table 2, the jet pump acts as a resistance in the flow path in contrast to an inductor for Case a. Therefore, the energy loss terms, ELF_{ij}, in Equations (b-3) through (f-4) are formulated as:

$$ELF_{ij} = K_{ij} \, f(\rho, V)$$

where K_{ij} are empirical constants and $f(\rho, V)$ are dynamic heads at the associated locations.

The equations of Table 3 and the correlations for ELF_{ij} represent the jet pump model for single-phase, steady-state conditions. This model will be generalized to two-phase conditions in the following section.

2.3 Extension to Two-Phase Conditions and Implementation

The jet pump model developed in the preceding section was extended to two-phase conditions by utilizing the following definitions and assumptions:

a. The density in the drive nozzle, suction nozzle and the throat region was defined as:

$$\rho_j = (\alpha_g \rho_g + \alpha_f \rho_f)_j$$

b. The velocity head at any location, j, was defined as:

$$1/2 \rho_j V_j = G_j^2 / 2 \rho_j$$

c. The local loss coefficients for the liquid and vapor phases were assumed to be equal.

d. The incompressible adiabatic flow assumption was removed.

These assumptions were used to generalize the correlations for ELF_{ij} to two-phase conditions. The generalized correlations were implemented into the RELAP5YA code in a new subroutine. This new subroutine determines the flow configuration in the jet pumps and evaluates the associated correlations. The momentum equation in the code is modified to account for these additional terms.

Specifically, for the flow configuration corresponding to normal operation, the correlations for ELF_{NT} and ELF_{ST} account for the momentum mixing effect in the throat region. The remaining ELF_{ij} terms represent the flow dependent form loss coefficients. The dissipative energy associated with mechanical energy loss terms is accounted for in the mixture energy equation.

3. MODEL ASSESSMENT

The RELAP5YA jet pump model was assessed against the 1/6 scale model jet pump data. That included both single-phase, steady-state and two-phase transient tests.

3.1 Single-Phase, Steady-State Tests

The steady-state tests yielded pressure differences between the locations shown in Figure 2 for predetermined drive and suction flow rates. The facility and test procedure details are reported in Reference 4. These tests were modeled in RELAP5YA as shown in Figure 3. The steady-state performance of the jet pump model was assessed by comparing several calculated pressure differences to data.

The pressure difference across the jet pump diffuser is a function of the diffuser flow rate. Figure 4 shows this pressure difference plotted against the dynamic head ($\rho V^2/2$) at the inlet of the diffuser. The fact that the data lie on straight lines indicates that the forward and reverse flow loss coefficients for the diffuser are well represented by constant values.

The pressure difference between the drive line and throat region and that between the suction plenum and throat region, are dependent upon the direction and magnitude of the drive flow rate and parameter M. Results from the forward drive flow are presented first and those from the reverse drive flow tests next.

For forward drive flow condition, the measured pressure difference between the drive line and throat exit is normalized by drive nozzle velocity head ($\rho V^2/2$) and plotted versus M in Figure 5. The RELAP5YA calculated values are shown by a solid line on the same figure. The normalization of the pressure difference by the drive nozzle velocity head collapses the data obtained for many different drive flow rates onto a definite trend curve as seen in the figure. This facilitates the comparison of RELAP5YA calculated values to the data in a codified manner. Similarly, the normalized pressure difference between the suction plenum and the throat exit is plotted versus M in Figure 6. Both the RELAP5YA calculated values and the data are shown in this figure.

The calculated pressure differences for the reverse drive flow condition are compared to data in Figure 7 and Figure 8. The normalized pressure difference between the drive line and throat region is plotted versus M in Figure 7 and that between suction plenum and throat region is plotted versus M in Figure 8.

The characteristics of a jet pump are usually indicated by performance (N versus M) curves, where:

N is defined as:

$$N = \frac{\text{Specific Mechanical Energy Change in Suction Flow}}{\text{Specific Mechanical Energy Change in Drive Flow}}$$

According to Figure 1:

$$N = \frac{\overline{P}_E - \overline{P}_A}{\overline{P}_D - \overline{P}_E}$$

The parameter M has been previously defined as the ratio of suction flow rate to drive flow rate. For the jet pump modeled in RELAP5YA, the predicted and experimental characteristic curves for the positive drive flow are shown in Figure 9 and that for negative drive flow rates are shown in Figure 10.

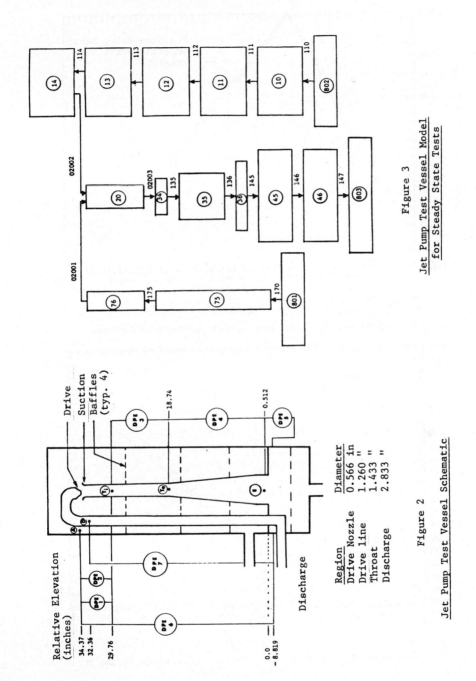

Figure 3

Jet Pump Test Vessel Model
for Steady State Tests

Figure 2

Jet Pump Test Vessel Schematic

Region	Diameter
Drive Nozzle	0.566 in
Drive line	1.260 "
Throat	1.433 "
Discharge	2.833 "

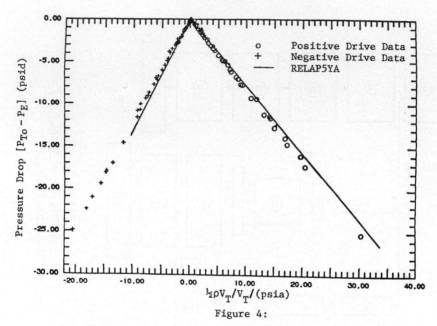

Figure 4:

Diffuser Pressure Drop versus Throat Dynamic Head

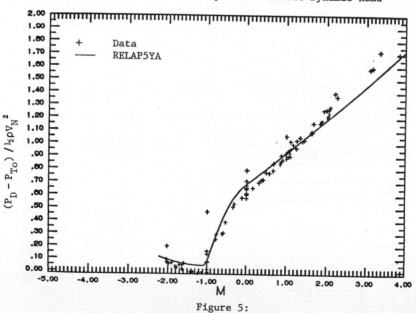

Figure 5:

Positive Drive; Normalized Pressure Difference Versus M
Drive Line to Throat

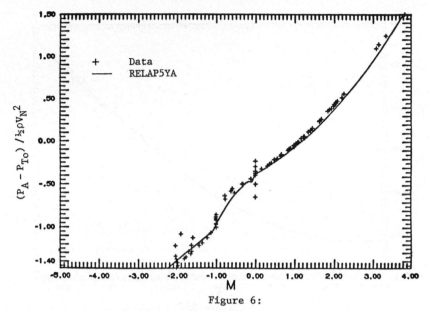

Figure 6:

Positive Drive; Normalized Pressure Difference versus M
(Suction Plenum to Throat)

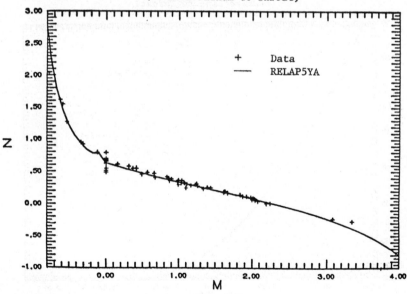

Figure 7:

Positive Drive; N versus M

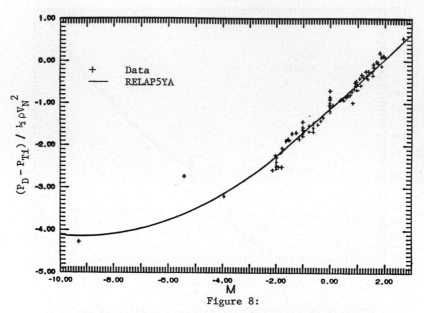

Figure 8:

Negative Drive; Normalized Pressure Difference Versus M
(Drive Line to Throat)

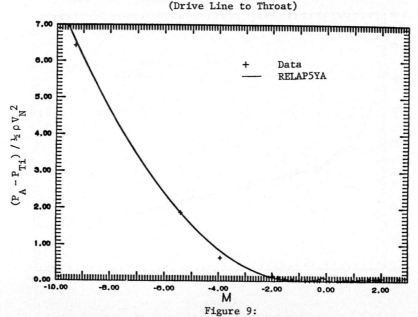

Figure 9:

Negative Drive; Normalized Pressure Difference versus M
(Suction Plenum to Throat)

The comparison of RELAP5YA results to data in Figure 4 through Figure 10 indicate that the jet pump model can predict the behavior of a jet pump well during steady-state, single-phase operation.

3.2 Two-Phase Transient Tests

To assess the performance of jet pump model under two-phase transient conditions, two blowdown tests were simulated [7]. Results from one of the tests are presented here.

The blowdown test loop is shown schematically in Figure 11. The nitrogen pressure in the blowdown vessel forced the water through the test assembly and jet pump. The valve lineup for this test produced a reverse flow through drive line. The suction and discharge flows were directed into the jet pump.

The blowdown test loop was modeled in RELAP5YA as shown in Figure 12. The blowdown pressure vessel was represented by a time-dependent volume, VOL-802. In this volume, pressure and temperature history was specified as that given in the data for PE-4 and TE-7 (see Figure 11). The system was connected to a time-dependent volume, VOL-800, which was maintained at atmospheric conditions.

Starting with water-filled stagnant conditions, the system blowdown was simulated. The calculated pressure histories in drive (P_{DR}) suction (P_{SU}) and discharge (P_{DG}) lines are compared to data in Figure 13. The substantial pressure drop from the suction and discharge lines to the drive line is due to critical flow at the drive nozzle.

The predicted and experimental volumetric flow rates through drive (Q_{DR}) suction (Q_{SU}) and discharge (Q_{DG}) lines are shown in Figure 14. The calculated values are reasonably close to data throughout the transient. The critical flow discharge coefficient for the drive nozzle was set to unity, however it appears that a value of 0.85 might have improved the results.

The calculated density in the drive line is compared to data in Figure 15. It can be seen that the calculated density matches the data very well.

The results from this blowdown test indicate that the RELAP5YA jet pump model can predict the behavior of a jet pump under transient two-phase conditions well. This blowdown test simulated a broken loop jet pump under approximate BWR LOCA conditions. The comparison of RELAP5YA results to data indicate that the RELAP5YA jet pump model can be reliably used for BWR safety analysis.

CONCLUSION

A mechanistic model for the jet pump was developed to represent the throat region of a jet pump. The model was derived by using conservation principles. The momentum mixing phenomena was captured by applying conservation of momentum equation to the throat region. The flow dependent form loss coefficients were identified by applying the mechanical energy equation to the throat region. The model was generalized to two-phase conditions and implemented into the RELAP5YA code.

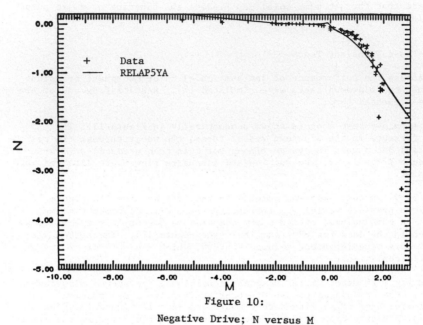

Figure 10:

Negative Drive; N versus M

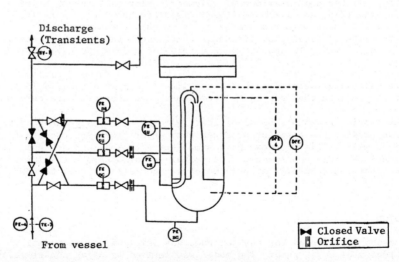

Figure 11:

Jet Pump Test Assembly Schematic

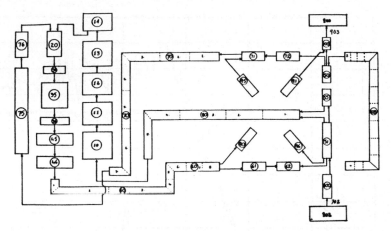

Figure 12:

RELAP5YA Model for Jet Pump Blowdown Test

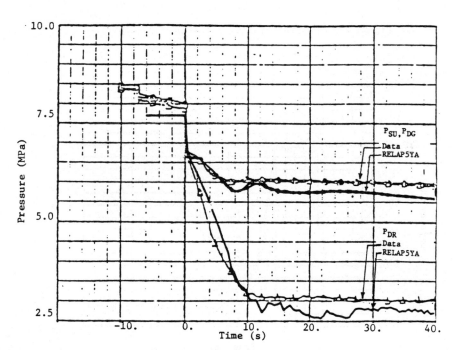

Figure 13:

Pressure in Jet Pump

44

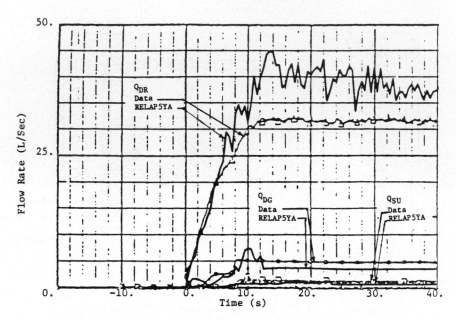

Figure 14: Volumetric Flow Rates in Jet Pump

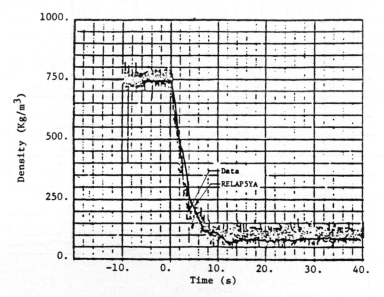

Figure 15: Density in Drive Line

The jet pump model was benchmarked against experimental data. The comparison of calculated results for steady-state tests to data indicate that the model can accurately predict the pressure differences within the jet pump. These results prove the veracity of the correlations for energy loss terms ELF_{ij}. The calculated pressure differences are a better measure of the performance of a jet pump model than the N versus M curves also presented in this paper.

The results for two-phase transient tests indicate that the RELAP5YA jet pump model can reliably predict the jet pump performance under two-phase transient conditions. Since the transient test simulated a broken loop jet pump under approximate BWR LOCA conditions, the results show that the jet pump model can be used for BWR safety analysis.

REFERENCES

1. Fernandez, R. T., et al., "RELAP5YA: A Computer Program for Light-Water Reactor System Thermal-Hydraulic Analysis," YAEC-1300 P, October 1982.

2. Ramsom, V. H., et al.,"RELAP5/MOD1 Code Manual Volume 1: System Models and Numerical Methods," NUREG/CR-1826, EGG-2070, September 1981.

3. Wilson, G. E., INEL 1/6 Scale Jet Pump Data Analysis, FG and G Report No. EGG-CAAD-5357, EG and G Idaho, Inc., February 1981.

4. Lahey, R. T., and F. J. Moody, The Thermal-Hydraulics of a Boiling Water Nuclear Reactor, ANS Monograph Series (1977).

5. Bird, R. B., W. E. Stewart and E. N. Lightfoot, Transport Phenomena, John Wiley and Sons, Inc., New York (1960).

6. Idel'chik, I. E., Handbook of Hydraulic Resistance, U.S. Atomic Energy Commission (1966).

7. Crapo, H. S., "BWR Jet Pump Report," EG and G Report No. P394, EG and G Idaho, Inc., November 1978.

Multi-Phase Flow and Heat Transfer III. Part B: Applications
edited by T.N. Veziroğlu and A.E. Bergles
Elsevier Science Publishers B.V., Amsterdam, 1984 — Printed in The Netherlands

ONCE-THROUGH STEAM GENERATOR MODULE FOR THE MODULAR MODELING SYSTEM

Wyn E. Bennett
Bechtel Power Corporation
San Francisco, California, U.S.A.

Jean-Pierre Sursock
Electric Power Research Institute
Palo Alto, California, U.S.A.

1. INTRODUCTION

In recent years, the nuclear industry has experienced a substantial shift
in the computational methods for thermal-hydraulic problems related to
safety. This shift is brought about by an increased emphasis on long duration
transients such as small-break loss of coolant accidents as opposed to design
basis accidents. Thus, along with the further refinement of large systems
codes such as RETRAN or RELAP 5, smaller, less accurate but faster running
codes are now being developed. Examples of these codes are SMABRE[1] and
DSNP[2] developed by Finland and by Argonne National Laboratory respectively. The Modular Modeling System[3] (MMS) is one such code. Its basic advantages are modularity and speed. User friendly features are based on the
application of two advanced simulator languages EASY 5[4] and ACSL[5].

The MMS code includes a large library of modules, each representing a
plant component such as pipes, pumps and heat exchangers. Both nuclear and
fossil power plants can be modeled by proper arrangement and coupling among
the modules. Extensive validation of the code has already been performed[6,7]
and additional work is now proceeding. More recently, the library has been
extended to include PWR primary loop components with two-phase flow
capability[8]. The objective is to analyze severe transients such as small-
break LOCA's and overcooling transients.

As part of this effort a Once-Through steam generator model has been
developed which addresses the major issues related to the Babcock & Wilcox
plant designs, including the so-called "Candy-Cane Physics." The objective of
this paper is to describe this model together with some ongoing verification.

The next section of the paper will describe very briefly the main
features of the two-phase modules and the role of this particular model in the
general library. The following section will discuss the technical detail of
the model and Section 4 will discuss current and future validation of this
work.

2. OVERVIEW OF THE MODULAR MODELING SYSTEM AND OF THE ONCE-THROUGH STEAM-GENERATOR MODULE

The Modular Modeling System (MMS) code was originally developed to study
operational transients in Nuclear and Fossil Power plants. Its major characteristics are reported elsewhere[3-6,7,8], but will be briefly summarized
here. The code is designed for use by utility engineers who want to perform

scoping studies of control systems or certain transient conditions (e.g., for procedures evaluation). The modular structure of the code allows a very flexible modeling of a power plant or parts of it (e.g., feedwater train). In addition to simulation, the ACSL and EASY5 languages can perform linear stability analysis (root locus, Bode and Nyquist plots) and sensitivity studies (e.g., optimization). The advanced languages which transform simple user input into FORTRAN statements, form the core of the software. Several integration routines for first order ordinary differential equations are also available (e.g., Gear algorithm, Adams-Moulton, Euler, etc.).

A library of components has also been developed. Each module represents a particular plant component such as a pump, a valve, or a steam generator. In order to simulate a Power Plant, the user must first input some information relative to the dimensions of each component to be modeled and its thermal-hydraulic characteristics. Next, he must specify the interconnection between the various components and finally he defines the boundary and initial conditions for his transient.

The level of modeling complexity can vary drastically from one module to another. Thus, if the user is unsatisfied with the accuracy of a given module he could design his own according to some clearly defined ground rules and insert it in his library.

Recently, a series of modules were developed to handle accident conditions in the primary side of Pressurized Water Reactors. Two-phase conditions can appear during small-break loss of coolant accidents (SBLOCA) and secondary induced transients. These two-phase conditions are modeled using the one-dimensional drift-flux approach and have yielded reasonably accurate and fast running models (close to real time simulation).

The steam generator model presented herein is a typical example of these two-phase modules. A trade-off is sought between accuracy (i.e., spatial resolution) and speed. The primary and secondary sides of the steam generator are divided each in several variable volumes. Inertia is neglected in the momentum equation and empirical models are often used. On the other hand, the level tracking approach adopted here and described below yields a fast running module with more than adequate accuracy.

The hot leg and candy-cane are also modeled as part of the module since usually during accidents, it is practically impossible to dissociate the processes developing in the candy-cane (e.g., steam bubble formation) from those in the tubes of the steam generator (e.g., condensation).

3. STEAM GENERATOR MODEL
3.1 Primary Side

The primary side of the Once-Through steam generator (OTSG) is divided into three or four volumes according to whether single-phase or two-phase conditions prevail in parts of the cooled length. There are also two fixed volumes corresponding to the upper and lower plena (cp. Figure 1). The tube region constitute a single volume if the liquid is subcooled. If, on the other hand, a mixture of steam and liquid enter this region, then a saturation level is computed below which the liquid is subcooled. Mass and energy balance equations are written in each volume:

$$V \dot{\rho} = V \left(\frac{\partial \rho}{\partial h} \dot{h} + \frac{\partial \rho}{\partial P} \dot{P} \right) = W_1' - W_2' - \rho \dot{V} \tag{1}$$

and

$$V\,(\rho\dot{h} - \dot{P}) = W'_1\,(h_1 - h) - W'_2\,(h_2 - h) - Q \tag{2}$$

where the dotted variables indicate the rate of change. The primed value indicate a flowrate relative to the moving boundaries of the volume V (if any) and Q is the energy out from the volume. The indices 1 and 2 refer to the inlet and outlet boundaries respectively.

The flowrate W' can be resolved into an absolute flow and boundary motion.

$$W' = W - \rho A \dot{Z} \tag{3}$$

The absolute value is computed using a steady-state orifice equation

$$W = KA\,(\Delta P - \rho g \Delta Z)^{1/2} \tag{4}$$

where ΔP and ΔZ are the pressure drop and elevation difference between two consecutive volumes. K is an empirical conductance (which includes the two-phase multiplier).

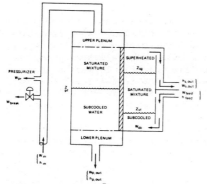

Figure 1: Schematic of the Once Through Steam Generator

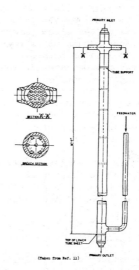

(Taken from Ref. 11)

Figure 2: IEOTSG Structure

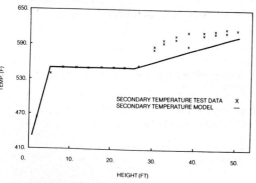

Figure 3: Steady State Secondary Temperature

The system of equations above can be solved for the rates of pressures and enthalpies if the primary were wholly in single-phase conditions (i.e., all volumes are fixed). Once the rates are known as function of the other variables, an implicit integrator (a version of the stiff Gear algorithm) is used to advance the system in time.

When two-phase conditions prevail, an additional assumption is needed to compute the saturation level (where $h = h_f$). This is done by coupling the heat transfer and the hydraulics as follows. The heat transfer above the level, Q_m, is assumed to be due entirely to condensation. Thus, the condensate flowrate is Q_m/h_{fg}. This flowrate is then used in the following quasi-steady-state momentum equation relating the pressures above and below the level.

$$P_m = P_\ell - \frac{1}{2} g \left[\rho_\ell z_f + \rho_m (L - z_f) \right] + \frac{fL}{2 \rho_f A^2 D} \left(\frac{Q_m}{h_{fg}} \right)^2 \tag{5}$$

Since P_m and P_ℓ are known, this equation is used to determine the level z_f. This procedure insures the consistency between the hydraulic and thermal processes.

3.2 Hot Leg and Candy-Cane

A similar approach is used in the hot leg and candy-cane. A single set of mass and energy balance equations is used in the hot leg. When two-phase flow occurs a level height is computed using the Sun-Duffey model[9]. The primary level in the steam generator can, of course, rise all the way up to the candy-cane. Thus, it is possible to simulate a bubble formation and collapse. It is also possible to observe reverse flow if the level on the steam generator side rises in the candy-cane faster than the hot leg level and overflows into the hot leg.

Flashing in the candy-cane is is also accounted for when the pressure there drop below the saturation pressure. The rate at which vapor is generated due to flashing is given by

$$W = \frac{\rho V}{h_{fg}} \left[x \left(\frac{1}{\rho_g} - \frac{dh_g}{dP} \right) + (1-x) \left(\frac{1}{\rho_f} - \frac{dh_f}{dP} \right) \right] \dot{P} \tag{6}$$

where V is the volume under consideration and x is the quality.

3.3 Secondary Side

The secondary side is divided into three variable volumes separated by the saturated level and the superheat level.

The mass and energy balance equations are written again, as in equations (1) and (2), but two additional assumptions are necessary:

1. The saturation level is computed using an incompressible flow approximation in the subcooled volume and

2. The rate of change of the pressure in the superheated and saturated regions are identical.

The boiling flowrate is obtained from the energy equation in the mixture region:

$$W_{boil} = \frac{1}{h_{fg}} [Q_m + K\overset{o}{P}]$$

where K is the "flashing" term.

$$K = \rho_m V_m [x_m (\frac{1}{\rho_g} - \frac{dh_g}{dP}) + (1-x_m) (\frac{1}{\rho_f} - \frac{dh_f}{dP})]$$

while the mixture level is derived from the continuity equation:

$$\overset{\cdot}{z}_m = \frac{1}{\rho_m A} [W_{dc} - W_{boil} - W_R] \tag{7}$$

where W_{dc} is the downcomer flowrate and W_R is the recirculation flowrate.

The mass and energy equations in the superheated volume are used to evaluate the secondary pressure and the secondary exit enthalpy. They are similar to equations (a-1) and (a-2).

The downcomer is treated as a single volume where the level is tracked according to:

$$\overset{\cdot}{z} = \frac{1}{\rho_f A_{DC}} (W_{feed} + W_R - W_{dc}) \tag{8}$$

where W_R is the recirculation flow rate and W_{dc} is the flow leaving the downcomer. The latter flow is computed according to a simple orifice formula, using the hydrostatic heads in the riser and in the downcomer to obtain the pressure drop.

The recirculation flowrate is computed such that a user specified enthalpy h_{dc} occur at the exit of the downcomer. A simple steady-state perfect mixing formula is applied.

$$W_R = W_{feed} \frac{h_{dc} - h_{feed}}{h_g - h_{dc}} \tag{9}$$

h_{dc} can be set to either h_f or h_{feed} corresponding to the B&W designs 177 and 205 respectively.

3.4 Heat Transfer and Tube Wall Temperature

The heat transfer is computed from the primary cooled regions to the tube walls and from the tube walls to the three secondary regions. The general form for the heat fluxes is:

$$q = U \Delta T$$

where U is an appropriate heat transfer coefficient (Dittus-Boelter, Schrock-Grossmann for boiling and Nusselt for condensation) and ΔT is a temperature difference between primary and wall or between wall and secondary.

The wall temperature T_w is computed according to:

$$\dot{T}_w = \frac{a}{MC_p} \left(q_p - q_s \right)$$

where a is a heat transfer area varying with levels position and MC_p is the thermal mass of the metal.

There are, in general, four heat fluxes to compute. These fluxes depends on the elevation of the primary (saturation) level relative to the elevation of the secondary levels. Thus, if the primary level lies between the saturation and superheat levels on the secondary side, as shown in Figure 1, the four nonzero heat fluxes are:

o   from primary saturated region to secondary superheat region

o   from primary saturated region to secondary saturated region

o   from primary subcooled region to secondary saturated region

o   from primary subcooled region to secondary subcooled region

4. COMPARISON AGAINST EXPERIMENTAL DATA

The model was compared against experimental data from a full height 19-tube steam generator tests performed at Babcock & Wilcox[10-11] (Figure 2). The primary pressure was 2210 psia. The initial steady-state temperature distribution on the secondary side is shown in Figure 3.

The transient test under consideration involved closing a valve in the steam line and, simultaneously, reducing the feedwater flowrate (Figure 4).

The resulting variations of secondary pressure and primary outlet temperature are shown in Figures 5 and 6, respectively. The agreement is seen to be very reasonable.

The data accuracy is ±6 psi for the pressure and about ±0.5°F for the temperature on the primary side.

The second transient analyzed with the model is a simulation of a loss of feedwater. The boundary conditions are shown in Figure 7. The secondary pressure and primary outlet temperature are compared against the experimental data in Figures 8 and 9.

The run time for both cases is on the order of 1/3 to 1/4 the real time on a Cyber 176. The model thus appears to perform efficiently in two fairly severe situations.

5. CONCLUSIONS

The model presented in the paper represents a Once-Through steam generator with potential two-phase conditions in the primary side. The model is based on the concept of variable volumes representing different thermodynamic conditions.

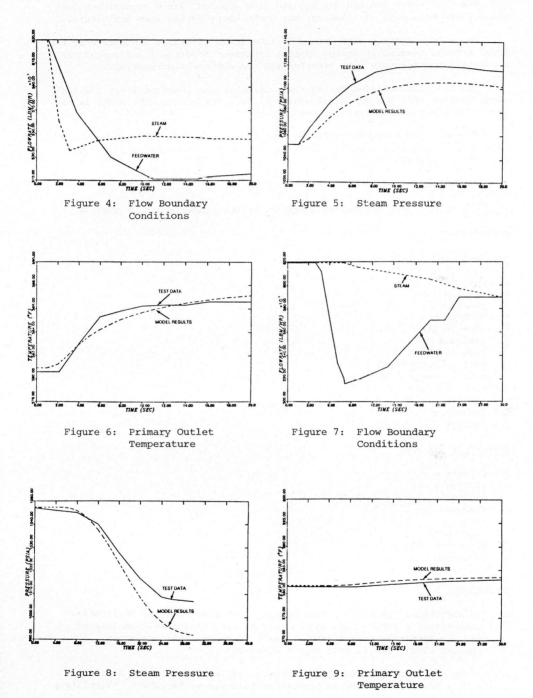

Figure 4: Flow Boundary
Conditions

Figure 5: Steam Pressure

Figure 6: Primary Outlet
Temperature

Figure 7: Flow Boundary
Conditions

Figure 8: Steam Pressure

Figure 9: Primary Outlet
Temperature

The candy-cane and the hot leg are also modeled, since significant two-phase phenomena, such as flashing, may occur there during some transients.

Comparison against data from a 19-tube steam generator indicate that the model behaves reasonably during typical transient conditions and reproduces accurately steady-state heat transfer and temperature distributions.

The run time achieved during these calculations (approximately 1/4 real time) implies that the numerical scheme is very efficient. The model is now part of the Modular Modeling System library.

Further qualification tests are expected as data for OTSG become available. The model will also be tested as part of a system analysis of Three Mile Island transients.

6. ACKNOWLEDGMENTS

One of us (WEB) acknowledges support by EPRI under contract RP694-9.

Nomenclature

a: heat transfer area
A: flow cross section
f: flow resistance
g: gravitational constant
h: enthalpy
K: conductance
L: tube length
P: pressure
q: heat flux
Q: heat transfer rate
U: heat transfer coefficient
W: mass flowrate
X: quality
Z: elevation
ρ: density

Subscripts

p: primary
s: secondary
ℓ: liquid
m: mixture
v: vapor
cc: candy-cane
f: saturated liquid
g: saturated vapor

REFERENCES

1. J. Miettinen, "SMABRE - A Fast Running Simulator Code for Small Break Analysis of a PWR." "ANS Specialists Meeting on Small-Break Loss of Coolant Accidents - Analyses in LWRs", Monterey, California (August 25-27, 1981), EPRI Report WS-81-201.

2. D. Saphier and J. T. Madell, "Analyzing Nuclear Power Plant Accidents with a New Special Purpose Simulation Language - The DSNP." "Simulation

Methods for Nuclear Power Systems." Conference-Tucson, Arizona – Proceedings May 1981, EPRI Report WS-81-212, (Ed. D. L. Hetrick and P. G. Bailey).

3. R. Dixon et al., "A Modular Modeling System for Power Plant Simulation." EPRI Report WS-81-212, (May 1981).

4. J. D. Burroughs, "Mainstream-EKS EASY5 Dynamic Analysis User's Guide." Boeing Computer Services Company, Report 10208-127, Revision 1, July 1981.

5. Mitchell and Gauthier Associates, "Advanced Continuous Simulation Language (ACSL), User Guide/Reference Manual." 2nd Edition 1981.

6. P. N. DiDomenico, "Power Plant Performance Modeling." EPRI Report CS/NP-2086 (November 1981).

7. A. B. Long and J. P. Sursock, "The Modular Modeling System Code: Review and Qualification." NRC 9th Water Reactor Safety Research Information Meeting, October 26-28, 1981.

8. J. P. Sursock and A. B. Long, "The Modular Modeling System: Two-Phase Flow Modules." 10th IMACS World Congress on System Simulation and Scientific Computation, August 8-13, 1982, Montreal, Canada Proceedings.

9. K. H. Sun, R. B. Duffey, and C. M. Peng, "The Prediction of Two-Phase Mixture Level and Hydrodynamically Controlled Dry-Out Under Low Flow Conditions." Int. J. of Multiphase Flow, Vol. 7, No. 5 (1981).

10. G. W. Loudin and W. J. Oberjohn, "Transient Performance of a 19-Tube Nuclear Integral Economizer Once-Through Steam Generator (OTSG)." Alliance Research Center Report 4679, Babcock & Wilcox (January 1976).

11. J. C. Lee et al., "Transient Modeling of Steam Generator Units in Nuclear Power Plants: Computer Code TRANSG-01." EPRI Report NP-1368 (March 1980).

Multi-Phase Flow and Heat Transfer III. Part B: Applications
edited by T.N. Veziroğlu and A.E. Bergles
Elsevier Science Publishers B.V., Amsterdam, 1984 — Printed in The Netherlands

57

A GENERALIZED FINITE DIFFERENCE MODEL FOR PHWR SYSTEM BLOW DOWN

S.K. Gupta, H.G. Lele and L.G.K. Murthy
Reactor Analysis and Studies Section
Bhabha Atomic Research Centre
Bombay 400 085, INDIA

ABSTRACT

This paper deals with the development of a computer code based on a generalized finite difference scheme for simultaneous solution of one dimensional, one fluid partial differential equations. The mass, momentum and energy equations at point nodes are used for this analysis. Comparison of this technique with contemporary mass centred control volume approach has been discussed. The method has been applied for validation to published experimental results. The results obtained with this method for analysis of a Loss of Coolant Accident (LOCA) for a single ended rupture of the inlet cold header of Rajasthan Atomic Power Plant (RAPP) are presented.

1. INTRODUCTION

The Rajasthan Atomic Power Plant is a CANDU type heavy water reactor. A rupture of the pressure coolant boundary would result in the rapid blow down of heavy water from the reactor coolant system. Consequently, a rapid degradation in the heat removal from the fuel rods would set in. This could result in the undesirable rise of fuel temperatures. The design basis accident for a reactor of this type is normally a rupture in the cold inlet header as it would lead to the worst consequences in terms of loss of coolant inventory from the system and rise of temperatures in the core. If not taken care of properly by installing emergency core cooling systems (ECCS) it could subsequently result in the melt-down of the core. Thus, it becomes important to accurately predict the system transients during blow down period so that proper ECCS could be designed.

2. OBJECTIVE

Most of the contemporary codes available for estimating system transients during blow down are displaced mesh control volume codes. The control volume approach is a lumped parameter approach, averaging thermodynamic properties in a control volume. For a reactor with pressure tube type of core, this approach becomes physically less realistic due to the complexity of the geometry. A code which assumes the variation of the properties between the nodes and with the ability to solve all the three conservation equations at the same point, would represent the system better. The code GFD1 has been developed with above mentioned principles.

3. DESCRIPTION OF CODES

Two computer codes GFD1 and NODE1 have been described. NODE1 is a control volume displaced mesh code while GFD1 is based on a generalized finite difference model, point code.

3.1 GFD1

The code is based on a set of one dimensional equations of conservation of mass, momentum and energy, assuming thermal equilibrium between vapour and liquid phases at a point. The governing equations are as follows:

$$A \frac{dP}{dt} + \frac{\partial (GA)}{\partial z} = 0 \tag{1}$$

$$A \frac{dh}{dt} + \frac{\partial (GAh)}{\partial z} = Q \tag{2}$$

$$A \frac{dG}{dt} + \frac{\partial (G^2 vA)}{\partial z} + \frac{AdP}{dz} + A\rho g \frac{dY}{dz} + F + \rho W = 0 \tag{3}$$

where A is flow area, F is frictional force per unit volume, G is mass velocity, h is enthalpy, P is pressure, Q is external heating rate per unit volume, t is time, v is mixture specific volume, W is shaft work per unit mass per unit length, Y is elevation, Z is space coordinate, e is mixture density. The equation of state used is in the form $P = P(h, \rho)$.

Martinelli and Nelson's factors are used as two phase friction factors with appropriate flow corrections as developed by Jones (Ref.1). Heat addition from the fuel rods and walls is calculated by solving the two dimensional conduction equation, using the generalized finite difference scheme. The fuel rod clad includes a Baker-just model for metal-water reaction if such a reaction occurs during the transient. The choked flows from the rupture are calculated using Moody's model as fitted into a formula as functions of pressure and enthalpy of the break node fluid[2].

The equations (1) to (3) are solved using the generalized finite difference scheme as described below.

3.2 The Generalized Finite Difference Scheme

The system is divided into a number of point nodes. Around, any node, all variables are assumed to vary in a parabolic form which is similar to the shape function as used in the finite element formulations

$$y = a_o + a_1 x + a_2 x^2 \qquad \qquad \ldots (4)$$

where y is the variable, a_o, a_1, a_2 are constants over a time-step and x is the space coordinate.

At any known instant designated 'n' variables at all the nodes are known from the earlier solution or from the initial conditions. For a node, the values of constants a_o, a_1 and a_2 for a time-step are calcu-lated by substituting the values of the variable in the neighbourhood, in equation (4) and solving it. Now, by differentiating this equation for each variable, the first and second order derivatives, to be used in equations (1) to (3), are estimated. And then, the equations are integrated over the time-step by any of the known standard techniques. The technique used in GFD1 is

$$\left(\frac{dF}{dt}\right)_n = \frac{F_{n+1} - F_n}{\triangle t} \qquad \qquad \ldots (5)$$

where F is the unknown variable, t is the $(n+1)^{th}$ time-step and $\triangle t$ is the time increment.

3.3 Prime and Minor Nodes

The GFD1 code divides the system into a number of prime nodes. The system between any two adjacent prime nodes can be divided into any number of minor nodes depending upon the space derivatives of different variables. The number of minor nodes can be changed during a transient. The variables at new nodes for the last known instant are estimated from the parabolic shape functions assumed in the region for the old minor nodes and then the transient is continued with the new nodes.

3.4 NODE1

The basic conservation equations for solution by the control volume approach are presented in Fig.1. The solution of this set of equations by the ordinary differential equation formulation and using Runge Kutta method has been presented in Ref.(3).

4. MODELLING OF SYSTEM

4.1 Nodalisation

The simulation of the experiment (4) and the actual RAPP primary heat transport system has been similar (Fig.2). The thermal input in the core region is an input as a function of time (Reactor kinetics is not considered in the code). The pumps on either side of the core perform in parallel. The inertia of the pumps is considered for estimating the pump coast down and homologous characteristics are used for single and two phase performance.

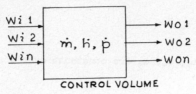

CONTROL VOLUME

1. MASS CONSERVATION $\quad \dot{m} = \mathcal{\Sigma} w$

2. ENERGY CONSERVATION $\quad \dot{h} = (\mathcal{\Sigma} wh + Q - h\dot{m} + \frac{v}{J}\dot{p})/m$

3. PRESSURE EQUATION $\quad \dot{p} = -\left[\frac{m}{\rho} + \frac{dv}{dh}\Big|_p (\mathcal{\Sigma} wh + Q - h\dot{m})\right]$

$$\left[m\,\frac{dv}{dp}\Big|_h + v\,\frac{dv}{dh}\Big|_p\right]^{-1}$$

4. MOMENTUM EQUATION $\quad \dot{W}_i = (p_i - p_{i+1} + \Delta p_{pump} + \rho\Delta h_i$

$$- K \cdot W_i^2)\,144\,g_c \left(\frac{A}{L}\right)_i$$

FIG. 1 $\quad$ EQUATION SET TO BE SOLVED

where P is pressure, h is enthalpy, e is density, v is specific volume of the mixture, m is the mass of the mixture in the central volume, W is the flow out of or into the control volume. J is a factor for conversion of units, g_c is gravitational constant and $\underset{A}{\underline{L}}$ is junction inertia.

4.2 Choice of Minor Nodes

For the region where the system is in a subcooled state, a smaller number of nodes are chosen. The liquid being highly incompressible, relatively small errors are introduced due to the fact that the spatial derivatives of the fluid velocity term is small in comparison to that in flashing and two phase flow regions. The number of nodes are increased as the system approaches saturation, crosses it and establishes itself in two phase region. After that, the number of nodes are decreased again with the assumption that again the spatial derivative of the fluid velocity does not vary severely. This can be substantiated by the fact that the void fraction versus quality curves flatten out beyond about 30 per cent quality. However, suitable criteria for adjusting the number of nodes by this method is yet to be established.

4.3 Subsequent Refining of Nodes

For adjusting the number of nodes, the spatial increment in the mesh size is altered and the same parabolic shape functions are interpolated for enlarging the number of nodes. However, for condensing the number of nodes, all that has to be done (apart from adjusting spatial increments) is just to ignore the intermediate nodes between two end nodes of the set of three nodes constituting the parabolic shape function definition. The repetition of this process will condense the nodes further in successive time-steps in a desired manner. The nodal distances (Δx's) in the adjacent nodes must not be allowed to change more than ten per cent otherwise numerical error could be introduced (5). However, a continuous

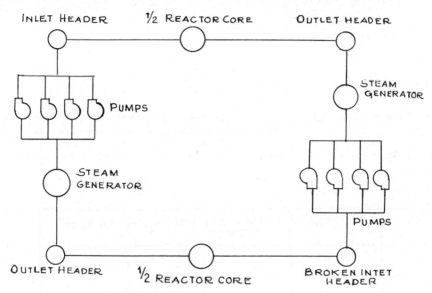

FIG. 2 NODAL DIAGRAM

gradation in the nodal distances could be resorted to.

4.4 Time-Step Control

The time-step during solution is varied depending upon the time derivatives of different variables. An estimate of the dynamic time constant of an equation can be obtained by dividing the variable value by its own derivative. At no instant, the incremental time-step is allowed to vary by more than 1/20th of the minimum dynamic time constant.

4.5 Fuel Rod Modelling

The fuel rod has been modelled using the generalized finite difference's shape function concept. The number of nodes are greatly reduced, once the initial large radial temperature gradient is reduced. A two dimensional conduction equation is used keeping in view the future modification of the code.

4.6 Break Node Modelling

The rupture nodes are defined in the input. At a ruptured node, the choked flow is calculated from Moody's model for the ruptured area. The shape functions for different flows around this area are chosen in the following manner (Fig.3).

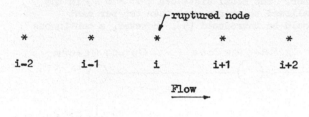

FIG. 3

If the i_{th} node in Fig.3 is the ruptured node, then based on the pressure and enthalpy of i_{th} node, choked flow from the break area (if found to occur) is calculated. Table 1 is then used for estimating (a_o, a_1, a_2) constants of equation (4) in the neighbourhood of the i_{th} node.

Table 1

NODE	VARIABLE	FLOWS FROM WHICH a_o, a_1 and a_2 ARE ESTIMATED
i-1	FLOW	W_{i-2}, W_{i-1}, $\left(W_i + W_{CHOKE}\right)$
i	FLOW	W_{i-1}, $\left(W_i + W_{CHOKE}\right)\left(W_i + W_{CHOKE}\right)$
i+1	FLOW	W_i, W_{i+1}, W_{i+2}

where W_{CHOKE} is the choked flow from the i_{th} node, W's represent the mass flows at various nodes.

The shape function constants for enthalpy for i_{th} node is also adjusted so as to account for different enthalpies of the choked flow and that of the $(i+1)^{th}$ node.

For plugging in a hydrodynamic line between any two nodes I and J, the pressures at points I and J are considered as boundary values and constants for the shape functions at the boundary nodes are estimated in a similar manner as described above.

4.7 Definition of Initial and Boundary Conditions

The initial conditions for GFD1 must be carefully prepared if the transient starts from subcooled state. Incompressibility of liquid may blow up the inaccuracies in the input or may demand extremely small time-steps thus consuming more computer time.

The GFD1 includes the provisions for estimating a transient when the boundary conditions in terms of pressures or flows are specified for a hydrodynamic line. In such cases for estimating the constants (a_o, a_1 and a_2) of equation (4) at the boundaries, the derivatives of the variables at the boundaries may be used.

5. RESULTS

The code GDF1 and NODE1 have been applied to a blow down experiment for comparing their predictions. The results of Ref.(4) which give experimental measurements of the RD-4 loop blow down in Canada have been chosen. The break node pressure with a 50% break area has been compared in Fig.4. It is noted that GDF1 overestimates the break pressure while NODE1 conservatively underestimates the break node pressure. Also, the kinks observed in the experimental results are better obtained with the NODE1 code. But the percentage variation from the experimental results is smaller in GDF predictions.

The heated section 2 outlet pressures (whose inlet header is rupturing) has been compared in Fig.5. Fig.6 illustrates the inlet pressure of the other heated section. The trend observed in Fig.4 is displayed here too. The GDF1 overestimating the pressures and NODE1 underestimating the pressures.

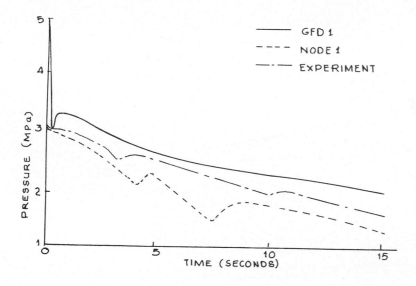

FIG.4 BREAK NODE PRESSURE

64

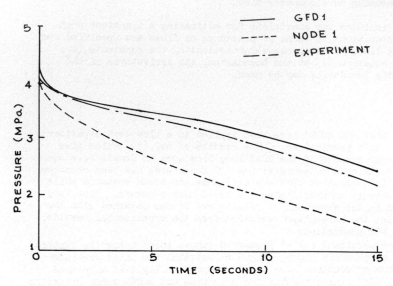

FIG.5 HEATED SECTION (BROKEN SIDE)

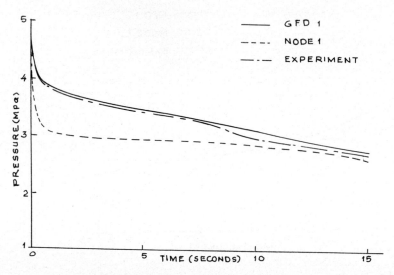

FIG.6 HEATED SECTION - INLET
(UNBROKEN SIDE)

Figs.7,8 and 9 describe the pressure transient curves for RAPP
primary heat transport system. The rupture sizes of 100, 75 and 50
per cent are assumed to take place at the inlet header of one half of
the core.

The flow through the core (near to broken point) has been displayed
in Fig.10. A sharp reversal in flow occurs which settles down to near
zero values.

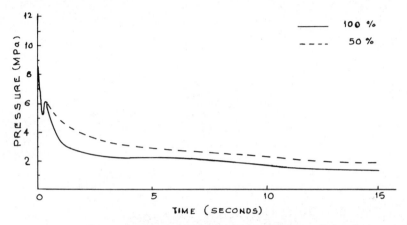

FIG.7 BREAK NODE PRESSURE (RAPP)

6. CONCLUSIONS

The results obtained indicate that the control volume code NODE1
is conservative in nature. The GDF1 overestimates the pressures and
thus delays the blow down period. It demands more work to be done in
the choice of prime nodes and selection of space meshes. For the code,
to be useful for small breaks, a multifluid model is also to be incorporated. However, the GFD method is flexible enough to handle two fluid
models.

REFERENCES

1. Jones, A.B., "Hydrodynamic Stability of a Boiling Channel", KAPL-
   2208, 1962.

2. Elliott, J.N., "RODFLOW, A Program for Studying Transients in a
   Power Reactor Coolant Circuit", TADI-11, Atomic Energy of Canada
   Limited, June 1968.

3. Lele, H.G., Gupta, S.K., and L.G.K. Murthy, "Validation and
   Comparison of Computer Codes for Reactor Safety Analysis for Loss
   of Coolant Accident", HMT-99-81, 6th National Heat and Mass
   Transfer Conference, 1981, Madras, INDIA.

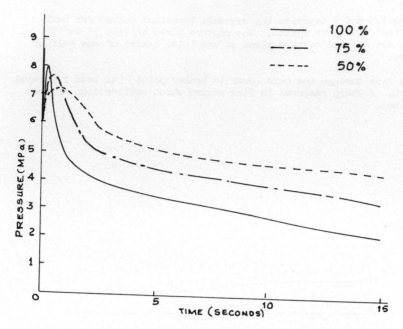

FIG.8   OUTLET PRESSURE-HALF CORE
(NEAR THE BREAK)

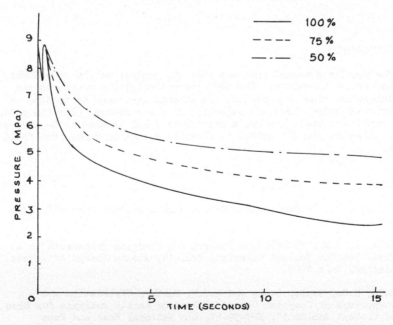

FIG.9   CORE INLET PRESSURE
(AWAY FROM BROKEN HEADER)

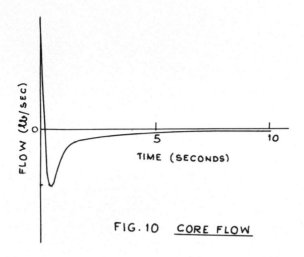

FIG. 10   CORE FLOW

4. Arrison, N.L., Hancox, W.T., Sulatisky, M.T., and Banerjee, S., "Blow down of a Recirculating Loop with Heat Addition", (202/77 BNES Conference, 1977).

5. Crowder, H.J., and Dalton, C., "Error in the Used of Non-Uniform Mesh System ", Vol.7, International Journal of Computational Physics, February, 1971.

6. Moody, F.J., "Maximum Two-Phase Vessel Blow Down from Pipes", GE, APED 4827, 1965.

7. Nahavandi, A.N., "Loss of Coolant Accident Analysis in PWR's", Nucl. Science.

8. Willie and Streeter, Fluid Transients , 2nd Edition, 1978, McGraw Hill.

9. "Loss of Coolant Accident and Emergency Core Cooling Models for General Electric Boiling Water Reactors", NEDO 10329, April, 1971.

Multi-Phase Flow and Heat Transfer III. Part B: Applications
edited by T.N. Veziroğlu and A.E. Bergles
Elsevier Science Publishers B.V., Amsterdam, 1984 — Printed in The Netherlands

A NEW APPROACH VIA A WEIGHTED RESIDUAL CRITERIA TO THE THERMAL HYDRAULICS
BEHAVIOUR OF THE REACTOR CORE AND APPLICATION TO TMI - 2 ACCIDENT

Tolga YARMAN and Beril TUĞRUL
Institute for Nuclear Energy
Technical University of Istanbul
Ayazağa, Istanbul - TURKEY

ABSTRACT

Space and time dependent thermal-hydraulics equations are written
for both fuel elements and coolant regions. Equivalent conduction equations
are proposed in the radial direction for the coolant channels.

As a first approximation the case of incompressible fluid is undertaken.

Space dependencies for various temperatures of interest have been
guessed. The application of a weighted residual criteria thus led to time
dependent equations involving only average fuel center line, fuel walls,
coolant at the walls of the fuel and coolant center line temperatures
embodying a simple picture of the overall thermal-hydraulics behaviour
of the entire core, or a specific portion of this to be handled.

The problem is further reduced to an eigenvalue problem if various
thermal hydraulics parameters can be considered as constant throughout
successive time steps.

An application of the technique is made to the first two minutes
of TMI-2 accident. The required amount of computation time is surprisingly
less (by   several orders of magnitude) than that generally required,
whereas the related predictions are very satisfactory.

1. INTRODUCTION

Under the actual computation circumstances it still is
very consuming to simulate a thermal-hydraulics behaviour of a nuclear
reactor system. The transient analysis of even a single fuel rod and surrounding coolant channel may be tiring. A global transient of the overall

reactor core, or of a portion of this, when required, thus can be hardly attacked.

There may further be convergence problems related to the equivalent space and time dependent finite difference scheme. Thus when the computation of results in question can be reached, confidence should be developed separately in order to believe them.

x

In this work we propose to guess space dependencies for various temperatures of interest throughout a thermal-hydraulics phenomena displayed by the overall reactor core or a portion of this.

The insertion of approximate temperature expressions into the original space and time dependent thermal-hydraulics equations written for various lumped regions of the nuclear system, and the application of a weighted residual criteria thus lead to time dependent equations involving only special definitions of "average fuel center line", "fuel walls", "coolant at the walls of the fuel" and "coolant center line temperatures", embodying a simple picture of the overall energetic evolution of the entire core.

As a first approximation the case of incompressible fluid is undertaken. This will eliminate from the space and time dependent thermal-hydraulics equations (written for both fuel and coolant regions), mass conservation, momentum conservation and fluid thermodynamics state equations.

2. MATHEMATICAL DESCRIPTION OF THE MODEL

We assume that the nuclear system we have to deal with is composed of fuel and coolant channel regions. We thus propose to write space and time dependent thermal-hydraulics equations for both fuel and coolant regions in question describing a thermal-hydraulics transient that takes place in our system. We propose equivalent conduction equations in cross flow direction for coolant channels. Thus, a global behaviour description of the space and time dependent thermal-hydraulics phenomena occuring within the reactor core is mathematically established, via coupling the mentioned equations written for various lumped regions through proper boundary conditions.

x

Below, we first give these equations with the mentioned boundary conditions. Then space dependencies of various temperatures of interest are guessed. The application of the weighted residual criteria will lead to time dependent equations only, which will be undertaken throughout Section 3, below (1,2,3,4).

2.1. Main Equations

Under the consideration of incompressible fluid, the space and time dependent thermal-hydraulics equations for both fuel and coolant regions can be written, with the familiar notation as follows:

In the fuel region :

$$\nabla \cdot k_f(x,y,z,T) \; \nabla T_{f,n}(x,y,z,t) + q_n(x,y,z,t)$$

$$= \frac{\partial \left[c_f(x,y,z,T) \, \rho_f(x,y,z,T) T_{f,n}(x,y,z,T) \right]}{\partial t} \qquad (1)$$

In the coolant region :

$$\nabla_{cf} \cdot K_a(x,y,z,T) \, \nabla_r T_{a,n}(x,y,z,t) - \frac{\partial \left[v_n(t) \, \rho_a(x,y,z,T) \, c_a(x,y,z,T) T_{a,n}(x,y,z,t) \right]}{\partial z}$$

$$= \frac{\partial \left[\rho_a(x,y,z,T) c_a(x,y,z,T) T_{a,n}(x,y,z,t) \right]}{\partial t} \qquad (2)$$

In these equations, $T_{f,n}(x,y,z,t)$ represents the temperature of the location (x,y,z) of the nth fuel element, at time t. $T_{a,n}(x,y,z,t)$ represents the temperature of the location $(x,\dot{y},z)$ in the nth channel, at time t.

Eq.(2) describes an equivalent conduction phenomena in the cross flow direction[x], assuming that the coolant flow occurs following the axial direction. The divergence and gradient operators in Eq. (2), for this reason,

---

[x] As is generally the case, it assumed that the system is cooled in the axial (z) direction. The heat transfer from fuel elements standing along this direction to the coolant is thus assumed to occur in the perpendicular direction to the flow direction. We call this the "cross flow" direction, and it takes place on the (x,y) plane.

do operate on the cross flow (or "cf") direction only.

$K_a(x,y,z,t)$ in Eq.(2) involves an appropriate definition to take care of the mentioned equivalent conduction phenomena within the coolant region. $K_a(x,y,z,t)$ is called "effective thermal conductivity" (5,6).

$v_n(t)$ in Eq.(2) is the flow velocity in the nth coolant channel, at time t.

For a further simplification we neglect conduction along the axial direction in fuel regions. (The axial dependencies of the fuel temperatures will be nevertheless induced by the axial dependencies of the fuel source term and the surrounding coolant temperatures.)

The following boundary conditions with the familiar notation shall be considered along Eqs. (1) and (2).

$$\left[-k_f(x,y,z,T) \, \nabla T_{a,n}(x,y,z,t)\right] . \, \vec{n} = h(T) \left[T_{w,n}(x,y,z,t) - T_{a,n}(x,y,z,t)\right] , \quad (3)$$

$$\left[-K_a(x,y,z,T) \, \nabla T_{a,n}(x,y,z,t)\right] . \, \vec{n} = h(T) \left[T_{w,n}(x,y,z,t) - T_{a,n}(x,y,z,t)\right] . \quad (4)$$

$\vec{n}$ in these equations is the outward looking unit vector perpendicular at the fuel wall.

$T_{w,n}(x,y,z,T)$ in Eqs. (3) and (4) refers to the fuel wall temperature within the nth coolant channel at the location (x,y,z) and time t. $T_{a,n}(x,y,z,t)$, on the other hand, represents the coolant temperature at the same location (to be more precise, at the coolant location pointed by $\vec{n}$ and immediately nearby the previous one) and at the same time.

One may recall that more notational elaboration should be required to describe $T_{w,n}(x,y,z,t)$ and $T_{a,n}(x,y,z,t)$ at "the wall"; for, more than one fuel element will deliver heat to a coolant channel. We will keep this in mind, but leave it out for the sake of simplification of our presentation.

2.2. Predictions of Space Dependenc   of Temperatures

For both fuel and coolant regions, a temperature space dependency   of the following form are assumed :

$$T(x,y,z,t) = A(y,z,t) \, x^2 + B(y,z,t) \, x + C(y,z,t) . \quad (5)$$

The quantities $A(y,z,t)$, $B(y,z,t)$ and $C(y,z,t)$ are determined in terms of the fuel center line, fuel walls, coolant at the fuel walls and coolant center line temperatures.

Thus, a parabolic description in the cross flow direction of both fuel and coolant regions for temperatures radial space dependencies has been adopted.

To reflect an easy picture of our proposition let us write the fuel and coolant temperatures, $T_f(r,z,t)$ and $T_a(r,z,t)$ according to Eq. (5), in the case of just one cylindrical fuel element of radius R_f, surrounded by a coolant channel of radius R_a, that provides thermal isolation at its outer boundary (cf. Fig.1):

$$T_f(r,z,t) = T_c(z,t)\ (1- \frac{r^2}{R_f^2}\ ) + T_w(z,t)\ (\ \frac{r^2}{R_f^2}) \quad , \qquad (6)$$

$$T_a(r,z,t) = T_{a,c}(z,t)\ \left[1- \frac{R_a - r}{R_a - R_f}\right]^2 + T_{a,w}(z,t)\left[\frac{R_a-r}{R_a-R_f}\right]^2 , \quad (7)$$

with the following definitions.

$T_c(z,t)$: fuel center line temperature at (z,t)

$T_w(z,t)$: fuel wall temperature at (z,t)

$T_{a,w}(z,t)$: coolant temperature at fuel wall and (z,t)

$T_{a,c}(z,t)$: coolant center line temperature at (z,t)

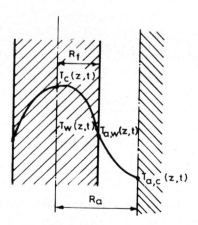

Figure 1. Fuel Rod With a Surrounding Channel
Showing The Temperatures of Interest

We will note that, because the heat conduction in the axial direction within fuel region is neglected we do not have to make a guess about the axial dependency of the fuel temperatures within the frame of our approach. Mathematically speaking there is no derivatives of the fuel temperatures with respect to the flow (z) direction.

This is however not the case for the coolant temperatures and we are bound to make an appropriate guess for the z dependencies of coolant temperatures, more precisely of those temperatures that enter in equations similar to Eq. (7) describing the cross flow direction dependencies of coolant temperatures.

We propose to adopt the simplest z dependency for the temperatures in question,i.e. they are assumed to vary linearly from **their** inlet to **their** outlet locations in our nuclear system.

If we pursue our simple example drawn along Fig.(1) we propose to write for $T_{a,w}(z,t)$ and $T_{a,c}(z,t)$ the following expressions :

$$T_{a,w}(z,t) = 2 \frac{\overline{T}_{a,w}(t) - T_g(t)}{H} z + T_g(t) \qquad , \qquad (8)$$

$$T_{a,c}(z,t) = 2 \frac{\overline{T}_{a,c}(t) - T_g(t)}{H} z + T_g(t) \qquad ; \qquad (9)$$

with the definitions,

$$\overline{T}_{a,w}(t) = \frac{1}{H} \int_0^H dz\, T_{a,w}(z,t) \qquad , \qquad (10)$$

$$\overline{T}_{a,c}(t) = \frac{1}{H} \int_0^H dz\, T_{a,c}(z,t) \qquad ; \qquad (11)$$

$T_g(t)$ being the coolant system inlet temperature at time t, and H the height of a fuel element.

One may notice that Eqs. (10) and (11) are definitions of an average coolant center line and coolant at the wall fuel temperatures.

2.3. Application of the Weighted Residual Criterion and Final Total
Differential Equations For The Average Temperatures

We now go to Eqs.(1), (2), (3) and (4) with expressions, similar to
those defined by Eqs.(6), (7), (8) and (9), that shall be written for
various fuel and coolant lumped regions of our system. Since the expressions
in question are only approximate, the equalities of the original equations
shall be broken and residues are obtained. We can repair the broken
equalities in a weighted integral sense. We thus integrate the residues
over appropriate spaces (in fuel, over fuel volume and in coolant region,
over coolant channel volume), and by reestablishing the original equalities,
obtain equations for the unknown quantities similar to those just defined;
i.e. (along our simple example), average temperatures, $\overline{T}_{a,w}(t)$, $\overline{T}_{a,c}(t)$,
together, with definitions similar to the following ones, drawn along
quantities that took place at the RHS of Eq.(6) :

$$\overline{T}_c(t) = \frac{1}{H} \int_o^H dz \ T_c(z,t) \qquad , \qquad (12)$$

$$\overline{T}_w(t) = \frac{1}{H} \int_o^H dz \ T_w(z,t) \qquad . \qquad (13)$$

We then come up with first order total differential equations with
respect time for the average temperatures similar to those of Eqs. (10),
(11), (12) and (13).

Within the frame of Fig. (1), we thus obtain four equations out of
which two are algebraic equations (corresponding to the boundary conditions between fuel and coolant) and the two others being first order total
differential equations with respect time. Thereby we write four equations
for the four unknowns $\overline{T}_c(t)$, $\overline{T}_w(t)$, $\overline{T}_{a,w}(t)$ and $\overline{T}_{a,c}(t)$.

It will be noticed that the complex time and space dependent thermal-hydraulics description of our system is now easily displayed in terms
of the average temperatures just introduced and other parameters of
interest; i.e. heat production term within fuel, coolant velocity in
channel, coolant system inlet temperature, etc.

It should be pointed out that, all of the thermal-hydraulics
parameters that enter Eqs.(1) and (2) and perhaps mainly the heat transfer
coefficient, are also space dependent (since they are dependent on tempe-

rature which varies from one location the another), and this should be given a consideration in a more detailed analysis. Yet for the purpose of our
presentation we find this forgivable.

3. SOLUTION OF THE FINAL COUPLED ALGEBRAIC AND FIRST ORDER TOTAL
   DIFFERENTIAL EQUATIONS

The system of equations obtained through the application of the
weighted residual criteria to the original thermal-hydraulics equations can
be reduced to an eigenvalue problem, if the parameters $k_f(x,y,z,T)$, $K_a(x,y,z,T)$, $c_f(x,y,z,T)$, $\rho_f(x,y,z,T)$, $c_a(x,y,z,T)$, $\rho_a(x,y,z,T)$, $h(T)$ can be kept
constant through a relatively short period of analysis, and both the time
dependencies of the source terms $q_n(x,y,z,t)$ and the coolant inlet temperature $T_g(t)$ can be, preferably, fit into an exponential form.

It can thus be checked that our average temperatures, unknowns of
the mentioned final system, can be expressed in terms of exponential functions whose arguments call upon the solution of a matrix eigenvalue problem.

It turns out, thereby, that the average temperatures $T_j(t)$'s, of
interest are written as :

$$\overline{T}_j(t) = \sum_{k=1} b_{j,k} e^{w_k t} \quad , \qquad j = 1, \ldots, \dot{j} \quad . \tag{14}$$

$\dot{j}$ is the number of unknowns, and K is the number of differential
equations together with the number of those bearing, thus an exponential
right hand side. Within the frame of Fig.(1), both $\dot{j}$ and K are 4.

4. APPLICATION OF THE TECHNIQUE

The technique presented has been applied to the first two minutes
(through which the coolant can be considered as incompressible) of TMI-2
accident (7).

Fifteen alined fuel elements of a central bundle are thus considered.

The time dependency of the source term composed of both neutronic and
residual heat terms is fit to an appropriate exponential form.

The inlet temperature of the system is found for the transient in
question and also fit into an exponential form.

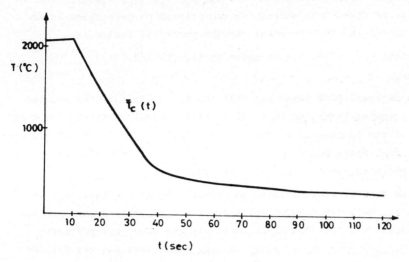

Figure 2. Typical Behaviour of Average Fuel Center Line Temperature With Respect To Time

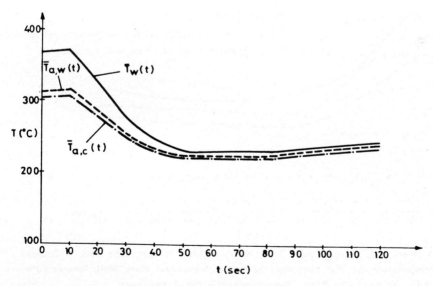

Figure 3. Typical Behaviour of Average Fuel Wall, Coolant At The Wall And Coolant Center Line Temperatures With Respect To Time

The thermal-hydraulics parameters k_f, K_a, c_a, ρ_a, c_f, ρ_f, h are assumed to be space independent for the present purpose, fixed in an appropriate manner and kept constant for the period of analysis.

A typical set of results is shown in Fig.(2) and (3).

5. CONCLUSION

It is noteworthy to point out that the calculation of the results through the proposed technique is extremly fast. Roughly speaking the speed of our predictions is usually a hundred times faster than those made through the conventional space and time dependent finite difference approach. Our results, in addition, seem to be very satisfactory.

For the sake of a comparison, our prediction of the behaviour of the average fuel wall temperature and that of a fuel wall temperature re-corded throughout the TMI-2 accident, with respect time, are presented in Fig.(4) (8,9,10). One should note that our prediction did not include the consideration of a fuel clad. Moreover, while fitting the behaviour of the source term of Eq.(1), with respect time, into an exponential form, an approximation has been made. A similar approximation is also made while fitting the coolant inlet temperature to an exponential form. It should be further recalled that the thermal-hydraulics parameters of interest are kept constant through the period of analysis.

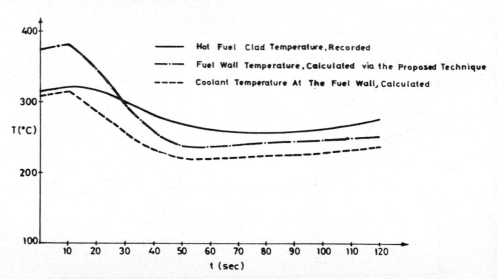

Figure 4. A Predicted Average, Fuel Wall (Without Clad) Temperature And A Coolant Temperature At The Fuel Wall; Versus A Measured Fuel Wall Temperature With Respect To Time, Throughout The TMI-2 Accident

These should be the primary reasons of the descrapencies between the calculated and the measured fuel wall temperatures shown in Fig. 4.

We thus think it would be interesting to add to Fig. 4, our prediction about the behaviour of the coolant temperature at the fuel wall, whose behaviour might then be expected to be closer to that of the measured fuel wall temperature under the mentioned circumstances.

x

Our prediction promises to be bettered by adopting a more realistic scheme of work, and it can yet be considered as satisfactory.

x

An effort is being made to generalize the proposed technique to the case of compressible fluid.

ACKNOWLEDGEMENT

This work is carried out under the auspices of the International Atomic Energy Agency (Research Contract No : 3214/RB).

REFERENCES

1. Yarman, T., Saldıray, E.S. and İşyar, A., "Soğutucu Kaybı Kazasına Analitik Bir Yaklaşım", Institute For Nuclear Energy, Technical University of Istanbul. Research Report, NEE-30 August, 1982.

2. Yarman, T., "A Weighted Residual Approach To The Solution of Thermal Hydraulics Problems", Turkish Journal Of Nuclear Sciences, Vol. 9, No.2, August, 1982.

3. Werner, W.F. "Weighted Residual Methods For The Solution Of Fluid Dynamics Problem", Proceedings Of The International Topical Meeting On Advances In Mathematical Methods For The Solution Of Nuclear Engineering Problems, ANS-ENS with IAEA, NEACRP of OECD, Munich, April, 27-29, 1981, Vol 1, 461-474.

4. Tuğrul, B., "Reaktör Isıl-Hidrolik Çözümlemesine Ağırlaştırılmış Ka-
   lıntılar Yoluyla Yeni Bir Yaklaşım, TMI-2 Kazası'na Uygulama" Ph.D.
   Thesis still being carried under the supervision of Professor T. Yarman, at the Institute For Nuclear Energy, Technical University of Istanbul.

5. Kakaç, S., "Transient Heat Transfer By Forced Convection In Channels",
   ASI Proceedings Of Turbulent Forced Convection In Channels And Rod
   Bundles, Istanbul, 20 July - 2 August 1978.

6. El-Wakil, M.M., <u>Nuclear Heat Transport</u> , International Textbook Company, 1971.

7. "Investigation Into The March 28, 1979 Three Mile Island Accident",
   Office of Inspection And Enforcement, U.S. Nuclear Regulatory Commission,
   NUREG-0600 Investigative Report No, 50-320/79-10, August 1979.

8. Ireland, J.R., Wehner, T.R. Kirchner, W.L., "Thermal-Hydraulic and Core
   Damage Analysis Of The TMI-2 Accident", <u>Nuclear Safety,</u> Vol. 22, No: 5,
   September - October 1981, 583-593.

10. Ireland, J.R., "TRAC Analysis Of Three Mile Island Accident", Transacti-
    ons, Vol. 34, TANSAO, American Nuclear Society Annual Meeting, Las Ve-
    gas, Nevada, 9-12 June 1980, 501-503.

Multi-Phase Flow and Heat Transfer III. Part B: Applications
edited by T.N. Veziroğlu and A.E. Bergles
Elsevier Science Publishers B.V., Amsterdam, 1984 — Printed in The Netherlands

ANALYSIS OF CORE PHYSICS AND THERMAL-HYDRAULICS RESULTS OF CONTROL ROD
WITHDRAWAL EXPERIMENTS IN THE LOFT FACILITY

D. J. Varacalle, Jr., T. H. Chen, E. A. Harvego, and H. Ollikkala[a]
EG&G Idaho, Inc.
P.O. Box 1625
Idaho Falls, Idaho 83415, U.S.A.

ABSTRACT

Two anticipated transient experiments simulating an uncontrolled control
rod withdrawal event in a pressurized water reactor (PWR) were conducted in
the Loss-of-Fluid Test (LOFT) Facility at the Idaho National Engineering Laboratory. The scaled LOFT 50-MW(t) PWR includes most of the principal features
of larger commercial PWRs. The experiments tested the ability of reactor
analysis codes to accurately calculate core reactor physics and thermal-
hydraulic phenomena in an integral reactor system. The initial conditions and
scaled operating parameters for the experiments were representative of those
expected in a commercial PWR. In both experiments, all four LOFT control rod
assemblies were withdrawn at a reactor power of 37.5 MW and a system pressure
of 14.8 MPa. In the first experiment, the average reactivity insertion rate
($\Delta\rho/\Delta t$) of 0.47 ¢/s resulted in a system scram due to high pressure (15.73 MPa)
at 58.5 s. The $\Delta\rho/\Delta t$ for the second experiment was 5.5 ¢/s, which
resulted in a reactor scram due to high power (48.8 MW) at 7.4 s. Pretest and
posttest RELAP5/MOD1 computer code calculations of thermal-hydraulic and core
neutronic parameters agreed reasonably well with the measured data. The LOFT
experiments demonstrated that state-of-the-art computer codes can calculate
the integral system response and reactor kinetics for anticipated transients
of this type.

1.   INTRODUCTION

Understanding the behavior of light water reactors during anticipated
transients is a major objective of the Nuclear Regulatory Commission (NRC)
Reactor Safety Research Program. The LOFT[1] facility is a major NRC sponsored testing facility that can simulate the response of a PWR over a wide
range of transient conditions. Since the LOFT facility is an integral[b]
nuclear facility, it can uniquely simulate the nuclear and thermal-hydraulic
processes that would occur in a commercial PWR during a control rod withdrawal
event.

---

a. Technical Research Centre of Finland.

b. The term integral is used to describe an experiment that characterizes the
most significant nuclear, thermal-hydraulic, and structural processes occurring in a complete system during a transient and differentiates it from the
separate effects, nonnuclear, small scale experiments conducted to address
specific areas of interest in transient analysis.

Two anticipated transient experiments (L6-8B-1 and L6-8B-2) were performed in the LOFT facility at the Idaho National Engineering Laboratory for the NRC in August 1982. The purpose of these experiments was to verify computer codes and analytical models and contribute to an understanding of symptoms, events, and plant conditions potentially leading to emergency or abnormal situations in a commercial PWR.

The basis for the need to consider anticipated transients in reactor safety evaluation is not that reactor protection and reactivity shutdown systems are unreliable, but that they may be challenged at a relatively high rate, therefore requiring a high reliability of the plant protection system (PPS). Although not as serious as the more severe design basis accidents, anticipated transients are a potential source of risk because of their relatively high probability. These risks can be particularly significant if anticipated transients are combined with equipment malfunctions or operator error.

The primary consideration in the design of the L6-8B experiments was to represent the integral behavior of a large PWR during a postulated control rod withdrawal transient, and to identify and investigate any unexpected events or thresholds in the large-plant response. The L6-8B experiments simulated the range of expected reactivity insertions resulting from an uncontrolled control rod assembly withdrawal accident in a commercial PWR. The experiments were designed to obtain integral plant response data and reactor kinetics data to evaluate the capabilities of RELAP5/MOD1, a state-of-the-art integral reactor computer code.[2]

To address issues relating to the identification of and recovery from anticipated transient events, the following objectives were defined for the LOFT L6-8B experiments:

1. Investigate integral plant response to a reactivity insertion event caused by the withdrawal of all four LOFT control rod assemblies.

2. Provide data to assess the applicability of point kinetics models for predicting transient reactor power.

In both L6-8B experiments, all four LOFT control rod assemblies were withdrawn with the core initially at 37.5 MW power. The average reactivity insertion rate was $4 \times 10^{-5} \Delta\rho/s$ (0.47 ¢/s) for experiment L6-8B-1 and $4 \times 10^{-4} \Delta\rho/s$ (5.5 ¢/s) for experiment L6-8B-2. In each experiment, the rod withdrawal rate was maintained until the reactor scram setpoints were reached. These experiment conditions bound a range of credible reactivity insertion accidents at power due to a control rod withdrawal in a PWR.

To characterize the thermal-hydraulic behavior of the system during the transients, measured parameters included coolant temperature and pressure, reactor power, fuel temperature, core reactivity, and pressurizer liquid level. These measured values also provided a means for comparison with values calculated by the RELAP5 computer code.

2. DESCRIPTION OF LOFT

The LOFT facility[1] includes a containment structure, support buildings, and a test assembly that houses a 50-MW(t) PWR that simulates the major behavioral aspects of generic four-loop 1000-MW(e) PWRs in carefully conducted experiments. The nuclear core is approximately 1.7-m long, 0.6-m in diameter, and contains 1300 fuel pins of 4.0 wt% U^{235}. The four control assemblies are of typical PWR design.

The primary coolant system (PCS) consists of an intact circulating loop simulating three loops of a PWR, and a second loop that can simulate a broken coolant pipe during large break loss-of-coolant experiments. The intact loop contains an operating steam generator, coolant pumps, and pressurizer. The broken loop piping was isolated for the L6-8B experiments.

The LOFT system is designed with a primary system volume-to-core power ratio similar to that in a commercial PWR. Primary system subvolumes, e.g., lower plenum, core region, upper plenum, outlet piping, steam generator, and inlet piping, also are designed with relative volumes equivalent to those in commercial PWRs. During power operation, heat is transferred from the LOFT reactor coolant system to the secondary side of a vertical, U-tube steam generator. Steam generated at a pressure of approximately 6.3 to 7.0 MPa is fed to an air cooled condenser and is condensed at approximately 2.0 MPa. A modulating steam flow control valve maintains secondary pressure.

The LOFT fuel rods are the type used in a typical PWR 15 x 15 rod array, except the active length of the LOFT rods is 1.68 m while that of a PWR 15 x 15 bundle fuel rod is 3.66 m. The center bundle fuel rods in the LOFT core were initially prepressurized with helium to 2.41 MPa, while the peripheral fuel rods were unpressurized.

Over 1000 channels of experimental data can be monitored during an experiment under computer direction. Fluid pressure, temperature, velocity, and density are monitored at key locations in the reactor coolant, emergency core cooling, blowdown, and secondary coolant systems by extensive instrumentation. Core thermocouples monitor fuel pin cladding and centerline temperatures. Four fixed nuclear detectors and a traversing in-core nuclear detector system determine core nuclear response and neutron flux shapes. The experiment assembly is shown in Figure 1.

3. EXPERIMENT DESCRIPTION

Using the American Nuclear Society (ANS) classification of plant conditions, which divides plant events into four categories in accordance with frequency of occurrence and potential radiological consequences to the public, an uncontrolled rod cluster control assembly bank withdrawal at power is classified as a Condition II event (fault of moderate frequency). This event, at worst, results in a reactor shutdown, with the plant still capable of returning to operation. By definition, this event does not initiate a multiple failure sequence, but does, however, result in an increase in core power. Since heat removal by the steam generators lags behind the core power generation, there is a net increase in the reactor coolant temperature and pressure. Unless terminated by manual or automatic action, the power mismatch and resultant coolant temperature rise would eventually result in departure from nucleate boiling (DNB). Therefore, in order to avert damage to the cladding, the PPS in a PWR is designed to terminate any such transient before the departure from nucleate boiling ratio (DNBR) falls below a designated value, typically in the range of 1.15 to 1.30.

The automatic features of a PWR PPS that prevent core damage following the postulated accident include actuation of a reactor trip if:

1. Two out of four power range neutron flux instrumentation channels exceed an overpower setpoint.

2. Two out of four pressurizer pressure instrumentation channels exceed a pressure setpoint (typically near 16.55 MPa).

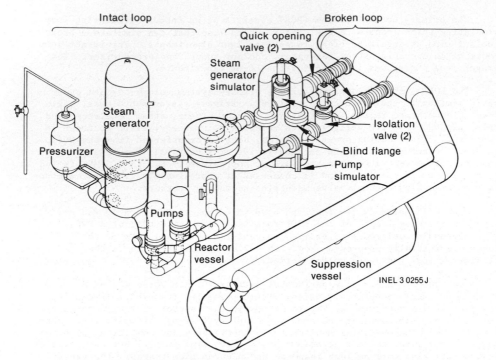

Fig. 1.   Axonometric Projection of LOFT System

For control rod withdrawal transients, the reactivity insertion rates investigated in the Westinghouse Trojan FSAR[3] ranged from 4 x 10^{-5} (slow rod withdrawal) to 8 x 10^{-4} Δρ/s (fast rod withdrawal). The predicted reactor trip on high neutron flux occurred shortly (within 3 s) after the start of the accident for the fast rod withdrawal, while the reactor trip on overpressure occurred after a longer period (32 s) for the slower rod withdrawal. The predicted rise in coolant temperature and pressure for the slow rod withdrawal was considerably larger than for the fast rod withdrawal.

The L6-8 experiments were initiated from operating conditions approximating those in a typical Westinghouse four-loop plant having an inlet temperature of 550 K and a core temperature rise of 36 K. The inlet temperature for the L6-8B experiments was 557 K. The LOFT core temperature rise was only 13.0 K, due to the safety requirements for a higher than scaled PCS mass flow rate (480 kg/s). The LOFT core average temperature, however, was 563.5 K, which corresponds closely to the commercial PWR average core temperature of 568 K.

For the L6-8B experiments, initial power was 37.5 MW rather than 50 MW (normal LOFT full power operation), to allow a longer time for the development of system dynamic characteristics in the tests. At this lower power, the maximum linear heat generation rates in the LOFT reactor (42.6 kW/m for the slow rod withdrawal experiment and 49.1 kW/m for the fast rod withdrawal experiment) are representative of beginning of life power levels in a commercial plant. These LOFT heat generation rates therefore provide a good approximation of PWR fuel temperature distributions and the correct Doppler contribution to the reactivity.

A steady-state primary coolant pressure of 14.7 MPa was chosen for the LOFT L6-8 experiments to maintain the normal LOFT overpressure scram point of 15.73 MPa, while still allowing the same potential for system overpressurization (1.03 MPa) that might occur in a commercial plant for this type of transient. The effect of this perturbation from normal commercial PWR coolant pressures (15.51 MPa) is minimal on the experiment and analytical results. In addition to initial conditions, the transient pressure is governed by the power-to-volume (P/V) ratio in PWRs. The P/V ratios for LOFT and a typical commercial PWR are 5.29 and 10.07 MW/m^3, respectively. Since the coolant expansion coefficient increases with increasing temperature, system pressure increases faster for larger P/V ratios and would therefore be expected to increase more rapidly in a commercial PWR.

The power-to-volume ratio in a PWR also affects an anticipated transient through its effect on the moderator reactivity feedback. As the coolant temperature rises, the moderator coefficient adds negative reactivity to the system, decreasing reactor power. The higher the P/V ratio, the greater the transient temperature rise rate and the greater the transient negative reactivity insertion rate due to moderator feedback. However, for the range of temperatures in the control rod withdrawal experiments, the LOFT moderator coefficient is more negative than that of a typical commercial PWR. This results in greater net negative moderator temperature feedback in LOFT than in a commercial PWR, and consequently a smaller net positive reactivity in the LOFT reactor due to the control rod withdrawal. Thus, the LOFT reactor would be expected to scram later than a commercial PWR for the same reactivity insertion rates.

The Doppler coefficient models the increased resonance absorption of neutrons in U^{238} with increasing fuel temperature. Since there is an increase in reactor power and fuel temperature during the L6-8B transients, the Doppler coefficient adds negative reactivity to the core. The LOFT Doppler reactivity coefficient is more positive than that in a commercial plant for the temperatures of interest in these experiments. Therefore, the LOFT Doppler contribution will be larger than that of a PWR for the same change in volume averaged fuel temperature.

The initial conditions for the LOFT experiments are shown in Table 1 and are representative of those in a commercial PWR at power. The experiments included a preconditioning phase, a testing phase, and a recovery phase. The transients were initiated by withdrawal of the control rods.

4. CALCULATIONAL TECHNIQUE

The RELAP5 computer code[2] was used for pretest[a] and posttest[b] calculations of system thermal-hydraulic response and core thermal and neutronic response for the LOFT experiments. RELAP5 is a reactor analysis code that can be used to predict the transient behavior of water cooled nuclear reactors (or simulators) subjected to postulated accidents such as those resulting from a loss-of-coolant accident (LOCA) or anticipated transient without scram (ATWS). It is a comprehensive program that predicts the interrelated effects of system

---

a. RELAP5/MOD1 Cycle 15, Idaho National Engineering Laboratory Configuration Control No. F00341.

b. RELAP5/MOD1 Cycle 18, Idaho National Engineering Laboratory Configuration control No. F00885.

Table 1. Measured Initial Conditions for the LOFT Experiments

EXPERIMENT	MLHGR[a] (kW/m)	SYSTEM PRESSURE (MPa)	COLD-LEG TEMPERATURE (K)	ΔT[b] (K)	CORE POWER[c] (MW)	PCS FLOW (kg/s)	CONTROL ROD POSITION[d] (m)
L6-8B-1	42.6	14.8	556.9	13.0	37.5	481.2	1.30
L6-8B-2	49.1	14.7	556.7	13.5	37.3	479.0	0.99

a. Maximum linear heat generation rate.

b. ΔT = core temperature rise.

c. Pretest power history included sufficient full power operation to obtain full fuel power preconditioning.

d. Distance above full-in position.

thermal-hydraulics, system heat transfer, core neutronics, and system component interactions. The code solves the governing conservation equations for mass, momentum, and energy using a one-dimensional nonhomogeneous, nonequilibrium hydrodynamic model. The required program input consists of a description of the system geometry, including fluid volumes, initial flows, pressure and temperature distributions, power generation, and material properties.

The code includes component process models for pipes, branches, abrupt flow area changes, choked flows, pumps, check valves, plant trips, control systems, heat structures, steam separator, and nuclear reactor core neutronics. The code also includes a point reactor kinetics model for core neutronics, a reactivity feedback control system model, and detailed secondary system component modeling.

The point kinetics model used in RELAP5 to calculate core power was considered to be adequate to simulate the L6-8B transients since changes in the core axial power shape were minimal during each of the two experiments. The driving function for the RELAP5 reactor kinetics equation is reactivity. Contributions to the reactivity include a time-dependent reactivity (control rod insertion) and individual reactivities due to feedback effects in each core region, including fuel temperature, coolant density, and coolant temperature feedback effects.

5. SYSTEM THERMAL-HYDRAULIC RESPONSE

The variables that affect system response, and thus core behavior, include coolant pressure, density, moderator temperature, and reactivity. The reactivity insertions in the L6-8 experiments provided the driving force that governed the system transient behavior by perturbating the core power and thus the coolant conditions.

In the following sections the experimental results are evaluated and discussed for each experiment and compared with the RELAP5 calculations based on the same initial conditions (Table 1).

Slow Rod Withdrawal Results (Experiment L6-8B-1)

The L6-8B-1 experiment involved an increase in primary coolant pressure that was sufficiently high to allow scram to occur prior to reaching the power trip setpoint. As the control rods were withdrawn at a rate of 0.19 cm/s (4.5 in./min) to initiate the transient, an average reactivity insertion of 0.47 ¢/s was imposed on the system causing the reactor power to increase. The reactor scrammed at 58.5 s on a high pressure trip signal (15.73 MPa, 2283 psia). Figure 2 illustrates the measured and posttest calculated core power. A steady-state power of 37.5 MW was established prior to the transient. With the initiation of the transient, the core power gradually increased resulting in a maximum power of 43.9 MW at 58.5 s. The RELAP5 posttest calculation for the power rate increase compares very well to the experimental data. Both the pretest and posttest calculated power increases were approximately 0.04 MW/s, which is slightly lower than the measured power increase of 0.11 MW/s.

Figure 3 illustrates the posttest calculated and experimentally derived total core reactivity for Experiment L6-8B-1. The calculated reactivity is slightly lower than the measured values. The RELAP5 calculation is the sum of the inserted positive reactivity and the negative feedback effects of the Doppler and moderator coefficients. The total inserted reactivity due to the control rod withdrawal was calculated to be 42.3 ¢.

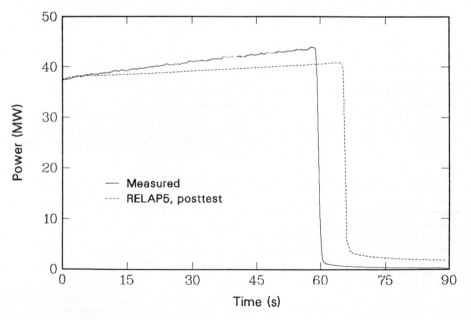

Fig. 2. Measured and Posttest Calculated Core Power for Experiment L6-8B-1

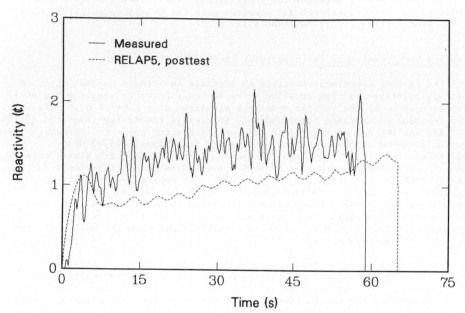

Fig. 3. Measured and Posttest Calculated Total Core Reactivity
for Experiment L6-8B-1

Figure 4 illustrates the RELAP5 posttest calculation of the Doppler and
moderator contribution. After the initial 5 s, the calculated moderator contribution dominated the Doppler contribution, and at the conclusion of the
calculation exceeded the Doppler contribution by a factor of 2. Just prior
to scram, the total integrated reactivity was calculated to be 1.3 ¢, with
Doppler feedback contributing a negative 13 ¢ and moderator feedback accounting for a negative 28 ¢. The difference between the calculated and measured
total reactivity is attributed to the RELAP5 code slight overprediction of the
negative reactivity due to moderator feedback.

The PCS fluid temperatures increased moderately as core power increased.
Since the main steam control valve (MSCV) did not open further to accommodate
the higher core power, steam generator heat removal capability was less than
the core heat generation rate. This resulted in a PCS energy imbalance and
an increase in PCS coolant temperatures. Figure 5 illustrates the measured
and posttest calculated cold- and hot-leg temperatures for the L6-8B-1 experiment. At scram (58.5 s), the measured hot- and cold-leg temperatures had
increased from the initial steady-state values to 575.5 and 559.4 K, respectively. The average measured PCS heatup rate during the experiment was
0.096 K/s which was calculated well by the RELAP5 code. The slight initial
offset in calculated temperature is due to code initialization limitations.
Due to the rise in coolant temperature, a significant amount of negative
reactivity was contributed by the moderator temperature coefficient thus
reducing the core power increase.

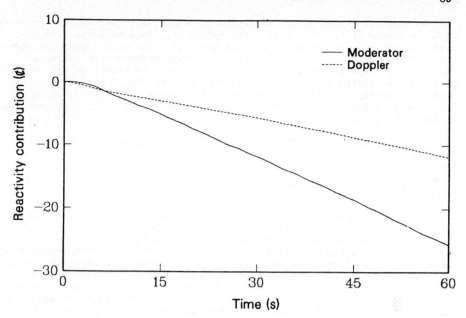

Fig. 4. RELAP5 Posttest Calculated Doppler and Moderator Reactivity
Contribution for Experiment L6-8B-1

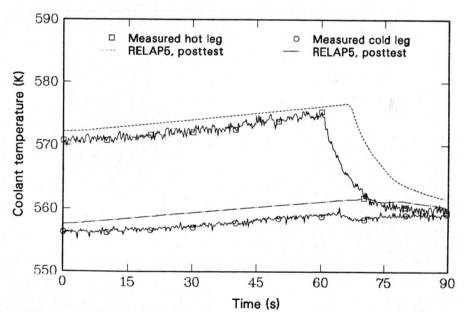

Fig. 5. Measured Posttest Calculated Cold- and Hot-leg Temperatures
for Experiment L6-8B-1

As shown in Figure 6, the measured fuel centerline temperatures increased 88 K, at a rate of 1.5 K/s, during the transient. This increase also contributed negative reactivity due to Doppler feedback. The RELAP5 posttest calculation trend of the rate of fuel centerline temperature change compares closely with the experimental data. The difference in magnitude between the measured and calculated fuel centerline temperature occurs because the measured value is representative of a hot bundle average temperature, while the RELAP5 calculation represents a core average temperature. The LOFT core hot bundle power is approximately 10% larger than the peripheral bundle power.

Figure 7 compares the measured and calculated pressure for the L6-8B-1 experiment. An initial steady-state pressure of 14.8 MPa was established prior to the experiment. With the initiation of the transient, the escalation in temperature of the PCS coolant inventory resulted in expansion of the fluid. The expansion of PCS fluid caused a rise in the pressurizer liquid level that in turn compressed the steam dome and increased the primary system pressure. Experiment L6-8B-1 achieved the high pressure scram setpoint (15.73 MPa) at 58.5 s after initiation. After scram, the primary coolant shrank due to the decreasing fluid temperature, and the pressurizer liquid level and PCS pressure decreased. The RELAP5 pretest calculated pressure rate increase was slower than measured, resulting in a predicted scram on high pressure at 105 s. Comparison of the pretest calculations with the measured data for reactor power, hot- and cold-leg temperatures, fuel centerline temperature, and reactivity indicate that these code pretest calculations compare closely with the measured data. The difference between the pretest calculation and the measured pressure is attributed to an input anomaly to the steam generator material properties. This anomaly increased the primary to secondary heat transfer and decreased the system pressurization rate. With the

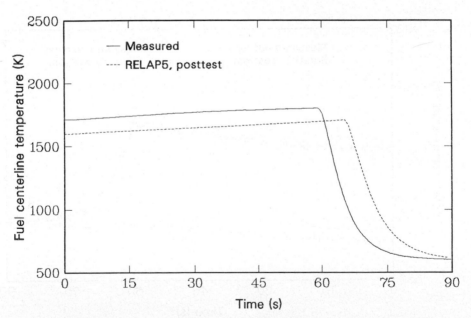

Fig. 6. Measured and Posttest Calculated Fuel Centerline Temperatures for Experiment L6-8B-1

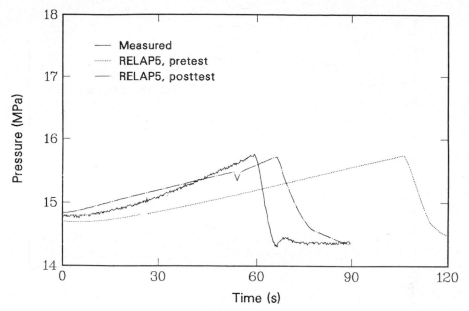

Fig. 7. Measured Pretest and Posttest Calculated Primary Coolant
System Pressure for Experiment L6-8B-1

proper input to the steam generator material properties, the posttest calculation for system pressure and reactor scram time (65.25 s) due to high pressure
shows a marked improvement over the pretest calculation.

5.2 Fast Rod Withdrawal Results (Experiment L6-8B-2)

In Experiment L6-8B-2, the control rods were withdrawn at a rate of
1.07 cm/s (25.3 in./min), more than five times faster than in Experiment L6-8B-1. In addition, in Experiment L6-8B-2, the rods were initially at
0.99 m (39 in.), which yielded a higher rod worth. Thus, the resultant average reactivity insertion rate during the fast rod withdrawal experiment was
5.5 ¢/s--12 times higher than that in the slow rod withdrawal experiment
(0.47 ¢/s). For the L6-8B-2 experiment, the total inserted reactivity due to
the control rod withdrawal was 40.4 ¢ according to the RELAP5 posttest calculation. Because of the rapid reactivity insertion rate, and the significant
time constant (approximately 20.1 s) for fuel heat conduction, the PCS coolant
did not have as much time to respond to the energy insertion as in Experiment L6-8B-1. Thus, PCS temperature and pressure did not increase as rapidly.

Figure 8 illustrates the calculated and measured core power. In the
experiment, an early reactor scram on high core average power (48.8 MW)
occurred at 7.4 s. The pretest and posttest calculated power rate increases
are almost identical to the measured power rate response. The differences at
scram are attributed to the slightly different trip setpoints for the two
calculations. The posttest calculation for the reactor scram time was 7.5 s
on high core average power (48.8 MW), which compares closely with the measured data. The pretest calculation predicted a scram at 13.2 s on a high
core average power of 51.5 MW in accordance with the pretest experiment
---cification.

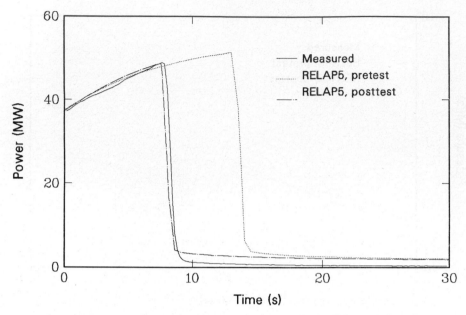

Fig. 8. Measured and Pretest and Posttest Calculated Core Power
         for Experiment L6-8B-2

Figure 9 illustrates the measured and calculated cold- and hot-leg temperatures for the L6-8B-2 experiment. The measured PCS temperatures had
increased approximately 2 K (3.6°F) at the time of reactor scram. The RELAP5
posttest calculations of coolant temperature compare well with the measured
data. The rise in the PCS temperature increased the system pressure. As
shown in Figure 10, the measured PCS pressure had increased 0.3 MPa (46 psia)
to 14.98 MPa at scram. The posttest calculation for pressure compares very
well with the measured data.

The fuel temperature rate increase was much more rapid in the fast rod
withdrawal experiment than in the slow rod withdrawal experiment. The measured fuel centerline temperature (Figure 11) in Experiment L6-8B-2 had
increased 155 K (279°F) at scram, at an average rate of 20 K/s (36°F/s), which
was approximately twice that observed in Experiment L6-8B-1. The RELAP5 posttest calculations for the fuel centerline temperature in Experiment L6-8B-2
compare very well with the measured data. The slight difference in initial
magnitude is attributed to the difference between the hot bundle and average
core power discussed above. Since the fuel temperature increase was so rapid,
Doppler feedback dominated the reactivity insertion of the control rods,
overshadowing the moderator feedback by a factor of almost 2, as shown in
Figure 12.

The posttest RELAP5 calculation for total reactivity compared very well
with the total measured core reactivity as shown in Figure 13. Just prior to
scram, the total integrated measured reactivity was calculated to be 11.4 ⊄,
with Doppler feedback contributing a negative 18 ⊄ and moderator feedback
accounting for a negative 11 ⊄.

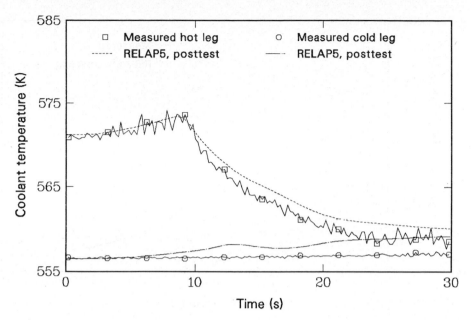

Fig. 9. Measured and Posttest Calculated Cold- and Hot-leg Temperatures
for Experiment L6-8B-2

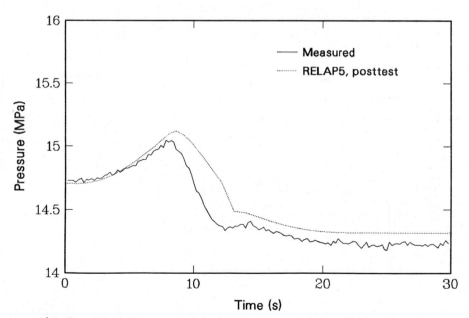

Fig. 10. Measured and Posttest Calculated Primary System Pressure
for Experiment L6-8B-2

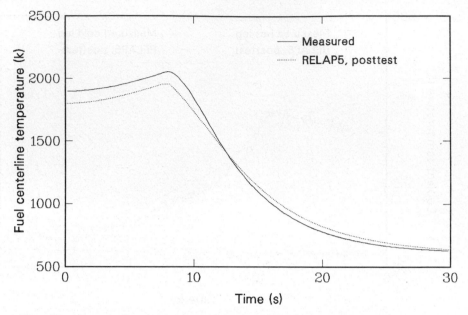

Fig. 11. Measured and Posttest Calculated Fuel Centerline Temperatures
for Experiment L6-8B-2

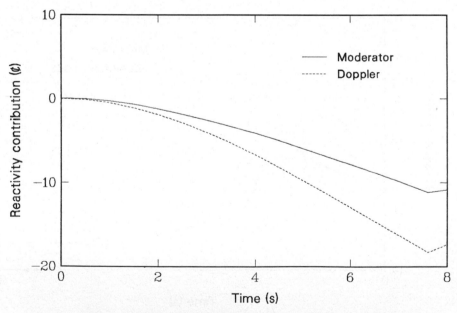

Fig. 12. RELAP5 Posttest Calculated Doppler and Moderator Reactivity
Contribution for Experiment L6-8B-2

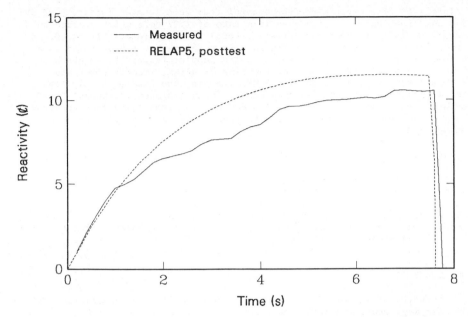

Fig. 13.   Measured and Posttest Calculated Total Core Reactivity

6.   CONCLUSIONS

The two LOFT anticipated transient experiments simulating an uncontrolled control rod withdrawal event in a PWR displayed the same core reactor physics and thermal-hydraulic trends as expected in a PWR, but magnitudes and response times were slightly different because of scaling.

Pretest and posttest calculations performed with the RELAP5/MOD1 code for these two experiments displayed trends in thermal-hydraulic and neutronic behavior similar to those observed in the experiments. The calculated net rate of energy deposition into the primary coolant, the calculated PCS pressure, and calculated power closely matched the experimental data. Total reactivity calculations matched the data very closely for both experiments. The calculations indicate negative reactivity feedback was dominated by moderator feedback for Experiment L6-8B-1 whereas the Doppler contribution dominated Experiment L6-8B-2.

In the L6-8B-1 experiment, the four control rod assemblies of the LOFT reactor were withdrawn from the initial 1.30-m position (from the bottom of the core) at a rate of 0.19 cm/s which yielded a $\Delta\rho/\Delta t$ of 0.47 ¢/s and a total inserted reactivity of 42.3 ¢. As PCS coolant temperature increased, the system scrammed on high PCS pressure (15.73 MPa) at 58.5 s into the test. The Doppler reactivity feedback and moderator feedback were of equal importance in regulating the reactor power until ∿5 s into the transient. After 5 s, the moderator feedback effect dominated the transient response. At the time of scram, reactor power was 43.9 MW. RELAP5 pretest and posttest calculations compared very well with the experiment data. Just prior to scram, the total integrated reactivity was calculated (posttest) to be 1.3 ¢, with Doppler feedback contributing a negative 13 ¢ and moderator feedback accounting for a negative 28 ¢.

In the L6-8B-2 experiment, the control rods were located initially at the 0.99-m position, and were withdrawn at a rate of 1.07 cm/s which yielded an average reactivity rate of 5.5 ¢/s and a total inserted reactivity of 40.4 ¢. Due to a faster reactivity insertion in this test, the system scrammed on a high power trip at 7.4 s into the test. The Doppler reactivity feedback dominated the entire transient. At the time of scram, reactor power was 48.8 MW and system pressure was 14.98 MPa. Again, the RELAP5 calculations compared very well with the measured data. The calculated (posttest) total integrated reactivity at scram was 11.4 ¢. Doppler reactivity feedback contributed a negative 18 ¢, while the moderator feedback reactivity produced a negative 11 ¢.

The two control rod withdrawal experiments performed in LOFT have provided valuable data for the assessment and evaluation of computer codes used in reactor safety research or reactor licensing. These experiments are particularly significant since they are the only control rod withdrawal experiments scheduled to be conducted in the LOFT facility and analyzed with the RELAP5 code. The data obtained from the tests in conjunction with the analyses conducted to date indicate point kinetics models are adequate for the prediction of transient reactor power for control rod withdrawal events.

ACKNOWLEDGMENTS

This work was supported by the U.S. Nuclear Regulatory Commisision, Office of Nuclear Regulatory Research, under DOE Contract No. DE-ACO7-76ID01570. The authors express their appreciation to J. E. Koske, A. H. Giri, K. J. McKenna, and P. Knecht for their assistance in this study.

NOTICE

REFERENCES

1. D. L. Reeder, "LOFT System and Test Description," NUREG/CR-0247, TREE-1208, EG&G Idaho, Inc., July 1978.

2. V. H. Ransom et al., "RELAP5/MOD1 Code Manual," NUREG/CR-1826, EGG-2070, EG&G Idaho, Inc., November 1980.

3. "Trojan Nuclear Plant Final Safety Analysis Report," USAEC Docket No. 50-344, Portland General Electric Company, February 1973.

Multi-Phase Flow and Heat Transfer III. Part B: Applications
edited by T.N. Veziroğlu and A.E. Bergles
Elsevier Science Publishers B.V., Amsterdam, 1984 — Printed in The Netherlands

EXPERIMENTAL STUDY OF PRE-CHF BOILING HEAT TRANSFER DURING BLOWDOWN

Saburo Toda and Akira Kumagai
Department of Nuclear Engineering
Tohoku University
Aramaki-Aoba, Sendai 980, Japan

ABSTRACT

Blowdown heat transfer experiments were performed in order to achieve better and detailed understanding of the pre-CHF boiling heat transfer in a PWR core which would be observed during a postulated loss of coolant accident. The experiments were carried out to obtain basic data for transient boiling heat transfer during blowdown in a boiling water loop provided with a transparent test section to observe phenomena in coolant behaviors, and also to examine the applicabilities of empirical correlations to blowdown transient which were recommended and/or used in actual analytical computer codes, however, were compared with not transient but steady data. Experiments of 41 runs were made with a single heater rod in an annular test section and with top discharge in the test section closely representing top blowdown in a reactor situation. These recommended correlations, however, could not satisfactorily predict the present data. Modified correlations were presented by authors, which were compared and correlated well with the present experimental data.

1. INTRODUCTION

Transient heat removal from a PWR core during a loss of coolant accident currently is an important and difficult problem in the field of reactor safety analysis. During the blowdown phase, the pressure of a PWR vessel first drops precipitously to the saturation pressure corresponding to the temperature of coolant. This is the subcooled depressurization period. The fuel rod heat transfer undergoes the several heat transfer regimes during blowdown. In this study we are concerned now with the pre-CHF boiling heat transfer taking place to subcooled liquid as well as to saturated two-phase mixture.

Subcooled nucleate boiling can occur during the normal steady state PWR operating as well as during the subcooled depressurization phase, which has been calculated by the Jen and Lottes [1], the Thom et al. [2] and the extended Chen [3, 4] correlations in RELAP 4/Mod 3 and Mod 5 codes, TRAC code, ALARM-P1 (JAERI) code and others.

During the saturated depressurization phase, the heat transfer to saturated two-phase mixture is occuring, where essentially two mechanisms are possible as follows, (1) saturated nucleate boiling in the low quality region and (2) forced convection vaporization in the high quality, high flow rate region. For determining this transfer, the Chen [3] and the Schrock and Grossman [5] correlations are recommended and used in the above-mentioned codes.

During the blowdown, the heat transfered from the fuel is an essential thermal quantity to the transient heat transfer and, in addition, heat removal from the fuel depends directly on the fluid flow and pressure transients of coolant within the core. The before-mentioned correlations are well supported by broad data base, which, however, were derived at the conditions of steady flow and steady pressure of fluid. Therefore, it will be required to be confirmed that these correlations can be applied with confidence for calculation of heat transfer in the phases of transient flow and pressure during blowdown. At present, few experimental data of forced convection boiling [6-8] are available at the condition of depressurization field in the single-phase or two-phase mixture fluid. During blowdown, flow patterns change very violently with transition from single phase flow to two-phase boiling flow at low or high quality, which give strong influences upon the heat transfer coefficient on the wall surface of the fuel rod. The exact magnitude of the heat transfer coefficient, which has been considered to be rarely important in the conservative calculation for safety analysis when it thoroughly high, is really important for the best or exact prediction of core heat transfer during blowdown to make it possible to show a correct history of blowdown transient.

The main purpose of this experimental study is to obtain basic data of transient boiling heat transfer during blowdown with transtion from subcooled fluid to saturated two-phase mixture, and to examine the applicabilities of correlations to blowdown transients which are used in actual computer program for safety analysis codes. Extensive comparisons of the present experimental data obtained are made with these correlations and, if they cannot explain the data well, new correlations are suggested by authors.

2. EXPERIMENTS

2.1. Experimental Apparatus and Procedure

The experimental apparatus will be described in detail. Fig. 1 shows a schematic flow diagram including the essential components of the boiling loop.

The test were carried out on a stainless steel tube of 10 mm O.D. x 0.5 mm thick x 450 mm long (effective heated length) mounted vertically in a test section of the boiling loop. Fig. 2 shows the test section in detail. Both upper and lower ends of test tube were welded to the electric copper terminals for heating.

The tube wall temperatures were measured at the outside surface. Two pieces of Chromel-Almel seath thermocouple of 0.34 mm O.D. were welded to the outside surface of the test tube for this purpose. These thermocouples were calibrated at the steady condition with the calculated outside surface temperature found by computing the temperature drop through the tube wall of which inner surface temperatures were measured with six pieces of Copper-Constantan thermocouple contacted rigidly with the inner surface of the test tube.

The test tube was vertically mounted at the center of the Pyrex glass tube of 28 mm I.D. which formed the outer wall surface of the testing annular flow channel as shown in Fig. 2. This glass tube channel was protected by a stainless steel pressure tube from the high pressure boundary, and the gap between the glass and stainless steel tubes was filled with stagnant water which also acted as thermal insulator.

Four pairs of optical observation windows of 20 mm wide x 265 mm long each

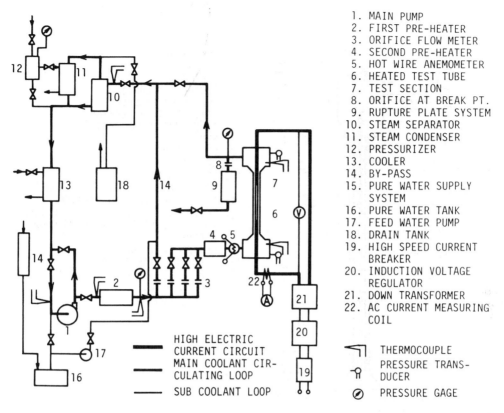

1. MAIN PUMP
2. FIRST PRE-HEATER
3. ORIFICE FLOW METER
4. SECOND PRE-HEATER
5. HOT WIRE ANEMOMETER
6. HEATED TEST TUBE
7. TEST SECTION
8. ORIFICE AT BREAK PT.
9. RUPTURE PLATE SYSTEM
10. STEAM SEPARATOR
11. STEAM CONDENSER
12. PRESSURIZER
13. COOLER
14. BY-PASS
15. PURE WATER SUPPLY
    SYSTEM
16. PURE WATER TANK
17. FEED WATER PUMP
18. DRAIN TANK
19. HIGH SPEED CURRENT
    BREAKER
20. INDUCTION VOLTAGE
    REGULATOR
21. DOWN TRANSFORMER
22. AC CURRENT MEASURING
    COIL

—— HIGH ELECTRIC
    CURRENT CIRCUIT
—— MAIN COOLANT CIR-
    CULATING LOOP
—— SUB COOLANT LOOP

⌐ THERMOCOUPLE
႙ PRESSURE TRANS-
   DUCER
⊘ PRESSURE GAGE

Fig. 1. Schematic Flow Diagram of Boiling Water Loop

Table 1. Experimental Runs and Range

P_0 = 0.588 MPa

RUN NO.	q_0 MW/m²	ΔT_{sub} °C	U_f m/s	RUN NO.	q_0 MW/m²	ΔT_{sub} °C	U_f m/s	RUN NO.	q_0 MW/m²	ΔT_{sub} °C	U_f m/s
1	1.16	70	0.5	15	1.74	70	3.0	29	2.09	30	1.0
2	1.16	70	1.5	16	1.74	50	0.5	30	2.09	20	3.0
3	1.16	70	3.0	17	1.74	50	1.5	31	2.09	20	2.0
4	1.16	50	0.5	18	1.74	50	3.0	32	2.09	20	1.0
5	1.16	50	1.5	19	1.74	30	0.5	33	2.09	20	0.5
6	1.16	50	3.0	20	1.74	30	1.5	34	2.09	10	3.0
7	1.16	30	0.5	21	1.74	30	3.0	35	2.09	10	2.0
8	1.16	30	1.5	22	1.74	30	4.0	36	2.09	10	1.0
9	1.16	30	3.0	23	1.74	15	0.5	37	2.09	10	0.5
10	1.16	15	0.5	24	1.74	15	1.5	38	2.09	5	2.0
11	1.16	15	1.5	25	1.74	15	3.0	39	2.09	5	1.0
12	1.16	15	3.0	26	1.74	5	0.5	40	2.33	50	3.0
13	1.74	70	0.5	27	1.74	5	1,5	41	2.33	50	3.0
14	1.74	70	1.5	28	1.74	5	3.0				

was made of platinum wire of 0.1 mm O.D. The fluid velocity was detected from electrical resistance of the platinum wire which was measured with a double bridge balance. This hot wire method was strongly affected with the fluid temperature variation. In the present experiment, however, the temperature of fluid at the entrance of the test section was found to be constant during the initial period of 50 ms. Therefore, this method received no influence of fluid temperature variations with time, and then could be calibrated at the condition of constant fluid temperature.

On the other hand, the method of (2) using the orifice was applied for the later long period of blowdown and, of course, for the period preceding to the test run to establish test conditions. Three kinds of orifice were used as shown in Fig. 1, of which aperture diameters were different, and could cover the full range of fluid velocity required in the present experiment.

To establish test conditions, the system water was circulated at the desired initial rate and two pre-heaters were operated to bring the water to the temperature approaching the saturation temperature of water at the exit of the test section. Then heat was applied to the test tube and the pressurizer was adjusted up to the initial system pressure until the desired initial conditions of heat flux, flow rate, pressure and temperature of water were attained finaly. After establishing initial test conditions, the Mylar film was ruptured by the needle driven simultaneously with the electrical pulse signal, while prior to this signal every recording instrument started to run.

Data were recorded for the sufficiently long period of blowdown transient. These analog deta were converted into digital data and analyzed numerically.

2.2. Scope of Experiments

Various series of runs and the range of variables at the initial time are shown in Table 1. Experiments of 41 runs were made at the test conditions of the initial system pressure at the exit of the test section, 0.588 MPa, the initial fluid velocity at the entrance to the test section, 0.5-4.0 m/s, the initial heat flux on the outside surface of the test tube, 1.16-2.33 MW/m^2, and the initial subcooling temperature of coolant at the inlet of the test section, 5-70 K. The aperture diameter of the orifice attached to the rupture plate system was selected to be 20 mm I.D. Recording time required for the present experiment was 20 seconds which was sufficiently long time in comparison with the order of 1 ms in the accuracy required for measuring the initial rapid transient.

3. EXPERIMENTAL RESULTS AND DISCUSSION

3.1. Experimental Results

Experimental data are first presented in the time history curves as shown typically in Figs.3, 4 and 5, each of which shows the time variations of coolant pressure and coolant temperature at the entrance and exit of the test section, respectively P_i and P_o, and T_{fi} and T_{fo}, the outside surface temperature of the test tube, T_w, and the coolant velocity, U_f, after the blowdown occurred. In the figures, the zero point at the time coordinate axis represents the blowdown-initiated time.

Both coolant pressures at the entrance and exit of the test section de-

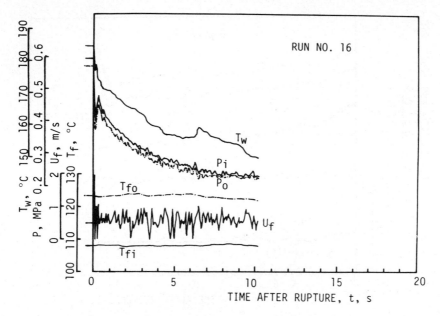

Fig. 3. Experimental Result of Run No. 16

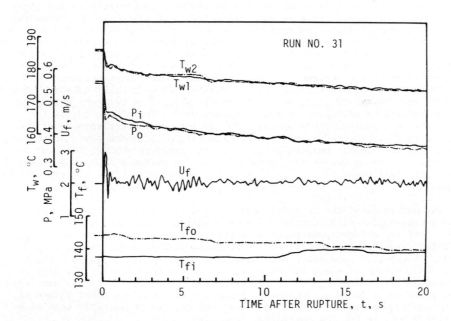

Fig. 4. Experimental Result of Run No. 31

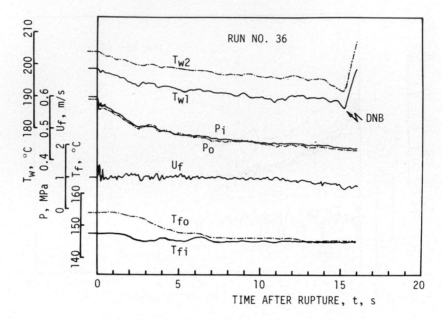

Fig. 5. Experimental Result of Run No. 36

crease rapidly after rupture and go down to the minimum pressure in the first period of 10 ms. Subsequently in the second period of 25–50 ms, the pressures tend to increase again to the maximum, the recovery peak pressure. These pressure behaviors are known to be the undershoot phenomena. After this period, the coolant pressure begins again to decrease gradually to the atmospheric pressure expected in infinite time elapsed. During the period from the minimum to recovery peak point of the coolant pressure, the exit coolant pressure exceeds in several cases the entrance value. Rapid bulk boiling of coolant caused by the pressure reduced brings the pressure recovery, while this recovery suppresses boiling bubbles to collapse and make pressure waves. Therefore, during the above-mentioned period, there appear very complex composite phenomena of the pressure recovery, the propagation and refraction of pressure waves which are peculiar to the present boiling loop. After the pressure reaches the peak of recovery, the normal tendency is attained that the entrance coolant pressure exceeds the exit pressure, but both pressures measured show the similar fluctuations.

The magnitude of the drop in coolant pressure during the first period is found to be smaller with decreasing the subcooling temperature of coolant, ΔT_{sub}, that is because spontaneous bulk boiling occurs in the low subcooling condition. When bubbles grow by coalescence and lead to slugs on the outside surface of the test tube under decreasing the pressure of subcooled coolant, rapid condensation of slugs occurs in the subcooled bulk flow at the exit of the test section and brings violent fluctuations in coolant velocity and pressure. Since the reduced pressure rate greatly depends on the total volume of discharged liquid coolant out of the loop, the coolant pressure is kept to be relatively high in the low subcooling condition.

Figs. 6 and 7 show typical comparisons of the coolant pressure histories

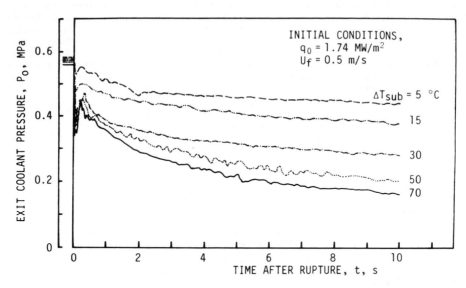

Fig. 6. Time Variations of Coolant Pressure at the
Exit of the Test Section

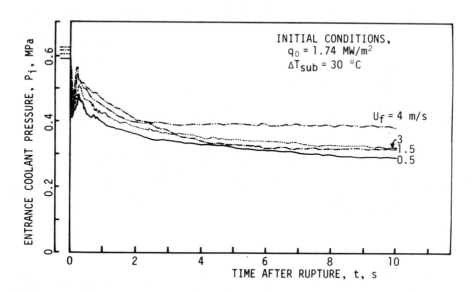

Fig. 7. Time Variations of Coolant Pressure at the
Entrance to the Test Section

i
l

m
d

t
m
s
T
i
t
t
t
D
c
s
l
b
c
f

t
F
c
w
t
t

f
i
t

w
T$_s$
a
f
T
i
r
n(

C
d
G$_f$
o

108

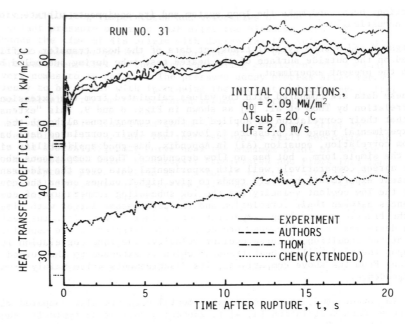

Fig. 8. Time Variation of Heat Transfer Coefficient on the Outside Surface of the Test Tube, and Comparisons between Experimental Data and Correlations (Run No. 31)

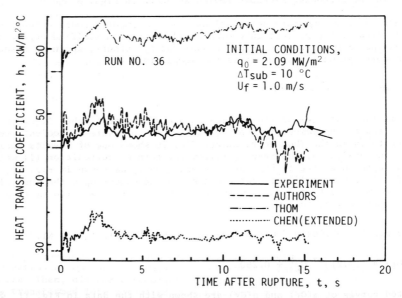

Fig. 9. Time Variation of Heat Transfer Coefficient on the Outside Surface of the Test Tube, and Comparisons between Experimental Data and Correlations (Run No. 36)

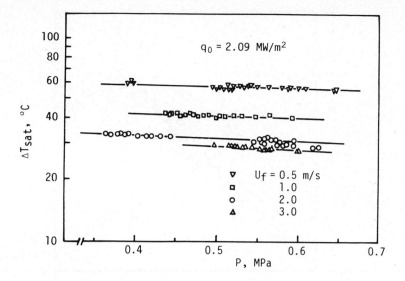

Fig. 10. Correlated Experimental Data during Subcooled Blowdown

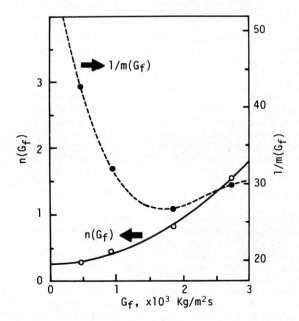

Fig. 11. Correlations of $m(G_f)$ and $n(G_f)$ with Experimental Data

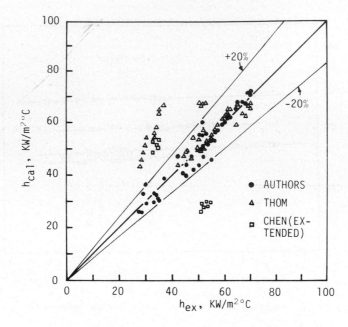

Fig. 12. Comparisons between Data and Correlations by
Thom et al, Chen and Authors in Subcooled
Blowdown Region

The authors'correlation (1) - (3) can predict well the subcooled blowdown
data in the range of the present experiment, having the maximum deviation of 20
percents, as shown in Fig. 12, where h_{cl} is the heat transfer coefficient calculated from the correlation and h_{ex} that obtained in the present experiment.
The figure also shows the comparison between the authors' the Thom and the Chen
correlations. In Figs. 8 and 9, the heat transfer coefficient values in transients predicted by the authors' correlation are also presented and show to be
much closer to the experimental data than those by other correlations, but show
a little large fluctuations with time as well as the Chen correlation, that is
because the flow rate dependence terms in these correlations are sensitive to
fluctuations of flow velocity measured as shown in Figs. 3 - 5.

3.3. Saturated Blowdown Region

In the saturated blowdown region, which follows the subcooled blowdown region, quality of fluid at the exit of the test section increases as the blowdown proceeds, and then the flow pattern changes from bubbly flow to slug and
annular flows. During this period, the violent fluctuations of the coolant
pressure and the flow quality observed in the subcooled or transition region are
suppressed, and then stable nucleate boiling and forced convection vaporizing
heat transfer are expected to be dominant. Coalescence of bubbles without condensing in the bulk flow and the discharged flow rate restricted by the critical
condition at the break point both suppress the thermal and hydrodynamic interactions to be fractional

Figs. 13 and 14 show typical experimental data of varied heat transfer coefficient with time, time history curves, which cover the wide range from sub-cooled blowdown to saturated blowdown. Here, let's pay major attention to the data in the saturated region in the figures. Experimental data are compared with the values calculated from the correlations by the Chen [3], equation (A3) in Appendix, and the Schrock and Grossman [5], equation (A2) in Appendix. The Chen correlation which is semi-empirical is known to work well for various fluids including water, and covers both the low and high quality regions, that is, the nucleate boiling and forced convection vaporization regions. There-fore, the Chen correlation is used as the best choiced correlation in the re-cent computer programs. The competing correlation is that by Schrock and Grossman. Their correlation principally is based upon the forced convection va-porization and includes the nucleate boiling effect represented by the boiling number, B_0.

As shown in Figs. 13 and 14, both correlations cannot predict satisfacto-rily the present data. The Schrock and Grossman correlation shows closer values to the experimental data at the high coolant velocity, however, tends to over-predict greatly with decreasing the coolant velocity which is approaching to a pool boiling condition. The Chen correlation shows relatively closer values to the present data than the Schrock and Grossman correlation, however, remains comparatively large differences with data. These are overpredicted at the low flow rate and underpredicted at the high flow rate.

A modified correlation for saturated blow down region is presented by au-thors, which is based upon the form of the Schrock and Grossman correlation,

$$N_u \ / \ (\ 0.023 \ Re^{0.8} \ Pr^{1/3} \ ) = (\ 360 \ B_0 + 2.08 \ ) + 5.33 \ (\ 1 \ / \ X_{tt} \ )^{2/3} \qquad (4)$$

where N_u is the Nusselt number, $Re = G_f(1-x)D_e/\mu_f$ the Reynolds number, Pr the Prandtl number and X_{tt} the Lockhart-Martinelli parameter. B_0 and X_{tt} are as follows,

$$B_0 = q \ / \ (\ G_f \ h_{fg} \ )$$

$$1 \ / \ X_{tt} = \{ \ x \ / \ (1-x) \ \}^{0.9} \ (\ \mu_g \ / \ \mu_f \ )^{0.1} \ (\ \rho_f \ / \ \rho_g \ )^{0.5} \qquad (5)$$

In the authors' correlation, the boiling effect represented by B_0 term is re-duced and, contrary to this, the forced convection effect represented by X_{tt} term is enhanced in comparison with the Schrock and Grossman correlation.

The authors' correlation (4) - (6) can predict well the present saturated data, having the maximum deviation of 20 percents, as shown in Fig. 15, where h_{cl} and h_{ex} are also the same definitions as in Fig. 12. In the figure, the comparisons between the authors, the Chen and the Schrock and Grossman correla-tions are presented. In Figs. 13 and 14, the predicted time history curves of heat transfer coefficient from the authors' correlation are also presented and found to be comparatively well in accordance with the experimental curves. The predicted curves show to be fluctuated largely because correlations are sensi-tive to variations of coolant velocity measured in the present experiment as shown in Figs. 3 - 5.

The authors' correlation is also compared with the data reported by Bennett et al. [9] as shown in Fig. 16. Their data are obtained at the steady condition. Compared with the present data, these experimental ranges are lower extremely in the heat flux and the mass flow rate both. The authors' correla-

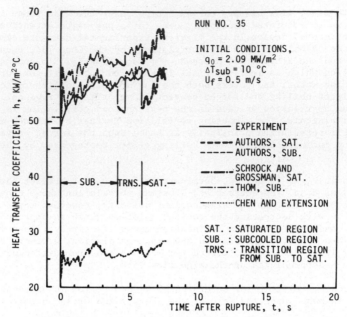

Fig. 13. Time Variation of Heat Transfer Coefficient on the
Outside Surface of the Test Tube, and Comparisons
between Experimental Data and Correlations (Run No. 35)

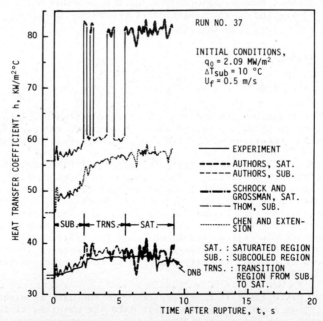

Fig. 14. Time variation of Heat Transfer Coefficient on the
Outside Surface of the Test Tube, and Comparisons
between Experimental Data and Correlations (Run No. 37)

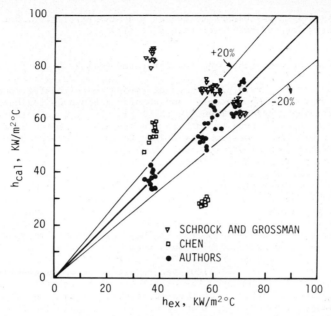

Fig. 15. Comparisons between Data and Correlations by Chen, Schrock and Grossman, and Authors in Saturated Blowdown Region

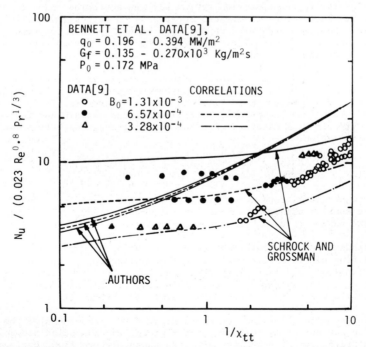

Fig. 16. Comparison of the Authors' Correlation, Equation (4), with Data by Bennett et al. [9]

tion predicts relatively higher values at high $(1/\chi_{tt})$ region, but correlates well with their data at the lower values of $(1/\chi_{tt})$.

3.4. Transition between Subcooled and Saturated Regions

Figs. 13 and 14 show typical data of varied heat transfer coefficient with time which cover both the subcooled and saturated blowdown regions, as was noted in the prceding sections. Then, in these figures, the correlations used in computer codes and suggested by authors are compared with the data in the wide range of blowdown transient from its origin to the end of experimental run. The composite correlation delived by Chen, which is a combination of the extended Chen correlation for subcooled liquid and the Chen correlation for saturated two-phase mixture, equations (A6) and (A3), can make a smooth transtion between subcooled and saturated blowdown modes. On the other hand, the Thom correlation for subcooled liquid cannot be applicable to quality region. Therefore, a switchover from Thom's correlation to the correlation of Schrock and Grossman, that is from equation (A1) to equation (A2), is required. The authors' correlations for the subcooled and saturated regions, equations (1) and (4), are also required for making a switchover between them.

Figs. 13 and 14 show the calculated results from these combined correlations. The Chen combination show the excellent smooth transtion between subcooled and saturated regions, but these predicted curves give comparatively large differences with the experimental data as was explained in detail in the preceding sections. The Thom − Schrock and Grossman combination, although it show exceptionally such a smooth switchover as shown in Fig. 13, gives an extraordinarily large stepwise jumping up or down in its calculated curve as shown in Fig. 14. In the trasition region between subcooled and saturated regions, quality in the flow fluctuates repeatedly to be positive or negative, that leads to repetitive jumping up and down of the curve calculated, as shown in Figs. 13 and 14. This behavior appears to be like a violent oscillation and is the major problem.

In the case of the authors' correlations, calculated curves show the same kind of stepwise oscillations in the transition region, but their magnitudes are relatively small as shown in Figs. 13 and 14. Therefore, it is concluded in the present study that the authors' correlations can explain and follow well the present experimental data and give the smooth transition between subcooled and saturated region.

4. CONCLUSION

Experiments of 41 runs were made with a single heated test tube (10 mm O.D.) in an annular channel test section (28 mm O.D.) and with top blowdown. Experimental ranges of variables for every run are as shown in Table 1.

Two kinds of measurements for the coolant velocity were applied. The anemometer of fine platinum wire was used for measuring the early stage of coolant velocity transient for about 50 ms after the break. For the later period of about 20 s up to the end of test run, the normal orifice flow meter was used. These composite measuring systems could operate satisfactorily.

Typical obtained data were shown in Figs. 3 − 7. Phenomena of the pre-CHF boiling were summarized as following two regions, the subcooled blowdown region and the saturated range of quality with two-phase mixture flow. During subcooled region, the coolant velocity and pressure showed violent variations induced by rapid condensation of bubbles, but they disappeared after transition to the saturated blowdown. Just after the break, vigorous bulk boiling and forced convection rapidly increased brought large decreases of the outside surface temperatures of the test tube. Thereafter, the surface temperature of the test tube continued to drop gradually until it started to rise again due to occurrence of DNB.

Transient data of the heat transfer coefficient on the outside surface of the test tube were compared with the values calculated from the extensions to the correlations by Thom et al. and Chen for subcooled region, and from the correlations by Schrock and Grossman, and Chen for saturated region. These compared results were shown in Figs. 8 - 15. The Thom correlation has good applicability since its form was very simple, but tended to give largely higher value at the low coolant velocity and low subcooling temperature. The correlation by Schrock and Grossman applied for saturated region agreed well with the experimental data in the high flow rate region, but overpredicted greatly with decreasing the coolant velocity. Therefore, it was concluded that a simple switchover from the Thom correaltion to that by Schrock and Grossman was not expected to be applicable directly.

The Chen correlation and its extension making a smooth transition between subcooled and saturated boilings showed comparatively large differences with the present data and were found to overpredict the data in the subcooled and high flow rate conditions.

Since above-mentioned correlations could not predict satisfactorily the data, modified correlations were presented by authors, equations (1) and (4) based on the Thom correlation for subcooled region and the Schrock and Grossman correlation for saturated region. They were compared and correlated well with the experimental data as shown in Figs. 12 and 15. The authors' correlations were applied for predicting transient time history curves of heat transfer coefficient as shown in Figs. 8 and 9 for subcooled region and in Figs. 13 and 14 for saturated region. Comparisons of the authors' correlations with the experimental data and other correlations showed that the authors' correlations had the good agreements with the data and the smooth transition between the subcooled and saturated regions.

NOMENCLATURE

B_0	boiling number
C_p	specific heat at constant pressure (J/kg K)
D_e	equivalent diameter (m)
G	mass flow rate (kg/m^2 s)
h	heat transfer coefficient (W/m^2°C)
h_{fg}	latent heat (J/kg)
k	thermal conductivity (W/m°C)
m	equation (2)
n	equation (3)
$Nu=hD_e/k_f$	Nusselt number
P	pressure
P_r	Prandtl number
q	heat flux (MW/m^2)
R_e	Reynolds number
t	time (s)
T	temperature (°C)
$\Delta T_{sat}=T_w-T_{sat}$	(°C)
$\Delta T_{sub}=T_{sat}-T_f$	(°C)

U   velocity (m/s)

$v_{fg}= \rho_g - \rho_f$ (m^3/kg)

x   quality

Greek

μ dynamic viscosity (kg/m s)

ρ density (kg/m^3)

σ surface tension (N/m)

X_{tt} Lockhart-Martinelli parameter

Subscripts

cl   calculated value

ex   experimental value

f   liquid

g   gas

i   at entrance

o   at exit

sat   at saturation condotions

w   at outside surface of test tube

0   at initial conditions

REFERENCES

1. Jens, W. H. and Lottes, P.A., "Analysis of Heat Transfer, Burnout, Pressure Drop and Density Data for High Pressure Water," USAEC Report ANL-4627, U.S.Atomic Energy Commission, 1951.

2. Thom, J. R. S., Walker, W. M., Fallon, T. A. and Reisting, G. F. S., "Boiling in Subcooled Water during Flow up Heated Tubes or Annuli," Proc. IME (London), Vol. 180, Pt 3C, 226-246 (1965-66).

3. Chen, J. A., "Correlation for Boiling Heat Transfer to Saturated Fluids in Convective Flow," I&EC Process Design Dev., Vol. 5, 322-329 (1966).

4. Collier , J. G., "Convective Boiling and Condensation," McGraw-Hill, New York, 1972.

5. Schrock, V.E. and Grossman, L.M., "Forced Convection Boiling Studies, Forced Convection Vaporization Project," USAEC Report TID-14632 (1959).

6. Lawson, C. G., "Heat Transfer from Electrically Heated Rods during a Simulated Loss-of-Coolant Accident," Chem. Eng. Progr. Sym. Seri., Vol. 67, 1-13 (1971).

7. Aoki, S., Inoue, A. and Kozawa, Y., "Transient Boiling Crisis during Rapid Depressurization," Heat Transfer 1974, Vol. 4, B6.3, 250-254 (1974).

8. Kumagai, A., Koibuchi, H. and Toda, S., "Transient Boiling Heat Transfer under Depressurization," Proc. 13th National Heat Transfer Symposium of Japan, A211 73-75 (1976). (Japanese)

9. Bennett, J. A. R., Collier, J. G., Pratt, R. C. and Thornton, J. D., "Heat Transfer to Two-Phase Gas-Liquid Systems, Part 1: Steam-Water Mixtures in the Liquid Dispersed Region in an Annulus," Trans. Inst. Chem. Engrs. (London), Vol. 39, 113 (1961).

10. Bjornard, T. A. and Griffith, P., "PWR Blowdown Heat Transfer," Thermal

and Hydraulic Aspect of Nuclear Reactor Safety, Vol. 1 (Light Water Reactors), ASME, 17-73 (1977).

APPENDIX

Correlations used and recommended in computer programs for safety analysis codes, which are refered in the present study as

1. The Thom correlation [2],

$$q = (1/22.7)^2 \, \Delta T_{sat}^2 \, \exp(0.230 \, P) \tag{A1}$$

2. The Schrock and Grossman correlation [5],

$$N_u \, / \, (0.023 \, Re^{0.8} \, Pr^{1/3}) = 7400 \, B_0 + 1.11 \, (1/X_{tt})^{0.66} \tag{A2}$$

3. The Chen correlation [3],

$$h = h_{FCV} + h_{NB} \tag{A3}$$

$$h_{FCV} = F \cdot 0.023 \, \{R_e(1-x)\}^{0.8} \, Pr^{1/3} \, (k_f/D_e) \tag{A4}$$

$$h_{NB} = S \cdot 0.00122 \, k_f^{0.79} \, C_{pf}^{0.45} \, \rho_f^{0.49} \, / \, (\sigma^{0.5} \, \mu_f^{0.29} \, h_{fg}^{0.24} \, \rho_g^{0.24})$$
$$\times \, \{h_{fg} \, / \, (T_{sat} \, v_{fg})\}^{0.75} \, \Delta T_{sat}^{0.99} \tag{A5}$$

where functions F and S are given in graphical form.

4. The extension to the Chen correlation [4],

$$q = h_{NB} \, \Delta T_{sat} + h_{FCV} \, (T_w - T_f) \tag{A6}$$

Multi-Phase Flow and Heat Transfer III. Part B: Applications
edited by T.N. Veziroğlu and A.E. Bergles
Elsevier Science Publishers B.V., Amsterdam, 1984 — Printed in The Netherlands

TRANSIENT TWO-PHASE FLOW WITH HEAT EXCHANGE IN A HORIZONTAL ANNULAR TUBE

A. A. Kendoush, S. A. K. Hamoodi
Nuclear Research Centre
Iraqi Atomic Energy Commission
Baghdad, Iraq

J. M. Hassan
College of Engineering
University of Baghdad
Baghdad, Iraq

ABSTRACT

An experimental system was designed and installed to investigate the effects of depressurisation in a horizontal annular test section; Freon-113 was used as a coolant in the experimental system with a maximum operating pressure of 194 kN/m^2 and the test conditions covered a range of initial circulating velocities up to 0.55 m/s . Thermodynamic non-equilibrium was found experimentally from measuring both the pressure and temperature of the coolant at the outlet of the test section during the blowdown . The void fraction and the local heat transfer coefficient were increasing throughout the blowdown while the local surface temperature of the inner heated pipe decreased mainly due to the turbulence and agitation in the bulk of the flowing coolant created by bubble formation . Nucleate boiling and flash evaporation are the two mechanisms responsible for bubble formation . Satisfactory agreement was found between the heat transfer coefficient results and Chen correlation in the later stages of the transient .

1. INTRODUCTION

Recently boiling heat transfer and two-phase flow have achieved world wide interest primerly because of their application in nuclear power reactors .The main important application of the present project, which investigate the effects of depressurisation in two-phase flow, is the thermal-hydraulic behaviour of the light water cooled nuclear reactors during a postulated loss of coolant accident (LOCA) .

120

Lawson[1] carried out a transient heat transfer tests and used heated rods
which were similar to the actual fuel rods geometrically . High heat transfer
coeficient were observed during blowdown . Aoki et al[2] have also obtained
high values of heat transfer coefficient at the depressurisation of the coolant.

Flash evaporation and thermodynamic non-equilibrium are associated with the
depressurisation of the flowing coolant as indicated by Simpson et al[3] and
Fauske[4] .

The subjects investigated in this work were the following:-

A. The effects of depressurisation on the thermal behaviour of the Freon-113
which was adopted as the flowing medium .

B. Thermodynamic non-equilibrium of the coolant during the transient .

C. The rate of convective heat transfer in the absence of the gravitational
forces as a horizontal test section was utilised .

2. EXPERIMENTAL SYSTEM

A full diagram of the experimental system is shown in Fig.1 . In the design
of the system, the following consideration were adopted to obtain controlled
sets of depressurisation tests :-

1. The provision of a primary cooling circuit which comprises the test section,
the expansion chamber, the circulation pump and the reservoir tank . This

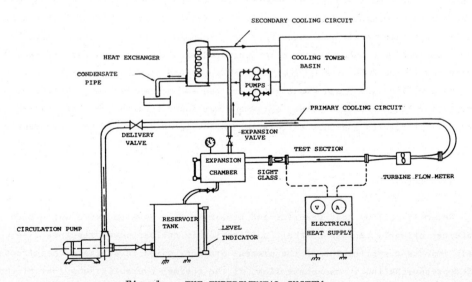

Fig. 1 - THE EXPERIMENTAL SYSTEM.

circuit enabled the investigator to control the system pressure and the flow
rate of the coolant .

2. Freon-113 was the coolant chosen for the primary cooling circuit . The low
boiling point of liquid Freon at a particular pressure, saves considerable
amount of thermal energy input . The use of liquid Freon for the modelling of
the thermal hydraulic behaviour of water in nuclear reactors, avoids working at
the conditions of elevated temperature and pressure[5],[6] .

3. A secondary cooling circuit was adopted to draw water from the basin of the
reactor cooling tower by pumps to the heat exchanger . The water takes away
the heat of condensation liberated by the Freon vapour .

4. Selection of a circulation pump for the primary circuit was made taking into
consideration the frictional and head losses as well as the balance between the
quantity of Freon remaining in the reservoir tank and the half-full level of
Freon in the expansion chamber . The reason of maintaining a half-full level
in the expansion chamber was to control the pressure rise of the system due to
the evaporation of liquid Freon in the test section .

5. The geometrical arrangment of the test section is illustrated in Fig.2 with
the inner pipe heated electrically . Chromel/Alumel thermocouples were spot
welded to outer surface of the inner pipe . The inlet and outlet temperatures
of the coolant were measured by silicon diode probes . Pressure transducers
were used for coolant pressure recording.   A sight glass was provided at the
outlet of the test section to visualize the flow pattern . A turbine flow meter

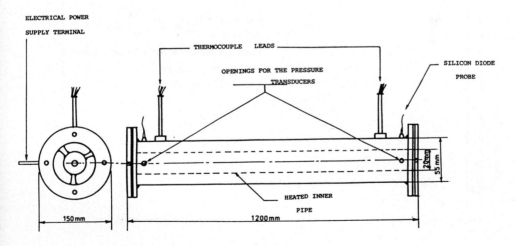

Fig. 2 - THE TEST SECTION

122

was mounted at the inlet of the test section for the measurement of the circulation velocity. Gamma-ray attenuation technique was used to measure the void fraction.

6. The resulting signals of all the instruments were fed to a U.V. recorder and continuously recorded during all transient tests.

More details on the experimental system, instruments distribution and methods of calibration may be found in reference (9).

3. DISCUSSION OF EXPERIMENTAL RESULTS

As a result of depressurisation the system pressure was reduced as shown in Fig.3. Measuring the absolute pressure and the local coolant temperature at the outlet of the test section, enabled the degree of non-equilibrium encountered during the transient to be estimated interms of the difference between the measured local and the saturation temperature corresponding to the measured local pressure as shown in Fig.4. The initial subcooled conditions at the start of the transient is also illustrated in Fig.4. In all cases it was found that the departure from thermodynamic equilibrium increases toward the

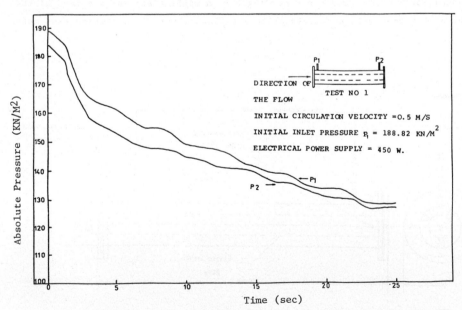

Fig. 3 — THE PRESSURE TIME HISTORY ALONG THE TEST SECTION

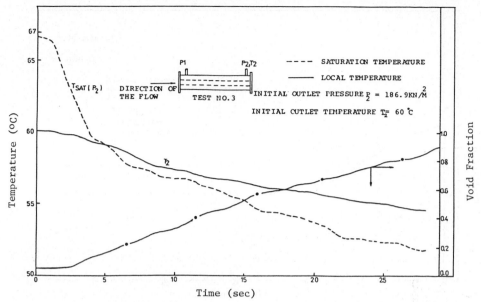

Fig. 4 -   THE VARIATION OF NON-EQUILIBRIUM DURING THE TRANSIENT

end of the blowdown while the void fraction increases which is in agreement with Zuber[7] who stated that low void fraction is associated with high thermodynamic equilibrium and vice versa.

The void formation during the transient is illustrated in Fig.5.   Flash

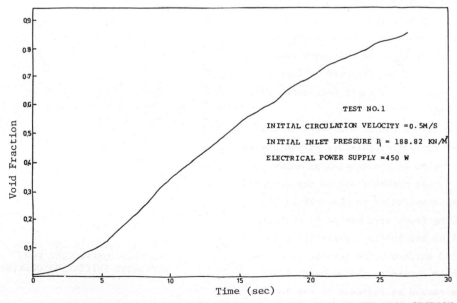

Fig. 5 -   THE VARIATION OF THE VOID FRACTION AT THE EXIT OF THE TEST SECTION.

124

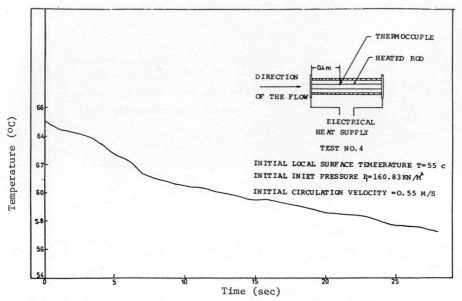

Fig. 6 - THE VARIATION OF THE SURFACE TEMPERATURE OF THE HEATED ROD.

evaporation in the bulk of the coolant and nucleate boiling on the surface of the heated pipe were the two causes responsible for void formation but it is not easy to interpret the proportion of the two causes. The void fraction results obtained in this work were similar to those reported by Edwards and Mather[8] in their horizontal test section.

The reason for the decrease in the local surface temperature of the heated inner pipe of the annulus as shown in Fig.6. was probably due to the enhanced flash evaporation in the bulk of the flowing Freon accompanied by nucleate boiling and bubble generation on the heated surface. The turbulence of the bubble motion in the coolant might have caused an increase in the local convective heat transfer coefficient during the transient as shown in Fig.7.

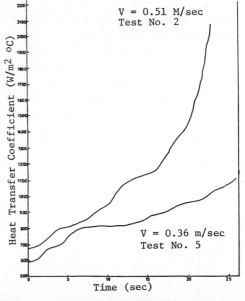

Fig. 7 - THE VARIATION OF THE HEAT TRANSFER COEFFICIENT FOR DIFFERENT INITIAL VELOCITIES.

Chen correlation[10], was used to compare the heat transfer coefficient results. Although the correlation was originally based on two-phase steady-state flow heat transfer conditions, it was applied at each successive time step during the transient. Satisfactory agreement was found between the present results and Chen correlation as shown in Fig.8 particularly during the later stages of the transient. This may be due to the predominance of the subcooled conditions at the earlier stages of the transient as illustrated in Fig.4.

Hamoodi[11] applied Chen correlation to transient two-phase vertically upward flow conditions. His data seemed to agree with the correlation at the downstream lengths of the test section.

ACKNOWLEDGMENT

This paper is published by permission of the Iraqi Atomic Energy Commission.

REFERENCES

1. Lawson,C.G. " Heat transfer from electrically heated rods during simulated loss of coolant accident. " Chem.Eng.Prog.Symp.Ser.67(1)1971 .

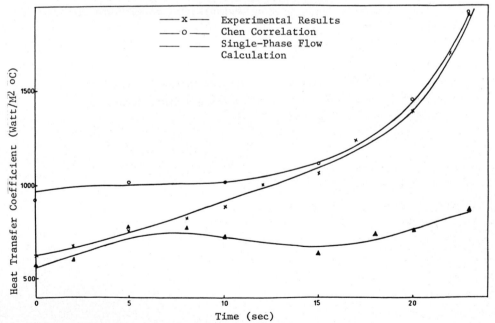

Fig. 8 - HEAT TRANSFER COEFFICIENT DURING TRANSIENT.

2. Aoki,S.,et al " Transient boiling crisis during rapid depressurisation "
Fifth International Heat Transfer Conference, Tokyo, Sep. 1974

3. Simpson, H.C., Rooney, D.R. and Callander, T.M.S. " Effect of rapid steam
off-take on natural circulation boilers "
Joint Symposium on Fluid Mechanics and Measurements in Two-Phase Flow Systems
Leeds, 1969

4. Fauske, H.K. , Chem. Eng. Prog. Symp. Ser. 61(210) 1965

5. Barnett, P.G. " The scaling of forced convection boiling heat transfer "
United Kingdom Atomic Energy Authority AEEW-R 134, 1963

6. Stevens, G.F. and Macbeth,R.V. " The use of Freon-12 to model forced convection burnout in water the restriction of the size of the model "
United Kingdom Atomic Energy Authority AEEW-R 683 , 1971

7. Zuber, N. , Discussion on Rapid Transients
Symposium on Two-Phase Flow Dynamics , EURATOM, Endhoven, Sep. 1967

8. Edwards, A.R. and Mather, D.J. " Some U.K. studies related to the loss of
coolant accident "
Symposium on Two-Phase Flow Group Meeting , Harwell, 1974

9. Hassan, J.M. " Transient heat transfer with phase change "
M.Sc. Thesis, College of Engineering
University of Baghdad , 1978

10. Chen, J.C. "A correlation of boiling heat transfer to saturated fluids in
convective flow "
Sixth Heat Transfer Conference , Boston, 1963
ASME paper No. 63-HT-34

11. Hamoodi, S.A.K. " Heat transfer in transient two-phase flow "
M.Sc. Thesis , Department of Thermodynamics and Fluid Mechanics
University of Strathclyde , 1978

Multi-Phase Flow and Heat Transfer III. Part B: Applications 127
edited by T.N. Veziroğlu and A.E. Bergles
Elsevier Science Publishers B.V., Amsterdam, 1984 — Printed in The Netherlands

EXPERIMENTAL SYSTEM DESCRIPTION FOR AIR-WATER CCFL TESTS OF THE 161-ROD
FLECHT-SEASET TEST VESSEL UPPER PLENUM

Stephen P. Fogdall and James L. Anderson
EG&G Idaho, Inc.
P.O. Box 1625
Idaho Falls, ID 83415, U.S.A.

ABSTRACT

A series of countercurrent flow limiting (CCFL) experiments has been performed by EG&G Idaho, Inc. in the Steam-Air-Water (SAW) test facility at the
Idaho National Engineering Laboratory on behalf of the U.S. Nuclear Regulatory
Commission (NRC). Tests were performed in a mockup of the vessel for the
161-Rod Systems Effects Test (SET) facility of the FLECHT-SEASET program, conducted by the Westinghouse Electric Corporation. Westinghouse and the NRC
will use the test results to provide a CCFL correlation to predict the flooding behavior in the upper plenum of the SET vessel. This paper presents a
description of the experimental system and the test conduct, including data
validation and uncertainty analysis.

The test objectives centered on experimentally obtaining coefficients in
the Wallis correlation for flooding with the specific vessel geometry. The
test conditions and vessel configuration are described and the design of the
test loop, instrumentation, and data acquisition are discussed. The estab-
lishment of a test point and the resultant data are described.

Data validation was a primary concern to ensure a successful test. The
discussion centers on the use of mass balances during a test point as a means
of establishing data validity. How this concern affected the design of the
test loop, instrumentation, and data acquisition is discussed, as are the
measurement uncertainties.

1. INTRODUCTION

The Westinghouse FLECHT-SEASET (Full Length Emergency Core Heat Transfer--
Separate Effects and System Effects Test) Natural Circulation and Reflux Test
Facility[a] in Pittsburgh, PA, is designed to provide experimental data from a
small-scale, simulated pressurized water reactor (PWR) during the natural circulation and reflux portions of a loss-of-coolant experiment. Prior to performing tests in the facility, Westinghouse and the U.S. Nuclear Regulatory
Commission (NRC) desired to have a countercurrent flow limiting (CCFL) correlation for use in predicting the flooding behavior in the FLECHT-SEASET test
vessel upper plenum. The CCFL testing reported in this paper was conducted in

———————————

a. This facility is also referred to herein as the System Effects Test (SET)
facility.

the Steam-Air-Water (SAW) test facility operated by EG&G Idaho, Inc. at the
Idaho National Engineering Laboratory, near Idaho Falls, ID. A series of
air-water tests were performed in the SAW loop to investigate the CCFL phenomena in a mockup of the FLECHT-SEASET vessel upper plenum. This paper presents the experimental system description and test conduct, including data
validation and uncertainty analysis. The CCFL test results are presented and
discussed in an accompanying paper [1], and the complete data set resulting
from these experiments is available in Reference [2].

2. TECHNICAL BACKGROUND

Countercurrent flow limiting occurs as a result of the interaction between
a gas updraft and a liquid downflow, which may be in the form of liquid droplets and/or liquid film. There is a limitation on the possible simultaneous
countercurrent flow rates of gas and liquid in a vertical conduit. Thus, CCFL
has been identified [3] as a mechanism that might reduce the core and upper
plenum refilling rate of a nuclear reactor during the reflood portion of a
loss-of-coolant accident (LOCA).

Data obtained from the CCFL tests in the SAW facility were applied to the
flooding correlations that have emerged from the experiments of different
researchers [4]. The correlations used were the Wallis correlation and the
Wallis-Kutateladze correlation.

The Wallis correlation for CCFL data from round tubes is given by

$$(J_g^*)^{1/2} + M(J_f^*)^{1/2} = C \tag{1}$$

where J_g^* and J_f^* are dimensionless mass fluxes of gas and liquid in the
countercurrent flow region defined as

$$J_i^* = \frac{\rho_i^{1/2} j_i}{[gw (\rho_f - \rho_g)]^{1/2}} = \frac{\dot{m}_i}{A[gw \rho_i (\rho_f - \rho_g)]^{1/2}} \tag{2}$$

where

j_i	=	superficial velocity of phase i (m/s)
ρ_f and ρ_g	=	phase densities of the liquid and gas, respectivelγ (kg/m^3)
ρ_i	=	density of phase i (kg/m^3)
g	=	gravitational acceleration (=9.78 m/s^2)
w	=	characteristic dimension (m)
$\dot{m}_i$	=	mass flow rate of phase i (kg/s)
A	=	minimum flow area (m^2).

Wallis recommended that for turbulent flow, the value of M in Equation (1) be
equal to unity. The value of C depends on the geometry and the manner in which

the liquid and gas are introduced into the system. A decision was made to use the hydraulic diameter as the characteristic dimension, w, in Equation (2), in view of the complex cross-sectional shape of the Westinghouse ground plate and the fact that the ground plate was the minimum flow area within the test vessel.

The Wallis-Kutateladze correlation is represented by

$$(K_g)^{1/2} + A(K_f)^{1/2} = B \quad . \tag{3}$$

The Kutateladze number for a phase, K_i, is given by

$$K_i = \frac{j_i \, \rho_i^{1/2}}{[g\sigma \, (\rho_f - \rho_g)]^{1/4}} = \frac{\dot{m}_i}{A[g\sigma \, \rho_i^2 \, (\rho_f - \rho_g)]^{1/4}} \tag{4}$$

where σ is the surface tension (N/m) and the other terms are as previously defined.

The gas Kutateladze number, K_g, expresses the balance between inertial forces in the gas, buoyancy forces, and surface tension forces. Application of this correlation does not require knowledge of any dimension or length characteristic of the system geometry.

3. TEST OBJECTIVES

The primary objective of the experimental program was to obtain sufficient CCFL data, for the specific vessel geometry, to determine the flooding characteristics and obtain flooding curves for the FLECHT-SEASET vessel upper plenum. Sufficient data were obtained to determine the values of the coefficients in the Wallis and the Wallis-Kutateladze correlations. Figures 1 and 2 are examples of CCFL curves obtained from the test data. Figure 1 is a CCFL curve of the Wallis correlation and Figure 2 is a CCFL curve of the Wallis-Kutateladze correlation. Secondary objectives of the test program were to (a) evaluate the relative merits of the two correlations, (b) evaluate the effect of various air injection nozzles on the CCFL curves, (c) determine the effect of liquid injection rate on the CCFL characteristics, and (d) determine the effect of different upper plenum internals on the CCFL characteristics of the upper plenum.

To accomplish the latter objective, CCFL curves were obtained during the air-water testing for three different sets of upper plenum internals: all internals in place (Figure 3); all internals, with the exception of the support columns; and support columns and the upper core support plate removed.

4. TEST CONDITIONS

The conditions maintained during the testing were

● Pressure: 170 kPa (25 psia)

● Temperature: Ambient (290 to 300 K)

Table 1. Primary Measurements

Measurement	Measurement Identifier	Location	Engineer Units	Range	Uncertainty (95% Conf.)
Differential pressure at venturi flowmeter	DPA*VEN*HI	FE-1 = Reference air, prior to injection	kPa	12.5-125	+0.5% FS
Differential pressure at venturi flowmeter	DPA*VEN-LO	FE-1 = Reference air, prior to injection	kPa	1.25-12.5	+0.5% FS
Absolute pressure	PA*VEN	FE-1 = Reference air, prior to injection	kPa	86-1100	+0.5% FS
Fluid temperature	TA*VEN	FE-1 = Reference air, prior to injection	K	280-300	+2 K
Differential pressure at orifice flowmeter	DPA*OUT*HI	FE-2 = Reference air, outlet of outlet surge tank	kPa	5-75	+0.5% FS
Differential pressure at orifice flowmeter	DPA*OUT*LO	FE-2 = Reference air, outlet of outlet surge tank	kPa	0.5-5	+1% FS +2% reading
Absolute pressure	PA*OUT	FE-2 = Reference air, outlet of outlet surge tank	kPa	86-775	+0.5% FS
Fluid temperature	TA*OUT	FE-2 = Reference air, outlet of outlet surge tank	K	280-300	+2 K
Volumetric flow rate	QW*INJECT	FE-3 = Prior to water injection into vessel	L/s	0.06-1.26	+0.5% FS
Fluid temperature	TW*INJECT	FE-3 = Prior to water injection into vessel	K	280-300	+2 K

Table 1. (continued)

Measurement	Measurement Identifier	Location	Engineer Units	Range	Uncertainty (95% Conf.)
Volumetric flow rate at turbine flowmeter	QW*CARRY	FE-5 = Exit of separator tank, carryover water out of hot legs	mL/s	6-775	+0.5% FS
Fluid temperature	TW*CARRY	FE-5 = Exit of separator tank, carryover water out of hot legs	K	280-300	+2 K
Volumetric flow rate at turbine flowmeter	QW*FALL	FE-4 = Fallback water, exit from bottom of vessel	mL/s	6-630	+0.5% FS
Fluid temperature	TW*FALL	FE-4 = Fallback water, exit from bottom of vessel	K	280-300	+2 K
Liquid level using differential pressure transducer	L*WCT	Water collection tank	cm	70-130	+1 cm
Absolute pressure	PV+0	Vessel at bottom of hot leg	kPa	86-775	+0.5% FS
Differential pressure	DP+0-83	Vessel, from hot leg elevation to above support plate	kPa	0.5-5	+1% FS +2% reading
Differential pressure	DP-83-101	Vessel, across support plate	kPa	0.5-5	+1 FS +2% reading
Differential pressure	DP-101-122	Vessel, across grounding plate	kPa	0.5-5	+1% FS +2% reading
Differential pressure	DP-122-171	Vessel, from below grounding plate to above grid spacer	kPa	0.5-5	+1% FS +2% reading
Absolute pressure	PV-171	Vessel, just above grid spacer	kPa	86-775	+0.5% FS

Fig. 6. Air injection--lower plenum drain assembly.

5.4 Data Acquisition System

Data acquisition and online analysis for the flooding tests were performed using the SAW loop data acquisition and analysis system (DAAS). This
system is configured around a Hewlett-Packard 21 MXE computer operating under
the HP-1000 RTE-IVB Operating System, and using three HP-91,000 A-to-D cards
for analog-to-digital conversion. The system is equipped with an HP-7925 disk
drive and an HP-7970E digital magnetic tape drive for extensive data storage.
Data were acquired over a 180-s time period, with 3680 samples per channel
being acquired, which is a sample frequency of 20 Hz per channel. After each
channel was sampled, mass flow rates were calculated and averaging buffers
updated. After all data for a test point were acquired, the average quantity
for each measurement and calculated parameter, and their uncertainties, were
output for a permanent record.

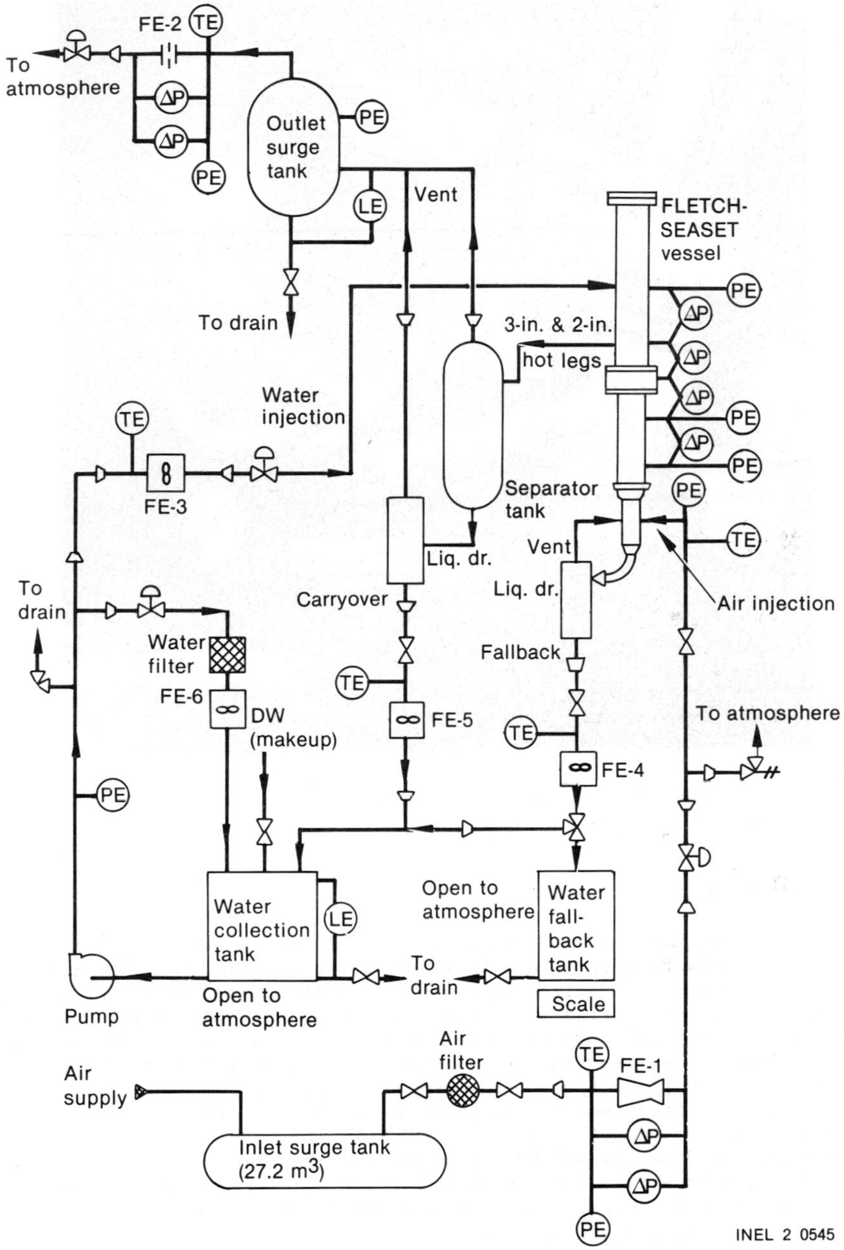

witł

thrc
watε
test

tran
poin

the
pres
valvι
tran

estał
estał

c omme

allow

progr

estab

was tɛ
the al
with ɛ
increɛ

point
hysteɪ

6.2 Ί

A
ing te
charac
acquis
series
nel of
period
time pι
floodiι

N:
first p
differε
orientɛ

Fig. 7. Piping and instrumentation diagram for SET facility Upper Plenum Flooding Test.

INEL 2 0545

Fig. 9. Air injection nozzles used for the scoping tests. The 2-in. end
cap (without extension) was used for the production testing.

cap as the production test series nozzle. The major reasons for this decision
were that (a) this nozzle produced data with the least amount of scatter as
measured with the standard error estimate, (b) the data showed no statistically
supported change in slope of a least squares linear fit, (c) the nozzle had no
orientation considerations, and (d) the nozzle injected air with a uniform
velocity profile through the core cross section.

Ten production test series were performed with the water injection rate
varied from 150 to 300 to 400 mL/s in each. Three were performed with all
upper plenum internals in place, three with the 10 support columns removed,
and three with the 10 support columns and the upper core support plate removed.
The tenth test series was performed at a lower test section pressure (125 kPa)
to obtain a CCFL correlation for comparison with the data obtained at 170 kPa.
At the lower test section pressure, a maximum of 285 mL/s of water would drain
from the test vessel. Therefore, the maximum water injection rate was lowered
to that value.

7. DATA VALIDATION

In order to be assured that valid data were obtained, the decision was
made to use mass balances of air and water as the major criterion for data
validation. The basic arguments for this decision are presented in the
following paragraphs along with mathematical expressions, results, and
discussion.

The data obtained during testing were applied online to the Wallis and
Wallis-Kutateladze correlations. In the Wallis correlation, given by Equation (1), J_g^* and J_f^* are dimensionless mass fluxes of gas and liquid in the
counter-current flow region, defined by Equation (2), and are functions of the
individual phase mass flow rates, fluid properties, and physical constants.
The fluid properties (phase density and surface tension) are dependent on

temperature and pressure. However, for a particular test point, the test section temperature and pressure were maintained constant (less than 1% variance
from the mean) and, therefore, the fluid properties were constant. Thus, the
Wallis dimensionless mass fluxes and (from similar reasoning) the Kutateladze

number are directly proportional to the mass flow rate, $\dot{m}$, as

$$J_i^* = A_1 \, \dot{m}_i \tag{5}$$

$$K_i = A_2 \, \dot{m}_i \tag{6}$$

where A_1 and A_2 represent, respectively, the portions of Equations (2) and
(4), that are constant during a test point.

The basis for data validation then rests on acceptance of the calculated
values of mass flow rate, $\dot{m}_i$. If, for both air and water, the absolute value
of the measured mass flow rate entering the vessel minus the measured mass flow
rate exiting the vessel was less than the sum of the individual measurement
uncertainties, then the mass flow rate values were considered valid and the
data point qualified. These criteria are expressed below.

For air flow rates:

$$\left| \dot{m}_{g-in} - \dot{m}_{g-out} \right| \leq \left| \Delta\dot{m}_{g-in} + \Delta\dot{m}_{g-out} \right| \tag{7}$$

For water flow rates:

$$\left| \dot{m}_{f-in} - \dot{m}_{f-out} \right| \leq \left| \Delta\dot{m}_{f-in} + \Delta\dot{m}_{f-out} \right| . \tag{8}$$

To apply the above criteria, the uncertainties in the individual mass flow rate
measurements are required. The derivation of these uncertainties is presented
next.

7.1 Uncertainties in Mass Flow Rates

The inlet gas mass flow rate, $\dot{m}_{g-in}$, was obtained with a venturi
flowmeter. The expression used [5,6] for mass flow rate using a venturi is
given by

$$\dot{m} = \beta^2 \, A_{pipe} \, Y \, \frac{C}{\sqrt{1 - \beta^4}} \, \sqrt{2 \, \rho_{gv} \, DP_{gv}} \tag{9}$$

where

β	=	ratio of minimum venturi ID to entrance pipe ID (d_1/d_2)
A_{pipe}	=	cross-sectional area of the entrance pipe (m^2)
Y	=	gas expansion factor
C	=	coefficient of discharge

ρ_{gv} =   air density in the venturi (kg/m^3), which, from the ideal gas law, is

$$\frac{R_G \, P_{gv}}{T_{gv}}$$

where

R_G =   gas constant for air (=3.483 kg/m^3-K/kPa)

P_{gv} =   gas static pressure at the venturi entrance (kPa)

T_{gv} =   gas temperature at the venturi entrance (K)

DP_{gv} =   pressure drop between the venturi throat and the upstream static pressure tap (kPa).

The expression, as applied to the 5-cm venturi used in this test, becomes

$$\dot{m}_{g-in} = C_1 \, Y \, \rho_{gv} \, DP_{gv} \tag{10}$$

where

$$Y \quad = \quad 1 - \frac{0.582 \, DP_{gv}}{P_{gv}}$$

DP_{gv} =   differential pressure across the venturi (kPa)

P_{gv} =   static pressure at the venturi entrance (kPa)

ρ_{gv} =   venturi gas density (kg/m^3)

C_1 =   1.547 x 10^{-2} m^2.

The method of analyzing the uncertainty of a computed result as presented by Kline and McClintock [7] may be summarized by

$$\Delta R = \left[\sum_{i=1}^{n} \left(\frac{\partial R}{\partial V_i} \, \Delta V_i \right)^2 \right]^{1/2} \tag{11}$$

where

R =   result

V_i = i^{th} independent variable

$\partial R, \partial V$ =   the uncertainty in the result and the independent variable, respectively.

Applying this method to the inlet mass flow rate, Equation (10) yields

$$\frac{\Delta \dot{m}_{g-in}}{\dot{m}_{g-in}} = \left[\left(\frac{\Delta C_1}{C_1}\right)^2 + \left(\frac{\Delta Y}{Y}\right)^2 + 1/4\left(\frac{\Delta \rho_{gv}}{\rho_{gv}}\right)^2 + 1/4\left(\frac{\Delta DP_{gv}}{DP_{gv}}\right)^2\right]^{1/2} \quad (12)$$

where the individual components in Equation (2) are as follows

$\dfrac{\Delta \dot{m}_{g-in}}{\dot{m}_{g-in}}$ = ratio of uncertainty to calculated value of inlet gas mass flow rate

$\dfrac{\Delta C_1}{C_1}$ = 0.0079 (see Footnote a below)

$$\frac{\Delta Y}{Y} = \left\{\frac{0.339\left[P_{gv}^2 (\Delta DP_{gv})^2 + DP_{gv}^2 (\Delta P_{gv})^2\right]}{P_{gv}^2 (P_{gv} - 0.582\, DP_{gv})^2}\right\}^{1/2} \quad \text{(see Footnote b below)}$$

$$\frac{\Delta \rho_{gv}}{\rho_{gv}} = \left[\left(\frac{\Delta P_{gv}}{P_{gv}}\right)^2 + \left(\frac{\Delta R_G}{R_G}\right)^2 + \left(\frac{\Delta T_{gv}}{T_{gv}}\right)^2\right]^{1/2} .$$

Outlet gas mass flow rate, $\dot{m}_{g-out}$, was obtained using a 4-in. orifice meter tube. Equations (9) and (10) are also applicable to an orifice, although the discharge coefficient and expansion factor are different.

Application of the uncertainty analysis to the discharge orifice measurement yields the uncertainty of the outlet gas mass flow rates as

$$\frac{\Delta \dot{m}_{g-out}}{\dot{m}_{g-out}} = \left[\left(\frac{\Delta C_2}{C_2}\right)^2 + \left(\frac{\Delta Y}{Y}\right)^2 + 1/4\left(\frac{\Delta P_{go}}{\rho_{go}}\right)^2 + 1/4\left(\frac{\Delta DP_{go}}{DP_{go}}\right)^2\right]^{1/2} \quad (13)$$

where the individual uncertainty components are

$\dfrac{\Delta \dot{m}_{g-out}}{\dot{m}_{g-out}}$ = ratio of uncertainty to calculated value of outlet gas mass flow rate

---

a. This value is the combined estimated uncertainties in the constants of Equation (9) and includes an uncertainty of 0.5% in the discharge coefficient.

b. The uncertainties for individual measurements are given in Table 1 and are based on manufacturers' data and engineering judgment.

$$\frac{\Delta C_2}{C_2} = 0.062$$

$$\frac{\Delta Y}{Y} = \left\{ \frac{0.118 \left[P_{go}^2 (\Delta DP_{go})^2 + DP_{go}^2 (\Delta P_{go})^2 \right]}{P_{go}^2 (P_{go} - 0.343 \, DP_{go})^2} \right\}^{1/2}$$

$$\frac{\Delta \rho_{go}}{\rho_{go}} = \left[\left(\frac{\Delta P_{go}}{P_{go}} \right)^2 + \left(\frac{\Delta T_{go}}{T_{go}} \right)^2 + \left(\frac{\Delta R_G}{R_G} \right)^2 \right]^{1/2}$$

and

$$T_{go} = \text{gas temperature at the orifice entrance (K)}.$$

The liquid mass flow rate measurement of the injected water, carryover water, and fallback water were made using turbine flowmeters. The expression for liquid mass flow rate using a turbine is

$$\dot{m}_f = Q_f \, \rho_f \tag{14}$$

where

$$\dot{m}_f = \text{liquid mass flow rate (kg/s)}$$

$$Q_f = \text{volumetric flow rate measured by the turbine (m}^3\text{/s)}$$

$$\rho_f = \text{liquid density (kg/m}^3\text{)}.$$

Applying Equation (11) to Equation (14), the measurement uncertainty in the liquid mass flow rate is given by

$$\frac{\Delta \dot{m}_f}{\dot{m}_f} = \left[\left(\frac{\Delta Q_f}{Q_f} \right)^2 + \left(\frac{\Delta \rho_f}{\rho_f} \right)^2 \right]^{1/2} . \tag{15}$$

Equation (15) can be used to obtain the uncertainty in inlet liquid mass flow rate, $\dot{m}_{f-in}$, the carryover mass flow rate, $\dot{m}_{f-carry}$, and the fallback mass flow rate, $\dot{m}_{f-fallbk}$. Equation (8), using these terms, becomes

$$\left| \dot{m}_{f-in} - (\dot{m}_{f-carry} + \dot{m}_{f-fallbk}) \right| \leq \Delta \dot{m}_{f-in} + \Delta \dot{m}_{f-carry} + \Delta \dot{m}_{f-fallbk} . \tag{16}$$

The above expressions were applied to the data during acquisition, to obtain the mass flow rates and their associated uncertainties. The mass balances, Equations (7) and (16), were then checked to ascertain the validity of the data.

Tables 2 and 3 present the mass flow rates and associated uncertainties for the gas and liquid data obtained during the test series in which all upper plenum internals were installed, the nominal water injection rate was 0.300 kg/s, and the air injection rate was varied from 0.13 to 0.35 kg/s. Of the 34 test points in this series, 5 failed the qualification criteria based on air mass summation (the largest deviation was 12%), and 11 failed the qualification criteria based on liquid mass summation (the criteria for 8 of these was within 20%). Test Points 3A01, 3A02, 3A03, 3A27, and 3A28 in Table 2 did not meet the criteria for data validation. Since these test points lie in a portion of the flooding curve which does not contribute to factors in the Wallis or Wallis-Kutateladze correlations, they were allowed to remain as trend data. A number of other test points, indicated in Table 3, did not meet the criteria for data validation, and were not used to determine correlation coefficients.

The same validation check was used for the other nine production test series and test points were eliminated from use if they failed to meet the mass balance criteria.

In addition to obtaining CCFL data, part of the responsibility in performing these CCFL tests was to ensure that the data obtained were valid. The mass balance criteria were based on theoretical considerations of the measurements and subsequent calculations. However, many other factors potentially contributed errors to the measurements. Some factors could be controlled and were eliminated or minimized, whereas others were uncontrollable.

7.2 Uncontrolled Factors

Some factors involved in testing were inherent and not removable. During some test points, as the air injection flow rate increased, a situation called storage occurred in the FLECHT-SEASET vessel upper plenum. At certain air flow rates, liquid tended to gather in certain local volumes until sufficient water gathered to change the air flow pattern through the upper plenum. A surge of water out of the vessel would then occur as carryover. This could cause the outlet liquid mass flow rate to be less than the inlet liquid mass flow rate, since some of the liquid inventory was being stored in the vessel. This discrepancy would not be accounted for in the uncertainty prediction. This storage and release occurs in a cyclic manner whose time period is dependent on the air injection rate and was not equal to the 180-s data acquisition time period. If the two time periods are not integral multiples of each other, the data could represent a net storage or a net release. This could also cause discrepancies between inlet and outlet liquid flow rates, greater than the predicted uncertainties.

Another possible error involves measurements of liquid mass flow rates with turbines at low flow rates. Turbines can indicate flow, even with zero actual flow, if vibration exists in the piping. The vibration can cause the turbine blades to vibrate, which the pickoff probe senses as blade motion and, thus, a liquid flow. Also, the liquid carryover and liquid fallback flow measurement turbines were mounted in a vertical position. At low flows, the water in droplet form could fall into the turbine blades with sufficient momentum to impart a spin to the blades. The turbine then could indicate an erroneous flow measurement.

Table 2. Gas Mass Flow Rate Balance ($\left| \dot{m}_{g-in} - \dot{m}_{g-out} \right| \leq \Delta \dot{m}_{g-in} + \Delta \dot{m}_{g-out}$)

Test Point	Inlet Mass Flow Rate (kg/s)		Outlet Mass Flow Rate (kg/s)		$\left\| \dot{m}_{g-in} - \dot{m}_{g-out} \right\|$ (kg/s)	$\Delta \dot{m}_{g-in} + \Delta \dot{m}_{g-out}$ (kg/s)
	$\dot{m}_{g-in}$	$\Delta \dot{m}_{g-out}$	$\dot{m}_{g-out}$	$\Delta \dot{m}_{g-out}$		
3A01	0.1310	0.0014	0.1162	0.0104	0.0148	0.0118[a]
3A02	0.1400	0.0015	0.1264	0.0105	0.0136	0.0120[a]
3A03	0.1510	0.0016	0.1382	0.0108	0.0128	0.0124[a]
3A04	0.1603	0.0016	0.1488	0.0111	0.0115	0.0127
3A05	0.1693	0.0017	0.1588	0.0115	0.0105	0.0132
3A06	0.1767	0.0017	0.1668	0.0118	0.0099	0.0135
3A07	0.1890	0.0018	0.1801	0.0124	0.0089	0.0142
3A08	0.1936	0.0052	0.1880	0.0128	0.0056	0.0180
3A09	0.2041	0.0049	0.1993	0.0134	0.0048	0.0183
3A10	0.2182	0.0046	0.2145	0.0142	0.0037	0.0188
3A11	0.2273	0.0045	0.2245	0.0148	0.0028	0.0193
3A12	0.2354	0.0044	0.2333	0.0153	0.0021	0.0197
3A13	0.2460	0.0042	0.2455	0.0160	0.0005	0.0202
3A14	0.1822	0.0018	0.1742	0.0121	0.0080	0.0139
3A15	0.2009	0.0052	0.1946	0.0131	0.0063	0.0183
3A16	0.2374	0.0045	0.2358	0.0154	0.0016	0.0199
3A17	0.2280	0.0045	0.2249	0.0148	0.0031	0.0193
3A18	0.2195	0.0046	0.2152	0.0143	0.0043	0.0189
3A19	0.2200	0.0046	0.2158	0.0143	0.0042	0.0189
3A20	0.2096	0.0048	0.2040	0.0136	0.0056	0.0184
3A21	0.2012	0.0050	0.1945	0.0131	0.0067	0.0181
3A22	0.1904	0.0019	0.1827	0.0126	0.0077	0.0145
3A23	0.1780	0.0018	0.1691	0.0119	0.0089	0.0137
3A24	0.1673	0.0017	0.1574	0.0114	0.0099	0.0131
3A24B	0.1682	0.0017	0.1589	0.0114	0.0093	0.0131
3A25	0.1594	0.0016	0.1494	0.0111	0.0100	0.0127
3A26	0.1479	0.0015	0.1364	0.0107	0.0115	0.0122
3A27	0.1397	0.0015	0.1272	0.0105	0.0125	0.0120[a]
3A28	0.1279	0.0014	0.1144	0.0105	0.0135	0.0119[a]
3A29	0.1762	0.0017	0.1680	0.0118	0.0082	0.0135
3A30	0.1804	0.0018	0.1724	0.0120	0.0080	0.0138
3A31	0.2322	0.0044	0.2317	0.0152	0.0005	0.0196
3A32	0.2405	0.0043	0.2411	0.0157	0.0006	0.0200
3A33	0.2326	0.0044	0.2325	0.0152	0.0001	0.0196

a. A test point in which the mass flow rate balance criterion was not met.

8. CONCLUSIONS

The following conclusions can be drawn as a result of the CCFL testing reported herein:

- CCFL data with small scatter and good repeatability can be obtained if care is taken in the design of the experimental system, including consideration of experimental uncertainties prior to testing.

Table 3. Liquid Mass Flow Rate Balance

$$\left| \dot{m}_{f\text{-}in} - (\dot{m}_{f\text{-}carry} + \dot{m}_{f\text{-}fallbk}) \right| \leq \Delta \dot{m}_{f\text{-}in} + \Delta \dot{m}_{f\text{-}carry} + \Delta \dot{m}_{fallbk}$$

| Test Point | Inlet Mass Flow Rate (kg/s) | | Outlet Mass Flow Rate (kg/s) | | | | $\left\| \dot{m}_{f\text{-}in} - \dot{m}_{f\text{-}carry} + \dot{m}_{f\text{-}fallbk} \right\|$ (kg/s) | $\Delta \dot{m}_{f\text{-}m} + \Delta \dot{m}_{f\text{-}carry} + \Delta \dot{m}_{f\text{-}fallbk}$ (kg/s) |
| | | | Carryover | | Fallback | | | |
	$\dot{m}_{f\text{-}in}$	$\Delta \dot{m}_{f\text{-}in}$	$\dot{m}_{f\text{-}carry}$	$\Delta \dot{m}_{f\text{-}carry}$	$\dot{m}_{f\text{-}fallbk}$	$\Delta \dot{m}_{f\text{-}fallbk}$		
3A01	0.3128	0.0071	0.0003	0.0029	0.3159	0.0032	0.0034	0.0132
3A02	0.3125	0.0071	0.0003	0.0029	0.3133	0.0032	0.0011	0.0132
3A03	0.3075	0.0071	0.0007	0.0029	0.3033	0.0032	0.0031	0.0132
3A04	0.3060	0.0071	0.0003	0.0029	0.2936	0.0032	0.0121	0.0132
3A05	0.3072	0.0071	0.0043	0.0029	0.2693	0.0032	0.0336[a]	0.0132
3A06	0.3084	0.0071	0.0094	0.0029	0.2256	0.0032	0.0734[a]	0.0132
3A07	0.3074	0.0071	0.1772	0.0029	0.1492	0.0032	0.0190[a]	0.0132
3A08	0.3082	0.0071	0.2077	0.0029	0.1171	0.0032	0.0166[a]	0.0132
3A09	0.3088	0.0071	0.2469	0.0029	0.0731	0.0032	0.0112	0.0132
3A10	0.3083	0.0071	0.2778	0.0029	0.0455	0.0032	0.0150[a]	0.0132
3A11	0.3078	0.0071	0.2871	0.0029	0.0277	0.0032	0.0070	0.0132
3A12	0.3085	0.0071	0.2947	0.0029	0.0155	0.0032	0.0017	0.0132
3A13	0.3096	0.0071	0.3011	0.0029	0.0046	0.0032	0.0039	0.0132
3A14	0.3031	0.0071	0.1445	0.0029	0.1726	0.0032	0.0140[a]	0.0132
3A15	0.3070	0.0071	0.2362	0.0029	0.0761	0.0032	0.0053	0.0132
3A16	0.3042	0.0071	0.2892	0.0029	0.0161	0.0032	0.0011	0.0132
3A17	0.3040	0.0071	0.2787	0.0029	0.0353	0.0032	0.0100	0.0132
3A18	0.3024	0.0071	0.2698	0.0029	0.0410	0.0032	0.0084	0.0132
3A19	0.3020	0.0071	0.2724	0.0029	0.0348	0.0032	0.0052	0.0132
3A20	0.3016	0.0071	0.2571	0.0029	0.0580	0.0032	0.0135[a]	0.0132
3A21	0.2984	0.0071	0.2254	0.0029	0.0894	0.0032	0.0154[a]	0.0132
3A22	0.2988	0.0071	0.1718	0.0029	0.1425	0.0032	0.0155[a]	0.0132
3A23	0.2992	0.0071	0.0982	0.0029	0.2143	0.0032	0.0133[a]	0.0132
3A24	0.3000	0.0071	0.0142	0.0029	0.2940	0.0032	0.0082	0.0132
3A24B	0.3104	0.0071	0.0325	0.0029	0.2842	0.0032	0.0063	0.0132
3A25	0.3116	0.0071	0.0010	0.0029	0.3250	0.0032	0.0144[a]	0.0132
3A26	0.3128	0.0071	0.0007	0.0029	0.3128	0.0032	0.0007	0.0132
3A27	0.3129	0.0071	0.0003	0.0029	0.3133	0.0032	0.0007	0.0132
3A28	0.3136	0.0071	0.0003	0.0029	0.3145	0.0032	0.0012	0.0132
3A29	0.3155	0.0071	0.0948	0.0029	0.2305	0.0032	0.0098	0.0132
3A30	0.3149	0.0071	0.1359	0.0029	0.1934	0.0032	0.0149[a]	0.0132
3A31	0.3135	0.0071	0.2924	0.0029	0.0342	0.0032	0.0131	0.0132
3A32	0.3131	0.0071	0.3006	0.0029	0.0080	0.0032	0.0045	0.0132
3A33	0.3141	0.0071	0.3003	0.0029	0.0107	0.0032	0.0031	0.0132

a. A test point in which the mass flow rate balance criterion was not met.

148

- Online data qualification can be an extremely useful tool for immediate verification of data integrity, thus preventing unrecognized problems from invalidating the data and requiring expensive and time-consuming repeat testing.

- CCFL phenomena are cyclic, requiring data acquisition over multiple cycles to obtain repeatable stationary data.

ACKNOWLEDGMENTS

The FLECHT-SEASET program is jointly funded and managed by the U.S. Nuclear Regulatory Commission, Electric Power Research Institute, and Westinghouse Electric Corp. This work was supported by the U.S. Nuclear Regulatory Commission, Office of Nuclear Regulatory Research, under DOE Contract No. DE-AC07-76IDO1570.

NOTICE

REFERENCES

1. Andreychek, T. S., Anderson, J. L., Fogdall, S. P., Rosal, E. R., "Counter Current Flow Limiting Phenomena in the 161-Rod FLECHT-SEASET Test Vessel Upper Plenum," Proceedings of the 3rd Multi-Phase Flow and Heat Transfer Symposium, Miami Beach, Florida, 18-20 April 1983.

2. Anderson, J. L. and Fogdall, S. P., "Experimental Data Report for Air-Water Flooding Tests of the FLECHT-SEASET Program SET Facility Vessel Upper Plenum," NUREG/CR-2636, EGG-2183, May 1982.

3. Yamanouchi, A., "Effects of Core Spray Cooling at Stationary State After Loss-of-Coolant Accident," Journal of Nuclear Science and Technology, 5, No. 10, October 1968, pp. 8-17.

4. Wallis, G. B., One-Dimensional Two-Phase Flow, McGraw-Hill, Inc., 1969.

5. Fluid Meters, Their Theory and Application, Report of ASME Research Committee on Fluid Meters, Fifth Edition, ASME, New York, NY, 1959.

6. Spink, L. K., Principles and Practice of Flow Meter Engineering, The Foxboro Co., 1967.

7. Kline, S. J. and McClintock, F. A., "Describing Uncertainty in Single-Sample Experiments," Mechanical Engineering, 75, No. 1, January 1953, pp. 3-8.

Multi-Phase Flow and Heat Transfer III. Part B: Applications
edited by T.N. Veziroğlu and A.E. Bergles
Elsevier Science Publishers B.V., Amsterdam, 1984 — Printed in The Netherlands

COUNTERCURRENT FLOW LIMITING PHENOMENA IN THE 161-ROD
FLECHT-SEASET TEST VESSEL UPPER PLENUM

Timothy S. Andreychek
Westinghouse Electric Corporation
Nuclear Technology Division
Pittsburgh, Pennsylvania 15230, U.S.A

James L. Anderson and Stephen P. Fogdall
EG&G Idaho, Inc.
Idaho Falls, Idaho 83415, U.S.A

E. R. Rosal
Westinghouse Electric Corporation
Nuclear Fuels Divison
Pittsburgh, Pennsylvania 15230, U.S.A

ABSTRACT

     Countercurrent flow limiting (CCFL) phenomena were experimentally studied
in a scaled model of a pressurized water reactor (PWR) upper plenum using air
and water as working fluids. The influence of rate of liquid injection, upper
plenum geometry, and system pressure on the flooding characteristics of the
scaled upper plenum assembly were studied. Test data were correlated using
the Wallis and the Wallis-Kutateladze correlations forms. Both correlation
forms yielded the same general flooding characteristics; the flooding curves
tended to be linear over the range of test parameters studied. The observed
flooding characteristics of the scaled upper plenum assembly were found to be
insensitive to both rate of liquid injection and upper plenum pressure. Upper
plenum assembly geometry, however, was found to significantly effect the
observed flooding behavior. A flooding correlation, based on the form suggested by Wallis, is proposed for use in evaluating the CCFL behavior of the
scaled model of the PWR upper plenum under simulated reflood conditions for a
LOCA event.

NOMENCLATURE

A   :   minimum flow area in FLECHT-SEASET SET Facility upper plenum (m^2)

B   :   intercept of the Wallis-Kutateladze Flooding correlation, deter-
            mined experimentally

C   :   intercept of the Wallis flooding correlation, determined experimentally

g   :   gravitational acceleration (9.78 m/s^2)

J*   :   dimensionless mass flux

K   :   Kutateladze number, ratio of inertial, buoyant, and surface tension forces

M   :   slope of the Wallis flooding correlation, determined experimentally

ṁ   :   mass flow rate (kg/s)

N   :   slope of the Wallis-Kutateladze correlation, determined experimentally

w   :   characteristic dimension for the Wallis correlation, taken to be the hydraulic diameter of the ground plate in the FLECHT-SEASET SET Facility upper plenum (m)

ρ   :   density (kg/m^3)

σ   :   surface tension of the liquid (N/m)

Subscripts:

f   :   liquid phase or component

g   :   gaseous phase or component

1. INTRODUCTION

Countercurrent flow limiting (CCFL) phenomena can occur due to the interaction between a vapor upflow and a liquid downflow that are coincidently attempting to traverse the same flow channel. The gaseous upflow can be either a condensible or a non-condensible vapor, and the liquid downflow can be either droplets or films. During the reflood portion of a postulated Loss-of-Coolant Accident (LOCA), steam generated in the reactor core flows up through the upper plenum carrying liquid water with it, and out through the primary system hot legs. This steam flow has been identified as a mechanism that could limit the fallback of stored and deentrained liquid water from the reactor upper plenum into the core and lower plenum [1].

CCFL phenomena can also occur in the upper plenum of an experimental facility that is simulating the reflood portion of a LOCA. One such experimental facility is the Systems Effects Test (SET) Facility, which is one of the test loops utilized in the FLECHT-SEASET Program [2], and consists of a 161-electrical-resistance-heater-rod core simulation, a downcomer, two steam generator simulators, associated piping, and accumulator/overflow tanks. The occurrence of CCFL phenomena within the upper plenum of the SET Facility during testing can influence the heat transfer process associated with the core simulation. Thus, knowledge of the CCFL characteristics of the SET Facility upper plenum was needed for both predicting and analyzing the thermal performance of the SET Facility core simulation during reflood testing. To develop this knowledge, an experimental program was undertaken.

It should be noted that the SET Facility was designed with a prototype to model ratio of 307 to 1, based on bundle flow area and volumes. The upper plenum was scaled to provide hydraulic performance typical of that of a pressurized water reactor (PWR). A detailed discussion of the scaling philosophy and methodology is discussed by Rosal et. al. [2].

2. TEST OBJECTIVES

The primary objective of the current test program was to obtain flooding curves for the actual SET Facility upper plenum assembly which consisted of ten support columns, an upper core plate, and a ground plate. Three configurations of the upper plenum assembly were tested:

1. All upper plenum internals in place.

2. The ground plate and upper core plate simulation in place and the ten support columns removed.

3. The ground plate in place with the upper core plate simulation and ten support columns removed.

The secondary objectives of the test program were:

1. Evaluation of effect of liquid injection rate on upper plenum flooding characteristics.

2. Measurement of collapsed liquid level in the upper plenum using differential pressure transducers.

3. Visual observation of flow conditions between the upper core plate and the ground plate, using an optical probe.

Using measured liquid and gas flows, dimensionless mass fluxes were calculated and plotted to develop flooding curves of each of the three SET Facility upper plenum assembly geometries tested.

3. TECHNICAL BACKGROUND

It is generally accepted that CCFL phenomena are governed by the interaction of shear, inertia, gravity, and surface tension forces of the two fluids, one liquid and the other a vapor, attempting to simultaneously traverse the same flow channel in opposite directions, as well as the geometry of the flow channel itself. Based on the experimental data of various researchers, essentially two flooding correlation forms have been proposed; the Wallis correlation and the Wallis-Kutateladze correlation [3]. The Wallis correlation accounts for flow channel geometry but neglects liquid surface tension, and the Wallis-Kutateladze correlation accounts for the liquid surface tension while neglecting flow channel geometry.

Wallis [3] correlated the CCFL data from round vertical tubes with:

$$(J_g^*)^{0.5} + M (J_f^*)^{0.5} = C \tag{1}$$

where J_g^* and J_f^* are dimensionless mass fluxes of gas and liquid, respectively, in the countercurrent flow region and are defined as:

$$J_g^* = \frac{\dot{m}_g}{A \, [gw\rho_g(\rho_f-\rho_g)]^{0.5}} \tag{2}$$

and

$$J_f^* = \frac{\dot{m}_f}{A \, [gw\rho_f (\rho_f - \rho_g)]^{0.5}} \tag{3}$$

The values of the coefficients M and C are evaluated from experimental data. Based on data from the aforementioned flooding experiments in round vertical tubes, Wallis recommended that, for turbulent flow, the value of M in Equation (1) is equal to unity [3], and the value of the intercept, C, is dependent upon the geometry of the flow channel and the manner in which the liquid and gas are introduced into the flow channel.

The Wallis correlation requires the definition of a characteristic dimension, w. It was reasoned that the CCFL phenomena would first begin to occur at that location or component in the SET Facility upper plenum for which the flow area was at a minimum. Such a component was the ground plate for the heater rod bundle. Thus, the hydraulic diameter of the ground plate was used as the characteristic dimension in evaluating the dimensionless mass fluxes for the Wallis correlation.

The Wallis-Kutateladze correlation can be written as:

$$(K_g)^{0.5} + N \, (K_f)^{0.5} = B \tag{4}$$

where K_g and K_f are the Kutateladze numbers for the gas and liquid flows, respectively, in the countercurrent flow region and are defined as:

$$K_g = \frac{\dot{m}_g}{A \, [g \, \sigma \, \rho_g \, (\rho_f - \rho_g)]^{0.25}} \tag{5}$$

and

$$K_f = \frac{\dot{m}_f}{A \, [g \, \sigma \, \rho_f \, (\rho_f - \rho_g)]^{0.25}} \tag{6}$$

The Kutateladze number represents a comparison among inertial, buoyant, and surface tension forces, and does not require the knowledge of any dimension or characteristic length of the flow channel geometry.

To simplify both the execution of the test program and the analysis of the resulting data, air and water were chosen as the working fluids. The use of these two fluids precluded the need to account for evaporation and condensation in evaluating the flooding performance of the SET Facility upper plenum assembly in the current test program.

4. EXPERIMENTAL SYSTEM AND MEASUREMENTS

The experimental facility to investigate the flooding characteristics of the SET Facility upper plenum hardware was designed, fabricated, and installed in the Steam-Air-Water (SAW) Loop at the Idaho National Engineering Laboratory by EG&G Idaho, Inc. The experimental facility modeled the SET Facility vessel, its upper plenum and all associated internals, and a short portion of the core. A detailed description of the experimental system, including a description of the test vessel, test loop, instrumentation, and the data acquisition and reduction syysystem has been reported by Fogdall and Anderson [4], and will not be repeated here. However, a schematic diagram of the test vessel in which the significant components are identified is provided for reference, Figure 1.

5. TEST CONDITIONS

The data from ten tests provided the basis for assessing the CCFL characteristics of the SET Facility upper plenum assembly. Three different upper plenum assembly configurations were tested for each of three different liquid injection rates. Air flow injection rate was varied for each upper plenum assembly configuration and liquid injection rate such that liquid carryover ranged from none to all of the injected liquid. All tests were performed at a system pressure of 170 kPa except for one; a reduced system pressure of 125 kPa was used to investigate the sensitivity of the observed flooding behavior of the upper plenum assembly to test vessel pressure. The test conditions studied are summarized in Table 1.

Each test was assigned a two-character designation based on water injection rate and upper plenum assembly configuration. The first character of the test designation was based on the water injection rate, and the second character was determined by the upper plenum assembly configuration. A test designation of 1A indicates a water injection rate of 150 mL/s with all internals installed, 3S indicates a 300 mL/s water injection rate with the support columns removed, and 4G indicates a 400 mL/s water injection rate with only the ground plate in place. The low pressure (125 kPa) test was assigned a test designation of LP.

TABLE 1. SUMMARY OF TEST CONDITIONS

Parameter	Value
Pressure	170 kPa (25 psia) 125 kPa (18 psia)
Temperature	Room Temperature
Air Injection Rate	0.13 to 0.35 kg/s (130 to 360 CFM at 170 kPa)
Water Injection Rate	400 mL/s (6.3 gpm) 300 mL/s (4.8 gpm) 150 mL/s (2.4 gpm)
Time Duration for Each Test Point	180 s

154

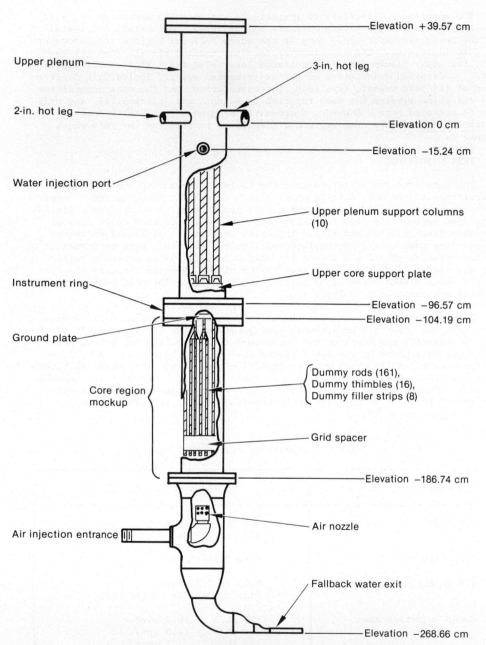

Elevation +39.57 cm

Upper plenum

3-in. hot leg

2-in. hot leg

Elevation 0 cm

Elevation −15.24 cm

Water injection port

Upper plenum support columns (10)

Upper core support plate

Instrument ring

Elevation −96.57 cm

Elevation −104.19 cm

Ground plate

Dummy rods (161),
Dummy thimbles (16),
Dummy filler strips (8)

Core region mockup

Grid spacer

Elevation −186.74 cm

Air nozzle

Air injection entrance

Fallback water exit

Elevation −268.66 cm

Figure 1.   Schematic Drawing of Test Section

6. DISCUSSION OF TEST RESULTS

The experimentally measured liquid and air flow rates were reduced to dimensionless mass fluxes and Kutateladze numbers. These dimensionless data were used to develope CCFL correlations for the various SET Facility upper plenum assembly configurations.

6.1 Comparison of Correlation Form

As previously stated, the Wallis and the Wallis-Kutateladze correlations are fundamentally different in that the former accounts for geometry and neglects liquid surface shear while the latter neglects geometry and accounts for liquid surface shear. To evaluate this fundamental difference in formulation on the reduced data, the root of the dimensionless mass fluxes of the Wallis correlation and the root of the Kutateladze numbers of the Wallis-Kutateladze correlation were plotted for test 4A as in Figures 2 and 3, respectively. Comparison of the data in these two figure show no significant difference between the two formulations; the resulting flooding curves tended to be linear over the range of test conditions studied. Thus, to facilitate relating the data to flow rates the Wallis correlation, being based on dimensionless mass fluxes, was used to evaluate the CCFL behavior of the SET-Facility upper plenum assembly.

6.2 Flooding Curves

The reduced CCFL data from the three tests conducted with all upper plenum internals installed, Test 1A, 3A, and 4A, are plotted in Figure 4. The data show that, the greater the liquid injection rate, the lower the air flow required to initiate flooding. Also, there is good agreement among the data for the three liquid injection rates tested, indicating that the liquid injection rate has little influence on the flooding characteristics of this SET-Facility upper plenum assembly configuration.

The data of Figure 4 tend to approximate a straight line having a slope of about −1.0 over the range:

$$0.04 < (J_f^*)^{0.5} < 0.30 \tag{7}$$

For values of the square root of J_f^* below 0.04, the slope of the data drastically increases, and a zero value for the fallback of the injected liquid is never measured. This behavior is attributed to the behavior of the turbine meter used to measure the liquid fallback. The minimum flow measurement rate of the turbine was 6 mL/S, which corresponds to a value of the square root of J_f^* of 0.04. Thus, these low mass fluxes were beyond the range of operation of the turbine meter, and were excluded from use in evaluating flooding correlations.

The reduced flow data expressed as dimensionless mass fluxes for the three tests performed with the ten support columns removed from the upper plenum assembly, Tests 1S, 3S, and 4S, are plotted in Figure 5. As with the previous case there is good agreement among the data over their common range of liquid fallback mass fluxes. This indicates that, as for the tests with all upper plenum internals installed, the rate of liquid injection has little influence on the flooding characteristics of the upper plenum assembly configuration tested. It is also noted that, unlike the data plotted in Figure 4, the data of Figure 5 do not exhibit an abrupt change in slope for values of the square root of J_f^* < 0.04. However, the liquid mass flow rates associated with

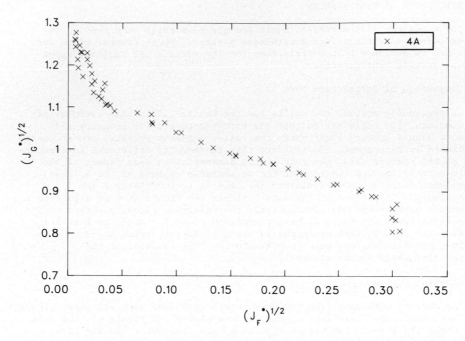

Figure 2. Wallis Correlation Flooding Curve, Test 4A

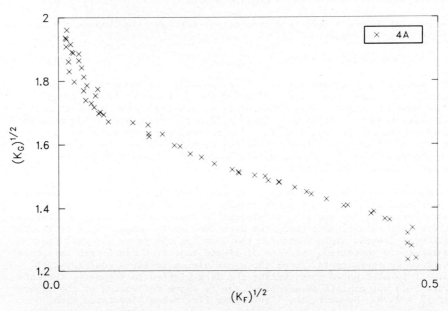

Figure 3. Wallis-Kutateladze Correlation Flooding Curve, Test 4A

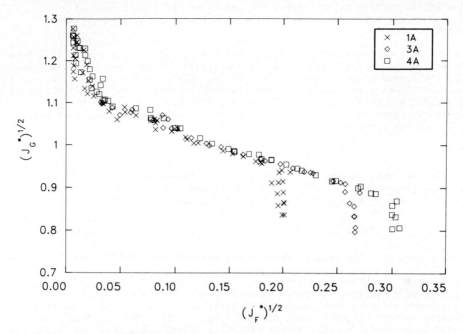

Figure 4. Wallis Correlation Flooding Curves, Tests 1A, 3A, and 4A

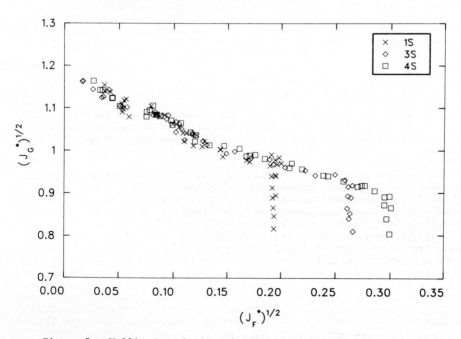

Figure 5. Wallis Correlation Flooding Curves, Tests 1S, 3S, and 4S

those values of dimensionless mass fluxes are still beyond the range of operation of the turbine meter. Thus, the values of the square root of
$J_f^* < 0.04$ were again excluded from use in evaluating flooding correlations.

The data of the flooding tests of the SET Facility upper plenum with both
the ten support columns and the upper core plate simulation removed, leaving
only the ground plate, Tests 1G, 3G, and 4G, show a sensitivity of the
observed flooding characteristics of the upper plenum asembly to liquid injection rate, Figure 6. The data of Test 1G, 150 mL/S liquid injection rate,
appears to approximate a straight line. In comparison, although they agree
well over the range of:

$$0.12 < (J_f^*)^{0.5} < 0.26 \qquad (8)$$

the data of Tests 3G and 4G, 300 mL/S and 400 mL/Sec liquid injection rate,
respectively, show an anomalous behavior in that they appear to fit a
quadratic form. It is also noted that, for values of the square root of
J_f^* of:

$$(J_f^*)^{0.5} < 0.12 \qquad (9)$$

there is considerable scatter between the data of Test 3G and that of Test 4G.

A second anomalous behavior of the data of Tests 3G and 4G is that complete or near complete carryover of the injected liquid was not attained at
the maximum air flow rate tested, 0.29 kg/S, as it had with other upper plenum
assembly configurations tested, Figures 7 and 8. This behavior could be the
result of the creation of a larger effective liquid storage volume in the SET
Facility upper plenum due to the removal of the core support plate simulation
aided in developing and maintaining flooding in the upper plenum.

The anomalous behavior of the data is shown in Figure 6 might also be due
to the interaction of the CCFL characteristics of the ground plate and the
upper core plate simulator in the SET Facility upper plenum assembly.
Although the flow area of the ground plate is the minimum flow area in the SET
Facility upper plenum assembly, the upper core plate simulation flow area
exceeds that of the ground plate by about only 10 percent. Thus, it is likely
that the flooding behavior shown in the data of Figures 4 and 5 is the result
of the coupled CCFL characteristics of the ground plate and the core support
plate simulation. It then follows that removing one of the two components
will alter the observed flooding behavior of the SET-Facility upper plenum
assembly.

One test was performed at a reduced system pressure to investigate the
influence of test section pressure on the observed flooding behavior of the
SET Facility upper plenum assembly. This test, designated Test LP, was performed with the same test parameters, other than system pressure, as Test 3A.
Although it was intended to perform the low pressure test at a liquid injection rate of 300 mL/s, the hydraulic characteristics of the experimental
system limited the maximum liquid injection rate about 300 mL/s. To assure a
stable and controllable liquid injection rate throughout the execution of Test
LP, a liquid injection rate of 285 mL/s was used. The reduced dimensionless
mass flux data from Tests 3A and LP are plotted in Figure 9, and the two sets
of data are seen to compare well over the entire range of mass fluxes
studied. It therefore appears that changes in system pressure of as much as
30% have negligable influence on the flooding characteristics of the SET

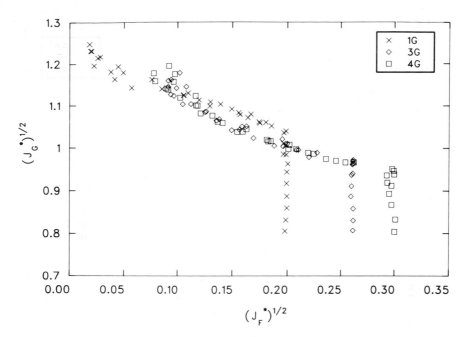

Figure 6. Wallis Correlation Flooding Curves, Tests 1G, 3G, and 4G

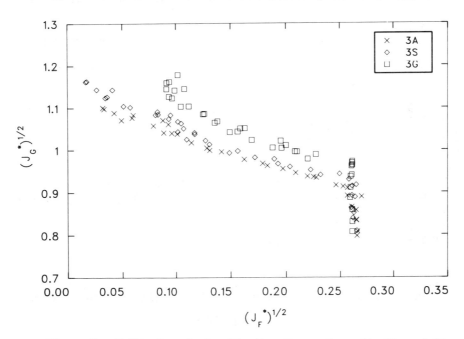

Figure 7. Wallis Correlation Flooding Curve, Tests 3A, 3S, and 3G

The data of Tests 1A and 1S were not used in the evaluation of the composite flooding correlation coefficients given in Equations 10 and 11. When viewed in plotted form, the data of Tests 1A and 1S appeared to agree well with data at larger liquid injection rates, Figures 4 and 5, respectively. However, the data of Table 2 show that the slope, M, calculated for the data of Tests 1A and 1S varies by about 10 percent or more from those values calculated for the other test data of Figures 4 and 5. The span of dimensionless liquid fallback mass fluxes over which the value of the slope, M, is evaluated for Tests 1A and 1S is small, therefore making the evaluation of the slope for those tests sensitive to small errors in either the experimentally measured air or fallback mass flow rate. By excluding the data of Tests 1A and 1S, this influence on the coefficients given in Equations 10 and 11 is minimized. It should be noted, however, that the flooding correlation defined by those two coefficients reasonably approximates the data from Tests 1A and 1S.

The data of Table 2 also support the observations made from the plotted data of Figures 6, 7, and 8; the flooding characteristics of the upper plenum assembly with only the ground plate installed are fundamentally different than with both the ground plate and upper core support plate simulation installed. The slope of the correlations fitted to the data of Tests 1G, 3G, and 4G is about 20 percent larger than the value given in Equation 10, the composite slope fitted from five tests. Furthermore, the calculated values for the fitted correlations to the flooding curve data of Tests 3G and 4G are approximately unity, the value recommended by Wallis as a result of his review of available experimental flooding data [3]. This observation tends to support the hypothesis advanced earlier by the authors that the flooding behavior shown in the data of Figures 4 and 5 is the result of the coupled CCFL characteristics of the ground plate and upper core plate simulation.

6.4 Pressure Drop Data

Differential pressure measurements were taken across four consecutive spans within the experimental facility, as identified in Table 3. The measurement designations given in Table 3 indicate the elevations spanned by the differential pressure transducers in units of centimeters, referencing the bottom of the hot log internal diameter as the 0.0 cm elevation. The particulars of the differential pressure measurements are discussed by Anderson and Fogdall [5]. The usefulness of these data in evaluating void fractions, and consequently mass storage, in the upper plenum assembly is limited as the range of the pressure transducers used in the test program was large compared to the elevation span over which the measurements were taken. Although the absolute accuracy of the pressure drop data is suspect, those data do show a sudden and significant change in value at the onset of flooding due to the storage of liquid mass in the upper plenum assembly, Figure 10.

6.5 Optical Probe Results

An optical probe system was installed to observe the flow conditions just above the ground plate during testing so as to detect the onset of flooding. A video system having a standard framing rate of 30 frames per second was used to monitor the output of the optical probe. At the lowest air injection rate, 0.13 kg/s, the monitor display showed only a grey blur. It was concluded that a high-speed movie camera would be reqired to record the flow conditions at this location for visual inspection. Utilization of such equipment, however, was beyond the scope of the current test program and no further attempts were made to visually observe the flow conditions above the ground plate in the upper plenum assembly during testing.

TABLE 3. DIFFERENTIAL PRESSURE MEASUREMENT LOCATIONS

Designation	Description
0–83	From the bottom of the hot leg internal diameter to 1.1 cm above the upper core plate simulation
–83–101	From 1.1 cm above the upper core plate simulation to 3.42 cm above the ground plate
–101–122	From 3.42 cm above the ground plate to the top of the core simulation, 17.48 cm below the top of the ground plate.
–122–171	From the top of the core simulation to 2.54 cm above the core simulation spacer grid.

7. FUTURE STUDIES

The anomalous behavior of Tests 1G, 3G, and 4G suggests that additional experimental studies be performed to evaluate the flooding characteristics of the upper core support plate simulation. From these studies an assessment may be made if the observed CCFL characteristics of the SET Facility upper plenum assembly with all internals installed is a result of the coupling of the flooding behavior of the ground plate and upper core support plate. Such

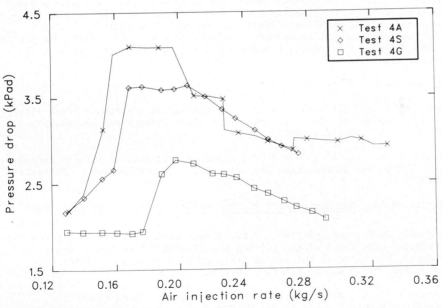

Figure 10. Upper Plenum Assembly Pressure Drop, 0 cm to –83 cm Elevation

164

information would provide a basis for understanding CCFL phenomena in other LOCA test facilities in which both a ground plate and upper core plate simulation are used.

8. SUMMARY

CCFL phenomena was experimentally studied in the SET Facility upper plenum using air and water as working fluids. In general, the flooding behavior was found to be independent of system pressure and liquid injection rate. The Wallis correlation form was used to correlate the experimental data, and a flooding correlation was recommended for characterizing the flooding behavior of the SET Facility upper plenum assembly with all internals installed.

Anomalous flooding behavior in the SET Facility upper plenum assembly when only the ground plate was installed was observed. Two possible causes of the anomalous behavior were advanced; one being the increase in liquid storage capability of the upper plenum assembly due to the removal of the upper core support plate simulation, and the other being the coupling of flooding behavior of the ground plate and upper core plate simulators due to their simularity in flow area. An experimental program that would assess the validity of the two proposed explanations for the anomalous behavior was proposed.

ACKNOWLEDGEMENT

The Full Length Emergency Core Heat Transfer-Systems Effects and Separate Effects Test (FLECHT-SEASET) Program is a five year test program whose objective is to examine the reflood heat transfer aspects of a postulated loss-of-coolant accident (LOCA) for a pressurized water reactor (PWR). The FLECHT-SEASET program is jointly funded and managed by the U.S. Nuclear Regulatory Commission (NRC), Electric Power Research Institute (EPRI), and Westinghouse Electric Corporation. The countercurrent flow limiting (CCFL) testing reported here was conducted in support of the FLECHT-SEASET program at the Idaho National Engineering Laboratory (INEL) by EG&G Idaho, Inc. at the request of the NRC.

REFERENCES

1. Yamanouchi, A., "Effects of Core Spray Cooling at Stationary State After Loss-of-Coolant Accident", Journal of Nuclear Science and Technology, 5, 10, pp. 8-17, October, 1968.

2. Rosal, E. R., et. al., "PWR FLECHT SEASET System Effects Natural Circulation and Reflux Condensation Task Plan Report," NUREG/CR-2401, February 1983.

3. Wallis, G. B., One Dimensional Two-Phase Flow, McGraw-Hill, Inc., 1969.

4. Fogdall, S. P., and Anderson, J. L., "Experimental System Description For Air-Water CCFL Tests of the 161-Rod FLECHT-SEASET Test Vessel Upper Plenum", Presented at the 3rd Multi-Phase Flow and Heat Transfer Symposium, Maimi Beach, Florida, 18-20 April 1983.

5. Anderson, J. L., and Fogdall, S. P., "Experimental Data Report for Air-Water Flooding Tests of the FLECHT-SEASET Program SET-Facility Vessel Upper Plenum," NUREG/CR-2636, May, 1983.

Multi-Phase Flow and Heat Transfer III. Part B: Applications
edited by T.N. Veziroğlu and A.E. Bergles
Elsevier Science Publishers B.V., Amsterdam, 1984 — Printed in The Netherlands

A CORRELATION FOR PHASE SEPARATION IN A TEE[a]

G. E. McCreery
EG&G Idaho, Inc.
Idaho Falls, Idaho 83415, U.S.A.

ABSTRACT

The flow of homogenized two-phase fluid from the main run of a tee out the side port, exhibits partial phase separation because the gas, having lower momentum than the liquid phase, preferentially exits the side port. A correlation is given for phase separation in a tee. The form of the correlation is derived by simplification of the continuity and momentum equations describing the two-phase flow out a horizontally oriented side port. The final equations employ a term that is correlated to experimental data. Calculations using the correlation show generally good agreement with data for bubbly, slug, and churn-turbulent flow and with limited data available for the annular flow regime.

NOMENCLATURE

A	Cross section area
$A_{3,1_{\ell g}}$	Two-phase capture flow area
$A_{3,1_g}$	Gas capture flow area
A_{31}	$A_{3,1_{\ell g}} + A_{3,1_g}$
D	Diameter
P	Pressure
t	Time
u	Axial velocity in x direction
v	Radial velocity in z direction

---

a. Work supported by the U.S. Nuclear Regulatory Commission, Office of Nuclear Regulatory Research under DOE Contract No. DE-AC07-76ID01570.

W	Mass flux
X	Quality
x	Axial distance
z	Radial distance from side port entrance
α	Void fraction
$\alpha*$	Capture area fraction
ρ	Density
$\rho*$	Modified density
ξ	Experimentally determined coefficient defined by Equation (18)

Subscripts

0	Initial
1	Upstream of side port
2	Downstream of side port
3	Side port
g	Gas
h	Homogeneous
i	Phase index
ℓ	Liquid

1. INTRODUCTION

Interest in phase separation in tees has been recently stimulated by the possibility of small break loss-of-coolant accidents in pressurized water reactor piping that joins with the primary coolant loop. The flow through a rupture in such a pipe or piping component, such as a valve, is represented by the flow out the side port of a tee. If the flow in the primary coolant loop is stratified, the flow out a horizontal break is influenced greatly by the elevation of the liquid-gas interface in relationship to the break. Flow of stratified fluid out a break is discussed by Zuber,[1] Crowley and Rothe,[2] and Reimann and Khan.[3] If the break is located below the interface, gas can enter the break by vapor pull-through, with or without formation of vortex. If the break is located above the interface, liquid may be entrained by the gas flow. Flow out the break is greatly influenced by its orientation with the horizontal plane, with more liquid exiting a bottom break and more gas exiting a top break.

If the flow in the primary coolant loop is more homogenized, such as for bubbly, slug, or churn-turbulent flow, the flow out the side port of a tee exhibits partial phase separation because the gas, having lower momentum than the liquid, more easily turns the corner into the side port than the liquid.

Well homogenized flow would occur, for instance, in the primary loop of a reactor during a small break accident with the pumps left running. An accident of this type, with the break situated in the intact loop cold leg, is simulated by LOFT Nuclear Loss-of-Coolant Experiment L3-6.[4] In addition, phase separation in tees for these flow regimes is of importance to the chemical processing industry for steam distribution lines, etc. Phase separation in tees has been experimentally investigated by Collier,[5] Honan and Lahey,[6] Saba and Lahey,[7] and Reimann, Seeger, and John.[8] In addition, Saba and Lahey have presented an analytical model capable of calculating phase separation.[7] The model requires the simultaneous solution of five continuity and momentum equations for subsonic flow through a tee with a horizontally oriented side port. The solution, which is typically obtained numerically, gives good agreement with Saba and Lahey's and Collier's data.[5]

This paper addresses phase separation of homogonized flow in a tee with a horizontal side port. I derive a correlation by first simplifying the conservation equations, which describe the flow sufficiently to permit simple solution. Because of the simplification, I employ a term correlated to the experimental data.

2. DERIVATION

The model formulates the void fraction or density in the side port of a tee as a function of upstream flow conditions and the ratio of the downstream to side port mass flux, W_2/W_3.

A. Initial Assumptions (Refer to Figure 1)

1. The flow is uniformly dispersed (bubbly or mist).

2. $A_1 = A_2$.

3. $A_3 \leq A_1$.

4. Buoyancy effects are neglected.

5. Both liquid and vapor are captured and diverted into the side port from upstream flow area $A_{3,1_{\ell g}}$.

6. Only gas is captured and diverted from upstream flow area $A_{3,1_g}$.

7. $A_{3,1_{\ell g}}$ is geometrically similar to flow area $A_{3,1_{\ell g}} + A_{3,1_g}$.

8. Pressure losses are small such that liquid and vapor densities do not vary

$$\rho_{3_g} = \rho_{1_g} = \rho_g$$

$$\rho_{3_\ell} = \rho_{1_\ell} = \rho_\ell .$$

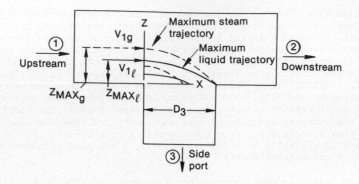

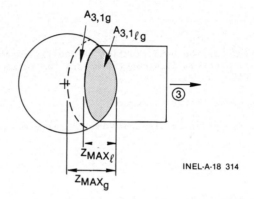

INEL-A-18 314

Figure 1. Flow geometry, top and end view of tee.

B.   General Approach

The solution for side port void fraction (α_3), or density (ρ_3) in the side port is obtained from continuity equations for the two phases once the capture area ratio defined as

$$\alpha^* = \frac{A_{3,1_{\ell g}}}{A_{3,1_{\ell g}} + A_{3,1_g}} \tag{1}$$

is determined. Since the gas phase has lower momentum than the liquid phase, it is collected and diverted into the side port from a larger upstream flow area

$$(A_{3,1_{\ell g}} + A_{3,1_g})$$

than is the liquid (collected in area = $A_{3,1_{\ell g}}$).

The collection areas are bounded by z_{max_g} and z_{max_ℓ} (Figure 1), which are the starting positions at $x = 0$ of the maximum trajectories of gas and liquid particles which enter the side port.

The solution for the area ratio proceeds by first simplifying the momentum equations to a form that may be solved for the distance ratio

$$z_{max_g} / z_{max_\ell} \quad .$$

The distance ratio is then used with an assumed flow geometry to find the area ratio. Because of the simplifications employed, a coefficient correlated to experimental data is necessary, and is incorporated in the solution.

C.   Particle Trajectories

Gas and liquid particle trajectories are solved from the momentum equations by first simplifying the two-dimensional forms of the equations such that the equation in the z direction is decoupled and may be solved separately.

In fixed coordinates, the two-dimensional steady state continuity and momentum equations for the ith phase are[9]

$$\frac{\partial}{\partial x} (\rho_i U_i) + \frac{\partial}{\partial z} (\rho_i v_i) = 0 \tag{2}$$

$$\frac{1}{\rho_i} \frac{\partial P}{\partial x} = -u_i \frac{\partial u_i}{\partial x} - v_i \frac{\partial u_i}{\partial z} + F_{x_i} \tag{3}$$

$$\frac{1}{\rho_i} \frac{\partial P}{\partial z} = -u_i \frac{\partial v_i}{\partial x} - v_i \frac{\partial v_i}{\partial z} + F_{z_i} \tag{4}$$

where u and v are the particle velocities in the x and z directions, and F_x and F_z are the forces per unit volume in the x and z directions, respectively.

If it is assumed for small to moderate values of (W_3/W_2) that

$$\frac{\partial P}{\partial z} = \text{constant}$$

$$\frac{\partial P}{\partial x} << \frac{\partial P}{\partial z}$$

$$u_i = u_{i_o} - \int_{x=o}^{x} \frac{\partial v_i}{\partial z} dx \approx u_{i_o}$$

and

$$\frac{\partial v_i}{\partial x} = 0 \quad .$$

Equation (4) then reduces to

$$\frac{1}{\rho_i} \frac{\partial P}{\partial z} = -v_i \frac{\partial v_i}{\partial z} + F_z \quad . \tag{5}$$

This may be written for the two phases as

$$(1 - \alpha) \rho_\ell v_\ell \frac{dv_\ell}{dz} = (1 - \alpha) \frac{dP}{dz} + \alpha F_d \tag{6}$$

$$\alpha \rho_g v_g \frac{dv_g}{dz} = \alpha \frac{dP}{dz} - \alpha F_d \tag{7}$$

where gravitational and wall friction forces are neglected and F_d is the drag force per unit gas volume in the z direction.

D.   Solution Using Modified Density Terms

z_{max_g} and z_{max_ℓ} may be obtained by setting $z_g = 0$ and $z_\ell = 0$ and using specific values of F_d, such as for bubbly flow. However, because of the many simplifications employed, this approach is not fruitful. Therefore, an alternative heuristic method is used in which the phasic densities in Equations (6) and (7) are replaced by actual densities plus a modified density term that accounts for the αF_d term. The modified density term is therefore a function of void fraction. The term is also assumed to be a function of W_2/W_3, since the flow ratio is a primary variable observed in the data.

Equations (6) and (7) are thus reduced to

$$z_g = \frac{-1}{2 \rho_g^*} \frac{dP}{dz} t^2 + z_{max_g} \tag{8}$$

$$z_\ell = \frac{-1}{2 \rho_\ell^*} \frac{dP}{dz} t^2 + z_{max_\ell} \tag{9}$$

where

$$\rho_g^* = \rho_g + f_g$$

$$\rho_\ell^* = \rho_\ell + f_\ell$$

$$f_g, f_\ell = \text{modification terms.}$$

The capture distance ratio at $x = 0$ is then, from Equations (8) and (9),

$$\frac{z_{max_g}}{z_{max_\ell}} = \frac{\rho_\ell^*}{\rho_g^*} \left(\frac{t_{max_g}}{t_{max_\ell}} \right)^2 \tag{10}$$

where

$$t_{max_i} = \text{transit time for maximum particle trajectories}$$

$$\approx \frac{D_3}{u_{1_i}} .$$

The form of the modified density terms is chosen to be compatible with the boundary conditions

(1) $\alpha \to 1$, $z_{max_\ell}/z_{max_g} \to 1$, $v_{1_\ell} \to v_{1_g}$, $\rho_g^* \to \rho_g$

(2) $\alpha \to 0$, $z_{max_\ell}/z_{max_g} \to 1$, $v_{1_g} \to v_{1_\ell}$, $\rho_\ell^* \to \rho_\ell$.

Also, because the interphase forces in the z direction are negative for the liquid phase and positive for the gas phase,

$$\rho_g^* \geq \rho_g$$

and

$$\rho_1^* \leq \rho_\ell .$$

This reasoning suggests modified density terms of the form

$$\rho_g^* = \rho_g + (1 - \alpha)^\xi (\rho_\ell - \rho_g) \tag{11}$$

$$\rho_\ell^* = \rho_\ell - \alpha^\xi (\rho_\ell - \rho_g) \tag{12}$$

where ξ is experimentally determined, and is a function of $\dfrac{W_2}{W_3}$.

The capture area fraction is given by

$$\alpha^* = \left(\frac{z_{max_\ell}}{z_{max_g}}\right)^n . \tag{13}$$

If the capture areas are circular, then n = 2; if the capture areas are annular, then n = 1. The actual case is intermediate and the final correlation agrees well with choice n = 1.5. Thus,

$$\alpha^* = \left(\frac{\rho_g^*}{\rho_\ell^*}\right)^{1.5} \left(\frac{v_{1_g}}{v_{1_\ell}}\right)^3 . \tag{14}$$

E. Homogeneous Flow Solution

The void fraction in the side port is solved from the one-dimensional continuity equations for the two-phases. Addition of the continuity equations for the gas and liquid mass flows that enter the side port yields

$$\left[\rho_\ell \, v_{3_\ell}(1 - \alpha_3) + \rho_g \, v_{3_g} \, \alpha_3\right] A_3$$

$$= \left[\rho_g \, v_{1_g} \, \alpha_1 + \rho_\ell \, v_{1_\ell}(1 - \alpha_1)\right] \alpha^* A_{31} + \rho_g \, v_{1_g}(1 - \alpha^*) A_{31} \tag{15}$$

where

$$A_{31} = A_{3,1_{\ell g}} + A_{3,1_g}.$$

The homogeneous solution is obtained by setting

$$v_{1_g} = v_{1_\ell}$$

$$v_{3_g} = v_{3_\ell}$$

$$v_1 A_{31} = v_3 A_3$$

in Equation (15), yielding the simple form,

$$\alpha_{3_h} = 1 - \alpha^*\left(1 - \alpha_{1_h}\right) \tag{16}$$

where

$$\alpha^* = \left(\frac{\rho_g^*}{\rho_\ell^*}\right)^{1.5}$$

and ρ_g^* and ρ_ℓ^* are given by Equations (11) and (12), and, the homogeneous void fraction is

$$\alpha_{1_h} = \left(1 + \frac{1 - x_1}{x_1} \frac{\rho_g}{\rho_\ell}\right)^{-1}.$$

The limits of Equation (16) at high and low upstream void fraction are

$$\alpha_3 \to 1, \, \alpha^* \to 1 \text{ as } \alpha_{1_h} \to 1$$

$$\alpha_3 \to 0, \, \alpha^* \to 1 \text{ as } \alpha_{1_h} \to 0.$$

Equation (16) may be reformulated in terms of density to yield

$$\rho_3 = [\rho_\ell(1 - \alpha_1) + \rho_g \alpha_1] \, \alpha^* + \rho_g(1 - \alpha^*)$$

or

$$\rho_3 = \rho_\ell \left(\frac{\rho_g^*}{\rho_\ell^*}\right)^{1.5} + \rho_g \left[1 - \left(\frac{\rho_g^*}{\rho_\ell^*}\right)^{1.5}\right] \, . \tag{17}$$

The value of ξ in Equations (11) and (12) is determined from a comparison of the homogeneous form of the model with data to be

$$\xi = 1.15\left(\frac{W_2}{W_3}\right)^{0.096} \qquad \text{for} \qquad \frac{W_2}{W_3} \geq \frac{30}{70} \tag{18}$$

and

$$\xi = 1.0 + \left(\frac{W_2}{W_3}\right) 0.14038 \qquad \text{for} \qquad \frac{W_2}{W_3} < \frac{30}{70} \, . \tag{19}$$

3. COMPARISON WITH DATA

This section presents a comparison of calculations using the homogeneous form of the model [Equations (16) through (19)] with available data. The various graph coordinates are, except for the data of Honan and Lahey, the coordinates used in the referenced work.

A comparison of calculations using the homogeneous form of the model with the data of Honan and Lahey[6] is shown in Figure 2. The data were taken at flow rates

$$W_1 = 1356.0, \ 2035, \ 0, \ 2713.0 \, \frac{\text{kg}}{\text{s-m}^2}$$

and flow ratios

$$\frac{W_2}{W_3} = \frac{30}{70}, \ \frac{50}{50}, \ \frac{70}{30} \, .$$

Qualities ranged from 0.1 to 1.0% and the flow regime was primarily slug with some data in the bubbly, churn, and wispy annular regimes as plotted on the map by Hewitt and Roberts.[10]

The calculated void fraction shows good agreement with the data and lies within the uncertainty of the measurements. These data are for a vertically oriented tee main run with a horizontal side port. Data taken for a horizontal tee by Saba and Lahey[7] is similar except that they show slightly higher separation at higher values of W_2/W_3.

Figure 3 compares calculations with Harwell data as reported in References 5 and 7. The data are for much higher qualities and lower mass

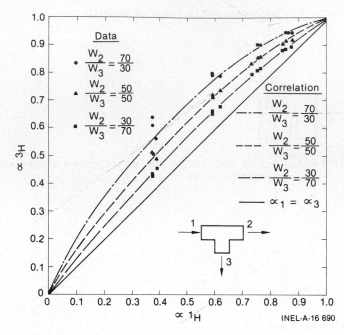

Figure 2. Calculated side port homogeneous void fraction compared with the data of Honan and Lahey.

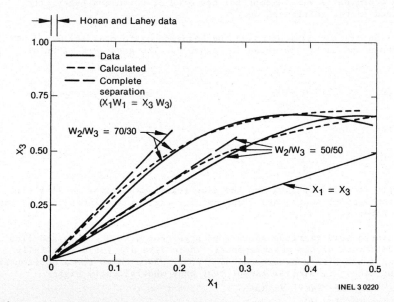

Figure 3. Calculated side port quality compared with Harwell data.

flow rate (W = 136.0 kg/s-m^2) than that of Honan and Lahey. The figure is
scaled with homogeneous quality since void fractions are compressed into a
narrow range close to α = 1.0. The agreement of the calculations with the
data is suprisingly good, considering that the flow regime is primarily
annular. The quality X_3 is predicted to within 10%.

Figures 4 and 5 compare calculations with air-water data taken at KfK[9]
Karlsruhe in using their graph coordinates. Data encompassed bubbly, slug, and
annular flow regimes. Calculations are compared only with data taken with the
tee oriented in the horizontal plane. Data were also taken with the side port
pointing vertically upward and downward. The data show marked variation from
the horizontal orientation data, with more separation exhibited for upward
orientation and less separation for downward orientation.

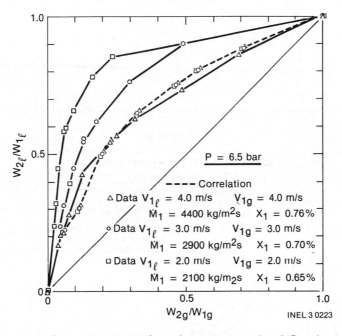

Figure 4. Comparison of correlation with KfK data for 50% void fraction.

Calculations generally agree well with the KfK data except at low
velocity, as shown in Figure 4. This is probably due to progressively more
stratification as velocity is lowered. Phase separation, as given by the
ratio $W_{2\ell}/W_{1\ell}$, is predicted to within 15% except for the lowest
upstream velocity (2 m/s) curve, which is underpredicted by as much as 35%.
The data in Figure 5, which exhibit high inlet slip, are predicted fairly well
despite the homogeneous flow assumption. This is probably because the tee
itself acts as a homogonizer, especially for the flow entering the side port.

As the final data comparison, Figure 6 compares calculations with density
data from LOFT Nuclear Small Break Loss-of-Coolant Experiment L3-6[4] during
the period 100 to 1000 s after experiment initiation. The experiment simulated
a 2.5% of intact loop cold leg flow area pipe break in a commercial pressurized

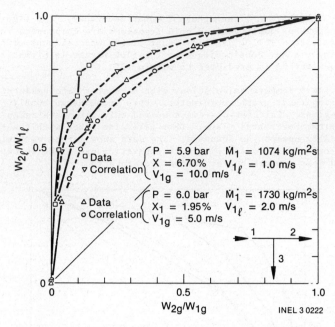

Figure 5.   Comparison of correlation with KfK data.

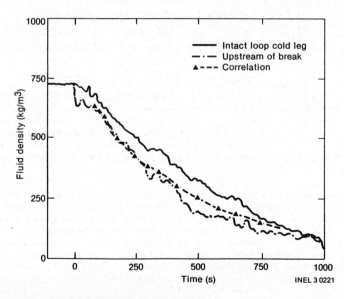

Figure 6.   Calculated side port density compared with LOFT nuclear
Experiment L3-6 data.

water reactor. The break was situated in the side port of a tee off the intact loop cold leg pipe. The side port flow area was 0.015 times that of the intact loop cold leg. The flow regimes in the intact loop cold leg were bubbly and slug. Flow was choked at an orifice placed several feet downstream of the tee junction in the side port.

Calculations show agreement with the LOFT data to within 10%. The generally good agreement indicates that phase separation is not strongly dependent on flow areas or on the ratio of side port flow area divided by upstream flow area. This area ratio independence is indicated in the model by the independence of the capture distance ratio [Equation (15)] on flow area. Calculations agree best with the LOFT data between $\sim$100 s and 280 s (calculations before 100 s were not performed because transient blowdown was underway). At 280 s, the flow within the pump transitioned from bubbly flow to partially stratified churn flow at a void fraction of $\sim$0.35.[11] As with the KfK data, partial stratification leads to an underprediction of phase separation.

4. CONCLUSIONS

A model of phase separation in a tee has been developed by simplification of the momentum equations that describe flow out the side port of the tee, with a horizontal side port, and by inclusion of modified density terms that contain an experimentally determined exponent. The homogeneous form of the model, as given by Equations (16) through (19), is for flow through a tee with a horizontally oriented side port.

The correlation shows agreement with the data of Honan and Lahey[7] to within experimental uncertainty and agreement with the much higher quality and lower flow rate data of Harwell to within 10%. Calculations agree with KfK Karlsruhe data to within 15% except at lower velocities, due to flow stratification. Calculations also agree with the data from LOFT Nuclear Loss-of-Coolant Experiment L3-6 to within 10%. This agreement gives preliminary evidence that phase separation in a tee is, at least to first order, flow area and flow area ratio independent.

The correlation compares best with data from the more homogonized flow regimes (bubbly, slug, and churn), though annular flow data is predicted fairly well. The correlation underpredicts phase separation when the flow is partially or completely stratified. Although more data is necessary to assess the accuracy and limits of the correlation with regard to flow regime, mass flux ratio (W_2/W_3), flow quality, and flow area ratio, calculations using the correlation show generally good agreement for a wide range of data.

NOTICE

REFERENCES

1. N. Zuber, "Problems in Modeling of Small Break LOCA," NUREG-0724, 1981.

2. C. J. Crowley and P. H. Rothe, "Flow Visualization and Break Mass Flow Measurements in Small Break Separate Effects Experiments," ANS Specialist Meeting on Small Loss of Coolant Accident Analyses in LWRs, Monterey, California, August 25-27, 1981.

3. J. Reimann and M. Khan, "Flow Through a Small Break at the Bottom of a Large Pipe with Stratified Flow, Second International Topical Meeting on Nuclear Reactor Thermal-Hydraulics, Santa Barbara, California, January 1983.

4. P. D. Bayless and J. M. Carpenter, Experiment Data Report for LOFT Nuclear Small Break Experiment L3-6 and Severe Core Transient Experiment L8-1, NUREG/CR-1868, EGG-2075, January 1981.

5. J. G. Collier, "Single-Phase and Two-Phase Flow Behavior in Primary Circuit Components, Proceedings of NATO Advanced Study Institute, August 1976, Istanbul, Two-Phase Flows and Heat Transfer, Volume 1, Hemisphere Publishing, Washington.

6. T. J. Honan and R. T. Lahey, Jr., The Measurement of Phase Separation in Wyes and Tees, NUREG/CR-0057, December 1978.

7. N. Saba and R. T. Lahey, Jr., Phase Separation Phenomena in Branching Conduits, NUREG/CR-2590, March 1982.

8. J. Reimann, W. Seeger, H. John, Erste Experimente zur Umverteilung einer Luft-Wasser-Strömung in einem T-Stück, Kernforschungszentrum Karlsruhe, 06-01-03P08A, August 1980.

9. K. H. Ardron, "One-Dimensional Two-Fluid Equations for Horizontal Stratified Two-Phase Flow, International Journal of Multiphase Flow, Vol. 6, 1980, pp. 295-304.

10. G. F. Hewitt and D. N. Roberts, "Studies of Two-Phase Flow Patterns by Simultaneous X-Ray and Flash Photography," AERE-M 2159, H.M.S.O. (1969).

11. G. E. McCreery, J. H. Linebarger, J. E. Koske, "Primary Pump Power as a Measure of Fluid Density During Bubbly Two-Phase Flow, Second International Topical Meeting on Nuclear Reactor Thermal-Hydraulics, Santa Barbara, California, January 1983.

Multi-Phase Flow and Heat Transfer III. Part B: Applications
edited by T.N. Veziroğlu and A.E. Bergles
Elsevier Science Publishers B.V., Amsterdam, 1984 — Printed in The Netherlands

COMPARISON OF ANALYTICAL AND EXPERIMENTAL RESULTS FOR A PWR PRESSURIZER
SAFETY VALVE DISCHARGE

J. C. Rommel and S. A. Traiforos
Bechtel Power Corporation
Gaithersburg, Maryland   20877, U.S.A.

ABSTRACT

In the aftermath of the TMI-2 accident, testing has been completed on
pressurizer safety and relief valves to demonstrate their safe operability.
With this testing, experimental data on the response of the discharge piping
to a safety valve actuation transient became readily available. This paper
presents a comparison of this data with analytical calculations. This comparison demonstrates that accurate predictions of the discharge piping fluid
properties and the fluid generated piping forces can be made using the computer codes RELAP5/MOD1 and REPIPE. The two-step process used for the analytical predictions allowed RELAP5/MOD1 to compute the time-dependent piping
thermal-hydraulic conditions, while REPIPE calculated the piping reaction
forces. For comparison with REPIPE, other methods of calculating the piping
loads were investigated and the results were compared with the REPIPE generated forces.

1.   INTRODUCTION

Protection against the overpressurization of the reactor coolant system
in pressurized water reactors is provided by safety and relief valves
(S/RV's). These valves discharge the pressurized fluid into a discharge tank
through a piping network containing many elbows and bends. When these valves
are actuated, not only does the pressurizer fluid create a thermal-hydraulic
transient in the downstream piping, but also a complex series of fluid induced
dynamic loadings are generated at each point in the piping network where a
change in flow direction occurs.

To ensure the adequate design of the piping and piping supports in the
discharge network, accurate methods of predicting the thermal-hydraulic response and loading histories must be available. One such method is a two-step
process utilizing the computer codes RELAP5/MOD1 Cycle 14 [1], hereafter noted
as RELAP5, and REPIPE [2], both available from CDC. This combination of the
RELAP series of computer codes and REPIPE has been successfully used in the
past for the study of piping transients [3].

The first step, the thermal-hydraulic analysis, makes use of RELAP5 to
determine the discharge piping fluid properties, while the second step, the
force calculation, has REPIPE computing the time-dependent dynamic loadings.
Using this methodology, it is the intent of the paper to present a comparison
of analytical calculations to available experimental data.

2. METHOD OF SOLUTION

2.1. RELAP5/MOD1

As stated previously to perform the thermal-hydraulic analysis, the computer code RELAP5 was selected. RELAP5 was developed to investigate the thermal-hydraulic behavior of pressurized water reactors under accident conditions. However, because of the advanced nature of the five-equation, two-fluid hydrodynamic model, the two-phase, nonhomogeneous, nonequilibrium flow conditions commonly encountered in pressure relief transients can be effectively analyzed using this code. The five equations utilized by RELAP5 are the two phasic continuity equations, the two phasic momentum equations and the mixture internal energy equation. The use of an additional energy equation is eliminated by the assumption that the least massive phase exists at the saturation state corresponding to the local pressure.

The input information required by RELAP5 completely describes the initial fluid conditions and the geometry of the system being analyzed. The input includes the initial physical fluid characteristics of temperature, pressure, mixture quality and noncondensible content and such geometrical information as flow area, volume length and vertical orientation. Other required input includes control actions such as valve motion.

Using this input, RELAP5 simultaneously solves the five field equations at each time advancement for such fluid properties as velocity, pressure, internal energy and mixture quality.

A more detailed discussion of the basic theory and inherent assumptions incorporated into RELAP5 is given in Reference 1.

2.2. REPIPE

Due to the fact that RELAP5 cannot directly output the hydraulic forces that result from fluid pressure and momentum variations, a post-processor is required to accept the RELAP5 output and translate this information into piping forces. The post-processor selected for this second step of the analysis was REPIPE. REPIPE computes the net force in each of the RELAP5 defined control volumes and distributes the forces along user defined piping segments. Though not investigated in this analysis, the REPIPE results are customarily used as input for piping structural analysis computer codes.

The basis for the theoretical development of REPIPE is the methodology developed by Moody [4] for the calculation of the time dependent loadings on piping networks. From Moody, for the bounded pipe segment analyzed in this paper, the reaction "wave force" is predicted by:

$$R = \sum_i A_i \left(\frac{L_i}{A_i}\right)\left(\frac{1}{g_c}\right)\left(\frac{dw_i}{dt}\right) \tag{1}$$

where:

R = net pipe reaction force on bounded segment

$\sum_i$ = summation over all volumes in bounded segment

A_i = volume area

$\dfrac{L_i}{A_i}$ = volume length to area ratio

g_c = gravitational constant

$\dfrac{dw_i}{dt}$ = derivative of average volume mass flow rate

Though not utilized during the course of this analysis, Moody's development also includes an equation for the calculation of the reaction "blowdown force" for unbounded pipe segments. This equation is also incorporated into REPIPE.

3. EXPERIMENTAL TESTING

As a consequence of a relief valve failing during the TMI-2 accident, the United States Nuclear Regulatory Commission requested that testing be completed on S/RV's to demonstrate the operability of these valves during both normal and accident conditions. As part of the response to this request, an extensive testing of spring-loaded safety valves was recently conducted at Combustion Engineering (C-E). A secondary objective of this testing was to experimentally measure the response of the discharge piping during an S/RV opening transient.

The experiments at C-E consisted of the testing of a matrix of different safety valves, initial pressurizer fluid conditions, and upstream piping configurations. A review of the experimental and analytical results indicated that the maximum downstream piping loads occurred when the discharge through the safety valve was a cold water slug, forming a loop seal, followed by high pressure steam. This loop seal arrangement, shown in Figure 1, is a typical design feature for a vendor of nuclear steam supply systems, for it minimizes the amount of valve steam leakage during normal operation. Due to the severity of the loads generated by this scenario, it was selected as a representative case for comparison.

In this test, the initial temperature distribution of the water seal varied from $100^{\circ}F$ at the valve inlet to $650^{\circ}F$ (saturation) at a point approximately eight feet upstream of the valve. The total volume of the slug was about one cubic foot. The valve used during this test was a Crosby 6M6 safety valve. The elevation change along the vertical downcomer was nearly twenty-five and one-half feet. Load cells were positioned at the bottom of this downcomer to measure the generated forces. Measurement devices were also placed at various locations along the piping network to determine the transient fluid pressure, temperature and mass flow rate. The location of three of the pressure sensors (those used for comparison with analysis) are given in Figure 1.

4. ANALYTICAL MODEL

Modelling piping systems in RELAP5 requires the use of a series of control volumes and flow paths between the volumes. The model of the experimental facility used for this analysis consisted of 124 volumes and 123 flow

182

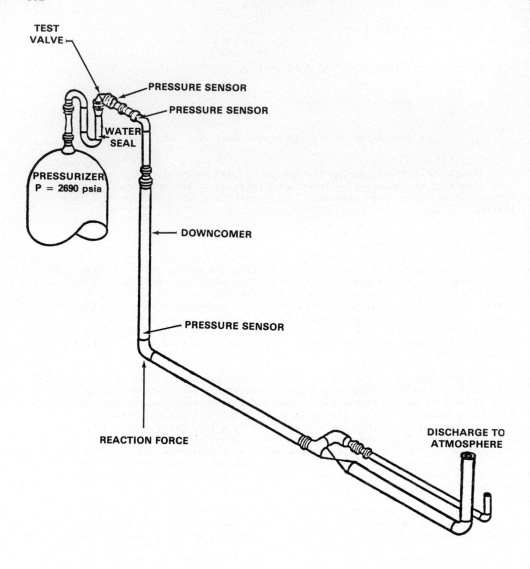

FIGURE 1. ISOMETRIC OF EXPERIMENTAL FACILITY

paths. The path lengths of the control volumes varied from one-half foot to approximately two feet. The smaller lengths were used in the nodalization of the vertical downcomer, where the largest loads would be generated. These lengths were later varied to assess the sensitivity of the force-time history on nodal length.

The safety valve was modelled as a RELAP5 "valve" component. The "valve" component is a special flow path that allows for a variable flow area. The flow area for the valve at full lift, as given by the manufacturer, was 0.0253 ft^2. However, this area was reduced slightly in the analysis to account for the discharge coefficient and thus allow agreement with the experimental flow rates. The special RELAP5 model that determines the existence of critical flow at flow paths was applied at the valve discharge plane.

In the experimental testing, the safety valve began to open when the pressurizer reached 2560 psia. However, the valve stem oscillated violently at high frequencies for about 900 milliseconds, after which the oscillations ceased and the valve reached full lift in 15 milliseconds. The pressurizer was then at 2690 psia. The modelling of the valve neglected the stem oscillations and opened the valve in 15 milliseconds, with the pressurizer at 2690 psia. This assumption resulted in a time lag appearing in the analytical results. This lag developed because the liquid slug was able to pass through the valve during the stem oscillations. Due to the uncertainty of the new slug position, and because as it is shown in the next section, this did not affect the shape and magnitude of the measured quantities, no attempt was made to account for the new position in the analytical model.

The effect of heat transfer between the fluid and the pipe walls was neglected. Due to the short duration of the transient (250 milliseconds) the effects of heat removal from the slug would be small. This assumption will lead to significantly smaller computer costs and slightly conservative piping loads.

5.   COMPARISON OF RESULTS

When the safety valve actuated, the high pressure source accelerated the liquid slug and steam through the valve and into the originally air-filled discharge piping. This sudden surge of liquid and steam created a pressure response in the downstream piping which RELAP5 was able to reproduce quite well. Figures 2, 3 and 4 [5] show comparisons of the experimental pressures at three separate locations (shown in Figure 1) to the calculated pressures. Keeping in mind the time lag discussed previously, it can be seen from these curves that the calculated trends and magnitudes of pressure at these locations were in good agreement with the experimental results. The only real discrepancy occurred at the inlet elbow to the downcomer. The experimental results indicated the presence of a pressure spike, whereas no spike was analytically predicted. This spike has been attributed to error in the pressure sensor response.

In addition to creating a pressure response, the fluid acceleration inside the piping generated dynamic loadings at each point in the piping network where a change in flow direction occurred. The REPIPE computed force for the vertical downcomer pipe segment is shown compared to the experimentally measured force in Figures 5 and 6 [5]. The negative sign signifies a force, applied by the fluid on the pipe, acting in the direction opposite the one labelled REACTION FORCE in Figure 1. Figure 5 gives both the force history calculated by REPIPE and that measured experimentally, while Figure 6 allows for a better comparison between analytical and experimental results by adding a time lag to the experimental results. From review of these plots, it can be concluded that the assumptions used for the valve modelling did not affect the shape and magnitude of the calculated pressures and forces.

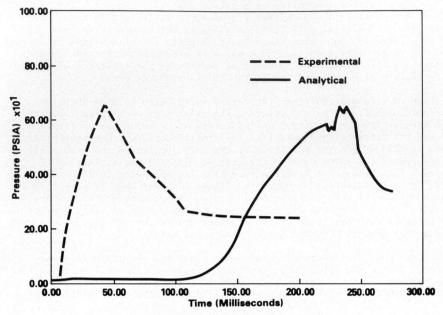

FIGURE 2. PRESSURE AT VALVE OUTLET

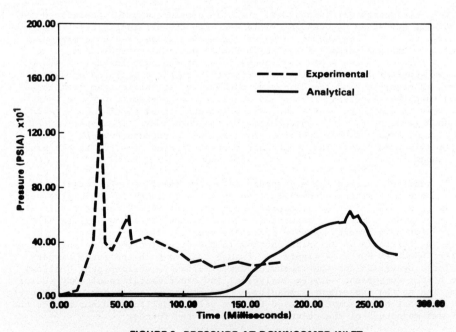

FIGURE 3. PRESSURE AT DOWNCOMER INLET

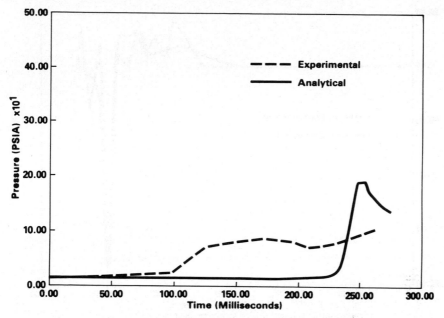

FIGURE 4. PRESSURE AT BOTTOM OF DOWNCOMER

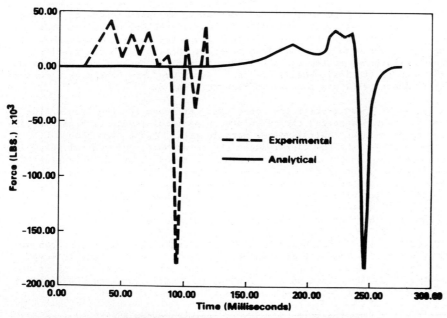

FIGURE 5. REACTION FORCE ON DOWNCOMER

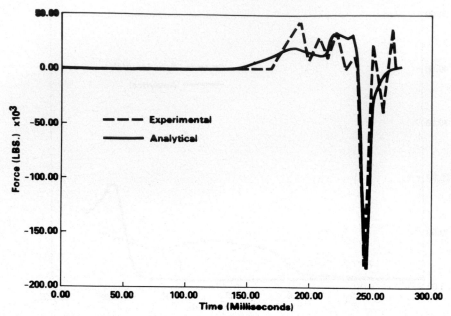

FIGURE 6. FORCE COMPARISON ACCOUNTING FOR TIME LAG

5.1. Nodalization Sensitivity Study

To estimate the importance of the control volume path length on the magnitude of the peak calculated hydraulic forces, the RELAP5 model was modified by increasing the size of the nodal lengths along the vertical downcomer. When these lengths were increased from one-half foot, used in the results discussed previously, to two feet, the magnitude of the peak force dropped from 183 kips to 66 kips. When a one-foot length was used, the peak force was 141 kips. These results are shown in Figures 7 and 8.

A review of these results indicates that, at least for the liquid slug followed by steam discharge, the calculated pipe loadings can be highly sensitive to the user-input control volume path lengths. Thus, even though by increasing path lengths the user reduces the total number of control volumes and consequently computer charges, care must be taken to avoid increasing the lengths beyond a point where the desired accuracy of the solution is being threatened. For all above analyses, adequately small time steps were used to ensure independence of the solution from the magnitude of the time step.

5.2. Alternate Methods of Calculating Dynamic Loadings

In addition to REPIPE, other methods of calculating dynamic loadings are presently available. To determine which, if any, of these methods is most suitable for the use in piping hydraulic calculations, a comparison of these different methods was completed. To conduct the comparison, the RELAP5 analysis using the one-foot path length nodalization scheme was used.

The first method investigated was the post-processor REFORC [6]. REFORC was developed for the Electric Power Research Institute (EPRI) specifically

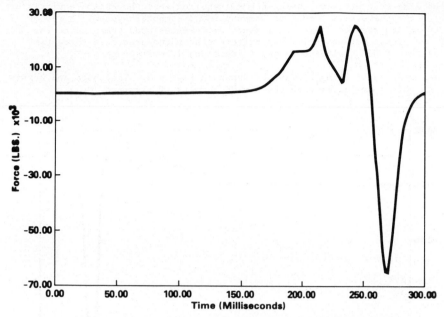

FIGURE 7. CALCULATED FORCE USING TWO FOOT NODAL LENGTH

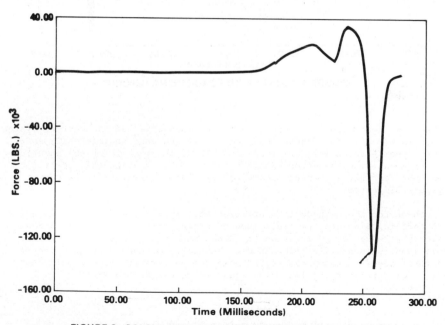

FIGURE 8. CALCULATED FORCE USING ONE FOOT NODAL LENGTH

for the S/RV test program. Using REFORC required that the RELAP5 analysis be rerun on the University Computing Company (UCC) computer system. A review of the UCC RELAP5 results showed only minor differences when compared to the CDC RELAP5 results. Consequently, any differences in the predicted force time histories must be attributed to the differences in post-processors.

The results of using REFORC are shown in Figure 9. The shape and magnitudes of this curve agree with the REPIPE results of Figure 8.

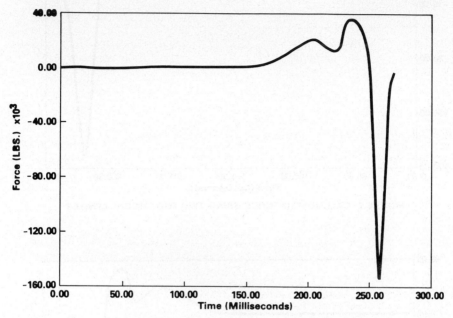

FIGURE 9. CALCULATED FORCE USING REFORC

The next two methods for determining loadings utilized the control system included in RELAP5. This control system allows the user to internally model algebraic and differential equations into RELAP5 and then print the equations' solutions along with the other RELAP5 output. Thus by using this control system, the need for a post-processor to calculate hydraulic forces can effectively be eliminated.

Each of the two methods that utilized this approach modelled an equation into RELAP5 to calculate the fluid induced piping forces. The first method modelled the "wave force" equation (1) discussed above, except that instead of volume, junction quantities were used. The second method calculated the net fluid force based on the pressure and momentum difference between both ends of the vertical downcomer section. The equation modelled into RELAP5 for this approach was:

$$R = \left(P + \frac{wV}{g_c}\right)_u A_u - \left(P + \frac{wV}{g_c}\right)_\ell A_\ell \qquad (2)$$

where:

R = net pipe reaction force on pipe segment

P = fluid pressure

w = fluid mass flow rate

V = fluid velocity

A = pipe area

g_c = gravitational constant

u = upper elbow of vertical downcomer segment

ℓ = lower elbow of vertical downcomer segment

Provided that the wall shear forces, including drag caused by orifices or other restrictions, are small, the two equations should give similar results.

The results derived from these two methods are presented in Figures 10 and 11. As expected the calculated forces were similar. The only major difference between these two curves occurred when the slug passed through a reducer (at approximately 0.2 seconds) and the shear effects became large.

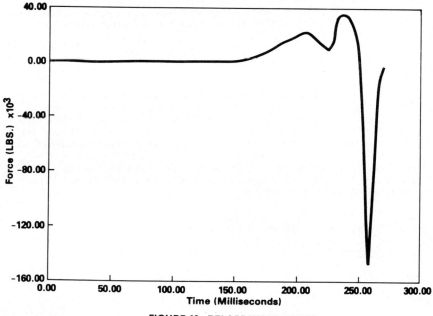

FIGURE 10. RELAP5 WAVE FORCE

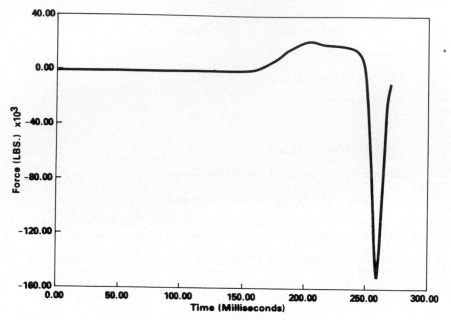

FIGURE 11. RELAP5 PRESSURE-MOMENTUM FORCE

A comparison of all four methods of determining the fluid generated pipe loadings is presented in Table 1. This comparison strongly suggests that each method was equally capable of translating the RELAP5 fluid properties into usable hydraulic force histories.

6. CONCLUSIONS

The results presented in this paper demonstrate that the RELAP5-REPIPE package can give accurate predictions of the piping loads as well as other fluid parameters resulting from a safety valve discharge of a cold water loop seal. Although results were not shown in this paper, the same conclusion was reached for the discharge of a two phase mixture through the safety valve.

Additionally, it was shown that other methods for calculating the fluid induced piping loads are available, and that these methods predicted forces similar in shape and magnitude to REPIPE.

Table 1. Comparison of Alternate Methods of Force Calculations

Method	Peak Force (kips)
REPIPE	141
REFORC	153
RELAP5 - "wave force"	148
RELAP5 - "pressure-momentum"	151

REFERENCES

1. "RELAP5/MOD1, Code Manual," Volumes I and II, NUREG/CR-1826, EGG-2070 Draft, Revision 1, EG&G Idaho, March 1981.

2. "REPIPE, Application Reference Manual," Cybernet Services, Control Data, 1980.

3. Traiforos, S. A., and Gilmer, J. M., "Dynamic Loading of a Feedwater Line Due to Check Valve Closure Using the RELAP4/MOD5-REPIPE Package," ASME Symposium on Fluid Transients and Structural Interactions in Piping Systems, Boulder, Colorado, June 22-24, 1981.

4. Moody, F. J., "Time Dependent Forces Caused by Blowdown and Flow Stoppage," ASME Paper 73-FL-23 (1973).

5. Wheeler, A. J., and Siegel, E. A., "Measurements of Piping Forces in a Safety Valve Discharge Line," ASME Paper 82-WA/NE-8.

6. "REFORC, A Computer Program for Calculating Fluid Forces Based on RELAP5 Results," Preliminary Draft Report, EDS Report No. 01-0650-1194, Revision 0, EDS Nuclear Inc., February 1982.

Multi-Phase Flow and Heat Transfer III. Part B: Applications 193
edited by T.N. Veziroğlu and A.E. Bergles
Elsevier Science Publishers B.V., Amsterdam, 1984 — Printed in The Netherlands

PWR RESPONSE TO AN INADVERTENT BORON DILUTION EVENT

James Adams
EG&G Idaho Inc.
P.O. Box 1625
Idaho Falls, Idaho 83415, U.S.A.

Dimitri Gidaspow, Vaikuntem Raghavan, and Hun-Der Sui
Department of Chemical Engineering
Illinois Institute of Technology
Chicago, Illinois 60616, U.S.A.

ABSTRACT

    Response of a pressurized water reactor (PWR) to an inadvertent boron
dilution event during refueling was studied experimentally in the Loss-of-
Fluid Test (LOFT) facility and in a plexiglass model of the LOFT reactor
vessel. These experiments showed that good mixing of the diluent and fluid
volumes occurred in the direct flow path volumes which are used in simulating
residual heat removal (RHR) in a PWR. In addition, other fluid volumes, not
in the RHR path, also mixed, delaying the time to criticality beyond that
expected based on RHR volumes only.

1. INTRODUCTION

    The U.S. Nuclear Regulatory Commission (NRC) requires the PWR response
to an inadvertent boron dilution event during refueling to be analyzed as
part of the PWR licensing safety analysis. The plant conditions assumed for
this analysis include: the plant is in cold shutdown; the reactor vessel
head is removed and the refueling tank is filled; the RHR system is being
used to remove core decay heat; and since the reactor vessel head is removed,
there are no control rods in the core, and the core reactivity is controlled
by dissolved boron in the coolant.

    In the licensing safety analysis, it is postulated that a diluent injec-
tion is inadvertently initiated and that the injection continues undetected
until an alarm setpoint is reached at one of the source range detectors. A
typical setpoint for these detectors is five times the background count rate.
Then, the NRC specifies [1], there must be a minimum of 30 min between the
time that this alarm setpoint is reached and the time that the core reaches
criticality. The method used by the PWR vendor to calculate the time-dependent boron concentration and the time to criticality includes an assumption
that the fluid volumes in the direct RHR flow path are instantaneously and
perfectly mixed (that is, that the boron concentration in these volumes is
uniform). The boron concentration, therefore, can be expressed as follows,
assuming that the boron concentration of the diluent is zero.

$$C(t) = C_o e - \frac{Qt}{V} \qquad (1)$$

where

    $C(t)$ =   time-dependent boron concentration (ppm)

C_o = initial boron concentration (ppm)

Q = diluent volumetric flow rate (m³/s)

V = effective (direct flow path) volume (m³)

t = time (s).

The effective volume includes the reactor vessel to the top of the nozzles and the RHR piping. Solving Equation (1) for the time at which the reactor becomes critical, t_c, results in:

$$t_c = \frac{V}{Q} \ln \frac{C_o}{C_c} \tag{2}$$

where

C_c = boron concentration at criticality.

To determine whether the assumption of perfect mixing is conservative, the NRC sponsored[a] a two-part nuclear boron dilution experiment, Experiment L6-6, in the LOFT facility [2] located at the Idaho National Engineering Laboratory and a series of dilution experiments in a scaled plexiglass model of the LOFT reactor vessel at the Illinois Institute of Technology. This paper reports the results from both sets of experiments.

2. EXPERIMENTAL FACILITIES

The LOFT facility is a 50-MW (thermal), volumetrically scaled PWR system, designed to reproduce both in time and approximate magnitude, the significant thermal, hydraulic, and nucleonic events expected to occur during off-nominal transients such as the postulated loss-of-coolant accident (LOCA). Figure 1 shows the LOFT primary system which consists of a reactor vessel, which houses the nuclear core; an intact loop consisting of active pumps, pressurizer, steam generator, and associated piping; a broken loop; and a blowdown suppression system. Figure 2 shows a cutaway view of the reactor vessel. The vessel was designed so that nominal PWR hydraulics, including flow resistances, would be preserved during transients. As shown, the coolant flows down the downcomer, into the lower plenum, up through the core (single pass) into the upper plenum, and out the nozzles.

Figure 3 shows a schematic of the LOFT primary system including the experimental RHR (recirculation line) and diluent injection systems, the letdown and sampling systems, and the boron sample locations used during Experiment L6-6. As shown in Figure 3, coolant samples were taken from four locations during Experiment L6-6. Batch or "grab" samples were taken from the lower plenum (BC-LP-001) at a location 200 degrees from the intact loop cold leg nozzle (that is, the fluid injection location), the RHR pump discharge (BC-P120-001), and downstream of the purification system (letdown system) ion exchanger (BC-P140-002). These samples were taken approximately every 5 min, and analyzed for boron concentration off-line. In addition, a continuous sample was taken at the letdown system pump discharge (BC-P140-001), and analyzed approximately every 5 min on-line. Due to the

a. Work supported by the United States Nuclear Regulatory Commission, Office of Nuclear Regulatory Research under DOE Contract No. DE-AC07-76ID01570.

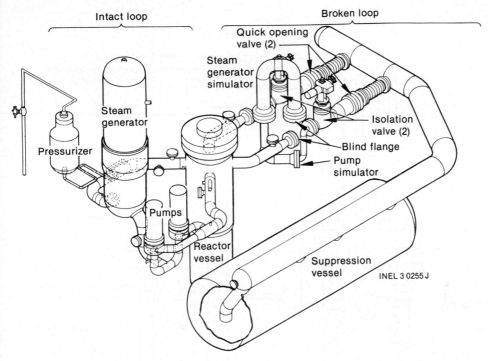

Fig. 1.   Axonometric Schematic of the LOFT Primary Coolant System

line delays involved, the results from the last two sample locations are not
reported here but are available in References [3] and [4]. Additional
information regarding the LOFT system and its scaling basis is found in
References [2] and [5].

    A clear plexiglass scaled model of the LOFT reactor vessel and RHR
(recirculation) line was fabricated and used to conduct mixing experiments at
the Illinois Institute of Technology. Figure 4 shows the reactor vessel
model, and Figure 5 is a schematic of the entire model. The volume scale
factor used was 166 (linear scale 5.5), and was applied to the downcomer,
lower plenum, core region, upper plenum, and the recirculation line. The
volumetric flow rates (recirculation and diluent) were also scaled down by
the same factor. This scaling basis (volume scaling for dimensions and flow
rate) is the same as that used in the scaling of the LOFT experiments to a
commercial PWR. Additional fabrication details include:

1.   The volume of the small annular downcomer (filler gap) in the LOFT
      system was added to the volume of the main annular downcomer and
      appropriately scaled down

2.   The core bypass volume in the LOFT reactor was added to the flow
      volume through the core and appropriately scaled down

3.   The distribution annulus volume above the top of the nozzle in the
      LOFT reactor was added to the annular volume in the upper plenum
      and appropriately scaled down

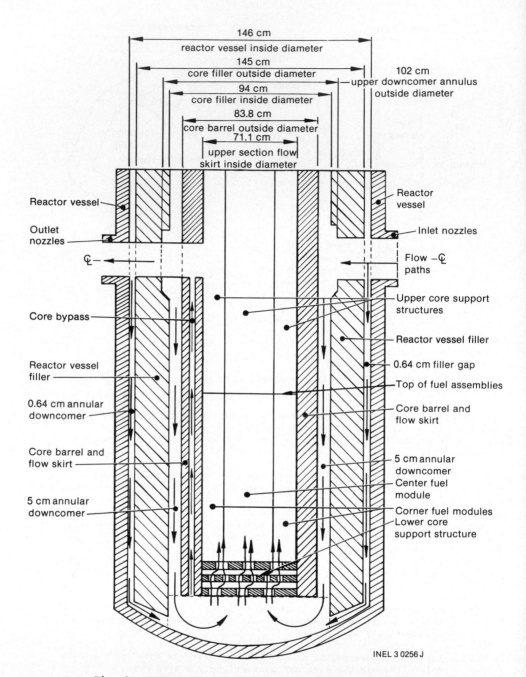

146 cm
reactor vessel inside diameter

145 cm
core filler outside diameter

102 cm
upper downcomer annulus
outside diameter

94 cm
core filler inside diameter

83.8 cm
core barrel outside diameter

71.1 cm
upper section flow
skirt inside diameter

Reactor vessel

Outlet
nozzles

Reactor
vessel

Inlet nozzles

Flow
paths

Core bypass

Upper core support
structures

Reactor vessel filler

Reactor vessel
filler

0.64 cm filler gap

Top of fuel assemblies

0.64 cm annular
downcomer

Core barrel and
flow skirt

Core barrel and
flow skirt

5 cm annular
downcomer
Center fuel
module

5 cm annular
downcomer

Corner fuel modules
Lower core
support structure

INEL 3 0256 J

Fig. 2. LOFT Reactor Vessel Cutaway To Show Flow Paths

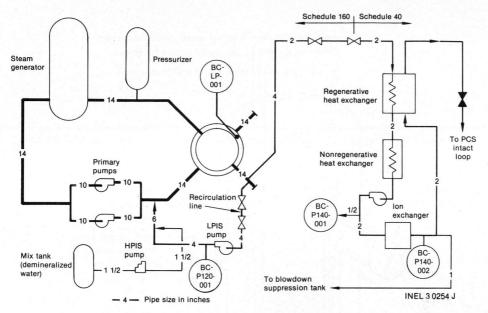

Fig. 3. LOFT System Schematic for Boron Dilution Configuration

4. The upper plenum above the nozzle was sealed off from the rest of the reactor (no upper plenum flow)

5. The core was simulated by acrylic rods

6. The instruments inside the LOFT reactor system were not simulated

7. The lower core support structure was eliminated for ease of fabrication

8. The reactor vessel filler in the LOFT reactor was eliminated in the model and the model reactor wall was chosen to be 0.953 cm (3/8 in.) thick to maximize a see-through capability.

Subsequent to fabrication, the model was filled and the water volume carefully measured to confirm the total volume of the model. The volume was 2.35×10^{-2} m^3 (0.830 ft^3) compared to the design value of 2.34×10^{-2} m^3 (0.828 ft^3), a difference of <1%. The flow rates for both the recirculation and diluent injection pumps were linear in the desired flow range. Flowmeters were installed in both lines to measure the flow. A flow-type conductivity cell was used to track the changing concentration. Calibration data for the conductivity cell were taken using standard solutions, and showed a linear conductivity-concentration relation in the desired concentration range of 500 to 4250 ppm. In the plexiglass model experiments, sodium chloride was used as the solute instead of boric acid. This change was made because of better stability and increased ease of preparation of sodium chloride compared to boric acid. Since the diffusivity of sodium chloride in water is 1.10×10^{-5} cm^2/s compared to 1.26×10^{-5} cm^2/s for boric acid, it was concluded that the substitution was acceptable.

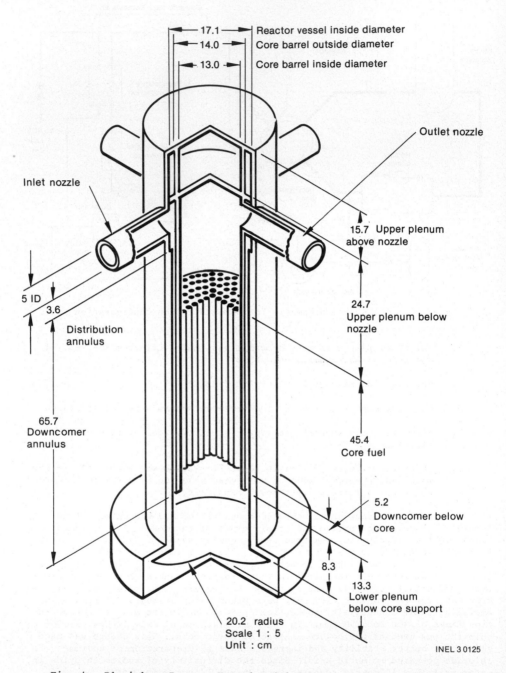

Fig. 4. Plexiglass Reactor Vessel Model Cutaway To Show Internals

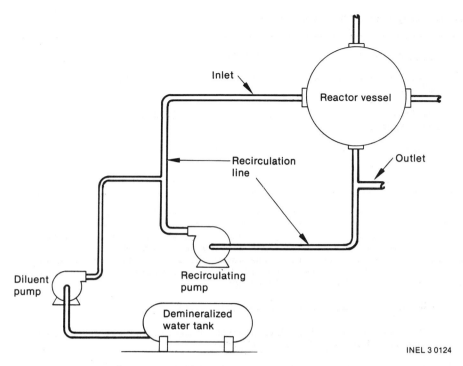

Fig. 5. Plexiglass Model Loop Schematic

3. EXPERIMENT L6-6

In order to reproduce the phenomena and the timing thereof expected to occur during an inadvertent boron dilution event in a PWR, the flow rates in Experiment L6-6 [6] were scaled, based on the following:

1. The rates of boron-concentration decrease in LOFT and in a reference PWR should be the same

2. The mixing time (that is, the time to completely exchange the mixed volume) should be the same.

Trojan, a Westinghouse-design four-loop PWR, owned and operated by Portland General Electric in Prescott, Oregon, was used as the reference plant in scaling the LOFT flow rates. Table 1 compares the assumed mixed volumes and the flow rates for the LOFT and Trojan plants. The Trojan values used to scale the LOFT values are based on the Trojan Final Safety Analysis Report [7].

Calculations made to estimate the mixing which would occur in LOFT, compared to that in Trojan, showed that, based on the scaled flow rates, the Reynolds numbers were higher in Trojan than in the LOFT experiment. Since mixing is dominated by turbulence (as opposed to molecular diffusion), mixing is thus expected to be better in Trojan than in LOFT, and Experiment L6-6 mixing results should be conservative when applied to Trojan. In order to

TABLE 2. Sequence of Events for Experiment L6-6

| Event | Time After Experiment Initiation[a] (s) | |
	L6-6A	L6-6B
Experiment initiated[a]	0	0
Source range alarm setpoint reached[b]	4 930 ± 50	5 380 ± 50
Criticality reached	7 416 ± 10	8 058 ± 10
Reactor scrammed	7 458.4 ± 0.2	8 093.2 ± 0.2
Diluent flow stopped	7 460 ± 1.5	8 096 ± 1
Experiment terminated[c]	10 160 ± 1.5	10 796 ± 1

a. Experiment initiation is defined as when the diluent injection flow was started [8].

b. Source range alarm setpoint is defined as five times the detector count rate at k_{eff} = 0.9. This was determined after the experiment and no alarm sounded.

c. Experiment termination occurred by definition [8] 45 min after diluent injection was stopped.

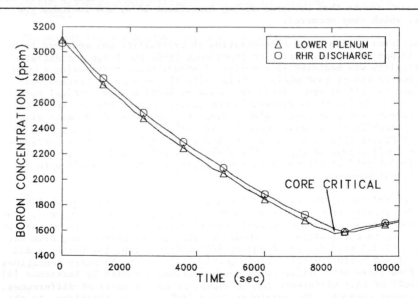

Fig. 7. Comparison of Lower Plenum and RHR Pump Discharge Boron Concentrations

3.2 Comparison with Preexperiment Calculation

Reference [9] reports the results of preexperiment calculations on Experiment L6-6. These calculations were performed using Equation (1) and an assumed value for the mixed volume of 3.98 m^3 (141 ft^3). This volume was the sum of the volumes for the recirculation line (including those portions of the primary coolant loops directly involved) and the reactor vessel up to an elevation corresponding to the top of the nozzles. This assumed volume is consistent with nominal PWR analysis of this type of event. The results of this calculation are shown in Figure 8, together with the lower plenum concentration measured for L6-6A and L6-6B. As shown, the boron concentration was calculated to decrease much more rapidly than measured. The calculated time to criticality was 5130 s (85.5 min), versus the measured times of 7416 s (123.6 min) for L6-6A and 8058 s (134.3 min) for L6-6B. Since it has been concluded above that the assumption of perfect mixing is adequate for calculations, it was decided that the large difference between measured and calculated times to criticality was caused by the assumed mixed volume, the actual effective mixed volume being ∿30% larger than assumed in the calculation. As a point of reference, the volume in the reactor vessel above the nozzles (not included in the assumed effective mixed volume) is ∿1/3 of this additional volume. Thus, significant volume outside the reactor vessel was also mixed.

4. POSTEXPERIMENT ANALYSES

A principal goal of the postexperiment analysis of Experiment L6-6 was to explain the difference in times to criticality between measured and predicted data. A key in determining the answer to this was provided by the pressurizer liquid level shown in Figure 9 for L6-6A. As shown, the level was held constant, within the specified ±0.025 m (±1 in.) throughout. In accomplishing this, however, slow, low-amplitude oscillations in the pressurizer liquid level occurred as the operators adjusted the letdown flow to maintain the constant liquid mass. These oscillations, shown in more detail in Figure 10, were within the specified pressurizer liquid level band, but, as discussed below, were apparently sufficient to cause significant mixing of the liquid in the pressurizer and intact loop hot leg. These liquid volumes were previously assumed to be stagnant.

The oscillating liquid level was the result of an oscillating fluid flow in these additional volumes. As illustrated in Figure 11, the oscillating fluid velocity between the up and down position at any one cycle is:

$$\text{velocity} = \frac{L_U - L_D}{t_U - t_D} \tag{3}$$

where

$L_{U,D}$ = level in the up, down position

$t_{U,D}$ = corresponding times.

In terms of a frequency, f, which is the average number of peaks and valleys per unit time, the average velocity for outflow is:

$$\text{average velocity} = (L_U - L_D)_{ave} \times f . \tag{4}$$

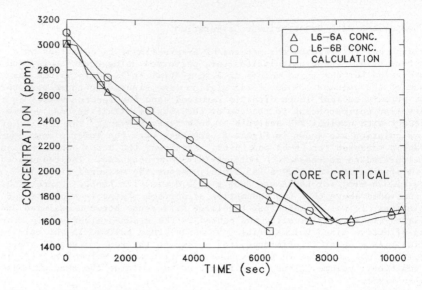

Fig. 8. Comparison of Lower Plenum Concentrations During L6-6A and
L6-6B with Perfect Mixing Model Calculations

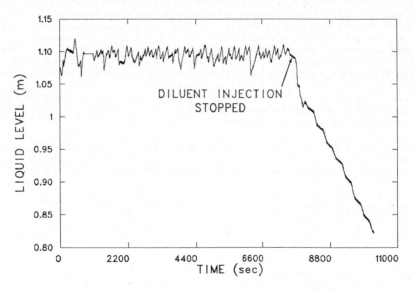

Fig. 9. Pressurizer Liquid Level During L6-6A

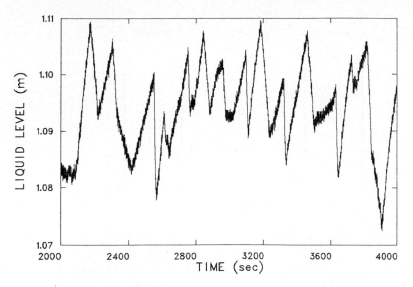

Fig. 10. Expanded Pressurizer Liquid Level During L6-6A

From Figure 10, we estimate:

$$f = 0.014 \ s^{-1}$$

and

$$(L_U - L_D)_{ave} = 0.015 \ m \ (0.59 \ in.).$$

The oscillatory flow rate, $\dot{Q}$, is:

$$\dot{Q} = A \times f \times (L_U - L_D)_{ave} \tag{5}$$

where

A = surface area of liquid in the pressurizer = 0.5 m^2 (5 ft^2).

Substituting the measured values into Equation (5) results in:

$$\dot{Q} = 1 \times 10^{-4} \ m^3/s \ (1.6 \ gpm).$$

This estimated value of 1 x 10^{-4} m^3/s compares with the diluent flow rate of nearly 5 x 10^{-4} m^3/s. Thus, the low-frequency, oscillating flow was of the same order of magnitude as the diluent flow rate. Since the oscillating fluid had a high (compared to that in the reactor vessel) boron concentration, it partially countered the dilution due to the diluent flow. The intact loop oscillation fluid velocity was 1.6 x 10^{-3} m/s, compared to 2 x 10^{-2} m/s in the reactor vessel due to RHR flow. The pressurizer fluid velocity was 1.8 x 10^{-4} m/s.

206

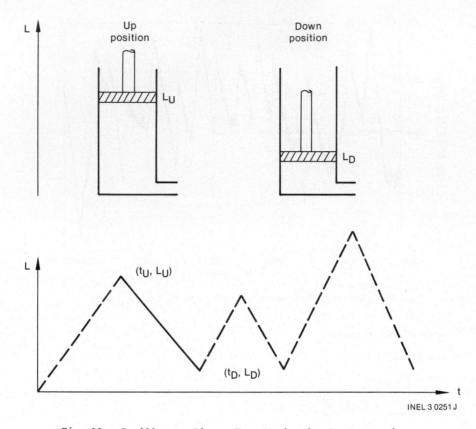

Fig. 11.   Oscillatory Piston-Type Motion in the Pressurizer

It can also be shown that the fluid in the intact loop and pressurizer did not merely move back and forth in a piston-type flow, but mixed. This mixing caused the addition of boron into the reactor vessel.

To see how this low velocity fluid flow results in good mixing, it is noted that mixing in a pipe can be described by the dispersion equation [10]:

$$\frac{\partial C}{\partial t} + U \frac{\partial C}{\partial x} = D_{eff} \frac{\partial^2 C}{\partial x^2} \tag{6}$$

where

$$D_{eff} = \frac{R^2 U^2}{48 \, D} \quad \text{for laminar fully developed flow}$$

$$C = \text{concentration (ppm)}$$

$$D = \text{molecular diffusivity of } H_3BO_3 = 10^{-9} \, m^2/s$$

R   =   pipe radius = 0.142 m (hot leg pipe)

U   =   average velocity = 1.6 x 10^{-3} m/s

t   =   time (s)

x   =   length dimension (m).

Using these values,

$$D_{eff} = 1 \frac{m^2}{s}.$$

The Peclet number, which provides a measure of how effectively mixing takes place, is defined as

$$\text{Peclet number} = Pe = \frac{L_U}{D_{eff}} \tag{7}$$

where L = pipe length.

For the 6-m length of the intact loop hot leg, Equation (7) yields Pe = 0.01, which indicates good mixing. Thus, according to this argument, oscillating fluid in the intact loop and its components should not just communicate in a piston-like fashion with the fluid in the reactor vessel, but should mix fairly effectively. This hypothesis is supported by two additional pieces of information, discussed in the following paragraphs.

When the diluent and letdown flows were terminated at 7460 s (124 min), the sample-line flows were continued which resulted in a decrease in pressurizer water liquid level as the water in the pressurizer replaced that lost from the intact loop. This is illustrated in Figure 8. Since the intact loop contained fluid with a higher concentration of boron than that in the reactor vessel, this influx from the loop resulted in an increase in reactor vessel boron concentration, as shown in Figures 6 and 7. Using the amount of influx, measured by the pressurizer liquid level decrease and the reactor vessel boron concentration increase, the pressurizer boron concentration at diluent flow cessation was calculated to be 2200 ± 230 ppm for L6-6A and 2100 ± 230 ppm for L6-6B [3]. Thus, the pressurizer boron concentration decreased ∿800 to 900 ppm in each case from the initial condition concentration of ∿3000 ppm.

An additional datum was provided by a one-time grab sample taken from an instrumentation line located near the steam generator inlet plenum. This sample, which was taken subsequent to L6-6B, was analyzed and found to contain 1800 ± 15 ppm boron. A similar sample taken at the primary coolant pump inlet was measured to contain 3070 ± 15 ppm boron. These two samples support the conclusions that (a) there was no backward flow in the loop, since if backward flow had existed, the pump inlet boron concentration would have decreased, and (b) the intact loop hot leg was nearly at the same boron concentration as the reactor vessel, indicating good mixing in this part of the intact loop.

4.1 Well-Mixed Model with Infinite Source

Oscillations in flow caused a more slowly decreasing boron concentration than expected, based on Equation (1) and the assumed direct flow path volume. An additional source for boron may be due to that contained in:

1. Intact loop hot leg, 14-in. Schedule 160 pipe (0.384 m^3, 13.6 ft^3)

2. Reactor vessel upper head (0.348 m^3, 12.3 ft^3)

3. Pressurizer filled to a level of 1.08 m (0.542 m^3, 19.1 ft^3)

4. Steam generator inlet plenum (0.335 m^3, 11.8 ft^3)

5. A portion of the inlet side of the steam generator tubes (volume undetermined).

As a first approximation, assume that the extra source of boron is infinite. Then, a material balance in boron results in (see Figure 12):

$$V \frac{dc}{dt} = -W_{DIL} \, C + \dot{Q}_{in} \, C_o \qquad (8)$$

where

C = boron concentration (ppm)

$\dot{Q}_{in}$ = oscillatory inflow rate (m^3/s)

W_{DIL} = dilution flow rate (m^3/s)

C_o = C (t = 0).

The solution to Equation (8) is

$$C = \frac{\dot{Q}_{in} \, C_o}{W_{DIL}} + \left(C_o - \frac{\dot{Q}_{in} \, C_o}{W_{DIL}} \right) \exp - \frac{W_{DIL}}{V} t \quad . \qquad (9)$$

If we use $\dot{Q}_{in}$ = 9.9 x 10^{-5} m^3/s (oscillatory flow over one full cycle) and take the other values to correspond to L6-6A conditions (W_{DIL} = 4.5 x 10^{-4} m^3/s, C_o = 3008 ppm, and V = 3.98 m^3), the boron concentration is as shown in Figure 13, where the measured L6-6A lower plenum concentration is also shown for comparison. The calculated concentration is ~5% higher than measured.

4.2   Finite Source Model

To take into account the finite sources of boron in the additional volumes (that is, the decreasing boron concentration in these volumes), the following differential equations can be set up for boron concentrations C_R and C_A in the reactor vessel and additional volumes, respectively, see Figure 14:

$$V_R \frac{dC_R}{dt} = -W_{DIL} \, C_R + \dot{Q}_{in} \, C_A \qquad (10)$$

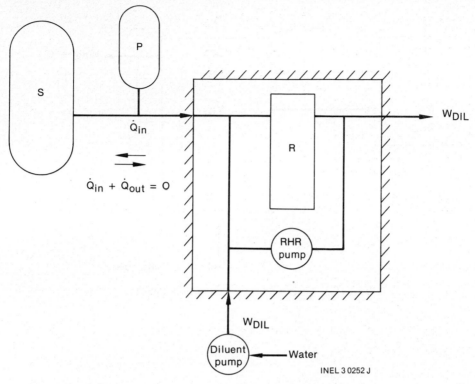

Fig. 12. Boron Dilution System Model with Oscillatory Flow Component

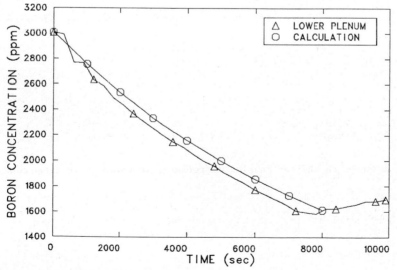

Fig. 13. Comparison of Lower Plenum Concentration in L6-6A with Infinite Additional Volume Model Results

210

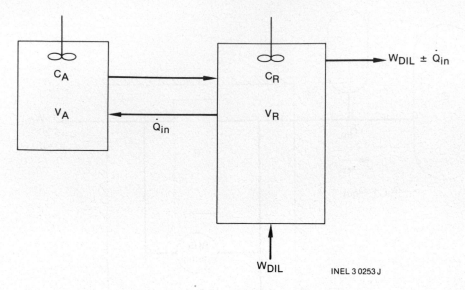

Fig. 14. Finite Additional Volume Model

$$V_A \frac{dC_A}{dt} = \dot{Q}_{in}\, C_R - \dot{Q}_{in}\, C_A \qquad\qquad (11)$$

with the initial conditions

$$C_R\,(o) = C_A\,(o) = C_o.$$

To solve Equations (10) and (11), first introduce the following dimensionless parameters:

$$\tau = \frac{W_{DIL}\,t}{V_R}, \quad \overline{C}_R = \frac{C_R}{C_o}, \quad \overline{C}_A = \frac{C_A}{C_o}$$

$$\alpha = \frac{V_A}{V_R}, \quad \beta = \frac{\dot{Q}_{in}}{W_{DIL}}, \quad \gamma = \frac{\alpha}{\beta}\,.$$

The solution to the two coupled differential equations, in dimensionless form, is:

$$\overline{C}_R = A\, e^{m_1 T} + B\, e^{m_2 T} \qquad\qquad (12)$$

where the roots m_1 and m_2 are:

$$m_{1,2} = \frac{-(1+\gamma) \pm \sqrt{(1-\gamma)^2 + 4\,\gamma\,\beta}}{2\,\gamma}$$

and the integration constants are:

$$A = \frac{m_2 + 1 - \beta}{m_2 - m_1}$$

$$B = \frac{\beta - 1 - m_1}{m_2 - m_1} \quad .$$

The dimensionless additional volume concentration is:

$$\bar{C}_A = \frac{m_1 + 1}{\beta} A\, e^{m_1 T} + \frac{m_2 + 1}{\beta} B\, e^{m_2 T} \quad . \tag{13}$$

As a check on the solution, Equation (12) reduces to Equation (9) when the V_A, and hence γ, become very large.

$\bar{C}_R$ was evaluated using the following values:

1. W_{DIL} equals 4.5×10^{-4} m^3/s

2. C_o equals 3008 ppm

3. $\dot{Q}$ equals 5.0×10^{-5} m^3/s (average flow over 1/2 cycle since we must now consider the effect of the backward flow on C_A)

4. V_A equals 1.61 m^3

which results in

$$\alpha = 0.403, \ \beta = 0.111, \ \text{and} \ \gamma = 3.679 \ .$$

Therefore,

$$\bar{C}_R = 0.81\, e^{-1.16 \times 10^{-4} t} + 0.186\, e^{-0.263 \times 10^{-4} t} \quad .$$

The calculated and measured boron concentrations in the reactor vessel are shown in Figure 15. The agreement, again, is within 5%; however, this time the calculated boron concentration is lower than measured.

5. PLEXIGLASS MODEL EXPERIMENTS

The procedure for conducting the plexiglass model boron dilution experiments was to:

1. Prepare a standard sodium chloride solution of known concentration (~40 L).

2. Fill the reactor vessel model and the loop with the sodium chloride solution free of air bubbles.

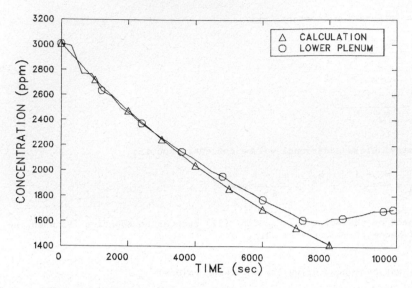

Fig. 15. Comparison of Lower Plenum Concentration During L6-6A with Finite
Additional Volume Model Results

3. Set the recirculating pump to the desired recirculation rate.

4. Inject the sodium chloride solution through the high-pressure
diluent pump at the desired dilution rate. Adjust outlet flowmeter
to match the dilution rate. Establish steady state (∿15 min) and
record the reactor vessel outlet concentration (i.e., the initial
concentration of the sodium chloride solution in the system).

5. Switch the injection of the sodium chloride solution to the deminer-
alized water, thus starting the dilution.

6. Trace the reactor vessel outlet concentration versus time until the
concentration drops well below half the initial value.

7. Discontinue the injection, end the recirculation, and drain the
system.

Data from two boron dilution experiments, which were replicates of each other,
are shown in Figure 16. The data are shown to be reproducible to within 1%
in the two experiments. Results from the plexiglass model experiments are
discussed in the following sections.

5.1 Effect of the Recirculation Rate

The effects of the recirculation rate on the mixing was measured in
three experiments which were conducted using the same diluent injection rate
(2.70×10^{-6} m^3/s, corresponding to the LOFT value scaled as discussed in
Section 2) and initial sodium chloride concentration, 2900 ppm. The recirculation flow rates used were 8.00×10^{-6} m^3/s, 1.65×10^{-5} m^3/s, and
2.76×10^{-5} m^3/s.

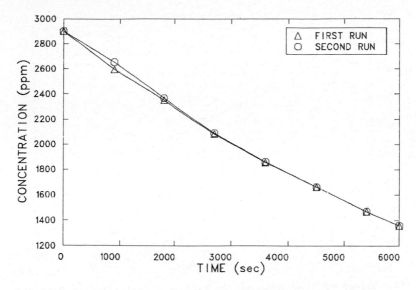

Fig. 16.   Results from Plexiglass Reproducibility Experiments

The last value corresponds to the LOFT L6-6A flow rate. The Reynolds numbers in the downcomer for these experiments were 90, 162, and 232, all of which indicate laminar flow.

In each of these experiments, the concentration decreased to 1400 ppm at ∿5700 s (95 min) after initiation, indicating no measurable effect of recirculation flow rate on the mixed volume for this range of Reynolds numbers.

5.2 Variance of Volume Mixing

The next five experiments were conducted using a common recirculation rate and different diluent injection flow rates to measure the variance of the mixed volume under these conditions. In the first three of these experiments (dilution Runs 1, 2, and 3), the initial concentration of sodium chloride was set at 2900 ppm and the recirculation rate was 2.49×10^{-5} m^3/s. Figure 17 shows the outlet concentration of sodium chloride as a function of time for Runs 1, 2, and 3 with diluent injection rates of 3.75×10^{-5}, 2.70×10^{-6}, and 1.63×10^{-6} m^3/s, respectively.

In the next two experiments (dilution Runs 4 and 5), the initial concentration of sodium chloride was set at 4570 ppm and the recirculation rate again was 2.48×10^{-5} m^3/s. Figure 18 shows the outlet concentration of sodium chloride versus time for Runs 4 and 5 with diluent injection rates of 3.75×10^{-5} and 2.63×10^{-6} m^3/s, respectively.

The linearity for all five experiments is clearly shown in Figure 19. The effective well-mixed volume (V_{eff}) was calculated from the slope of the well-mixed formula using the relation:

$$V_{eff} = \frac{\text{Injection rate}}{\text{Slope}}. \tag{14}$$

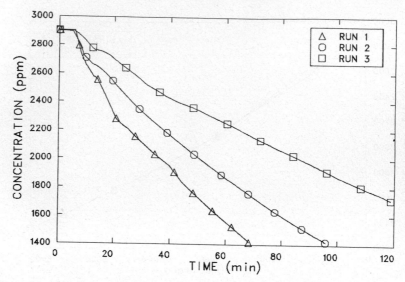

Fig. 17. Effect of Diluent Injection Rate at a Recirculation Rate of
2.49 x 10^{-5} m^3/s and an Initial Sodium Chloride Concentration of 2900 ppm

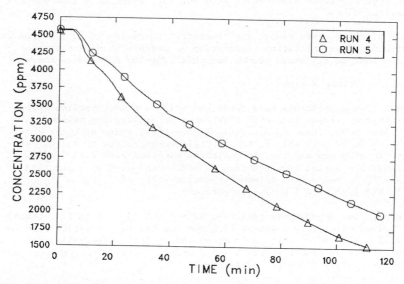

Fig. 18. Effect of Diluent Injection Rate at a Recirculation Rate of
2.49 x 10^{-5} m^3/s and an Initial Sodium Chloride Concentration of 4570 ppm

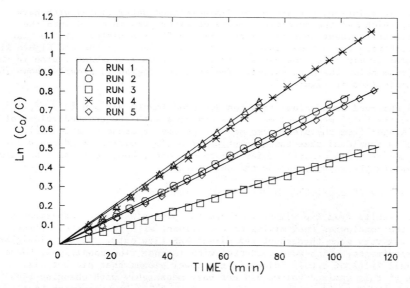

Fig. 19. Liniarity of Plexiglass Model Dilution Runs 1 Through 5 at a
Recirculation Rate of 2.49 x 10^{-5} m^3/s

The results are summarized in Table 3. The effective volume is 21.3 ± 0.2 x
10^{-3} m^3. This is ~10% lower than the measured total volume of 23.5 x
10^{-3} m^3.

Table 3. Effective Plexiglass Experiment Volume Determination

Dilution Run	Injection Rate (x 10^{-6} m^3/s)	Recirculation Rate (x 10^{-5} m^3/s)	Initial Concentration (ppm)	V_{eff} (x 10^{-4} m^3)
1	3.75	2.49	2900	21.20
2	2.70	2.49	2900	21.26
3	1.63	2.49	2900	21.10
4	3.75	2.49	4570	21.65
5	2.63	2.49	4570	21.45

5.3 Location of Stagnant Volumes

In addition to the dilution experiments, a dye experiment was conducted to
visually confirm the lack of total mixing and identify the location of
stagnant volumes. The recirculation flow rate was 2.28 x 10^{-5} m^3/s and
the diluent injection flow rate was 2.28 x 10^{-6} m^3/s.

218

Fig. 22. Closeup of Plexiglass Model 20 min After Dye Injection Showing
Stagnant Volume in the Lower Plenum

agency thereof, or any of their employees, makes any warranty, expressed or
implied, or assumes any legal liability or responsibility for any third
party's use, or the results of such use, of any information, apparatus, product or process disclosed in this paper, or represents that its use by such
third party would not infringe privately owned rights. The views expressed
in this paper are not necessarily those of the United States Nuclear
Regulatory Commission.

REFERENCES

1. United States Nuclear Regulatory Commission, Office of Nuclear Regula-
    tion, Standard Review Plan, Section 15.4.6, "Chemical and Volume Control
    System Malfunction that Results in a Decrease in Boron Concentration in
    the Reactor Coolant (PWR)," April 1975.

2. Reeder, D. L., "LOFT System and Test Description (5.5-Ft Nuclear Core 1 .
    LOCEs)," NUREG/CR-0247, TREE-1208, July 1978.

3. Adams, J. P., and Berta, V. T., "Quick-Look Report on LOFT Boron Dilu-
    tion Experiment L6-6," EGG-LOFT-5867, May 1982.

219

4. Stitt, B. D., and Divine, J. M., "Experiment Data Report for LOFT Boron Dilution Experiment L6-6," NUREG/CR-2733, EGG-2197, June 1982.

5. Ybarrondo, L. J., et al., "Examination of LOFT Scaling," 74-WA/HT-53, Proceedings of the Winter Meeting of America Society of Mechanical Engineers, New York, November 17-21, 1974, CONF-741104.

6. Fullmer, K. S., and Koske, J. E., "LOFT Experiment Definition Document, Boron Dilution Experiment L6-6," EGG-LOFT-5730, April 1982.

7. Portland General Electric, "Trojan Final Safety Analysis Report," Section 15.2.4, "Uncontrolled Boron Dilution".

8. Fullmer, K. S., and Koske, J. E., "LOFT Experiment Operating Specification, Boron Dilution Experiment Test L6-6," EGG-LOFT-5729, March 1972.

9. Adams, J. P., and Girard, C., "LOFT L6-6 Experiment Prediction," EGG-LOFT-5803, March 1982.

10. Levich, B. G., Physicochemical Hydrodynamics, Englewood Cliff: Prentice-Hall, 1962, pp. 116-120.

Multi-Phase Flow and Heat Transfer III. Part B: Applications
edited by T.N. Veziroğlu and A.E. Bergles
Elsevier Science Publishers B.V., Amsterdam, 1984 — Printed in The Netherlands

PRESSURIZER AND STEAM GENERATOR BEHAVIOR UNDER PWR TRANSIENT CONDITIONS[a]

A. B. Wahba[b] and V. T. Berta
EG&G Idaho, Inc.
P.O. Box 1625
Idaho Falls, ID   83415, U.S.A.

W. Pointner
Gesellschaft fuer Reaktorsicherheit (GRS) mbH
8046 Garching, West Germany

ABSTRACT

Experiments have been conducted in the Loss-of-Fluid Test (LOFT) pressurized water reactor (PWR), at the Idaho National Engineering Laboratory, in which transient phenomena arising from accident events with and without reactor scram were studied. The LOFT PWR is a fully operational, 50-MW(t) facility and is related to a commercial PWR through volumetric scaling principles. The main purpose of the LOFT facility is to provide data for the development of computer codes for PWR transient analyses. Significant thermal-hydraulic differences have been observed between the measured and calculated results for those transients in which the pressurizer and steam generator strongly influence the dominant transient phenomena. Pressurizer and steam generator phenomena that occurred during four specific PWR transients in the LOFT facility are discussed. Two transients were accompanied by pressurizer inflow and a reduction of the heat transfer in the steam generator to a very small value. The other two transients were accompanied by pressurizer outflow while the steam generator behavior was controlled.

The RELAP5/MOD1 code did not calculate the measured rapid pressure decrease which resulted from the pressurizer spray initiation. Also, differences between measured and calculated heat transfer in the steam generator were sufficiently large to necessitate improvements in heat transfer regimes and in the way steam generators are modeled.

Calculations with DRUFAN-02 (a lumped-parameter, thermal-hydraulic code with a thermodynamic nonequilibrium model) have shown that temperature and mass flow rate out of the pressurizer spray line greatly influence the pressure behavior. Very good agreement between measured and calculated pressure was found under the assumption that the spray water was subcooled.

----

a. This work was supported by both the U.S. Nuclear Regulatory Commission, Office of Nuclear Regulatory Research, under DOE Contract No. DE-AC07-76ID01570, and the Federal Ministry of Research and Technology in West Germany.

b. German representative to the LOFT Program.

1. INTRODUCTION

The pressurizer and steam generators are major components of a pressurized water reactor (PWR). During normal operation of a nuclear power plant, the pressure of the primary coolant system is controlled by the pressurizer, which normally includes spray nozzles, valves, and a heater. Steam condensation phenomena, caused by the actuation of the spray system, control small system pressure increases. In case of a large increase in system pressure, the pressure operated relief valve (PORV) automatically opens to limit the pressure. When the system pressure decreases, the pressurizer heater is initiated and the subsequent vaporization phenomena will increase the pressure. Correct prediction of reactor system behavior must therefore include the correct prediction of such phenomena in the pressurizer. Also, the steam generators (SG) have a control function on the behavior of the primary system which is associated with the energy balance. In case of operational transients, such as loss of feedwater, the SG secondary coolant boils off and the system behavior depends, to a great extent, on changes in the heat transfer between primary and secondary sides of the SG. Due to the high surface-area-to-volume ratio in the SG, detailed knowledge of the heat transfer phenomena is needed.

This paper discusses the pressurizer and SG phenomena that occurred during specific PWR transients in the Loss-of-Fluid Test (LOFT) facility. The principal purpose of the LOFT facility is to provide data for the development of computer codes for PWR transient analyses. Significant thermal-hydraulic differences have been observed between measured and calculated results for the following transients in which the pressurizer and SG strongly influenced the dominant transient phenomena:

1. Loss-of-feedwater without reactor scram

2. Loss-of-offsite power without reactor scram

3. Rapid cooldown

4. Steam generator tube rupture.

The pressurizer and steam generator phenomena characteristics during the above transients were in the same order:

- 1 and 2   Pressurizer inflow; steam generator heat transfer reduction to very small values

- 3 and 4   Pressurizer outflow; steam generator heat transfer varying.

2. LOFT FACILITY

The LOFT facility is a fully operational, volumetrically scaled (LOFT/TROJAN ~1/50), PWR facility. The experimental facility, as shown in Figure 1, includes six major subsystems which have been instrumented such that system variables can be measured and recorded during transient operation. The major subsystems of LOFT are the reactor vessel, intact loop, broken loop, blowdown suppression system, emergency core cooling system, and secondary coolant system. The reactor vessel includes the 50-MW(t) nuclear core, with 1300 fuel rods arranged in five square (15 x 15) and four triangular (corner) fuel modules. The fuel assemblies are identical to the 15 x 15 fuel rod arrays used in commercial PWRs, except that the active fuel length is 1.68 m compared with 3.7 m in a commercial reactor.

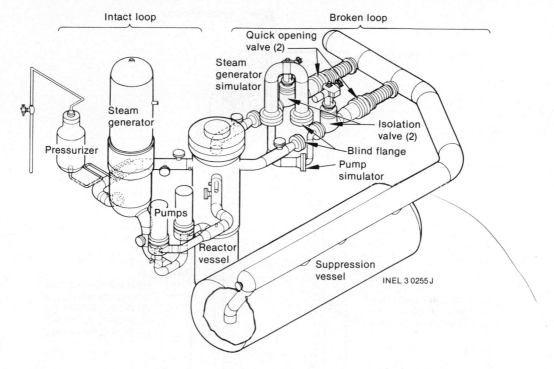

Fig. 1. LOFT major components.

The intact loop simulates three loops of a commercial, four-loop PWR and contains a steam generator (about half the height of a commercial SG), two primary coolant pumps in parallel, and a pressurizer. The pressurizer, as shown in Figure 2, includes electrical heaters at the bottom and spray nozzles at the top. When the spray system is in operation, water from the cold leg flows through a long spray line to the nozzles. The spray line is a stainless steel tube about 15 m long, 33 mm in outer diameter, and 6 mm thick, and is insulated with a 38-mm layer of calcium silicate. The spray line is not instrumented.

The broken loop, designed for loss-of-coolant accident simulations, consists of hot and cold legs that are connected to the reactor vessel and the blowdown suppression tank header. The broken loop hot leg contains a simulated steam generator and a simulated pump. These components were not attached during the transients discussed in this paper but were replaced with a blind flange as shown in Figure 1. A complete description of LOFT is given in Reference [1].

3. LOFT TRANSIENTS

The LOFT transients identified in Section 1 were initiated from the operating conditions listed in Table 1. Those conditions for Transients 1, 2, and 3--loss-of-feedwater without scram, loss-of-offsite power without scram, and

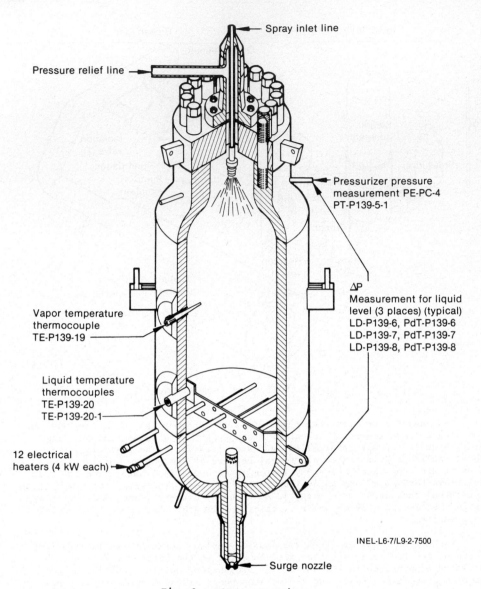

Fig. 2. LOFT pressurizer.

the rapid primary system cooldown, respectively--are typical of operating conditions in commercial PWR plants. The operating conditions for Transient 4, steam generator tube rupture, are typical of hot standby conditions in commercial PWRs. Transients 1, 2, and 3 [2-7] are termed "anticipated transients," in which the initiating event is not a loss-of-coolant accident condition in the primary coolant system. Transient 4 [8,9] is classified as a small-break accident which, in this case, involves a coolant break path between the primary and secondary coolant systems of a PWR.

Table 1. Plant Operating Conditions at Transient Initiation

	Range	
	---	---
Condition	Transients 1, 2, and 3	Transient 4[a]
Pressure, MPa	14.75 to 15.0 (2140 to 2177 psia)	15.5 (2248 psia)
Cold leg temperature, K	556 to 557 (542 to 544°F)	562 (552°F)
Core axial temperature differential, K	19 to 20 (34 to 36°F)	$\sim$0
Core maximum linear heat generation rate, kW/m	50.2 to 53.1 (15.3 to 16.2 kW/ft)	--
Core power/system volume, MW/m^3	6.67 to 6.92 (189 to 196 kW/ft^3)	--
Core decay heat/system volume at 1000 s after transient initiation, kW/m^3	102.4 to 123.6 (2.9 to 3.5 kW/ft^3)	--[a]
Loop mass flow/system volume, $kg/s \cdot m^3$	46.9 to 66.1 (3.93 to 4.13 lbm/s-ft^3)	65.2 (4.07 lbm/s-ft^3)

a. Transient 4 was initiated from the listed conditions 5040 s after reactor scram at operating conditions similar to those listed for Transients 1, 2, and 3. Decay heat/system volume at 5040 s was 41 kW/m^3 (1.16 kW/ft^3).

The response of the primary coolant system during these kinds of transients is strongly dependent on the pressurizer and steam generator. Transients 1 and 2 were initiated, in effect, by a large reduction in primary-to-secondary-system heat transfer in the steam generator. This led to increased energy deposition in the primary system, which in turn caused coolant inflow to the pressurizer due to coolant swell. The pressure and temperature response of the primary system is therefore dependent on (a) the heat transfer mechanisms in the SG as the secondary side changes from normal operating conditions to a highly voided state, and (b) the thermodynamic state and the effectiveness of controls (pressurizer spray system and PORV venting) in the pressurizer.

Transient 3 was the opposite of Transients 1 and 2, in that the initiating event was a rapid increase in SG secondary energy demand accompanying a steam line rupture, which caused a cooldown of the primary system. The cooldown in turn causes a coolant volumetric shrink and resultant outflow from the pressurizer to the primary system piping. During this transient, the SG heat transfer rate increased, but the heat transfer regimes remained essentially unchanged. The thermodynamic state and control systems (spray and heaters) in the pressurizer influenced the pressure response, or control, of the primary system up to the time the pressurizer emptied.

Transient 4 was initiated by the rupture (simulated) of a single steam generator tube. A recovery procedure was used which combined continued SG operation with pressurizer spray actuation to mitigate coolant flow from the primary system to the secondary system. This transient was similar to Transient 3 in that both the steam generator and pressurizer thermodynamic characteristics were similar; that is, SG heat transfer regimes were essentially unchanged, and there was coolant outflow from the pressurizer to the primary coolant system piping.

The four transients ended when stable conditions were achieved with the system under operator control. The stable conditions were, in all cases, a hot standby condition, pressurizer control reestablished, and the system proceeding to cold shutdown conditions under operator control.

4. COMPUTER CODES

The LOFT transients discussed in the previous section were calculated in pre- and postexperiment modes using the RELAP5/MOD1 computer code [10]. The RELAP5 code is designed to analyze the thermal-hydraulic behavior of a light water reactor (LWR) during LOCA and non-LOCA transients. The code is based on a one-dimensional, nonhomogeneous, nonequilibrium, hydrodynamic model. A point kinetics model is used to calculate the power transient in the nuclear core.

Postexperiment calculations of the steam generator tube rupture transient (Transient 4) were carried out at the Gesellschaft fuer Reaktorsicherheit (GRS) mbH in West Germany with the DRUFAN-02 code. The DRUFAN code [11] was developed at the GRS to analyze blowdown and refill phases of LOCA transients, as well as non-LOCA transients in LWRs. The code is based on the lumped-parameter approach, which allows the flexible configuration of control volumes. The physical model is based on the separate field equations for steam and liquid phases and overall field equations for energy and momentum of the mixture. The thermodynamic nonequilibrium assumptions are based on the mass transfer rate between phases, which is calculated by means of correlations for the growth and shrinkage of steam bubbles and liquid droplets, respectively. Drift flux models and a mixture level tracking model are implemented in the DRUFAN-02 version [12], and a heat conductor model is available for simulating structures, electrical heater rods, and fuel rods. A point kinetics model is used to calculate fuel rod powers, and a comprehensive package of flow-regime-dependent heat transfer and critical heat flux correlations is implemented to calculate the heat transfer between solid structures and the fluid.

5. PRESSURIZER PHENOMENA

Transients 1 and 2 (loss-of-feedwater and loss-of-offsite power) were both accompanied by a degradation of the heat sink. Primary fluid expansion due to the increase in primary coolant temperature caused inflow in the pressurizer during both transients. The rate of pressure increase during the initial 60 s of Transient 1 was controlled by the pressurizer spray, as shown in Figure 3. After that time, continued degradation of the SG heat transfer, as well as the continuing decrease of the pressurizer vapor volume, caused a rapid increase in pressurizer pressure. The system pressure was then limited by the PORV and safety relief valve.

System pressure during the first 70 s of the transient is shown in more detail in Figure 4. The pressurizer spray was actuated twice during this period at a pressure setpoint of 15.35 MPa. The experimental results showed a

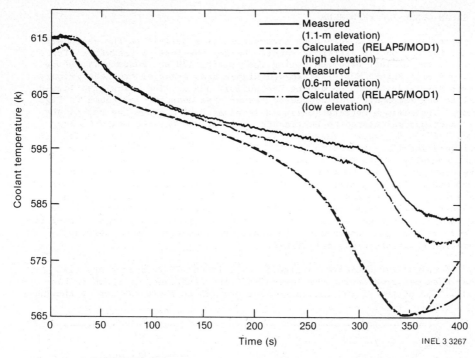

Fig. 8. Fluid temperatures at two elevations in the pressurizer during
Transient 3 compared with RELAP5/MOD1 postexperiment calculations.

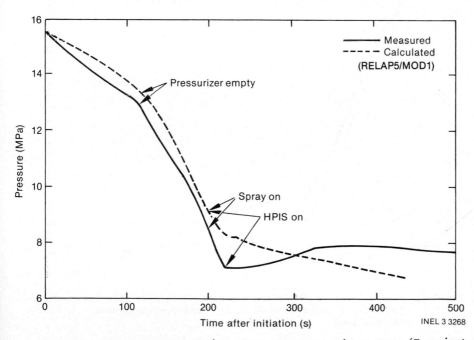

Fig. 9. System pressure during steam generator tube rupture (Transient 4)
compared with RELAP5/MOD1 preexperiment calculations.

6. STEAM GENERATOR PHENOMENA

Transients 1 and 2 were both accompanied by a boiloff in the secondary side of the SG, but at different rates. During the loss of feedwater (Transient 1), the boiloff occurred during the initial 120 s. Postexperiment calculations with RELAP5/MOD1 [13] showed higher heat transfer from the primary to secondary side of the SG during the boiloff phase. The calculated liquid level in the secondary side decreased faster than measured, as shown in Figure 10. The minimum detectable liquid level at 0.25 m above the top of the tube sheet was calculated to be reached at 80 s, whereas the measurements showed 120 s. Secondary pressure (Figure 11) started to decrease rapidly as the liquid level passed the top of the U-tubes. The main steam valve was closed manually at about 70 s and the pressure recovered again. The bypass valve was cycled twice to keep the secondary pressure below 6.5 MPa. All RELAP5/MOD1 calculations, typical of the postexperiment calculations shown in Figure 11, do not contain cycling of the valve as was observed during the transient. Measured and calculated fluid temperature at the bottom of the SG downcomer are compared in Figure 12. Very good agreement was found during the boiloff process, but superheated steam was measured after 90 s, whereas saturation temperature was calculated.

The calculated decrease in liquid level in the SG secondary was also faster than measured during the loss-of-offsite transient, as shown in Figure 13. The bottom of the instrument's range (25 cm above the top of the tube

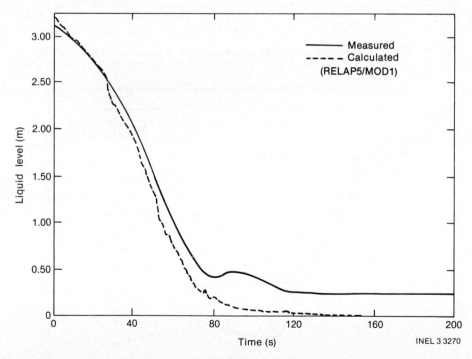

INEL 3 3270

Fig. 10. Liquid level in the steam generator secondary side during
Transient 1 compared with RELAP5/MOD1 postexperiment
calculations.

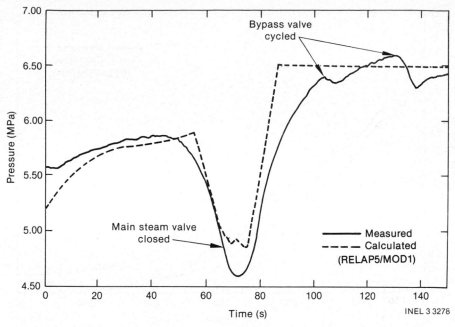

Fig. 11.   Steam generator secondary pressure during Transient 1 compared with RELAP5/MOD1 postexperiment calculations.

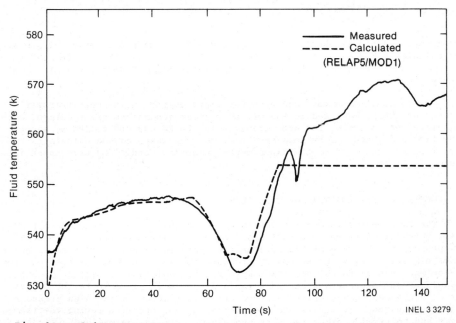

Fig. 12.   Fluid temperature at the bottom of the steam generator downcomer during Transient 1 compared with RELAP5/MOD1 postexperiment calculations.

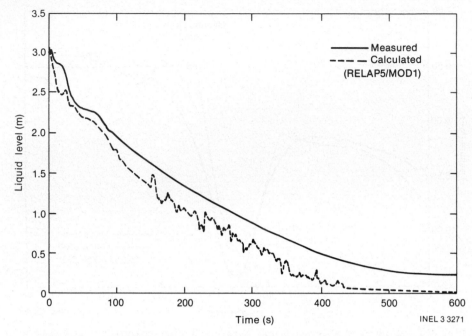

Fig. 13. Liquid level in the steam generator secondary side during
Transient 2 compared with RELAP5/MOD1 postexperiment
calculations.

sheet) was reached at about 500 s, compared with a calculated value of 370 s.
The calculated higher heat sink in the SG between 50 and 350 s is thought to
be one of the reasons for the lower calculated system pressure shown in
Figure 14.

These results indicate the need for improvements in the heat transfer
package of RELAP5/MOD1 and in the way that steam generators are modeled. During Transients 3 and 4, the heat transfer in the SG was controlled by secondary feed and bleed. Very good agreement between measured and calculated
variables was found after the experimental boundary conditions were taken into
account [15].

7. POSTEXPERIMENT ANALYSIS OF TRANSIENT 4

The main objective of Transient 4 was to examine a new procedure for PWR
plant recovery after rupture of a small number of steam generator tubes. The
new procedure makes use of the pressurizer spray to decrease the primary pressure to a value below the secondary pressure in order to minimize outflow from
the primary system to the secondary system. As mentioned in Section 5, the
spray-induced condensation caused a rapid decrease in pressure to the target
value, but a repressurization occurred after initiation of the high pressure
injection system. To understand the phenomena that occurred during the transient and to examine other recovery procedures using a thermal-hydraulic system code, the phenomena must be accurately calculated. The German thermal-
hydraulic code DRUFAN-02 was extensively used to examine the influence of the
following parameters on the results of the transient:

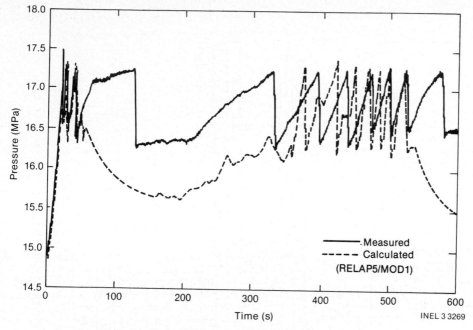

Fig. 14. System pressure during loss-of-offsite power (Transient 2) compared with RELAP5/MOD1 postexperiment calculations.

1. Temperature of the water in the spray line

2. Mass flow rate out of the spray line

3. Initial water level in the pressurizer

4. Mass flow rate out of the break

5. Cooldown rate of the primary fluid

6. Heat losses from the pressurizer vessel.

The DRUFAN-02 nodalization used for this study is shown in Figure 15. The results of a sensitivity study on the first two parameters is discussed in detail because these parameters were not specifically measured during the transient. Parameters 3, 4, and 5, which were measured during the transient however, do have a large influence on the pressure behavior during the initial 200 s and were incorporated in the sensitivity study. As an example, an increase in the pressurizer water level by 0.1 m causes a delay in the initiation of spray by about 40 s. Further, increasing the break flow by about 25% causes spray initiation about 30 s earlier. Also, the 20% increase in cooldown rate causes an earlier initiation of the spray by about 30 s. Neglecting the heat losses (Item 6 above) from the pressurizer vessel has a negligible influence on the pressure behavior.

236

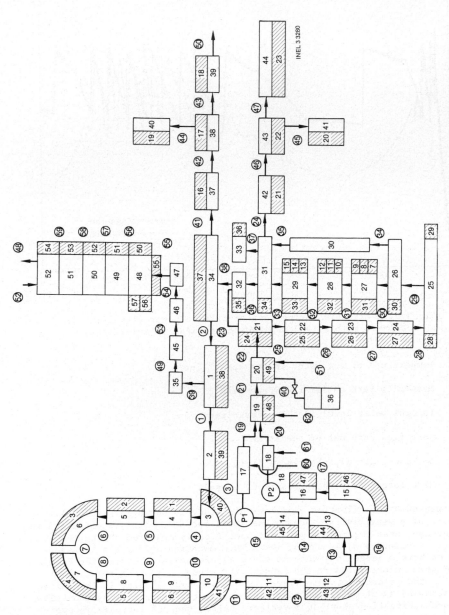

Fig. 15. DRUFAN-02 nodalization for LOFT steam generator tube rupture (Transient 4).

7.1 Influence of Spray Temperature

As mentioned in Section 2, the spray line is 15 m long and insulated with calcium silicate about 4 cm thick. The heat loss was calculated to be about 4 kW, which amounts to a cooldown of the spray water by about 1 K for a mass flow rate of 1 kg/s. The initial spray temperature was assumed to be 380 K for a period of 5 s, which is estimated to be the travel time through the spray line. The spray temperature was then allowed to increase rapidly to about 1 K lower than the cold leg temperature to account for heat loss to the ambient during the spray line travel time. These were the assumptions for the best-estimate calculations shown in Figure 16. The spray temperature during cold spray was kept constant at 380 K. During hot spray, the initial cold period (5 s) was neglected. The fourth curve in Figure 16 shows the influence of decreasing the spray temperature to about 10 K lower than the best-estimate value. Figure 17 shows the influence of these parameters on pressurizer liquid level.

7.2 Influence of Spray Mass Flow Rate

The mass flow rate used for the best-estimate calculations was that determined from flow characteristics of the spray valve. A sensitivity study was performed by changing this value by ±50%. The high mass flow rate value caused a decrease in pressure by about 0.1 MPa, as shown in Figure 18, and an increase in water level by about 4 cm, as shown in Figure 19. The low value of the mass flow rate had the same influence in the opposite direction, as shown in Figures 18 and 19.

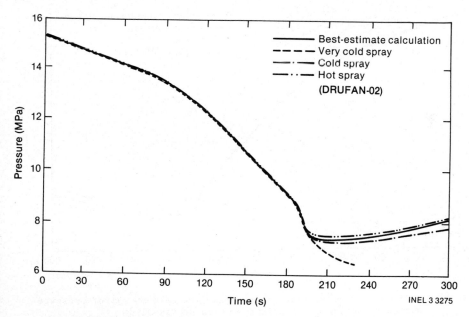

Fig. 16.   Influence of spray temperature on system pressure during Transient 4 (calculations with DRUFAN-02).

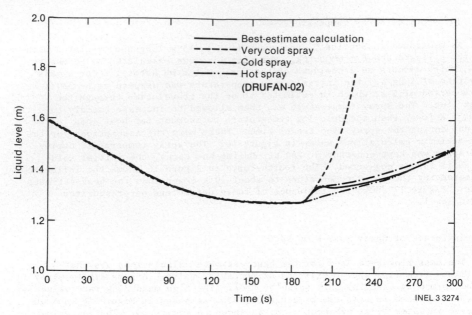

Fig. 17. Influence of spray temperature on pressurizer level during Transient 4 (calculations with DRUFAN-02).

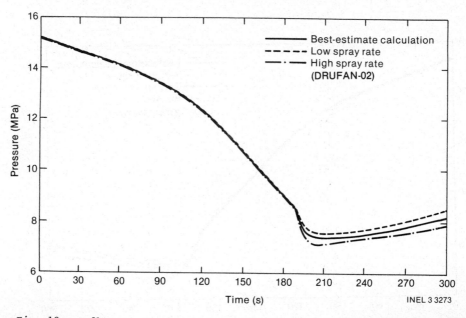

Fig. 18. Influence of pressurizer spray mass flow rate on system pressure during Transient 4 (calculations with DRUFAN-02).

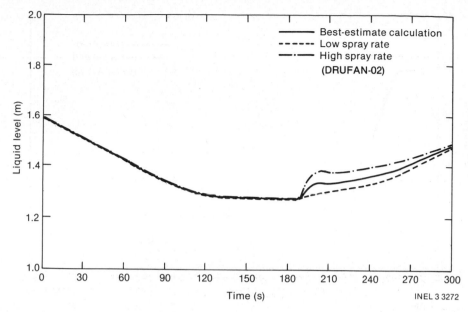

Fig. 19. Influence of pressurizer spray mass flow rate on pressurizer liquid
level during Transient 4 (calculations with DRUFAN-02).

7.3 Results of Postexperiment Calculations with DRUFAN-02

The break flow during Transient 4 was adjusted to preexperiment
calculated values to simulate a single tube rupture in the SG of a large PWR.
The measured break flow given in Reference [9] was used in the postexperiment
calculations presented in this section. The ±0.07-L/s uncertainty of the
measurements and a system leakage of 0.012 kg/s were taken into account. The
mass flow out of the pressurizer spray line was assumed to be 0.9 kg/s. The
water temperature out of the spray line was assumed to be 380 K for 5 s after
spray initiation, and was then allowed to heat up rapidly to about 1 K below
the cold leg temperature. These assumptions were in accordance with the heat
losses from the spray line to the environment. Under these assumptions, the
agreement between measured and calculated system pressure was very good, as
shown in Figure 20.

The calculated water level in the pressurizer agrees very well with the
measured value until 350 s, as shown in Figure 21. The difference between the
measured and calculated water level between 350 and 900 s (58 kg) can be
accounted for by a 0.1-kg/s larger break flow during this period.

8. CONCLUSIONS

Long-term transients in PWR systems, such as those discussed in this
paper, wherein the pressurizer and steam generators strongly influence the
dominant transient characteristics, have been shown to contain thermodynamic

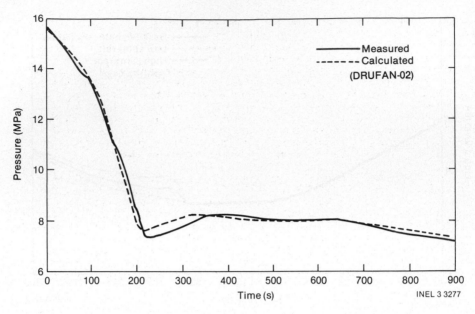

Fig. 20. System pressure during Transient 4 compared with DRUFAN-02 post-experiment calculations.

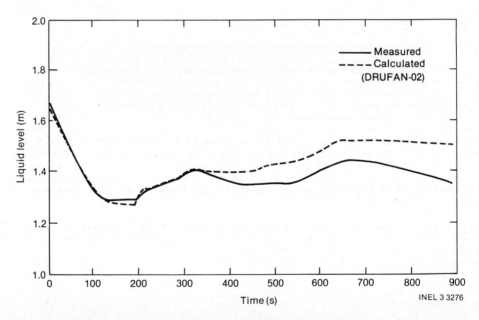

Fig. 21. Pressurizer liquid level during Transient 4 compared with DRUFAN-02 postexperiment calculations.

nonequilibrium phenomena associated with those components. The presence of these phenomena is observed in comparisons of measured transient characteristics with calculations from current-generation systems codes based on either equilibrium or nonequilibrium assumptions. The differences between measured and calculated characteristics of the LOFT transients are consistent for pressurizer inflow and outflow conditions and for steam generator rapid heat transfer degradation or varying heat transfer conditions. The magnitude of the differences is sufficiently large to warrant developing the ability of the systems codes to calculate the nonequilibrium phenomena. In addition, heat transfer modes and condensation phenomena, which may also be associated with thermodynamic nonequilibrium and which are not fully understood or calculable, are occurring in the pressurizer and steam generator secondary.

The RELAP5/MOD1 code, as currently developed, is not capable of correctly calculating the spray-induced condensation in the pressurizer. Also, other thermodynamic nonequilibrium phenomena, such as temperature stratification in the pressurizer, cannot be calculated. The limitations of RELAP5/MOD1 in calculating transients that contain nonequilibrium phenomena have been recognized, and development of a new version of RELAP5 to handle these phenomena is underway [16]. The DRUFAN-02 code, with a more sophisticated thermodynamic nonequilibrium model than in RELAP5/MOD1, is shown to be capable of correctly calculating such phenomena as spray-induced condensation in the pressurizer.

9. REFERENCES

1. Reeder, D. L., "LOFT System and Test Description," NUREG/CR-0247, TREE-1208, July 1978.

2. Adams, J. P., "Quick-Look Report on LOFT Nuclear Experiment L9-3," EGG-LOFT-5848, April 1982.

3. Bayless, P. D., and Divine, J. M., "Experiment Data Report for LOFT Anticipated Transient Without Scram Experiment L9-3," NUREG/CR-2717, EGG-2195, May 1982.

4. Sanchez-Pope, A. E., "Quick-Look Report on LOFT Nuclear Experiment L9-4," EGG-LOFT-6071, October 1982.

5. Batt, D. L., Divine, J. M., McKenna, K. J., "Experiment Data Report for LOFT Anticipated Transient Without Scram Experiment L9-4," NUREG/CR-2978, EGG-2227, November 1982.

6. Adams, J. P., "Quick-Look Report on LOFT Nuclear Experiment L6-7/L9-2," EGG-LOFT-5526, August 1981.

7. Stitt, B. D., and Divine, J. M., "Experiment Data Report for LOFT Anticipated Transient Experiment L6-7," NUREG/CR-2277, EGG-2121, September 1981.

8. Adams, J. P., "Quick-Look Report on LOFT Nuclear Experiment Series L6-8," EGG-LOFT-6031, September 1982.

9. Jarrell, D. B., Divine, J. M., McKenna, K. J., "Experiment Data Report for LOFT Anticipated Transient Experiment Series L6-8," NUREG/CR-2930, EGG-2219, November 1982.

10. Ransom, V. H., et al., "RELAP5/MOD1 Code Manual, Volume 1: System Models and Numerical Methods; Volume 2: Users Guide and Input Requirements," NUREG/CR-1826, EGG-2070, March 1982.

11. Wolfert, K., "Die Beruecksichtigung Thermo-Dynamischer Nichtgleichts-zustaende Bei Der Simulation Von Druckabsenkungsvorgaengen," Dissertation der Technische Universitaet Muenchen, 1979.

12. Steinhoff, F., "DRUFAN-02 Interim Program Description, Part 1," Gesellschaft fuer Reaktorsicherheit (GRS) mbH Forshungsgelande, 8046 Garching, West Germany, GRS-A-685, March 1982.

13. Koizumi, Y., Grush, W. H., Behling, S. R., "Steam Generator Behavior During LOFT Loss-of-Feedwater ATWS Experiment L9-3 and Loss-of-Offsite Power ATWS Experiment L9-4," paper to be presented at the ASME Winter Annual Meeting, Boston, MA, November 1983.

14. Griffith, P., "Pressurizer Modeling" Preliminary Report, August 1982, Department of Mechanical Engineering, MIT, Cambridge, MA 02139.

15. Condie, K. G., "RELAP5 Reference Calculation for LOFT Experiment L6-7/L9-2," EGG-LOFT-6014, January 1983.

16. Ransom, V. H., "RELAP5/MOD2: A Generic PWR System Transient Analysis Code," to be published.

Multi-Phase Flow and Heat Transfer III. Part B: Applications
edited by T.N. Veziroğlu and A.E. Bergles
Elsevier Science Publishers B.V., Amsterdam, 1984 — Printed in The Netherlands

A FIRST ORDER MODEL FOR NON-EQUILIBRIUM ECC PRESSURE SUPPRESSION

Robert W. Lyczkowski
Components Technology Division
Argonne National Laboratory
Argonne, Illinois 60439, U.S.A.

ABSTRACT

Closed form analytical solutions have been obtained which account for non-equilibrium pressure suppression during emergency core cooling (ECC) injection. The mathematical model uses the two-fluid theory of two-phase flow[1,2]. Many assumptions have to be made in order to obtain the analytical solutions; however, they appear to be reasonable since agreement with experimental data is possible. These assumptions can be lifted but then numerical methods would have to be employed to solve the equations. The objective of this paper is to identify the controlling mechanisms involved in the non-equilibrium pressure suppression process and to describe basic trends. If all the assumptions were to be lifted, the full two-fluid theory would result. In other words, the model can be made as sophisticated as one wants by merely making fewer assumptions.

It has been found that the heating of the jet is a first order effect relatively insensitive to assumptions and that the pressure suppression effect is second order and relatively sensitive to the assumptions made. The so-called pressure undershoot is predicted.

1. HIGH SPEED JET HEATING MODEL

Consider a high speed jet of cold water flowing through a relatively stagnant hot steam environment as shown in Fig. 1. The inlet jet flow rate, v_{zo} , is so fast that the average vapor velocity, v_z^g , gravity and all transverse velocities can be considered negligible. Steam supply is more than sufficient to make up for the amount condensed on the cold jet and dragged along by interphase frictional effects and so the system pressure is nearly constant. Since the liquid jet velocity is high, it will not break up but retain a nearly cylindrical shape, growing slightly in diameter as the steam condenses on its surface.

Under these assumptions, the governing field equation for the water column is[3]

$$r\rho_\ell C_\ell \left(\frac{\partial T_\ell}{\partial t} + v_z^\ell \frac{\partial T_\ell}{\partial z} \right) = k_\ell \frac{\partial}{\partial r} \left(r \frac{\partial T_\ell}{\partial r} \right). \tag{1}$$

The average bulk temperature of the jet is defined as[3]

244

$$\hat{T}(t,z) = \frac{\int_0^{2\pi} \int_0^{R_\ell(t)} T_\ell(t,z,r,\theta) \ r dr d\theta}{\int_0^{2\pi} \int_0^{R_\ell(t)} r dr d\theta} \tag{2}$$

Equation (1) is integrated over (r, θ), as

$$\int_0^{2\pi} \int_0^{R_\ell(t)} \rho_\ell C p_\ell \left(\frac{\partial T_\ell}{\partial t} + v_z^\ell \frac{\partial T_\ell}{\partial z} \right) r dr d\theta = k_\ell \int_0^{R_\ell(t)} \frac{\partial}{\partial r} \left(r \frac{\partial T}{\partial z} \right) dr \int_0^{2\pi} d\theta \ . \tag{3}$$

Assuming that the jet radius is constant and interchanging the order of differentiation and integration and use of the definition of the bulk temperature, Eq. (2) causes Eq. (3) to assume the form

$$\rho_\ell C p_\ell \left(\frac{\partial \hat{T}_\ell}{\partial T} + v_z^\ell \frac{\partial \hat{T}_\ell}{\partial z} \right) \left(2\pi \frac{R_{\ell o}^2}{2} \right) = k_\ell \left(r \frac{\partial T_\ell}{\partial r} \Big|_0^{R_{\ell o}} \right) 2\pi \ . \tag{4}$$

The term on the right hand side of Eq. (4) is the rate of heat added per unit length of the jet, q'_{cond}. If it is assumed that the condensation takes place in a thin region on the surface of the jet as shown in Fig. 1, then the rate of condensation can be given by the Nernst film approximation,

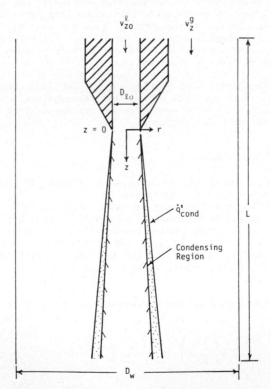

Figure 1.   High Speed Jet Heating Geometry

$$\dot{q}'_{cond} = h\left(T_g - \hat{T}_\ell\right) 2\pi R_{\ell o} \xi \ , \tag{5}$$

where ξ is a correction factor to account for non-cylindrical heat transfer area. This factor is difficult to measure directly and is usually lumped together with the "true" heat transfer coefficient h as

$$\dot{q}'_{cond} = h_{eff}\left(T_g - \hat{T}_\ell\right) \cdot 2\pi R_{\ell o} \ , \tag{6}$$

where h_{eff} is the effective overall heat transfer coefficient which may be much higher than h.

Equation (6) is used to simplify Eq. (4) since the heat flux per unit length entering the surface of the jet at r=R is given by

$$-k_\ell \left(r \frac{\partial T_\ell}{\partial r}\bigg|_{R_{\ell o}}\right) \cdot 2\pi = h_{eff}\left(T_g - \hat{T}_\ell\right) \cdot 2\pi R_{\ell o} \ . \tag{7}$$

We now assume that the gas temperature is at saturation at the average system pressure $\overline{P}$ as

$$T_g = T_{sat} \ (\overline{P}) = \overline{T}_{sat} \tag{8}$$

and that the jet is symmetric. Combination of Eqs. (4), (7), and (8) produces

$$\frac{\partial \hat{T}_\ell}{\partial t} + v_z^\ell \frac{\partial \hat{T}_\ell}{\partial z} + \frac{2h_{eff}}{\rho_\ell C_\ell R_{\ell o}}\left(\hat{T}_\ell - \overline{T}_{sat}\right) = 0 \ . \tag{9}$$

Equation (9) may easily be integrated if ρ_ℓ, C_ℓ, v_z, h_{eff} and $\overline{T}_{sat}$ are assumed to be constant. Introduce the auxiliary equation

$$d\hat{T}_\ell = \frac{\partial \hat{T}\ell}{\partial T} dt + \frac{\partial \hat{T}\ell}{\partial z} dz \tag{10}$$

to reduce Eq. (9) to a system of ordinary differential equations. This system is give by

$$\frac{dt}{1} = \frac{dz}{v_z^\ell} = \frac{d\hat{T}_\ell}{\dfrac{2h_{eff}}{\rho_\ell C_\ell R_{\ell o}}\left(\hat{T}_\ell - \overline{T}_{sat}\right)} \tag{11}$$

which is equivalent to

$$\frac{d\hat{T}_\ell}{dt} + \frac{2h_{eff}}{\rho_\ell C_\ell R_{\ell o}}\left(\hat{T}_\ell - \hat{T}_{sat}\right) = 0 \tag{12}$$

along the characteristic direction

$$\frac{dz}{dt} = v_z^\ell. \tag{13}$$

246

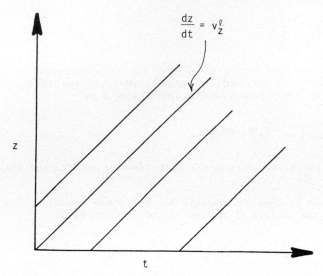

$$\frac{dz}{dt} = v_z^\ell$$

z

t

Figure 2.   Integration Path for Transient Jet Heating Model

The ordinary differential equation given by Eq. (12) or equivalently

$$v_z^\ell \frac{d\hat{T}}{dz} + \frac{2h_{eff}}{\rho_\ell C_\ell R_{\ell o}} \left(\hat{T}_\ell - \overline{T}_{sat} \right) \tag{14}$$

may be integrated along the flow path as shown in Fig. 2.   It is clear that the group multiplying $\hat{T}_\ell - \overline{T}_{sat}$ in Eq. (12) can be thought of as an inverse time constant for the liquid jet heating, $\tau_{T_\ell}^{-1}$ as

$$\tau_{T_\ell}^{-1} = \frac{2h_{eff}}{\rho_\ell C_\ell R_{\ell o}} . \tag{15}$$

The solution of Eqs. (12) or (14) and (13) with

$$\hat{T}_\ell = \hat{T}_{\ell o} \quad (z=0, \ t>0), \ (t=0, \ z>0) \tag{16}$$

is

$$\frac{\hat{T}_\ell - \overline{T}_{sat}}{\hat{T}_{\ell o} - \overline{T}_{sat}} = \exp \left(-t/\tau_{T_\ell} \right) \tag{17}$$

where

$$t = z/v_z^\ell . \tag{18}$$

It is clear then that t plays the role of a "residence" time equal to the average volume of the jet divided by the average volumetric flow rate as

$$t = \frac{V_\ell}{Q_\ell} = \frac{A_\ell \rho_\ell z}{A_\ell \rho_\ell v_z^\ell} = \frac{z}{v_z^\ell} . \tag{19}$$

Therefore, we can treat the process as "quasi-steady" with the understanding that we move along with the jet at velocity v_z^ℓ .

An estimate of the time constant τ may be made for ECC using a value of $\overline{h}_{eff}$ = 100 BTU/(s-ft^2 °F) which is typical for condensing heat transfer coefficients on fast jets and in steam-jet ejectors[4,5,6,7] as

$$\tau_{T_\ell} = \frac{(60)(1)(1/2)}{(2)(100)} = 6.67 \text{ s.} \qquad (20)$$

The jet heating model was compared with the steady-state data of Zinger[4] at two jet velocities, 14 and 25 m/s (45.9 and 82 ft/s). The physical parameters of the experiment were:

$$\overline{P} = 1.7 - 2.0 \text{ atm } (24.97 - 29.38 \text{ psia})$$
$$2R = D = 15 \text{ mm } (4.92 \times 10^{-2} \text{ft})$$
$$\hat{T}_{\ell o} = 20°C \ (5=68°F)$$
$$v_z = 14 - 25 \text{ m/s } (45.9 - 82.0 \text{ ft/s})$$

There is some confusion in the paper as to the pressure. At one point it was given as 1.2 - 1.4 atm and in another as 1.7 - 20 atm. We chose the latter pressure range.

The average value of $\overline{T}_{sat}$ was chosen to be 250°F at 30 psia. The time constant τ_{T_ℓ} was calculated from the analysis of Levy and Brown who obtained good agreement between his analysis and his steams injector data using a value of h_{eff} of 80-100 BTU/(s-ft^2°F) with an injection velocity of 117 ft/s (35.7 m/s).[7] As can be seen in Fig. 3 reproduced from Zinger's paper this value is in the range of his high speed jet data. It also appears to be a good length average for the 25 m/s curve as shown in Fig. 4. The time constant τ_{T_ℓ} calculated for this experiment is 7.375x10^{-3} s or 7.375 ms. The temperature of the jet then becomes

$$T_\ell(z) = 250 - 182 \ \exp\left[\left(-136 \ z/v_{zo}^\ell\right)\right] \qquad (21)$$

where v_{zo}^ℓ = 82 ft/s. Equation (21) has been plotted in Fig. 5 where it is compared to the data. As can be seen, good agreement has been achieved with the basic trend established. The mysterious 50 mm off-set of the temperature profile was not explained in the paper. Coincidentally, the diameter of the water pipe is also 50 mm. It also appears that the established temperature is roughly 221°F which would correspond to about 20 psia (1.4 atm.). On the other hand, this pressure may be the equilibrium value produced by the cold jet pressure suppression. Considering all the vagaries of the experiment, this agreement is remarkable, proving that the basic mechanism has been identified to be interphase heat transfer.

The 14 m/s (45.9 ft/s) run was computed by calculating a value of τ_{T_ℓ} which would force the model to agree with the data of 600 mm (1.968 ft)

248

in Fig. 5. The value of τ_{T_ℓ} computed in this manner is 3.6374×10^{-2} s which would correspond to a value of h_{eff} of approximately 20 BTU/$(s-ft^2-°F)$ which is in remarkable agreement with h_{eff} from Fig. 3. The computations are plotted in Fig. 5 and compared to the data. Once again, the trend has been established and agreement with data is good.

The model can be made more sophisticated by allowing h_{eff} to be a function of $\hat{T}_\ell - \bar{T}_{sat}$ as well as v_z. Correlations of this type are obtainable both experimentally and theoretically.[4] Two such models are plotted in Fig. 3. The curve marked "1" appears to be empirical and is given by

$$h_{eff} = 5.608 \; v_z^\ell \; \Phi \; (R), \; BTU/ \left(s-ft^2- °F \right) \tag{22}$$

where

$$R = C_\ell \left(T_{sat} - \hat{T}_\ell\right)/\left(h_{gs} - h_{\ell s}\right), \; \text{dimensionless} \tag{23}$$

and Φ is given in tabular form in reference[4].

It appears this correlation overestimates the data. The curve given by "2" in Fig. 3 is theoretical and very complicated. It appears to agree with the data at very low flow rates.

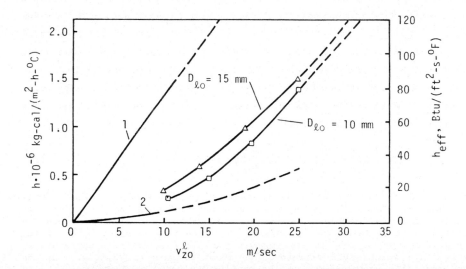

Figure 3. Variation of the Coefficient of Heat Transfer with the Velocity of Flow According to Experimental and Calculated Data (L = 800 mm). (4) 1 – from the Formula of G. N. Abramovich and A. P. Proskuryakov, (4); 2 – from the Formula of S. S. Kutateladze and V. M. Borishansky. (4)

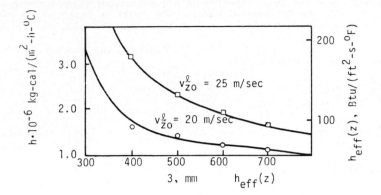

Figure 4. Variation of h, the Mean Coefficient of Heat Transfer, with z, the Height of the Jet, According to Experimental and Calculated Data ($D_{\rho o}$ = 15 mm); a) v_{zo}^{ℓ} = 20 m/ sec; b) v_{zo}^{ℓ} = 25 m/sec. (4)

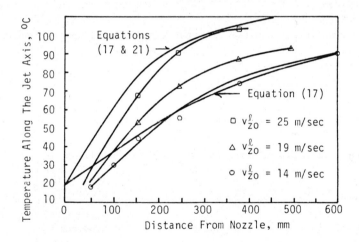

Figure 5. Change of Temperature Along the Axis of the Jet ($D_{\rho o}$ = 15 mm; L = 800 mm; p = 1.2-1.4 Atmosphere Absolute) (4) and Comparisons with Analysis.

2. PRESSURE RELAXATION MODEL

Having been able to explain the first order behavior of the high speed jet heating in a nearly constant pressure field, the next step is to identify the mechanisms governing the depressurization using steam jet ejector pressure data. These devices have been analyzed by a number of investigators[5, 6, 7].

We will treat the steam injector as a quasi-steady-state process similar to that done for the jet heating model. The model is based on the two-fluid theory and so six field equations need to be accounted for[1, 2].

The first field equation needed to describe the process in the vapor continuity equation is given by

$$\frac{dW_z^g}{dz} = -\dot{m}'_{cond} \tag{24}$$

where W_z^g is the vapor mass flow rate and $-\dot{m}'_{cond}$ is the vapor condensation rate per unit length.

The second field equation is the liquid continuity equation given by

$$\frac{dv_z^\ell}{dz} \simeq 0 \; . \tag{25}$$

Equation (25) implies that the fractional change of liquid flow is negligible with respect to the fractional change of the vapor flow. This assumption was made implicitly in order to develop the jet heating model. Equation (25) can be integrated to obtain

$$v_z^\ell = v_{zo}^\ell = \text{inlet liquid velocity} \; . \tag{26}$$

The third field equation is that of the liquid energy equation and is given by

$$\frac{d}{dz}\left(W_z^\ell h_\ell\right) - h_{\ell s}\dot{m}'_{cond} = \dot{q}'_{cond} \tag{27}$$

where kinetic energy, frictional heating, potential and other effects have been neglected.

The constitutive relation for $\dot{q}'_{cond}$ is given by[1, 5, 6, 7]

$$\dot{q}'_{cond} = \left(h_{gs} - h_{\ell s}\right)\dot{m}'_{cond}. \tag{28}$$

Expansion of Eq. (27) and combination of Eq. (28) produces

$$W_z^\ell \frac{dh_\ell}{dz} + h_\ell \frac{dW_z^\ell}{dz} + h_{\ell s}\dot{m}'_{cond} = \left(h_{gs} - h_{\ell s}\right)\dot{m}'_{cond} \; . \tag{29}$$

Although we set $dv_z^\ell/dz \simeq 0$ for purposes of the liquid velocity calculation, the relation

$$\frac{dW_z^\ell}{dz} = \dot{m}'_{cond} \tag{30}$$

still remains true. Combination of Eqs. (30) and (29) and simplification produces

$$W_z^{\ell} \frac{dh_{\ell}}{dz} + \left(h_{\ell} - h_{\ell s} \right) \dot{m}'_{cond} = \dot{m}'_{cond} \left(h_{gs} - h_{\ell s} \right). \tag{31}$$

Since $\left(h_{\ell} - h_{\ell s} \right) \ll \left(h_{gs} - h_{\ell s} \right)$, Eq. (30) becomes $\tag{32}$

$$W_z^{\ell} \frac{dh_{\ell}}{dz} = \dot{m}'_{cond} \left(h_{gs} - h_{\ell s} \right). \tag{33}$$

Equation (33) was solved in Section 2 with constant gas temperature $T_g = \overline{T}_{sat}$. This can be seen when the following identities are made:

$$W_z^{\ell} = \rho_{\ell} A_{\ell} v_z^{\ell} = \rho_{\ell} \pi R_{\ell o}^2 v_z^{\ell} , \tag{34}$$

$$\frac{dh_{\ell}}{dz} = \frac{dh_{\ell}}{dT} \frac{d\hat{T}_{\ell}}{dz} = C_{\ell} \frac{d\hat{T}_{\ell}}{dz} \tag{35}$$

and

$$\dot{q}'_{cond} = h_{eff} \left(T_g - \hat{T}_{\ell} \right) \cdot 2\pi R_{\ell o} = \left(h_{gs} - h_{\ell s} \right) m'_{cond} \tag{36}$$

Substitution of Eqs. (34)-(36) into Eq. (33) produces Eq. (14). Equation (36) can be used to obtain the condensation rate as

$$\dot{m}'_{cond} = \frac{\pi D_{\ell o} h_{eff} \left(T_g - \hat{T}_{\ell} \right)}{h_{gs} - h_{\ell s}}. \tag{37}$$

Since we have assumed that the liquid velocity is constant, the fourth field equation given by the liquid momentum equation is not required.

The fifth field equation is given by the vapor momentum equation as

$$\frac{d}{dz} \left(W^g v_z^g \right) + \hat{v} \dot{m}'_{cond} = -A_g \frac{dP}{dz} \tag{38}$$

where $\hat{v}$ is the velocity at which condensate is transferred to the liquid. Interphase friction has been assumed zero. Assuming that the vapor momentum flux is negligible with respect to the other terms in Eq. (38) produces:

$$A_g \frac{dP}{dz} = -\hat{v} \dot{m}'_{cond} \tag{39}$$

Assuming, as did Levy and Brown[7] that

$$\hat{v} = v_{zo}^{\ell} \tag{40}$$

causes Eq. (39) to assume the form

$$A_g \frac{dP}{dz} = -v_{zo}^\ell \dot{m}'_{cond}. \tag{41}$$

Combination of Eq. (41) with Eq. (37) produces

$$\frac{dP}{dz} = - \frac{v_{zo}^\ell \, h_{eff} \, \pi D_{\ell o} \left(T_g - \hat{T}_\ell \right)}{A_g \left(h_{gs} - h_{\ell s} \right)}. \tag{42}$$

The assumptions used to arrive at Eq. (42) seem reasonable for the rapid jet heating model, but are suspect for the steam jet ejector having appreciable vapor velocity. This was confirmed during the course of the steam jet ejector analysis. The major difference between the Zinger high speed jet heating experiment and the steam-ejector experiments in that the steam has an appreciable velocity in the latter experiments. A more complete model for the vapor stagnation pressure is given by Eq. (12) in the Levy and Brown paper[7].

The final (sixth) field equation is given by

$$T_g = T_g(P) = T_{sat} \tag{43}$$

which may be linearized about the equilibrium pressure, P_{eq}, as

$$T_{sat} = T_{sat_{eq}} + \frac{dT_g}{dP} \Big|_{eq} \left(P - P_{eq} \right). \tag{44}$$

Combination of Eqs. (43) and (45) results in

$$A_g \frac{dP}{dz} + \frac{v_{zo}^\ell \, \pi D_{\ell o} \, h_{eff}}{\left(h_{gs} - h_{\ell s} \right)} \frac{dT_g}{dP} \Big|_{eq} \left(P - P_{eq} \right) = - \frac{v_{zo}^\ell \, \pi D_{\ell o} \, h_{eff}}{\left(h_{gs} - h_{\ell s} \right)} \left(T_{sat_{eq}} - \hat{T}_\ell \right). \tag{45}$$

The jet heating model from the previous section is

$$\hat{T}_\ell = \overline{T}_{sat} + \left(T_{\ell o} - \overline{T}_{sat} \right) \exp \left[\left(- \frac{z}{v_{zo}^\ell \tau_{T_\ell}} \right) \right] \tag{46}$$

where $\overline{T}_{sat}$ was held constant. Define the inverse pressure time constant as

$$\tau_P^{-1} = \frac{\left(v_{zo}^\ell \right)^2 \, \pi D_{\ell o} \, h_{eff}}{g_c \left(h_{gs} - h_{\ell s} \right) A_g} \left(\frac{dT}{dP} \Big|_{eq} \right). \tag{47}$$

Combination of Eqs. (45)–(47) assuming $\overline{T}_{sat} = T_{sat_{eq}}$ produces

$$\frac{dP}{dz} + \frac{1}{\tau_P v_{zo}^\ell} \left(P - P_{eq} \right) = \left(\frac{\frac{dP}{dT} \Big|_{eq}}{\tau_P v_{zo}^\ell} \right) \left(T_{\ell o} - T_{sat_{eq}} \right) \exp \left(- \frac{z}{v_{zo}^\ell \tau_{T_\ell}} \right). \tag{48}$$

Equation (48) may be integrated with the initial condition

$$P(z=0) = P_o \tag{49}$$

as

$$P = P_{eq} + \left(P_o - P_{eq}\right) \exp\left(-\frac{z}{v_{zo}^{\ell} \tau_P}\right) - \left[\frac{T_{sat_{eq}} - T_{\ell o}}{\frac{dT}{dP}\bigg|_{eq}\left(1 - \frac{\tau_P}{\tau_{T_\ell}}\right)}\right]$$

$$\left[\exp\left(-\frac{z}{v_{zo}^{\ell} \tau_{T_\ell}}\right) - \exp\left(-\frac{z}{v_{zo}^{\ell} \tau_P}\right)\right]. \tag{50}$$

Equation (50) will be referred to as "Model I."

The values of τ_{T_ℓ} and τ_P were estimated for the $v_{oz}^{\ell} = 117$ ft/s liquid injection rate run of Levy and Brown[7] using $h_{eff} = 100$ BTU/(s-ft^2°F) and a constant diameter jet. The other parameters are

$$D_{zo}^{\ell} = 3.75 \times 10^{-2} \text{ft}$$

$$D_w = 0.112 \text{ ft}$$

$$A_g = \pi\left(D_w^2 - D_{zo}^{\ell}\right)^2 = 3.5 \times 10^{-2} \text{ ft}^2$$

$$\frac{dT}{dP} = 3\text{°F/psi at approximately 20 psia}$$

The two time constants came out to be

$$\tau_P = 2.23 \times 10^{-3} \text{ s}$$

$$\tau_{T_\ell} = 5.625 \times 10^{-3} \text{ s}$$

The run chosen for analysis is shown in Fig. 6 as the bottom-most curve labelled "A". Estimation of the remainder of the necessary input to Eq. (50) are

$$P_o = 12 \text{ psia}$$

$$P_{eq} = 1.5 \text{ psia}$$

$$T_{\ell o} = 40\text{°F}$$

$$T_{sat_{eq}} = 101\text{°F}$$

Substitution of all the parameters into Model I, Eq. (50) produces

$$P = 1.5 + 10.5\, e^{-3.825z} - 33.7\left(e^{-1.5192} - e^{-3.825z}\right)$$

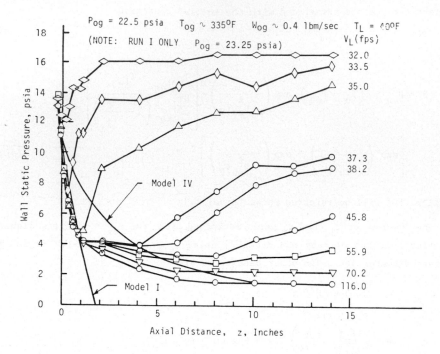

Figure 6. Effect of Inlet Liquid Velocity on Axial Pressure Distribution (BPVO) and Comparisons with Analysis, Adapted from Reference 7.

where z is in feet. The results are plotted in Fig. 6. It is clear that this simple Model I is adequate to analyze steam injector only for very short distances since the computed pressure becomes negative. The linearization assumption, Eq. (44), probably breaks down. It might be adequate for the high speed jet and ECC, however. The very interesting aspect of the result of Model I is that it predicts an "undershoot" due to non-equilibrium heating as shown in Fig. 7. This undershoot results in momentarily superheated liquid at the local pressure. It is obvious that other mechanisms come into play since the data at this flow rate does not have the undershoot. However, there are undershoots at lower liquid flow rates as shown in Fig. 6.

A second model, called Model II set the time constants equal

$\tau_{T_\ell} = \tau_P = \tau$ in Eq. (50). The result is

$$ P = P_{eq} + \left[\left(P_o - P_{eq} \right) - \frac{\left(T_{sat_{eq}} - T_{\ell o} \right)}{\left. \frac{dT}{dP} \right|_{eq}} \left(\frac{z}{v_{zo}^\ell \tau} \right) \right] exp \left[- \left(\frac{z}{v_{zo}^\ell \tau} \right) \right]. \quad (52) $$

The results for $\tau = \tau_{T_\ell}$, and $\tau = \tau_P$ are plotted in Fig. 7. When $\tau = \tau_{T_\ell}$, the results are above Model I but still produces a negative pressure undershoot although not as low. The recovery from the undershoot to

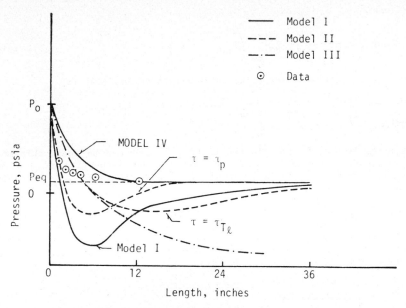

Figure 7. Comparison of Various Models with Data[7] v_{zo}^{ℓ} = 117ft/s

equilibrium takes longer because the time constant is higher. When $\tau = \tau_p$, the pressure again lies above Model I and the undershoot is just slightly less than that produced by using $\tau = \tau_{T_\ell}$. The undershoot recovery to equilibrium is faster than Model I.

The third model, called Model III, was obtained by using an average constant temperature difference between the liquid and vapor in Eq. (45) as

$$\frac{dP}{dz} + \frac{1}{\tau_p v_{zo}^{\ell}} \left(P - P_{eq} \right) = \frac{\left.\frac{dP}{dT}\right|_{eq} \left(\widehat{T_\ell} - T_{sat_{eq}} \right)}{\tau_p v_{zo}^{\ell}} \ . \tag{53}$$

The average temperature difference is obtained by integrating Eq. (46) between 0 and L as

$$\overline{\widehat{T_\ell} - T_{sat_{eq}}} = \int_0^L \left(T_o - \overline{T}_{sat_{eq}} \right) \exp \left(-\frac{z}{v_{zo}^{\ell} \tau_{T\ell}} \right) dz \ / \int_0^L dz$$

to obtain

$$\overline{T_\ell - T_{sat_{eq}}} = \left(T_o - \overline{T}_{sat_{eq}} \right) \left(\frac{v_{zo} \tau_{T\ell}}{L} \right) \left[1 - \exp \left(-\frac{L}{v_{zo} \tau_{T\ell}} \right) \right]. \tag{54}$$

Equation (53) is integrated with $P(z=0)=P_o$ to obtain

$$P = P_{eq} + \left[P_o - P_{eq} + \left.\frac{dP}{dT}\right|_{eq} \left(\overline{T_{sat_{eq}} - \widehat{T}_\ell} \right) \right] \qquad X$$

$$\exp\left(-\frac{z}{v_{zo}^{\ell}\tau_P}\right) - \frac{dP}{dT}\bigg|_{eq}\left(T_{sat_{eq}} - \hat{T}_{\ell}\right) . \tag{55}$$

The average temperature differencing using L=1 ft in Eq. (54) is

$$\overline{T_{sat_{eq}} - \hat{T}_{\ell}} = 31.4°F \tag{56}$$

The results of Model III are plotted in Fig. 7. As can be seen, this model reaches a steady state well below the data. The assumption of a finite average temperature difference is good for short distances but not longer distances.

Model IV is Model III with the average temperature difference set to zero. The results are plotted in Fig. 7. The undershoot is eliminated and the equilibrium pressure is approached from above at about the right time but the pressure is above the data. A switch from Model I to Model IV at one inch would produce good agreement with the data.

3. EFFECT OF NONCONDENSABLES

It is well known that the effect of noncondensable gases considerably lowers condensation rates (reference [8], for example). This considerable lowering would serve to increase the estimated time constants, significantly decreasing the jet heating and pressure suppression processes.

4. CONCLUSIONS

As more data is found, the conclusions may change. However, based on the analysis of two relatively simple experiments, the following conclusions have been deduced.

The description of the jet heating is a first order effect which is relatively independent of the pressure field. The rate of the jet heating is primarily a function of the interphase heat transfer coefficient which is a function of the relative velocity and the difference between phase temperatures. An average overall h_{eff} which is only a function of jet inlet flow rate is adequate to establish the correct trends.

The pressure suppression model is second order in that this phenomenon is a function of the jet heating model. The description of this process is more sensitive to assumptions and sub-processes. It appears that for rapid jet injection, the phase temperatures equilibrate very rapidly and so the heat transfer drops to zero in a very short distance. A combination of Models I or II and IV would yield the initial rapid pressure decrease and the secondary slower approach to equilibrium. However, as the steam jet injector data indicates at lower liquid flow rates, the possibility does exist for pressure undershoots because of suppression. The pressure falls quickly, causing a superheated liquid situation. Because the vapor is hotter than the liquid, the jet continues to heat up. As the jet temperature approaches saturation, the pressure recovers to the final equilibrium value.

The overall pressure equilibration time can be estimated a priori quite accurately using Model IV, but the computed pressure will be too high for short times.

NOMENCLATURE

A_g = Area of vapor, ft^2

A_ℓ = Area of liquid, ft^2

C_ℓ = Liquid heat capacity, $BTU/(lb_m°F)$

$D_{\ell o}$ = Inlet jet diameter, ft

D_w = Injector diameter, ft

g_c = Conversion factor $(ft-lb_m)/(lb_f - s^2)$

h = "True" heat transfer coefficient, $BTU/(s\ ft^2-°F)$

h_{eff} = Effective heat transfer coefficient, $BTU/(s-ft^2-°F)$

h_{gs} = Saturation vapor enthalpy, BTU/lb_m

h_ℓ = Liquid enthalpy, BTU/lb_m

$h_{\ell s}$ = Saturation liquid enthalpy, BTU/lb_m

k_ℓ = Liquid thermal conductivity, $BTU/(s-ft-°F)$

L = Liquid jet length, ft

$\dot{m}'_{cond}$ = Condensation rate on the jet per unit length, $lb_m/(s-ft)$

P = Pressure, psi

P_{eq} = Equilibrium pressure, psi

P_o = Inlet pressure, psi

$\dot{q}'_{cond}$ = Heat transfer rate on the jet per unit length, $BTU/(s-ft)$

Q_ℓ = Volumetric flow of jet, ft^3/s

r = radial coordinate, ft

$R_{\ell o}$ = Inlet jet radius, ft

t = time

T_ρ = Local liquid temperatures, °F

$\hat{T}_\ell$ = Bulk liquid temperature defined by Eq. (2)

$T_{\ell o}$ = Inlet liquid temperature, °F

T_{sat} = Saturation temperature, °F

$\overline{T}_{sat}$ = Average saturation temperature, °F

$T_{sat_{eq}}$ = Equilibrium saturation temperature, °F

V_ℓ	=	Volume of jet, ft^3
$\hat{v}$	=	Velocity associated with phase change, ft/s
v_{zo}^ℓ	=	Liquid velocity, ft/s
W_z^g	=	Vapor mass flow, lb/s
W_z^ℓ	=	Liquid mass flow, lb/s
z	=	Axial coordinate direction, ft
Θ	=	Azimuthal coordinate direction, radians
ξ	=	True to nominal area factor for heat transfer, dimensionless
ρ_ℓ	=	Liquid density, lb_m/ft^3
τ_p	=	Relaxation time for pressure defined by Eq. (47)
τ_{T_ℓ}	=	Relaxation time for jet temperature defined by Eq. (15)

REFERENCES

1. R. W. Lyczkowski, Dimitri Gidaspow, and C. W. Solbrig, "Multiphase Flow Models for Nuclear Fossil and Biomass Energy Conversion," in Advances in Transport Processes, pp. 198-351, A. S. Mujumder and R. A. Mashelkar, Eds., Wiley Eastern Ltd., New Delhi, 1982.

2. M. Ishii, Thermo-Fluid Dynamic Theory of Two-Phase Flow, Eyrolles Paris, 1975.

3. R. B. Bird, W. E. Stewart, and E. N. Lightfoot, Transport Phenomena, third printing, John Wiley, New York, 1960.

4. N. M. Zinger, "Heating a Jet of Water in a Vapor-Filled Space," in Problems of Heat Transfer During a Change of State, S. S. Kutateladze, ed., pp 75-85, AEC-tr-3405, 1953.

5. M. A. Grolmes, "Steam-Water Condensing-Injector Performance Analysis with Supersonic Inlet Vapor and Convergent Condensing Section," Argonne National Laboratory, ANL-7443, May 1968.

6. J. H. Linehan and M. A. Grolmes, "Condensation of a High Velocity Vapor on a Subcooled Liquid Jet in Stratified Flow," in Heat Transfer 1970, paper Cs 2.6, Vol. VI, Elsevier Pub. Co., 1970.

7. E. K. Levy and G. A. Brown, "Liquid-Vapor Interactions in a Constant-Area Condensing Ejector," J. Basic Eng., pp 169-180, March 1972.

8. H. K. Al-Diwany and J. W. Rose, "Free Convection Film Condensation of Steam in the Presence of Non-Condensing Gases," Int. J. Heat Mass Transfer, pp 1359-1369, 1973.

Multi-Phase Flow and Heat Transfer III. Part B: Applications   259
edited by T.N. Veziroğlu and A.E. Bergles
Elsevier Science Publishers B.V., Amsterdam, 1984 — Printed in The Netherlands

THERMAL-HYDRAULICS OF THE PRECURSORY COOLING DURING BOTTOM
REFLOODING

Mario De Salve and Bruno Panella
Dipartimento di Energetica
Politecnico di Torino
10129 Torino,ITALY

ABSTRACT

The thermal-hydraulics of the precursory cooling phase during the bottom
flooding of a channel is analysed. In the inverted annular flow regime a void
fraction model and a method to estimate the thermal non-equilibrium for both
liquid and vapor phases are presented and a criterion for the stability of
the liquid core is derived;some heat transfer correlations are discussed.
In the dispersed flow film boiling regime fluiddynamics and heat transfer are
analysed and some models,which take into account the thermal non-equilibrium,
are tested for the typical bottom flooding conditions. The precursory cooling
wall temperatures,predicted by an overall calculation code,are compared with
a single tube bottom flooding experimental data.

1.   INTRODUCTION

The thermalhydraulics ahead of the quench front during the emergency cooling by bottom reflooding is very important for the wall temperature history,
so it is essential to study the precursory cooling region flow patterns and
heat transfer. The flow patterns which may occur downstream of the quench front
are the inverted annular flow (at low void fraction,i.e. less than 40%),the
dispersed flow (at high void fraction,i.e. higher than 80%) and the single
phase (vapor only) flow if the channel is long enough. The inverted annular
flow (which may not occur at low flow rate) eventually develops into dispersed flow regime through a transition region with few large drops within a continuous vapor medium. The heat transfer regimes in the precursory cooling region are film boiling regimes,that is inverted annular film boiling and liquid
deficient regimes,characterized by a strong thermal non-equilibrium. Many a-
nalytical studies have been done in recent years on the thermal-hydraulics
of bottom reflooding [1,2,3,4] :models of the inverted annular flow have been
more recently presented by Elias and Chambré [5] and by Fung and Groeneveld[6];
several authors have studied the dispersed film boiling regime [7,8,9,10,11].
The contribution of the present paper is an analysis of the thermal-hydraulics during the rewetting of a hot tube by bottom flooding: a theoretical
void fraction-flow quality correlation is presented for the inverted annular
flow regime and a criterion for the stability of the liquid core is derived

from the conservation equations;thermal non—equilibrium for both liquid and vapor phases is estimated;some film boiling heat transfer correlations are a-nalysed and tested. In the dispersed film boiling region some models,taking into account thermal non-equilibrium,are compared and the droplets evolution is evaluated by fragmentation criteria and evaporation rate. Several heat tra-nsfer correlations in the inverted annular film boiling and in the dispersed film boiling regimes are discussed. The models are adopted in an overall cal-culation code which predicts the wall temperature history during the bottom flooding of tubular channels,in order to compare the precursory cooling tempe-ratures predicted with single tube bottom flooding experimental results [12].

2. INVERTED ANNULAR FLOW HYDRODYNAMICS

In the inverted annular flow (IAF) pattern a low viscosity vapor layer separates a continuous liquid core,which may contain some entrained bubbles, from the heater surface. The heat transfer mechanism (IAFB) is stable film bo-iling and the heat is transferred from the wall to the vapor and subsequently from the vapor to the liquid core by forced convection evaporation;there is moreover some radiation heat from the wall to the liquid. IAF occurs in bottom reflooding downstream of the quench front (QF) position if the flow rate and the void fraction upstream of the QF don't allow the occurrence of the annular flow pattern.   The transition from bubble—churn flow to annular flow may be evaluated by the Wallis criterion[13]. In the IAF a double thermal non—equi-librium may occur as the liquid core may be still subcooled and the vapor la-yer is superheated,as the vapor temperature increases with the distance from the QF.

2.1. IAF void fraction

Yadigaroglu and Arrieta suggest to use the drift—flux model with V_{gj} =0 and C_o =3 ,which yielded reasonable good agreement with the data examined in their paper[14] . Elias and Chambré [5] analysis is restricted to subcooled re-gion assuming a constant vapor film thickness(that is a constant void fraction) which is correlated as a function of the fluid equilibrium quality at the QF po-sition,by using the heat transfer experimental data of Fung et al.[15] :

$$\alpha = 0.0521 \exp (20.327 \ x_e) \ ; \tag{1}$$

when the liquid surface starts to boil it is easily entrained by the vapor and the heat transfer mechanism can be characterized as dispersed flow film boi-ling (DFFB) rather than IAFB. Void fractions predicted by the relation (1) are strongly underpredicted even for positive values of x_e . Fung and Groene-veld [6] present a two—fluid model by considering the developments of the ther-mal boundary layers both in the liquid core and in the vapor film; several pa-rameters are left indeterminate and they are adjusted to fit their hot patch tecnique data. The predicted void fractions are usually smaller than the mea-sured values:according to ref.[6] this can be attributed to the entrainment of the vapor in the liquid core.

In the present paper the cross sectional averaged void fraction as a fun-ction of the flow quality in the IAF pattern is derived by the mass conservation equations for the liquid and the vapor, which can be equated:

$$- \frac{\partial}{\partial t} [\rho_\ell (1 - \alpha)] - \frac{\partial}{\partial z} [\rho_\ell u_\ell (1 - \alpha)] = \frac{\partial}{\partial t} (\rho_v \alpha) + \frac{\partial}{\partial z} (\rho_v u_v \alpha) \quad (2)$$

The following assumptions are made in order to obtain a simplified form of the continuity equations:

1. the flow is steady state;
2. the liquid and vapor densities are constant;
3. the liquid and vapor velocity profiles are uniform over the cross section, that is the cross-sectional averaged values are assumed in eq.(2);
4. the axial gradient of the average liquid velocity is assumed negligible.

Then the following relation is obtained from eq.(2):

$$\frac{d\alpha}{\alpha} = - \frac{\rho_v \, du_v}{\rho_v u_v - \rho_\ell u_\ell} \quad (3)$$

Integrating the eq. (3) from the QF position (that is from the onset of the IAF) the cross-sectional averaged void fraction is expressed by:

$$\alpha = \alpha_o \frac{\rho_v u_{vo} - \rho_\ell u_\ell}{\rho_v u_v - \rho_\ell u_\ell} \quad (4)$$

where $\alpha = \alpha_o$ and $u_v = u_{vo}$ are the values at the QF.

If eq. (4) is substituted in the definition of the flow quality x, the following relation is obtained between the cross-sectional averaged void fraction and the flow quality:

$$\alpha = x + \alpha_o (1 - x) [1 - \frac{x_o (1 - \alpha_o)}{\alpha_o (1 - x_o)}] \quad (5)$$

where x_o is the flow quality at the onset of the IAF; α_o can be evaluated by the well known Zuber and Findlay [16] drift flux correlation with the Dix[17] expression for the distribution parameter. The thickness of the vapor layer δ can be evaluated from the fraction of the channel cross section occupied by the vapor annulus α_{IA}:

$$\alpha_{IA} = 1 - [1 - 2 \frac{\delta}{D}]^2 \quad (6)$$

If the fraction of the liquid core cross section, occupied by the vapor in the liquid column is assumed constant along the channel and equal to α_o, the ratio α_{LC} between the vapor cross section in the liquid core and the channel cross section is given by:

$$\alpha_{LC} = \alpha_o (1 - \alpha_{IA}) \quad (7)$$

As α_{IA}, α_{LC} and α are correlated by:

$$\alpha = \alpha_{IA} + \alpha_o (1 - \alpha_{IA}) \quad (8)$$

equations (5),(6) and (8) yield:

$$\alpha_{IA} = x - \frac{x_o}{1 - x_o} (1 - x) \tag{9}$$

$$\delta = \frac{D}{2} \left[1 - \sqrt{\frac{1 - x}{1 - x_o}} \right] \tag{10}$$

According to the analysis of Dougall and Rohsenow [18] for the stable film boiling the ratio between the vapor film thickness and the channel diameter is
normally less than 0.1 ,which corresponds to $\alpha_{IA} < 0.36$. Plummer et al.[19]
suggested that at void fraction of about 0.4 the IAF regime changes into DFFB.
Limiting values of about 0.3 and 0.5 for the void fraction are suggested by
Groeneveld and Rousseau[20] and by Yadigaroglu and Arrieta[14] .

In fig.1 the void fraction is plotted versus the flow quality:it can be
seen that downstream of the QF position (where $x=x_o$ and $\alpha =\alpha_o$ are evaluated by
the drift flux model) α increases very slowly with x and depends strongly
on the values of x_o and α_o at the transition between bubble flow and IAF. According to eq.(9)$\alpha_{IA}$is a linear function of the flow quality. This model underpredicts the axial gradient of the data obtained by Fung and Grœneveld[21]
by the hot patch tecnique;this can be due to entrainment of the vapor from the
annulus to the liquid core but such a phenomenon will accelerate the instability of the IAF:probably that's why void fraction increases along the channel more than one would expect for an IAF pattern like that above defined.

2.2 Inverted annular flow stability

The onset of liquid entrainment and carryover depends on the extension
of the IAF region,that is on the liquid core stability. According to Arrieta and
Yadi garoglu[2] the onset of entrainment criterion is obtained prescribing a
Weber number,as the Jensen's analysis [22] on the stability of the interface
shows that the Weber number based on the channel diameter is the critical parameter;they assumed that all the liquid was entrained above We=10 . Andersen

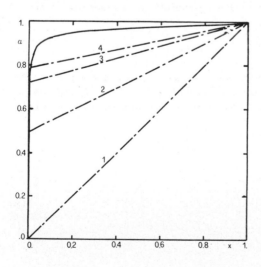

Fig.1. Typical void fraction versus flow quality for the
bubble–churn flow (dark
line) and for the inverted
annular flow (dashed lines)
at p = 1 bar.
1. $x_o= 0$;   2. $x_o= 0.001$
3. $x_o= 0.005$ 4. $x_o= 0.01$

[23]considered the stability of a liquid jet,on the basis of the Raileigh theory, and obtained the following expression of the jet length L_j :

$$\frac{L_j}{d_j} = \frac{7.85 \sqrt{\rho_\ell / \rho_v} \sqrt{We_j}}{[A + \sqrt{0.5 + A^2}][0.5 + 3/128 \; We_j^2 + We_j/8\sqrt{0.5 + A^2}]^{1/2}} \tag{11}$$

where $A = We_j 3/16$ and We_j is the Weber number based on the jet diameter d_j.

The following IAF stability criterion is derived by the continuity equation (2) and by the momentum equations for liquid and vapor phases:

$$\frac{\partial}{\partial t}[(1 - \alpha)\rho_\ell u_\ell] + \frac{\partial}{\partial z}[(1 - \alpha)\rho_\ell u_\ell^2] + \frac{\partial}{\partial z}[(1 - \alpha)p_\ell] = B \tag{12}$$

$$\frac{\partial}{\partial t}(\alpha\rho_v u_v) + \frac{\partial}{\partial z}(\alpha\rho_v u_v^2) + \frac{\partial}{\partial z}(\alpha p_v) = -\alpha\rho_v g + p_{vi}\frac{\partial\alpha}{\partial z} \tag{13}$$

where

$$B = -(1 - \alpha)\rho_\ell g - p_{\ell i}\frac{\partial\alpha}{\partial z}$$

with the following hypothesis:
1. the interface pressures $p_{\ell i}$ and p_{vi} are equal;
2. $p_v = p_\ell + \sigma/R$,where R is the liquid core radius;
3. the liquid and vapor densities don't depend on z and t.

By using the method presented in ref.[24] the eqs.(2,3,12,13) can be written under the following matrix form:

$$M'\frac{\partial \overline{X}}{\partial t} + M''\frac{\partial \overline{X}}{\partial z} = C \tag{14}$$

where $\overline{X}$ is the solution vector:

$$\overline{X} = [\alpha \; u_\ell \; u_v \; p_\ell] \tag{15}$$

M' and M'' are 4x4 matrix and C is a column vector.
Stability is ensured if the s values of the characteristic equation

$$\det[M'' - sM'] - 0 \tag{16}$$

are real. In terms of Weber number such a condition gives:

$$We < 2[1 + \frac{\rho_v}{\rho_\ell}\frac{1 - \alpha}{\alpha}][\frac{1}{\sqrt{1 - \alpha}} + \frac{\alpha}{2\sqrt{(1 - \alpha)^3}}] \tag{17}$$

Figure 2 shows the critical Weber number,predicted by the eq.(17), as a function of the void fraction at 1 bar.

2.3. Thermal non-equilibrium and heat transfer in inverted annular flow boiling

Upstream of the quench front position subcooled nucleate boiling may occur and according to Zuber et al. [25] a profile fit model can be assumed for the enthalpy of the liquid phase H_ℓ. In the IAFB the liquid core may be subcooled too and the same profile fit correlation can be used to predict

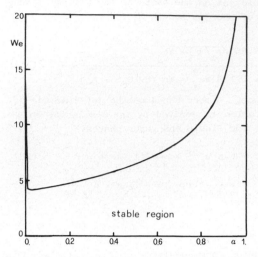

Fig.2. Variation of the critical Weber number with the void fraction at p = 1 bar.

the liquid enthalpy;the vapor layer is superheated and the following profile fit correlation is here suggested for the cross sectional averaged vapor enthalpy:

$$H_v = H_{vs} + (H_{vmax} - H_{vs}) \left[1 - \exp\left(\frac{H - H_q}{H_{vmax} - H_q}\right) \right] \tag{18}$$

where H_{vmax} is the maximum vapor enthalpy value attainable (corresponding to the wall temperature) and H_q is the enthalpy of the mixture at the onset of IAFB. The IAF actual quality is expressed by:

$$x = \frac{H - H_\ell}{H_v - H_\ell} \tag{19}$$

Typical actual quality and equilibrium quality against the mixture enthalpy are shown in fig.3;also the liquid and vapor enthalpies are herein plotted (fig.3a).

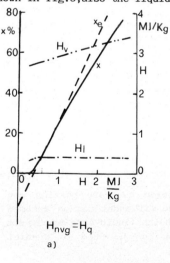

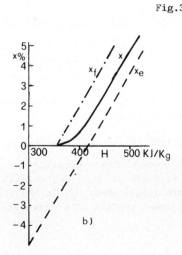

Fig.3.Flow and equilibrium quality,liquid and vapor enthalpy for the inverted annular flow versus the enthalpy of the mixture at p =1 bar and ΔT_{sub} = 20 °C.
b) Quality variation at low values of the enthalpy.

The enthalpy H_q is here assumed equal to the liquid enthalpy at the net vapor generation point with a subcooling of 20°C (typical of bottom flooding conditions). From the definition of the actual and equilibrium quality the following relation can be written:

$$\frac{x}{x_e} = \frac{1 + \Delta H_{sub}/(H - H_{\ell s})}{1 + \Delta H_{sub}/H_{fg} + \Delta H_{sup}/H_{fg}} \qquad (20)$$

where ΔH_{sub} is the subcooling along the channel and ΔH_{sup} is the superheating of the vapor at the same elevation. It can be seen the variation of the flow quality,which is higher than the equilibrium quality at low enthalpy and becomes lower as the enthalpy of the mixture increases;when the liquid is still subcooled the condition that the flow quality is lower than the "frozen" flow quality (that is with no variation in subcooling and no superheating) is satisfied.

Most of the experiments in IAFB have been carried out on cryogenics and refrigerants due to the high heater wall temperature required to maintain film boiling especially at low qualities in water. Some experiments have been carried out also in water,because of the interest in LOCA heat transfer (see the review of Groeneveld and Gardiner [26]),but in temperature controlled systems or from transient tests;more recently the hot patch tecnique [15] has been used for the analysis of steady state subcooled film boiling. As expected the heat transfer coefficient increases with mass flux and with the subcooling;moreover the data show that the axial location has a strong effect on the film boiling temperature and on the heat transfer coefficient,which decays with distance downstream and levels off within 25 diameters from the hot patch(or from the QF). This effect can be attributed either to the strong evaporation near the QF,which considerably disturbs the flow,or to the increasing thickness of the vapor layer or to the increasing vapor superheat. Data,obtained with a tubular test section reflooded internally[27,12] as well as FLECHT data show an exponential variation of the heat transfer coefficient with distance from the QF.

The classical correlations developed for pool boiling conditions are not appropriate to predict the film boiling heat flux under the reflooding conditions;none of them considered the effects of subcooling and mass flux. For saturated conditions the correlations of Ellion[28] and Hsu[29] are the most applicable:the first one depends on the distance from the QF but this dependence is too weak $(z^{-1/4})$ to account for the data effect,the second one was derived on the basis of post-CHF FLECHT reflood data[30]and consists of the sum of the transition boiling and film boiling components:the last term is given by the modified Bromley correlation [31]. The Ellion correlation can be regarded as a lower bound as it works better at lower flow rate and subcooling;the Hsu correlation does not predict correctly the trend of the data due to the lack of flow rate and subcooling effects. Yadigaroglu and Arrieta [14] used a correlation of the form:

$$h = h_q \exp(-a \Delta z_q) \qquad (21)$$

where Δz_q is the distance from the QF position and a and h_q are empirical coefficients depending on the QF local conditions. Eq.(21) gives values in agreement with data of ref. [27] downstream of the QF,both in the IAFB regime and in DFFB regime. Also Elias and Chambré [5] derive an exponential variation of the heat flux with the distance from the QF in the region downstream of the QF position where they assume that no evaporation from the liquid occurs and the heat transport through the vapor layer is by molecular conduction: their model requires the vapor gap thickness, which has be chosen for each case to fit ref. [15] data. Another correlation with an exponential decay with the distance from the QF was derived by Sun et al.[32] for falling film rewetting:

$$h = \frac{h_b}{N} \exp(-0.05 \, \Delta z_q) \tag{22}$$

where h_b is the average boiling heat transfer coefficient for the sputtering region and N is a coefficient depending on the flow rate which characterizes the magnitude of the precursory cooling and can be correlated by the empirical relation(where Ψ is the spray flow rate per unit perimeter):

$$N = 800 \, \Psi^{-1/4} \tag{23}$$

Fig. 4 compares the hot patch technique data [15] with the heat flux predicted by

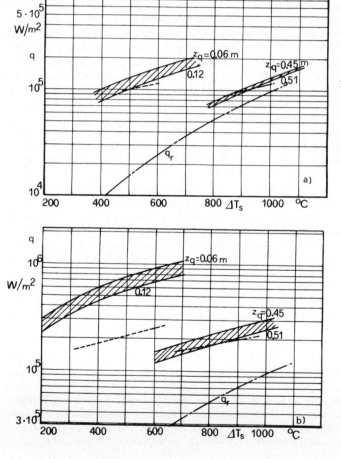

Fig.4.Comparison of the heat flux predicted by eq.(22) with the hot patch technique data, at p = 1 bar
a)G = 100 kg/m^{2}s
ΔT_{sub} = 60 °C
b)G = 500 kg/m^{2}s
ΔT_{sub} = 20 °C

Dashed lines are for data, solid lines for eq.(22).

the eq.(22),added to the radiative heat flux from the wall to the liquid;h_b is been evaluated by the Ponter and Haigh [33] critical heat flux correlation. The comparison is made at two mass velocities (100 and 500 kg/m^{2}s),at two values of subcooling at the QF (20 and 60 °C) and at two distances from the QF (0.06–0.12 m and 0.45–0.51 m;the double value of each distance is due to the uncertainty of the QF position in the hot patch). It can be seen that there is a fairly good agreement at the lower flow rate,but the correlation (22) overpredicts the data at the higher flow rate for the lower distance from the QF,due to the empirical character of eq.(23),derived for different conditions. However the trend seems correct and the subcooling effect is correctly evaluated.

3.   DISPERSED FLOW FLUIDDYNAMICS AND HEAT TRANSFER

In dispersed flow regime the flow pattern is characterized by small liquid droplets flowing in a continuous vapor stream with a strong thermodynamic non-equilibrium,at void fraction usually above   0.8;the heat transfer regime is called dispersed flow film boiling (DFFB). There is a transition region between IAFB and DFFB because the liquid core breaks up into liquid slugs ,which form large drops before further breaking down into smaller drops.

3.1. DFFB void fraction

According to Andersen [23] the fragmentation of the liquid core gives drops of diameter

$$\delta_d = d_j [\frac{3\pi}{\frac{3}{8} We_j + \sqrt{(\frac{3}{8} We_j)^2 + 2}}]^{1/3} \tag{24}$$

where d_j and We_j are the liquid core diameter and Weber number respectively. If the void fraction at the upstream boundary of DFFB is known the drops number density is given by

$$N = \frac{6(1-\alpha)}{\pi \delta_d^3} \tag{25}$$

Downstream of the onset of DFFB each drop may break up in n smaller droplets

$$n = [\frac{1}{2\xi} + \frac{\ln(\xi^{-3/2} + \sqrt{1/\xi^3 - 1})}{2\sqrt{\xi^{-3} + 1}} \sqrt{\xi}]^3 \tag{26}$$

where ξ is the drop deformation ratio given by the equation:

$$\xi^{1/2} + \xi^{-5/2} -2\xi^2 = C_D We_\delta \tag{27}$$

We_δ is the droplet Weber number and C_D is the drag coefficient,which can be evaluated by the Ingebo's [34] corrrelation:

$$C_D = 27/ Re_\delta^{0.84} \tag{28}$$

The droplet velocity is evaluated by the equation of motion for a droplet moving in an accelerated vapor medium; by assuming that all droplets acquire immediately their terminal velocity at constant mass flux the steady state form of the motion equation yields:

$$u_v - u_\ell = [4/3 \frac{(\rho_\ell - \rho_v)g \delta}{C_D \rho_v}]^{1/2} \tag{29}$$

The droplet diameter depends on the fragmentation and on the droplet evaporation. The initial droplet diameter δ_o is given by the relation:

$$\delta_o = \delta_d \ (1/n)^{1/3} \tag{30}$$

The droplet diameter δ changes along the channel due to the droplet fragmentation as n (eq.26) changes and δ can be evaluated by an equation similar to the eq.(30); δ changes also due to the droplets evaporation and a relationship between the actual vapor quality and the droplet size can be written as:

$$\delta = (\frac{1 - x}{1 - x_o})^{1/3} \delta_o \tag{31}$$

Equations (26,27,28,29,30,31) give the phase velocities and then the void fraction,which are necessary to calculate the DFFB local heat transfer.

3.2. Thermal non–equilibrium and heat transfer in DFFB

Several methods (based on empirical correlations or on mechanistic models) have been proposed to predict the actual quality and the heat transfer in DFFB. The main heat transfer mechanisms from the wall to the fluid are the following:
  1. wall to vapor convective heat transfer
  2. wall to droplets radiative heat transfer
  3. wall to droplets direct contact heat transfer
The vapor generation rate depends essentially on the heat transfer from the superheated vapor to the entrained liquid droplets interface but often 2. and 3. mechanisms cannot be neglected. The vapor temperature is controlled both by wall to vapor and vapor to liquid heat exchange. Only some methods which take in account thermodynamic non–equilibrium are here considered. Groeneveld and Delorme [8] derived a correlation(based on local information) to predict the actual vapor enthalpy and the actual quality from post–dryout data,by assuming that the heat transfer from the wall to the vapor can be predicted from conventional superheated steam heat transfer correlations with the actual vapor velocity;mechanisms n.2.and 3. are assumed negligible and the data pressure range was 7–215 bar,the mass velocity range was 130–4000 kg/m^2 s ,the equilibrium quality range was –0.12–1.6 . Chen et al. neglect the mechanisms n.2. and 3. and develop a phenomenological model from momentum transfer analogy to obtain the convective heat transfer coefficient [9] . The non–equilibrium vapor temperature and the actual quality prediction require the evaluation of an empirical parameter derived for the following film data:pressure ranges from 4.2 to 195 bar, mass velocity ranges from 16.5 to 3000 kg/m^2 s and flow quality ranges from 0.5 to 1.7;also this model is based on local information and removes the need for knowledge of detailed thermal and dynamic behaviour of droplets along the channel. As Nijhawan et al.[35] pointed out it is not surprising that both correlations (Groeneveld and Chen) are not able to predict non–equilibrium superheat for vapor;this is due to the empirical constants and to the almost comple-

te lack of experimental superheat data for the developments of such correlations. Saha[7] proposed a model based on a profile fit method for the actual quality, which is correlated to the equilibrium quality by an exponential function:such a method requires a vapor generation rate to be postulated,and the quality and the phases velocities at the upstream boundary of the DFFB region are required as boundary conditions. In the Delale model [10] the actual quality at every axial location is obtained as a function of the actual quality at the onset of the DFFB and of a parameter K,which depends on the heat transfer coefficient from the vapor to the droplets,on the droplet diameter,on the void fraction and on the equilibrium quality,if mechanisms n.2. and 3. can be neglected;two methods to calculate K are presented in ref.[10] .

On the basis of the Mastanaiah and Ganic[11] model,if the mechanisms n.1., 2. and 3. are taken into account,the actual quality axial change is given by:

$$\frac{dx}{dz} = \frac{1}{G\,H_{fg}} \left[-\frac{4}{D} (q_{w-d} + q_r) + q^v_{vd} \right] = \frac{\Gamma}{G} \tag{32}$$

where q_{w-d} and q_r are the heat flux corresponding to the mechaninisms 3.and 2. respectively and q^v_{vd} is the heat transfer from the vapor to the droplets per unit volume of the channel. The vapor temperature is given by:

$$\frac{dT_v}{dz} = \frac{1}{x\,Gc_v} \left[-\frac{4\,q_{w-v}}{D} - q^v_{vd} - \Gamma c_v (T_v - T_s) \right] \tag{33}$$

heat transfer from the vapor to the droplets can be evaluated by the Rane et al. [36] correlation,which gives the Nusselt number for a single droplet moving in a gas stream;the droplets density number and diameter are evaluated by eqs. (25,26,30 and 31). The heat transfer from the wall to all the impinging droplets can be calculated by the relation:

$$q_{w-d} = K_d\, c\, f\, H_{fg}\, \exp\left(1 - \frac{T}{T_s}\right) \tag{34}$$

where K_d is the droplet deposition velocity and f is the deposition factor; the bulk droplets concentration c can be evaluated by the relation:

$$c = \left[\frac{1}{\rho_\ell} + \frac{x}{1-x}\,\frac{1}{\rho_v} \right]^{-1} \tag{35}$$

Such a model,which requires a proper choice for the parameters K_d and f,can be applied also at negative equilibrium quality and in the transition region from IAFB to DFFB.

Several heat transfer correlations have been proposed to predict the DFFB heat transfer;between them the following can be properly adopted:the wall to vapor heat transfer coefficient can be evaluated by the Heinemann [37] correlation. Radiative heat transfer between the wall and the liquid is accounted for by the method adopted by Arrieta and Yadigaroglu [2],based on the model proposed by Sun et al. [38] . The wall to droplet heat transfer mechanism is the least understood of the above mentioned mechanisms;in the present work the correlations (34,35) are adopted. The total heat transfer from the wall is obtained by superposition of the convective,radiative and collision components.

4. PREDICTION OF SINGLE TUBE BOTTOM REFLOODING EXPERIMENTS

The aforementioned models to evaluate the precursory cooling ahead of the QF during the bottom reflooding of a single tube are adopted in an overall calculation code (RIBFLOW),which predicts the temperature distribution and the rewetting velocity along the channel. The code is based on a mechanistic model and the heat transfer coefficients,depending on the local flow conditions, provide the boundary conditions for the transient conduction equation solution. The philosophy and the structure of the code are reported in ref.[3] . The precursory cooling fluiddynamics is modeled by the method proposed in section 2.1. for the IAF and in section 3.1.for the DFF;thermal non-equilibrium is evaluated by the correlations (18,20) in the IAF region and on the basis of Mastanaiah and Ganic [11] model in the DFF region;heat transfer is predicted on the basis of Sun et al. [32] correlation for the IAFB and by the Heineman [37] correlation for the DFFB,taking into account the wall to droplet interaction (eq.(34)); radiative heat transfer to the liquid is taken in account too [2,38] . If the heat transfer coefficient predicted in the IAFB region,which decreases exponentially with the distance from the QF,becomes lower than that one evaluated for the DFFB regime,the last is assumed. Other models to predict thermal non-equilibrium in the DFFB region [7,8,9,10],the IAF stability [2,23] and the droplets fragmentation are provided to this modified version of the code.

The predictions of the code are compared with the data performed on a facility which consists of 1 m long,0.01 m inner diameter,1 mm thick stainless steel bare tube,heated electrically by a DC source. Details of the experimental setup and procedure are given in ref.[12] . The reported results are typical samples of prediction for the following run conditions(pressure p = 1 bar; subcooling at the inlet ΔT_{sub}= 80°C):

- Run 1 T_{wo} = 750°C; W = 20.5 g/s
- Run 2 T_{wo} = 750°C; W = 13.5 g/s
- Run 3 T_{wo} = 750°C; W = 27   g/s
- Run 4 T_{wo} = 570°C; W = 10   g/s

Figs.5 present the comparison between the code prediction and the experimental results of the wall temperature histories at the 20,40,60,80 cm elevations for Run 1. A proper choice for the value of the droplet deposition velocity seems to be K_d = 0.3 and a value of 1 for the deposition factor has been assumed. As regards the precursory cooling phase the agreement is good especially at 40 and 60 cm elevations where there are no inlet and outlet effects. The effect of the "quench temperature" T_q ,here defined as the value which corresponds to the transition from stable to unstable film boiling ,is shown:the quench time increases as T_q decreases (from fig.5a to fig.5b) and also the wall temperatures during the precursory cooling phase are affected because T_q is an important parameter to evaluate the lower boundary of the precursory cooling region. In the present study T_q is assumed as an input parameter. In figs.6 the effect of the flooding flow rate is presented for Run 2 and 3. At higher flow rate the heat transfer for the IAFB is overpredicted:this can be explained by the overestimation of the heat transfer coefficient predicted by the eq.(22) near the QF at higher flow rate and confirms the results shown in fig.4;on the other hand the heat transfer for the DFFB seems underpredicted,probably because the assumed value of the droplet deposition velocity K_d is too low (see also fig.8) at high flow rate .

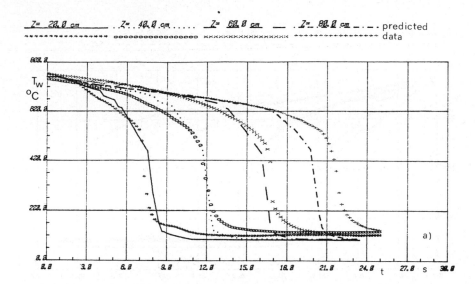

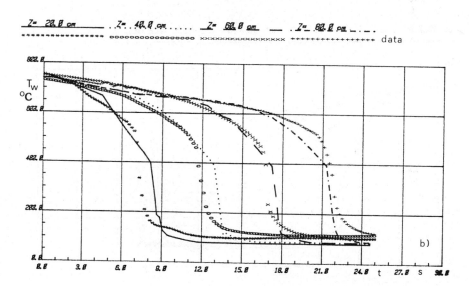

Fig.5. Comparison of experimental and predicted wall temperature
histories during the bottom reflooding of a single tube.
Run 1. K_d= 0.3 m/s; a) T_q= 450 °C ; b) T_q= 400°C

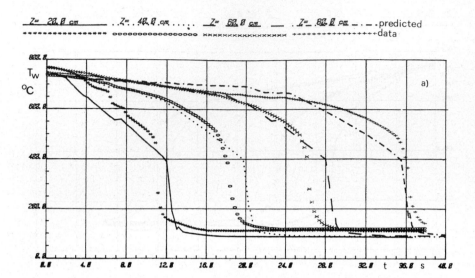

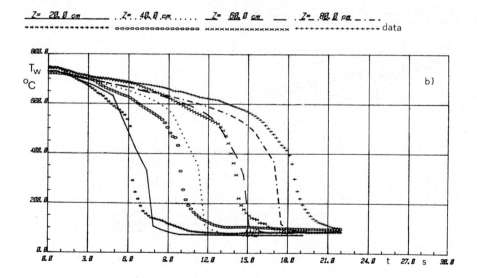

Fig.6. Comparison of experimental and predicted wall temperature
histories during the bottom reflooding of a single tube.
K_d= 0.3; a) Run 2, W = 13.5 g/s; b) Run 3, W = 27 g/s .

Fig. 7 presents the effect of the droplet fragmentation for Run 1 at the 60 cm elevation:the prediction with no fragmentation (dashed line) slightly underpredicts the heat transfer as the vapor liquid interface is lower than when the droplets fragmentation occurs.

The models analysed in section 3.1. for the DFFB have been compared for the experimental runs conditions:there are not significative difference between them,but the heat transfer is strongly underpredicted if such models are adopted from the onset of the DFFB,where the equilibrium quality is negative for the run conditions examined,at high inlet subcooling;instead they are in better agreement with the data if they are used when the equilibrium quality is positive and the Mastanaiah model is adopted for negative quality.

Figs.8 present the effect of the droplet deposition velocity K_d for Run 4, characterized by a low initial wall temperature and a low flooding rate:values of K_d =0.15 m/s and K_d = 0.3 m/s are assumed in fig.8a and 8b respectively. It can be seen the strong effect of the direct wall to droplets interaction mechanism on the heat transfer during the precursory cooling phase in bottom reflooding;if this mechanism is neglected the quench time is strongly overpredicted for the run conditions. It is therefore very important to express in a correlation like eq.(34) appropriate coefficients as a function of the wall temperature and flow rate for the conditions occurring during the rewetting by bottom reflooding.

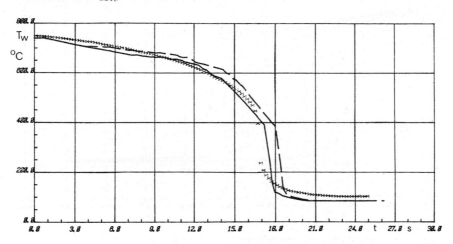

Fig.7. Comparison of experimental and predicted wall temperature history during the bottom reflooding of a single tube at z = 60 cm. Run 1. Effect of the droplet fragmentation: dashed line is for no fragmentation. K_d= 0.3 .

274

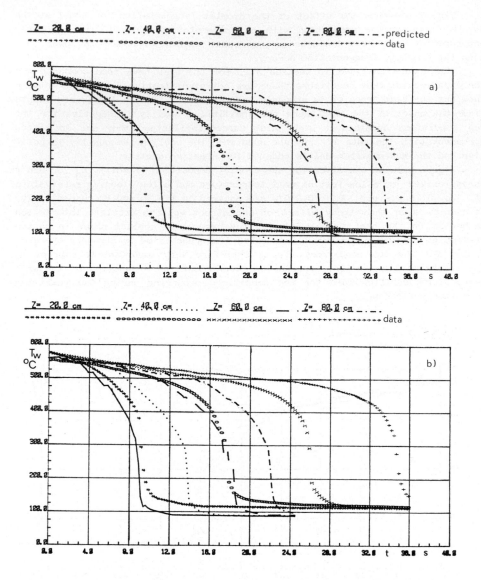

Fig. 8. Comparison of experimental and predicted wall temperature
histories during the bottom reflooding of a single tube.
Run 4 . Effect of the droplet deposition velocities K_d :
a) K_d= 0.15 ; b) K_d= 0.3

5. CONCLUSIONS

A comprehensive analysis of the precursory cooling phase during bottom flooding is presented and the comparison of the models with experimental data led to the following conclusions:

- the models presented for the thermal-hydraulics in the inverted annular flow region provide simple correlations between the void fraction, flow quality and equilibrium quality, and a formulation of the flow stability, but appropriate experiments on the IAF are needed;

- the Sun heat transfer correlation predicts well the wall temperatures for the inverted annular film boiling regime except at high flow rate near the quench front;

- in the dispersed flow film boiling region the post-dryout correlations analysed, which take into account the thermal-nonequilibrium, can be adopted only if the equilibrium quality is positive; the predictions obtained by a model based on the Mastanaiah study are in fairly good agreement with the data also if the equilibrium quality is negative;

- the direct wall to liquid droplet interaction seems to be a very important heat transfer mechanism in the precursory cooling phase during bottom reflooding, also at high wall temperature; experiments in such conditions are needed;

- several models presented or adopted in the present paper will need experimental testing and some adjustament of their parameters.

ACKNOWLEDGEMENTS

The assistance of G.Falletti for modifications of the code is gratefully acknowledged.

This work has been sponsored by the Italian National Committee for Nuclear and Alternative Energies (ENEA).

NOMENCLATURE

c_v	vapor specific heat	W	Flow rate	
C_D	Drag coefficient	We	Weber number $(\rho_g D[u_g - u_\ell]^2/\sigma)$	
d_j	Liquid core diameter	x	Flow quality	
D	Channel diameter	x_e	Equilibrium quality	
f	Droplet deposition factor	x_f	Frozen flow quality $(\frac{H - H_{\ell o}}{H_{vs} - H_{\ell o}})$	
G	Mass flux	z	Axial coordinate	
h	Heat transfer coefficient	α	Void fraction	
H	Enthalpy	Γ	Vapor generation rate per unit volume	
H_{fg}	Specific latent heat of vaporization			
K_d	Droplet deposition velocity	δ	Vapor thickness or droplet diameter	
L_j	Liquid core height			
n	Number of fragmentation droplets	ρ	Density	
N	Droplet number density	σ	Surface tension	
p	Pressure	ξ	Droplet deformation ratio	
q	Heat flux	Ψ	Spray flow rate per unit perimeter	
Re	Reynolds number			
t	Time			
T	Temperature			
u	Phase velocity			

Subscripts

d,δ	Droplet	vd	Vapor to droplets
i	Liquid—vapor interface	w	Wall
j	Liquid core	w—d	Wall to droplets
l	Liquid	w—v	Wall to vapor
LC	Liquid core		
NVG	Net vapor generation		Abbreviations
o	Initial or quench front		
q	Quench front	DFFB	Dispersed flow film boiling
r	Radiation	IAF	Inverted annular flow
s	Saturated	IAFB	Inverted annular film boiling
sub	Subcooled	QF	Quench front
sup	Superheated		
v	Vapor		

REFERENCES

1. De Benedetti,A;,and Martini,R.,"TRAFEM—A code for Predicting Transient Heat Transfer during Emergency Core Reflooding",European Two—Phase Flow Group,Grenoble,1977.

2. Arrieta,L.,and Yadigaroglu,G.,EPRI NP 756,1978.

3. De Salve,M.,and Panella,B.,Heat Transfer in Nuclear Reactor Safety (Ban-Koff and Afgan Ed.),Hemisphere Publishing Company,p.747,1982.

4. Kim,A.K.,and Lee,Y.,7th International Heat Transfer Conference,München, Vol.4,p.181,1982.

5. Elias,E.,and Chambré,P.,Nucl.Eng.and Des.,Vol.64,n.2,p.249,1981.

6. Fung,K.K.,and Groeneveld,D.C.,7th International Heat Transfer Conference, München,Vol.4,p.381,1982.

7. Saha,P.,Shiralkar,B.S.,Dix,G.E.,ASME paper n.77—HT—80,1977.

8. Groeneveld,D.C.,Delorme,G.G.J.,Nucl.Eng.and Des.,Vol.36,n.1,p.17,1976.

9. Chen,J.C.,Ozkaynak,F.T.,Sundaram,R.F.,Nucl.Eng.and Des.,Vol.51,p.143,1979.

10 Delale,C.F.,ASME Journal of Heat Transfer,Vol.102,n.4,p.501,1980.

11. Mastanaiah,K.,and Ganic,E.N.,ASME Journal of Heat Transfer,Vol.103,n.2, p.300,1981.

12. De Salve,M.,and Panella,B.,NUREG/CP-0014,Vol.2,p.1196,1980.

13. Wallis,G.B., One Dimensional Two—Phase Flow ,McGraw Hill,New York,1960.

14. Yadigaroglu,G.,and Arrieta,L.A.,NUREG/CP-0014,Vol.2,p.1173,1980.

15. Fung,K.K.,Gardiner,S.R.M.,Groeneveld,D.C.,Nucl.Eng.Des.,Vol.55,n.4,p.51,1979.

16. Zuber,N.,and Findlay,J.A.,ASME Journal of Heat Transfer,Vol.87,p.453,1965.

17. Dix,G.E.,"Vapor Void Fraction for Forced Convection with Subcooled Boiling at Low Flow Rates",Ph.D.Thesis,Univ.California,Berkeley,1971.

18.   Dougall,R.S.,and Rohsenow,W.M.,MIT Report,n.9079-26,1963.

19.   Plummer,D.N.,et al.,MIT Report n.72718-91,1974.

20.   Groeneveld,D.C.,and Rousseau,J.C.,"CHF and Post-CHF Heat Transfer",NATO Advanced Research Workshop,Spitzingsee,1982.

21.   Fung,K.K.,and Groeneveld,D.C.,Int.J.Multiphase Flow,Vol.6,p.357,1980.

22.   Jensen,R.,"Inception of Liquid Entrainment during Emergency Cooling of Pressurized Water Reactors",Ph.D.Thesis,Utah State University,1972.

23.   Andersen,J.G.M.,RISO Report n.296,1973.

24.   Delhaye,J.M., Two-Phase Flow and Heat Transfer in the Power and Process Industry ,Hemisphere Publishing Company,p.89,1981.

25.   Zuber,N.,et al.,3rd International Heat Transfer Conference,Chicago,Vol.5, p.24,1966.

26.   Groeneveld,D.C.,and Gardiner,S.R.M.,"Post-CHF Heat Transfer under Forced Convective Conditions",ASME Winter Annual Meeting,Atlanta,1977.

27.   Seban,L.,et al.,EPRI Report NP-743,April,1978.

28.   Ellion,M.E.,CALTEC Report JPL-MEMO -20,88,1954.

29.   Hsu,Y.Y.,"Proposed Heat Transfer Best Estimate Packages",1977.

30.   Cadek,F.F.,et al.,PWR FLECHT Final Report Supplement",W CAP-7931,1972.

31.   Bromley,L.A.,et al.,Industrial Eng. Chem.,Vol.45,p.2369,1953.

32.   Sun,K.H.,Dix,G.E.,Tien,C.L.,ASME J.of Heat Transfer,p.360,August,1975.

33.   Ponter,A.B.,Haigh,C.P.,International J.of Heat and Mass Transfer,Vol.12, n.4,p.429,April,1969.

34.   Ingebo,R.D.,NACO Technical Note 3762.

35.   Nijhawann,S.,et al.,ASME J.Heat transfer,Vol.102,n.3,p.465,August,1980.

36.   Rane,et al.,ASME Paper n.79-WA/HT-10,1979.

37.   Heineman,J.B.,ANL 6213,1969.

38.   Sun,K.H.,et al.,ASME J.of Heat Transfer,n.98,p.414,1976.

Multi-Phase Flow and Heat Transfer III. Part B: Applications
edited by T.N. Veziroğlu and A.E. Bergles
Elsevier Science Publishers B.V., Amsterdam, 1984 — Printed in The Netherlands

THE ROLE OF FLOODING PHENOMENA IN REFLUX CONDENSATION IN A VERTICAL
INVERTED U-TUBE

Pak T. Wan, René Girard and Jen-Shih Chang
Department of Engineering Physics and
Institute of Energy Studies
McMaster University
Hamilton, Ontario L8S 4M1, Canada

ABSTRACT

Reflux condensation in a vertical inverted U-tube inside steam generators form an important heat removal mechanism in loss of coolant accidents in nuclear power plants. The role of flooding on reflux condensation in a single inverted U-tube was studied. The range of reflux condensation covered in the present paper is the "complete reflux condensation" range, which consists of an oscillating single-phase water column sustained by a condensing two-phase countercurrent churn-annular flow region. The analysis based on the Wallis type of flooding correlation was carried out by postulating that the formation of the single-phase water column on top of the two-phase region was governed by "flooding" conditions at the entrance. The analytical results were compared with the experimentally measured rate of condensation for the range of operating conditions.

1. INTRODUCTION

The Three Mile Island small-break loss-of-coolant accident (LOCA) demonstrated the necessity for long term unpowered cooling schemes for removing decay heat from the nuclear reactor core. Under these accident conditions, the steam generators are expected to be the major heat sinks. Depending on the particulars of the accident scenario [1,2], such as the location and size of the break, and the details of the emergency core cooling system (ECCS), different heat removal mechanisms may be occurring in the steam generator tubes. These include reflux condensation, single- and two-phase natural circulation. To evaluate the potential consequences of a small-break LOCA, each of these cooling modes must be investigated in detail. The present paper concentrates on the reflux condensation cooling mode.

Figure 1 illustrates four observed flow regimes in a vertical steam generator tube as the steam injection rate is gradually increased [5-7]. Three modes of reflux condensation can be distinguished: filmwise reflux condensation, churn-annular reflux condensation without water column, and churn-annular reflux condensation with water column. When the steam injection rate is small, the condensate flows down the wall of the tube as a smooth or slightly wavy film. Calia and Griffith [3] and Sun et al. [4] have analyzed the heat removal capability of steam generators for this mode of reflux. As the steam injection rate is increased further, waves at the tube inlet bridge across the tube and an intermittent churn-annular flow is established; however, all the up-flow of steam is still completely condensed.

280

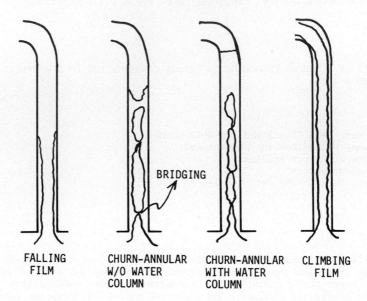

FALLING   CHURN-ANNULAR   CHURN-ANNULAR   CLIMBING
FILM   W/O WATER   WITH WATER   FILM
                 COLUMN   COLUMN

Figure 1. Various Flow Patterns in a Vertical Tube for Increasing
          Steam Flow Rate.

As the steam injection rate is further increased and if the vertical tube is
long enough, a water column is formed on top of the countercurrent churn-
annular region. This mode is termed complete reflux condensation by Banerjee
et al. [5]. The basic phenomenon and heat removal capability of the last two
reflux condensation modes have been studied experimentaly by Chang et al.
[6,7]. As the steam injection rate is increased further, the water column
lengthens until it reaches the top of the U-tube and eventually carryover
occurs, leading to climbing film flow and transition to two-phase natural
circulation. The present paper analyzes the role of flooding in the complete
reflux condensation mode.

2. ANALYTICAL MODEL

     The basic postulate of the model is that the formation of a single-phase
region (water column) on top of the two-phase countercurrent churn-annular
region corresponds to the attainment of the flooding point at the tube inlet
[5]. The flooding limit, also known as the countercurrent flow limit (CCFL),
refers to the phenomenon in which the downward flow of water is limited by the
upward steam flow. The flooding condition has been studied extensively
analytically and experimentally in recent years [8,9]. Analytical modelling
of the basic mechanisms is still in its infancy. In the absence of a
satisfactory analytical understanding, a large number of empirical flooding
correlations have been published. However, these correlations were formulated
from experimental data for a limited number of fluids and channel geometries.
Parametric dependence of the flooding correlations is not well established.
Investigators have reported conflicting findings regarding the geometric and

fluid property effects, e.g., entrance geometry effects, tube diameter and tube length effects, viscosity and surface tension effects, on the flooding characteristics. The reader is referred to the reviews by Deakin [8] and Tien et al. [9] for a more comprehensive treatment. The most widely used flooding correlations are of the Wallis [10] or the Kutateladze [11] types. For a Wallis type flooding correlation, it is expressed as

$$j_g^{*1/2} + n\, j_f^{*1/2} = c \qquad (1a)$$

where n and c are empirically determined constants. Wallis et al [10] found that most of his data for air-water and steam-water systems can be correlated by n ≃ 1, and c = 0.69 to 0.8, depending on the entrance geometries. For a Kutateladze type flooding correlation, it is given by

$$K_g^{1/2} + n'\, K_f^{1/2} = c' \qquad (1b)$$

where n' and c' are empirically determined constants. Jacoby et al. [12] found that their extensive data yielded values of n' = 0.689 – 2.16, and c' = 1.31 – 2.04 for air-water systems.

In the case of complete reflux condensation, all the injected steam is completely condensed; therefore, at steady state, conservation of mass yields,

$$m_g = m_f = m_c \qquad (2)$$

Using the definitions of j_g^*, j_f^*, K_g and K_f, it can be shown that

$$j_g^* = \left(\frac{\rho_f}{\rho_g}\right)^{1/2} j_f^* \qquad (3a)$$

$$K_g = \left(\frac{\rho_f}{\rho_g}\right)^{1/2} K_f \qquad (3b)$$

According to the basic postulate, for the case of complete reflux condensation, Eqs. (1a) and (3a), or Eqs. (1b) and (3b) must be satisfied simultaneously. The Wallis type flooding correlation yields:

$$m_c = \frac{c^2\, A\, \sqrt{g\, D\, \rho_g(\rho_f - \rho_g)}}{[1 + n(\frac{\rho_g}{\rho_f})^{1/4}]^2} \qquad (4a)$$

The Kutateladze type flooding correlation yields:

$$m_c = \frac{c'^2\, A\, \rho_g^{1/2}\, [g\sigma(\rho_f - \rho_g)]^{1/4}}{[1 + n'(\frac{\rho_g}{\rho_f})^{1/4}]^2} \qquad (4a)$$

Using Eqs. (4a) or (4b), the rate of condensation in complete reflux can be determined once the geometrical and fluid properties of the system are specified.

3. EXPERIMENTAL APPARATUS AND RESULTS

3. EXPERIMENTAL APPARATUS AND RESULTS

For completeness, the experimental apparatus and results obtained by Chang et al. [6,7] are repeated. The apparatus used is shown in Figure 2. The objective of these experiments was to study single tube complete reflux condensation in isolation from possible interactions with other system components, i.e., the boundary conditions at each end of the condenser tube were controlled to give constant plenum pressures.

The apparatus consisted of a series of eight consecutive, concentric pyrex glass sections, linked together by Teflon spacers. The Teflon pieces allowed measurement of pressure, temperature at the centre and near the wall inside the inner tube. The inner tube was connected at the bottom to a steam inlet plenum and at the top to an outlet plenum by a small section of copper tubing in the form of an inverted U. The void fraction was measured along the length of the tube by ring type capacitance transducers [13].

Steam was passed through a measurement station and an orifice meter before entering the bottom plenum on the side. The steam flowed upward inside the inner tube, where reflux condensation occurred. The steam flow rate was also determined by collecting condensate from the bottom plenum over a known time under constant operating conditions. The upper plenum was mounted with a combination of a pop and standard relief valve, which was used for pressurizing the system. The inlet and outlet plenum pressures were kept constant during a run. Cooling water was circulated in the annulus of each test section. The flow rates, inlet and exit temperatures of the cooling water were monitored. Forty-five channels of information (5 void, 12 pressure, 28 temperature) were measured along the length of the tube and at the plenums. The experimental data was acquired and analyzed by a NOVA-III mini-computer via an RTP analog to digital converter.

The test matrix explored the effects of system pressure, secondary cooling water temperature (i.e., wall heat flux), the length of the single-phase region and the inner tube diameter on the rate of condensation for complete reflux.

As long as the flow regime in the tube is that of complete reflux condensation, i.e., a single-phase region on top of a countercurrent churn-annular flow region, the following results were obtained:

(1) For the tube sizes from 1/2" to 1" I.D., significant tube size effect was observed on the rate of condensation.

(2) For the range of secondary cooling water temperature ($7^{\circ} \leq T \leq 45^{\circ}C$) in the experiments, no significant variation in the rates of condensation was observed.

(3) In the range of system pressure (1.0 bar $\leq p \leq 1.3$ bar) in the experiments, the condensation rates were approximately constant.

(4) For the range of single-phase region lengths (0.0 m $\leq L \leq 2.0$ m) in the experiments, no significant variation in the condensation rates was found.

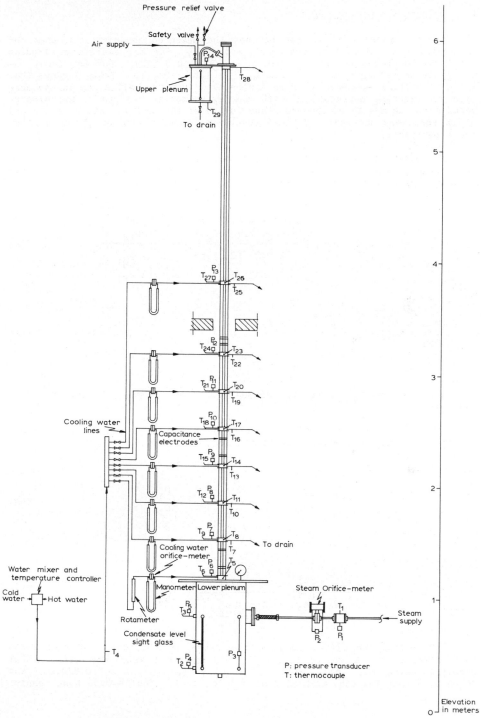

Figure 2. Schematic of Experimental Set-Up.

4. RESULTS AND DISCUSSION

Table 1 shows a comparison between the measured condensation rates and the calculated values from Eqs. (4a) and (4b). For Eq. (4a), the flooding coefficients chosen were: $n = 1.0$, $c = 0.69$ $_7/_9.8$ [12]. For Eq. (4b), the coefficients selected were: $n' = 1.0$, $c' = 3.2^{1/2}$ [4,14]. Table 1 shows that there is partial agreement between the results from the Wallis and Kutateladze type of flooding correlations with the experimental results. The discrepancies may be due to an inappropriate choice of the coefficients n, n', c and c' in Eqs. (4a) and (4b). Table 2 shows the typical range of j_g^* and j_ℓ^* from the experiments.

To recast the comparison in a form that is less sensitive to the flooding coefficients of n, n', c and c', the ratio of the condensation ratio at two flooding points are compared. Eqs. (4a) and (4b) yield:

$$\frac{m_{c1}}{m_{c2}} = \left(\frac{c_1}{c_2}\right)^2 \left(\frac{A_1}{A_2}\right) \sqrt{\frac{D_1\,\rho_{g1}(\rho_{f1} - \rho_{g1})}{D_2\,\rho_{g2}(\rho_{f2} - \rho_{g2})}} \left[\frac{1 + n_2\left(\frac{\rho_{g2}}{\rho_{f2}}\right)^{1/4}}{1 + n_1\left(\frac{\rho_{g1}}{\rho_{f1}}\right)^{1/4}}\right]^2 \tag{5a}$$

$$\frac{m_{c1}}{m_{c2}} = \left(\frac{c'_1}{c'_2}\right)^2 \left(\frac{A_1}{A_2}\right) \sqrt{\frac{\rho_{g1}}{\rho_{g2}}} \cdot \left[\frac{\sigma_1(\rho_{f1} - \rho_{g1})}{\sigma_2(\rho_{f2} - \rho_{g2})}\right]^{1/4} \cdot \left[\frac{1 + n'_2\left(\frac{\rho_{g2}}{\rho_{f2}}\right)^{1/4}}{1 + n'_1\left(\frac{\rho_{g1}}{\rho_{f1}}\right)^{1/4}}\right]^2 \tag{5b}$$

Generally, n_1 and n_2, c_1 and c_2, n'_1 and n'_2, c'_1 and c'_2 may not necessarily be the same.

To analyze the effect of tube size on the relative condensation rates, and for the same inlet steam conditions, Eqs. (5a) and (5b) reduce to:

$$\frac{m_{c1}}{m_{c2}} = \left(\frac{c_1}{c_2}\right)^2 \left[\frac{1 + n_2\left(\frac{\rho_g}{\rho_f}\right)^{1/4}}{1 + n_1\left(\frac{\rho_g}{\rho_f}\right)^{1/4}}\right]^2 \left(\frac{D_1}{D_2}\right)^{2.5} \tag{6a}$$

$$\frac{m_{c1}}{m_{c2}} = \left(\frac{c'_1}{c'_2}\right)^2 \left[\frac{1 + n'_2\left(\frac{\rho_g}{\rho_f}\right)^{1/4}}{1 + n'_1\left(\frac{\rho_g}{\rho_f}\right)^{1/4}}\right]^2 \left(\frac{D_1}{D_2}\right)^{2.0} \tag{6a}$$

Theoretical normalized condensation rates as a function of normalized tube diameter is shown in Figure 3, and compared with the experimental results, using Eq. (6a), with $n_1 = n_2 = 1$, $(c_1/c_2) = (0.8/0.69)$ and $(0.69/0.8)$, and Eq. (6b) with $n'_1 = n'_2 = 1$, $c'_1 = c'_2$. The 1" O.D. tube results

Table 1. Comparison of measured and calculated condensation
rates at atmospheric pressure.

1" O.D. tube:

    W* – calculated condensation rate = 76 – 103 gm/min.
    K** – calculated condensation rate = 179 gm/min.
    experimentally measured condensation rate = 100 $\pm$ 10 gm/min.

3/4" O.D. tube:

    W* – calculated condensation rate = 52 – 70 gm/min.
    K** – calculated condensation rate = 131 gm/min
    experimentally measured condensation rate = 70 – 105 gm/min.

1/2" O.D. tube:

    W* – calculated condensation rate = 11 – 15 gm/min.
    K** – calculated condensation rate = 38 gm/min.
    experimentally measured condensation rate = 35 – 43 gm/min.

W* – based on Wallis flooding correlation with n = 1, c = 0.69 – 0.80.

K** – based on Kutateladze flooding correlation with n' = 1, c' = $3.2^{1/2}$.

Table 2. Typical j_g^* and j_ℓ^* values for complete reflux condensation.

---

	j_g^*	j_ℓ^*
1" O.D. tube	~ 0.42 – 0.51	~ 0.010 – 0.013
3/4" O.D. tube	~ 0.48 – 0.72	~ 0.012 – 0.018

were used for normalization. Figure 3 shows that there is qualitative agreement on the tube size effect between the theoretical and experimental results; however, quantitative agreement is not very good.

For the effect of system pressure on the relative condensation rates, and for the same tube size, Eqs. (5a) and (5b) become:

$$\frac{m_{c1}}{m_{c2}} = \left(\frac{c_1}{c_2}\right)^2 \sqrt{\frac{\rho_{g1}(\rho_{f1} - \rho_{g1})}{\rho_{g2}(\rho_{f2} - \rho_{g2})}} \cdot \left[\frac{1 + n_2\left(\frac{\rho_{g2}}{\rho_{f2}}\right)^{1/4}}{1 + n_1\left(\frac{\rho_{g1}}{\rho_{f1}}\right)^{1/4}}\right]^2 \tag{7a}$$

$$\frac{m_{c1}}{m_{c2}} = \left(\frac{c_1'}{c_2'}\right)^2 \sqrt{\frac{\rho_{g1}}{\rho_{g2}}} \cdot \left[\frac{\sigma_1(\rho_{f1} - \rho_{g1})}{\sigma_2(\rho_{f2} - \rho_{g2})}\right]^{1/4} \cdot \left[\frac{1 + n_2'\left(\frac{\rho_{g2}}{\rho_{f2}}\right)^{1/4}}{1 + n_1'\left(\frac{\rho_{g1}}{\rho_{f1}}\right)^{1/4}}\right]^2 \tag{7b}$$

Figure 4 shows a comparison of the theoretical and experimental normalized condensation rates as a function of system pressure. The calculated condensation ratio shows an increasing trend with increasing pressure. For the range of experimental conditions considered, the experimental results are consistent with the theoretical results.

The effect of cooling water temperature on the condensation rates was previously examined by Banerjee et al. [5]. The first order effect of a change in the cooling water temperature is to leave the inlet steam conditions unchanged. Consequently, no change in Eqs. (5a) or (5b) results. Therefore, changes in the heat flux do not affect the steam condensation rates. Experimentally measured rate of condensation as a function of temperature difference between steam and average cooling water temperature is shown in Figure 5 to confirm this theoretical result. It should be noted that changes in the cooling water temperature affect the length of the two-phase region. However, as long as the tube is long enough so that complete reflux condensation occurs, no change in the condensation rate results.

Consider a small increase in the inlet steam flow rate in a tube undergoing complete reflux. This would initially lead to a temporary imbalance at the tube inlet, i.e., the steam flow rate exceeds the flooding limit. This would then lead to a net transfer of liquid up the vertical tube. Consequently, the single-phase region lengthens, which in turn pressurizes the inlet conditions slightly, and thereby shifting the point of flooding to a slightly higher pressure. The rate of change of the condensation rate with respect to a change in the single-phase length can be evaluated by:

$$\frac{\partial m_c}{\partial L_{1\phi}} = \frac{\partial m_c}{\partial p} \cdot \frac{\partial p}{\partial L_{1\phi}}$$

Figure 6 shows a comparison between the experimental and calculated rate of condensation as a function of the single-phase region length. Again, the experimental measured data fall within the bounds of the computed values. It is interesting to note that the model predicts a slight increase of rates of

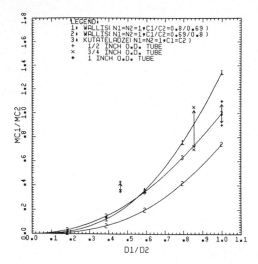

Figure 3. Tube Size Effects: Comparison of Experimental and Theoretical
Condensation Ratios. 1" O.D. Tube Results Used as Base Case.

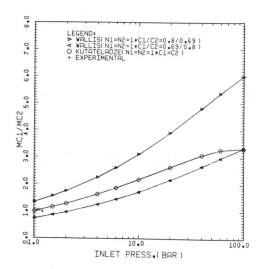

Figure 4. System Pressure Effects: Normalized Condensation
Rates vs. Pressure for 1" O.D. Tube.

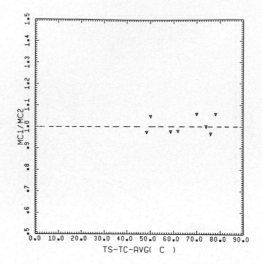

Figure 5. Wall Heat Flux Effects: Normalized Condensation Ratios vs. $(T_s - \bar{T}_c)$ for 1" O.D. Tube.

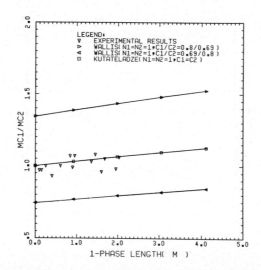

Figure 6. Effect of the Single-Phase Region Length: Normalized Condensation Rates vs. Single-Phase Region Length for 1" O.D. Tube.

condensation with increasing single-phase region length.

Figures 7a and 7b show two flow regime maps based on the present inter-pretation of complete reflux condensation. Boundary A is based on the rela-tionship of $j_g^* = j_f^*$ and $K_g = K_f$. Since $\rho_f > \rho_g$, and for the case of reflux condensation, it follows that $j_g^* > j_f^*$ and $K_g > K_f$. Boundary B represents the flooding limit as given by the Wallis or Kutateladze type flooding correla-tions. The solid line shows the typical locus of a vertical tube under reflux condensation. For the path OF, the flow regime inside the tube is that of filmwise reflux and the churn-annular reflux without water column. Upon reaching the point F, flooding conditions at the tube inlet is met and a water column begins to form on top of the countercurrent two-phase region. With increasing steam flow rates, the length of the single-phase region is increased, causing the tube inlet to be pressurized gradually and thereby shifting the flooding point from F to G. Eventually, the combined length of the single-phase and two-phase regions exceeds the length of the tube and carryover occurs, resulting in climbing film flow or two-phase natural circulation. The above picture is oversimplified in that the single-phase region is not stationary, but it oscillates on top of the two-phase region. Therefore, the oscillation amplitude of the single-phase region will also play a role in determining the transition conditions from reflux condensation to the climbing film flow regime.

If the above model is accepted, the present analysis provides a simple framework for scaling low pressure laboratory results to other inlet steam conditions or geometries of interest, e.g., in small-break LOCA analysis. Future experiments can be designed to test the validity of the scaling law.

One interesting implication can be seen using the above analysis. Typically, the steam generator tubes are very long. The rates of condensation corresponding to the point F (Figures 7a or 7b), when the single-phase region is formed, may be much smaller than that corresponding to the point G, when transition to climbing film flow occurs. The earlier model by Banerjee et al. [5] did not make such a distinction.

5. CONCLUDING REMARKS

A simple analytical model was presented to explain the effects of tube size, cooling water temperature, system pressure and single-phase region length on the rate of condensation in a vertical tube undergoing complete reflux condensation. There appears to be general agreement between the calculated and experimental results. More precise experiments under a wider range of conditions need to be performed to confirm the present model.

ACKNOWLEDGEMENTS

The authors wish to express their sincere thanks to P. Sergejewich, R.E. Pauls, W.I. Midvidy, and S.T. Revankar for valuable discussions and comments. This work was partially supported by the National Sciences and Engineering Research Council of Canada, Ontario Government BILD, and Ontario Hydro.

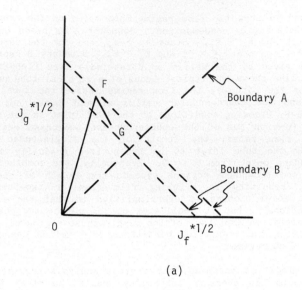

(a)

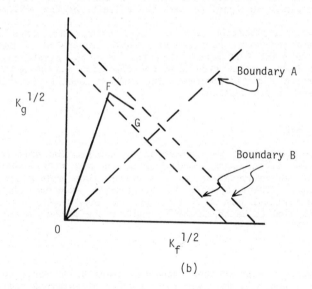

(b)

Figure 7. Proposed Flow Regime Maps for Reflux Condensation Based on:
(a) Wallis Type Flooding Curve;
(b) Kutateladze Type Flooding Curve.

REFERENCES

1. W.E. Burchill, "Physical Phenomena of a Small-Break Loss-of-Coolant Accident in a PWR", Nuclear Safety, 23, No. 5, 525-536, Sept.-Oct. 1982.

2. V.S. Krishnan, J-S. Chang and S. Banerjee, "Two-Phase Natural Circulation in a Horizontally Boiling-Vertical U-Tube Condensing Thermal Hydraulics Loop", Proc. of the 3rd Annual Conference of the Cdn. Nucl. Soc., A1-A6, 1982.

3. C. Calia and P. Griffith, "Modes of Circulation in an Inverted U-Tube Array with Condensation", Thermal Hydraulics in Nuclear Power Technology, Sun et al., eds., ASME HTD-vol. 15, 35-44, New York, 1981.

4. B.K-H. Sun, M. Toren and S. Oh, "Reflux Condensation in a Vertical Channel Flow", 2nd Int. Topical Meeting on Nucl Reactor Thermal-Hydraulics, Santa Barbara, CA, 1983.

5. S. Banerjee, J-S. Chang, R. Girard and V.S. Krishnan, "Reflux Condensation and Transition to Natural Circulation in a Vertical U-Tube", ASME 81-WHA/HT-59, 1981.

6. J-S. Chang and R. Girard, "Reflux Condensation Phenomena in a Vertical Tube Side Condenser", Advancement in Heat Exchangers, G.F. Hewitt, ed., Hemisphere Pubs., New York, 1981.

7. J-S. Chang, R. Girard, S. Revankar, P.T. Wan, W.I. Midvidy, R.E. Pauls and P. Sergejewich, "Heat Removal Capability of Steam Generators Under Reflux Cooling Modes in a CANDU-PHT System", submitted to the 4th Annual Conference of the Cdn. Nucl. Soc. 1983.

8. A.W. Deakin, "A Review of Flooding Correlations for Reflux Condensers", AERE-M2923, Oct. 1977.

9. C.L. Tien and C.P. Liu, "Survey on Vertical Two-Phase Countercurrent Flooding", EPRI-NP-984, Feb. 1979.

10. G.B. Wallis, P.C. deSieyes, R.J. Roselli and J. Lacombe, "Countercurrent Annular Flow Regime for Steam and Subcooled Water in a Vertical Tube", EPRI-NP-1336, 1980.

11. S.S. Kutateladze, "Elements of the Hydrodynamics of Gas-Liquid Systems", Fluid Mechanics - Soviet Research, 1, 29, 1972.

12. J.K. Jacoby and C.M. Mohr, Final Report on 3-D Experiment Project Air-Water Upper Plenum Experiments, EG & G Idaho, 3DP-TR-001, Nov. 1978.

13. G.A. Irons and J-S. Chang, "Particle Fraction and Velocity Measurement in Gas Powder Streams by Capacitance Transducer", Int. J. Multiphase Flow, 8, 1982.

14. C.L. Tien, "A Simple Analytical Model for Counter-Current Flow Limiting Phenomena with Vapour Condensation", Letters Heat Mass Transfer, 4, 231-238, 1977.

NOMENCLATURE

A	cross-sectional area of tube.
c	coefficient (intercept) in the Wallis flooding correlation.
c'	coefficient (intercept) in the Kutateladze flooding correlation.
D	tube diameter.
g	gravitational acceleration.
j	volumetric flux.
j*	nondimensional volumetric flux.
K	nondimensional Kutateladze number.
L	length.
m	mass flow rate.
n	coefficient (slope) in the Wallis flooding correlation.
n'	coefficient (slope) in the Kutateladze flooding correlation.
p	pressure.
ρ	density.
σ	surface tension.
T	temperature.

Subscripts

c	condensation.
f	liquid.
g	vapour.
1, 2	cases 1, 2
1, 2ϕ	single-, two-phase.

Multi-Phase Flow and Heat Transfer III. Part B: Applications
edited by T.N. Veziroğlu and A.E. Bergles
Elsevier Science Publishers B.V., Amsterdam, 1984 — Printed in The Netherlands

293

A STUDY ON THE RETENTION OF FISSION PRODUCTS BY REACTOR COMPONENTS DURING HYPOTHETICAL ACCIDENTS

W. W. Yuen
Associate Professor
Department of Mechanical and
 Environmental Engineering
University of California
Santa Barbara, California 93106, U.S.A.

K. H. Sun
Nuclear Safety and Analysis Department
Electric Power Research Institute
Palo Alto, California 94303, U.S.A.

ABSTRACT

The existing approaches utilized in the analysis of fission product transport in reactor components are evaluated. Based on a consideration of the thermal-hydraulic conditions existed within the component, the accuracy of a number of assumptions made in the existing approaches is demonstrated to be uncertain. The flow within a reactor component (upper plenum, pressurizer) is assumed to be fully developed turbulent pipe flow in the existing computer code. The relevant thermal-hydraulic conditions, however, indicate that the flow should be more appropriately described as a developing turbulent boundary layer. The existing models neglect concentration gradient existed within the steam/nuclide vapor mixture. Calculations show that this simple model will fail to predict the important delay of fission product release and underestimate the amount of retention. In the pool scrubbing analysis, the present approach assumes constant bubble radius and bubble velocity. A more reaistic model indicates that this assumption is too simplistic. The correlations currently used to predict particle deposition in a reactor component are also demonstrated to have uncertain accuracy. Alternate relation is proposed.

Using a one-dimensional model, the retention of fission product by a reactor component is calculated for three scenarios. They are the retention of fission product from a flow of steam/nuclide vapor mixture, the retention of fission product from a flow of steam/nuclide vapor and particulate mixture and pool scrubbing. For all three cases, the important dimensionless parameters which are relevant to the retention process are identified and their effects are illustrated quantitatively.

1. INTRODUCTION

The understanding of the various physical phenomena and mechanisms affecting the release of radioactive materials during postulated reactor accidents is currently a problem of considerable practical importance for the nuclear industry. This knowledge is one of the most important elements based on which the risk to the public of various postulated accidents in the operation of nuclear power plants is determined.

In a recent report (NUREG-0772) [1], a survey of the best technical information currently available for the above task was presented. The report clearly demonstrates that the current level of understanding of the various physical processes relating to this phenomenon has greatly improved over the

c_s^w = equilibrium vapor concentration of the nuclide at the temperature of the wall surface (g/cm^3)

A_w = area of wall surfaces (cm^2)

V = volume of the element (cm^3)

k_w = deposition velocity for mass transfer (cm/sec)

Assuming a fully developed turbulent pipe flow existed within the controlled volume, k_w is evaluated by the following correlation [5]

$$Sh = 0.023 \, Re^{0.83} Sc^{0.33} \tag{2}$$

where Sh, the Sherwood number Sc, the Schmidt number and Re, the Reynolds number are defined by

$$Sh = \frac{k_w D_H}{D} \tag{3}$$

$$Sc = \frac{\mu}{\rho D} \tag{4}$$

$$Re = \frac{\rho U D_H}{\mu} \tag{5}$$

with

D_H = Hydraulic diameter of the tube (cm)
ρ = density of the fluid (g/cm^3)
D = diffusion coefficient of nuclide vapor in steam (cm^2/sec)
μ = viscosity of the fluid $(g/(sec\text{-}cm))$
U = bulk mean velocity of the fluid within the control volume (cm/sec)

For the TMIB' accident which proceeds slowly, TRAP-MELT further assumes that the flow is natural-convection driven. The Reynolds number is generated from the Grashof number, Gr, by the relation

$$Re = \left(\frac{Gr}{70}\right)^{\frac{1}{2}} \tag{6}$$

where

$$Gr = \frac{g(1-T/T_w)L^3}{\nu^2} \tag{7}$$

with

g = gravitational constant $(980 \; cm/sec^2)$
T = bulk fluid temperature (°K)
T_w = temperature of the wall of the control volume (°K)

L = length of the control volume (cm)
ν = μ/ρ (cm^2/sec)

Two difficulties can be readily identified with the above characterization of the flow. They are:

1.   The Reynolds number and Grashof number as defined by equations

(5) and (7) respectively utilize different characteristic

lenghts. Equation (6) is thus inconsistent.

2.   The assumption of a fully developed turbulent flow within a

control volume is often inaccurate in a TMLB' accident.

As demonstrated in ref. 25 using some typical thermal-hydraulic data computed
for a TMLB' accident, the flow within the different control volumes should be
more accurately described as a developing, natural-convection-driven,
turbulent boundary layer. A more accurate estimate of the mass transfer
coefficient is given by [5]

$$Sh = Sc^{1/3}[0.037 \ Re_{eq}^{0.8} - 850] \tag{8}$$

with

$$Re_{eq} = (Gr)^{1/2} \tag{9}$$

$$Sh = k_w \ L/D \tag{10}$$

It is important to note that for a developing turbulent boundary layer, the
characteristic length utilized in the definition of Gr, Re and Sh is the
length of the control volume. The correction factor introduced in equation
(A1) for fully-developed turbulent pipe flow is thus not necessary in
equation (9).

3.2 Evaluation of particle deposition from the flow onto the wall of a control volume of the primary system.

Because of the existence of debris and aerosol within the reactor during
a hypothetical accident, the deposition of particles onto the wall of a control volume is an important process affecting the eventual release of fission
products into the environment. If particles are radioactive, their release
will add directly to the inventory of released fission products. For non-
radioactive particles, they provide addtional surfaces for the removal of
nuclide vapor (by vapor absorption and condensation onto the particle's surface) and their retention will affect indirectly the overall fission product
release.

In nearly all of the reported works in this area (an excellent reveiw is
given by Sehmel in reference [6]), the particle disposition process is des-

cribed by a deposition velocity, V_d. For a tube, V_d is defined by the expression

$$\frac{dN}{dt} = - V_d N(\frac{A_w}{V}) \tag{11}$$

where

N = the average concentration of particles in a circular cross section

of the tube (cm^{-3}).

In TRAP-MELT, the deposition velocity is calculated by two separate correlations depending on the size of the particle.

For supermicron particles (particles with radius greater than 1 micron), V_d is written as

$$V_d = K^+ U_f \tag{12}$$

where K^+ is a dimensionless parameter given by

$$K^+ = 1.47 \times 10^{-16} \rho_p^{1.01} \bar{R}^{2.1} Re^{3.02} \tag{13}$$

with

U_f = friction velocity (cm/sec)

ρ_p = particle density (g/cm^3)

$\bar{R}$ = ratio of particle diameter to pipe diameter

Re = Reynolds number

Consistent with the assumption of fully developed turbulent flow within the pipe, TRAP-MELT assumes that the friction velocity is given by

$$U_f = U(f/2)^{\frac{1}{2}} \tag{14}$$

with

$$f = 0.0014 + 0.125 \, Re^{-0.32} \tag{15}$$

It is important to note that equation (13) is a purely empirical relation proposed by Sehmel [6] based on his survey of available experimental data.

For submicron particles (particles with radius less than 1 micron), TRAP-MELT uses equation (12) but with

$$K^+ = \frac{(D_p/\upsilon)^{2/3}}{14.5 \frac{1}{6} \ell n \frac{(1+\phi)^2}{1-\phi+\phi^2} + \frac{1}{3^{\frac{1}{2}}} \tan^{-1} \frac{2\phi-1}{3^{\frac{1}{2}}} + \frac{\pi}{6(3)^{\frac{1}{2}}}} \tag{16}$$

where

$$\phi = \frac{1}{2.9(D_p/\nu)^{1/3}} \qquad (17)$$

and

D_p = Brownian diffusivity of particle (cm^2/sec).

Based on a consideration of the typical particle size and the flow condition within the control volume as described in ref. 25, it is the opinion of the present authors that equations (12) through (15) are not applicable for reactor conditions encountered in a TMLB' accident. Equations (16) and (17) can be utilized for all particle size and the friction factor should be given by

$$f = 0.0738 \ Re_{eq}^{-0.2} \qquad (18)$$

where Re_{eq} is as defined by equation (9).

3.3 Modeling of pool scrubbing.

One of the most important processes affecting the retention of fission product in the containment is the removal of nuclide vapor from steam/nuclide bubbles rising in the suppression pool. In ref. 25 of NUREG-0772 [1], the existing modeling approach to this phenomenon is described. Specifically, the model assumes that the bubble has a constant radius and constant fission product concentration as it rises. Using a constant overall mass transfer coefficient between the bubble and the pool, the scrubbing factor SF, which is defined as the fraction of nuclide vapor scrubbed from the bubble, is written as

$$SF = 1 - \exp(- \frac{3k_2\tau}{a}) \qquad (19)$$

where

k_2 = overall mass transfer coefficient for vapor transport (cm/sec)

τ = rise time of the bubble (sec)

a = radius of bubble (cm)

Using "typical" values for rise velocity (40 cm/sec), bubble radius (1.5 cm) and diffusivities of nuclide vapor in steam and water (0.07 and 2×10^{-3} cm^2/sec respectively), k_2 is estimated to be about 0.182 cm/sec based on a standard heat transfer correlation for turbulent flow around a solid sphere. Equation (19) becomes

$$SF = 1 - \exp(- \frac{0.546\tau}{a}). \qquad (20)$$

For a given rise time and bubble radius, SF is determined from the above equation.

As discussed in ref. 25 and demonstrated by a more detailed mathematical model in the next chapter, the above simple model for pool scrubbing can be quite inaccurate. Both the bubble radius and velocity can change significantly as the bubble rises. The accuracy of k_2 utilized in equation (20) is also uncertain.

3.4 Evaluation of thermodynamic properties and others.

Even with sound and realistic physical modellings, the prediction of fission product retention for a given hypothetical accident can still remain inaccurate because of the uncertainty in values of the various thermodynamic properties. A small change in the mass diffusivity of nuclide vapor in steam, for example, can lead to a substantial change in the prediction of the overall retention. Currently, many of the thermodynamic properties of fission products (such as partition coefficient, latent heat of vaporization, mass diffusivity, etc.) utilized in the various physical models are only known approximately. Results generated from the different models are thus accurate only in predicting qualitative trend. For accurate quantitative results, efforts to measure thse parameters accurately under accident conditions are needed.

A number of other areas where improvement on its physical modelling are possible can still be identified. Most of them, however, are either related to the previously mentioned phenomena or of less importance. In the evaluation of thermophoretic deposition of particles, for example, TRAP-MELT uses an expression developed by Brock [7] in which the temperature gradient at the wall is required. Currently, the Dittus-Boelter correlation for fully developed turbulent heat transfer in a pipe is used to evaluate this temperature gradient. Based on the previous discussion on flow characteristics within a control volume, a more accurate correlation for this purpose should be that of a turbulent boundary layer flow. In any case, the effect of this discrepancy is probably insignificant because thermophoretic deposition is not expected to be a dominant mechanism of particle deposition for most hypothetical accidents.

3. ANALYSIS

Even with accurate models developed for the different physical phenomena, these models must be combined in a mathematically and physically consistent manner in order that the amount of fission products released from the reactor during hypothetical accidents can be predicted accurately. The objective of this section is to present some example analysis based on a one-dimensional model. For simplicity, only one or two mechanisms of mass transfer will be included in each analysis. The physical models are thus idealistic. Nevertheless, the present results should be useful in identifying relevant dimensionless parameters for the different mass transfer mechanisms and obtaining qualitative trend on how these parameters affect the overall fission product release. Since different components of the reactor are modelled as one-dimensional control volume in the existing computer codes [2,3], some results of this chapter can also be applied directly to improve the accuracy of these codes in its prediction of fission products release.

4.1 Mass transfer from a steam/nuclide vapor mixture flowing in a circular pipe.

The geometry and coordinate system are shown in Figure 1. A steam/nuclide single-phase vapor mixture is flowing through the system. The initial mass fraction of nuclide vapor for the entering flow at x = 0 is known. The objective of the analysis is to determine the mass fraction of nuclide vapor for the exiting flow at x = L. The relevant mass, momentum, energy and species conservation equations for the present analysis can be readily obtained from standard textbook [8]. An exact solution to these equations, however, is clearly quite complicated and beyond the scope of the present work. It will also not serve the main purpose of this study which is to identify important dimensionless parameters and to develop qualitative trend. The present work thus simplifies the analysis by the following assumptions:

1. The temperature, pressure, density and velocity of the entering fluid, the exiting fluid and the fluid within the pipe are uniform at some prescribed average bulk values. (In actual analysis of a hypothetical accident, these thermodynamic properties for a given control volume, together with the wall temperature, will be predicted by computer codes such as MARCH. While they are generally functions of time, they can be considered as constant over a short time period for accidents which proceed slowly (such as TMLB'). Results of the present study can thus be applied to a given control volume separately over separate time periods.)

2. The mass fraction of fission products within a pipe is a function only of the axial coordinate x and time. Mass diffusion in the radial direction is described by a "convective-like" mass flux expression given by

$$q_m = \rho \, k_w (m_x - m_{x,w}) \tag{21}$$

where

q_m = mass flux deposited on the wall of the control volume

(g/cm^2)

ρ = density of the mixture (g/cm^3)

m_x = bulk mass fraction of fission products in the fluid

$m_{x,w}$ = equilibrium mass fraction of fission products adjacent

to the wall

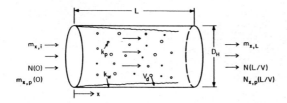

Fig. 1.   Geometry and coordinate system for a one-dimensional pipe-flow mass transfer analysis. (k_w, k_p and v_d stand for deposition velocities of the three mechanisms described in the text.

It is interesting to note that if k_w is calculated from heat and mass transfer analogy and $m_{x,w}$ is the saturated mass fraction of fission products evaluated at the wall temperature, equation (21) describes the rate of condensation of fission products on the control volume's wall. If k_w is tabulated from vapor absorption data [9] and $m_{x,w}$ is set to be zero, on the other hand, equation (21) describes the rate of absorption of fission products from the vapor mixture by the control volume's wall.

3. Mass diffusion in the axial direction is negligibly small compared to the axial convective mass transfer due to the bulk flow of the mixture.

Based on the above assumptions, the relevant mass transfer equation for the present problem becomes

$$\rho A \left[\frac{\partial m_x}{\partial t} + U \frac{\partial m_x}{\partial x} \right] = -\rho \left(\frac{4A}{D_H} \right) k_w (m_x - m_{x,w})$$ (22)

The initial and boundary conditions for the above equation are

$$m_x(x,t) = m_{x,0} \qquad (t \leq 0)$$ (23)

and

$$m_x(0,t) = m_{x,i} \qquad (t \geq 0)$$ (24)

with $m_{x,0}$ and $m_{x,i}$ being the initial mass fraction of the fission product within the pipe and the mass fraction of the fission product for the entering flow respectively.

Assuming that $m_{x,i}$ is constant, exact solution to equation (22) can be readily obtained. Details of the solution method and the expression for m_x for arbitrary values of x and t are given in ref. 25. Physically, the most important results for the present consideration is $m_{x,L}$, the mass fraction of the fission product of the exiting fluid at x = L. For a control volume with a uniform initial concentration of fission product vapor, results in ref. 25 yields

$$m_{x,L}(t) = m_{x,0} \exp\left(- \frac{4k_w t}{D_H}\right) + m_{x,w}\left(1 - \exp\left(- \frac{4k_w t}{D_H}\right)\right)$$ (25)

for t < L/U and

$$m_{x,L}(t) = m_{x,i} \exp\left(- \frac{4k_w L}{UD_H}\right) + m_{x,w}\left(1 - \exp\left(- \frac{4k_w L}{UD_H}\right)\right)$$ (26)

for t > L/U.

Equation (25) indicates that for t < L/U, the mass fraction of fission products in the exiting stream is a function only of $m_{x,0}$, $m_{x,w}$ and the dimensionless parameter of $k_w t/D_H$. There is a time delay of L/U before the effect of the incoming stream is felt at the exit. For slowly developing accidents, this delay can be quite substantial. Because of the possibility of human action and automotive control feature within the reactor that can halt an accident, this time delay can have significant impact on the overall estimated risk for a hypothetical accident.

It is important to note that in the existing analysis such as TRAP-MELT, each control volume is assumed to be well-mixed and hence no time delay for the release of fission products can be predicted. Indeed, if the governing transport equation in TRAP-MELT (equation (6) in reference [3]) is applied to the present situation in which only one mass-transfer mechanism is considered, it becomes

$$\rho \frac{dm_x}{dt} = -\rho(\frac{4A}{D_H})k_w(m_x - m_{x,w}) + \frac{\rho U}{L}(m_{x,i} - m_x) \tag{27}$$

The solution to the above equation is

$$m_x(t) = m_{x,0} e^{-t/t_c} + \frac{Ut_c}{L} m_{x,R}(1 - e^{-t/t_c}) \tag{28}$$

where

$$t_c = \frac{L}{U[1 + \frac{4K_w L}{UD_H}]} \tag{29}$$

and

$$m_{x,R} = m_{x,i} + (\frac{4k_w L}{UD_H})m_{x,w} \tag{30}$$

Equation (28) is not a function of position due to the well-mixed assumption and it yields a physically unrealistic prediction that the effect of the incoming stream is felt immediately at the exit.

For $t > L/U$, equation (26) indicates that the mass fraction of fission product at the exiting stream is constant and depends only on the incoming mass fraction $m_{x,i}$, the equilibrium mass fraction at the wall $m_{x,w}$ and the parameter $k_w L/(UD_H)$. Physically, this result is quite reasonable as the parameter $k_w L/(UD_H)$ can be interpreted as the ratio of the characteristic time for mass convection in the axial direction (L/U) to that of mass diffusion in the radial direction (D_H/k_w). When $L/U \gg D_H/k_w$, the radial mass diffusion effect dominates and the amount of retention is large. When $L/U \ll D_H/k_w$, on the other hand, the axial convective effect is large and the amount of retention is thus small.

As an example of application of the present analysis, the mass fraction of fission product of the vapor stream exiting from the pressurizer as a function of time for a TMLB' accident with thermal-hydraulic conditions is calculated based on equation (26) and presented in Figure 2. The value of k_w used in the calculation is taken from Table A2 assuming that natural convection is the dominant mode of heat and mass transfer. The initial mass fraction is assumed to be zero. The equilibrium mass fraction at the wall of the control volume is also assumed to be negligible compared to the bulk mass fraction. (This is reasonable assumption if the mass transfer process is assumed to be condensation and the partial pressure of the fission product (say CsI) at the wall is assumed to be the saturation pressure at the wall temperature given by standard reference [10] .) As a basis for comparison, results calculated with the TRAP-MELT approach (equation 28) are also presented in the same figure. It is interesting to note that equation (28) not only fails to predict the initial delay in the release of fission product, but also over-estimates the magnitude of release. Mathematically, it can be readily shown that the present approach is identical to the TRAP-MELT approach in the limit of

$\dfrac{k_w L}{UD_H} \to 0$. Since this parameter is not negligible (because of small U) in a

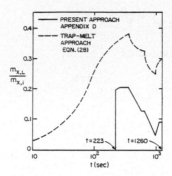

Fig. 2. Prediction of fission product
release from the pressurizer in
a TMLB' accident based on the
one-dimensional model with no
particulate flow.

TMLB' accident, the TRAP-MELT approach thus over-estimates the amount of fission product release. The amount of fission product release predicted by both approaches is less than the reported value in reference [1]. This is probably because only one mechanism of fission product mass transfer (say condensation) is assumed in the calculation.

4.2 Mass transfer from a mixture of steam/nuclide vapor and particles
flowing in a circular pipe.

The geometry and coordinate system are again as shown in Figure 1. General conservative equations for the present problem can be obtained from standard reference [8]. The same set of simplifying assumptions used in the previous case are again utilized to reduce the required mathematical complexity. In addition, the mixture is now assumed to consist of particles flowing with the same average bulk velocity of the mixture. All particles are assumed to have the same size and coagulation of particles is neglected. Physically, the assumption of constant particle size is not restrictive as the present analysis can be readily generalized to cases with some arbitrary particle size distributions. Similar to chapter 6 in reference [1], the size of particles is assumed to be 10μm or less. An order-of-magnitude estimate in the same reference shows that coagulation is relatively ineffective in changing the size of particles and thus can be ignored in the calculation.

As in the previous case, a time delay of L/U must elapse before the effect of the incoming mixture is felt at the exit. After such delay, the mass fraction of fission product of the vapor and the particle concentration leaving the control volume are constants independent of time. Because of the increased mathematical complexity, the present work develops solution only for time greater than L/U. It is also convenient to consider the problem in a reference frame moving with the same velocity as the entering mixture. The governing equations for the mass fraction of fission product in the vapor stream $m_x(t)$ and the particle concentration $N(t)$ are

$$A \frac{dN}{dt} = -V_d \left(\frac{4A}{D_H}\right) N \tag{31}$$

and

$$\rho A \frac{dm_x}{dt} = - \frac{4k_w \rho A}{D_H}(m_x - m_{x,w}) - N(t)A4\pi a^2 k_p (m_x - m_{x,p}) \tag{32}$$

where

k_p = deposition velocity for the condensation of fission product onto the surface of the particle (cm/sec)

$m_{x,p}$ = equilibrium mass fraction at the surface of the particle

As noted in the previous chapter, V_d in equation (31) is best evaluated by equation (12), together with equations (16), (17) and (18). The Brownian diffusivity of particles in steam D_p is related to the particle's mobility B by the well known relation [11]

$$D_p = BkT \tag{33}$$

with

k = Boltzman constant $(1.38 \times 10^{-23}$ Joule/°K)

T = absolute temperature (°K)

For the mobility of a particle of radius a, it is customary to use the empirical relation [12,13]

$$B = \frac{1 - X\ell/a + (Q\ell/a)e^{-ba/\ell}}{6\pi\mu a} \tag{34}$$

where

μ = viscosity of the carrier gas (g/(sec-cm))

ℓ = mean free path of the gas molecules (cm)

The constants X, Q and b depend on the carrier gas and particles. While data are available for these parameters for the flow of oil droplets in air [13] (X = 0.864 and Q = 0.29), they are not available for a particle/steam mixture at reactor conditions. For typical hypothetical accident (such as TMLB'), however, it can be readily shown that $\ell/a \ll 1$. The detailed assumption and calculation for this order-of-magnitidue estimate is given in ref. 25. Assuming that X and Q for a particle/steam mixture at reactor conditions are of the same order of magnetude as those for oil in air, it is reasonable to write

$$B = \frac{1}{g\pi\mu a} \tag{35}$$

Assuming that all thermodynamic properties are constant, equation (31) can be readily integrated to yield.

$$N = N_0 e^{-\beta t} \tag{36}$$

with N_0 being the intial particle concentration and

$$\beta = \frac{4V_d}{D_H} \tag{37}$$

Integration of equation (32) yields

$$m_x = m_{x,p} + (m_{x,i} - m_{x,p})\exp(-Et)\exp[-\frac{G}{\beta}(1 - \exp(-\beta t))]$$

$$+ E(m_{x,w} - m_{x,p})F(t)\exp(-Et)\exp[\frac{G}{\beta}\exp(-\beta t)] \tag{38}$$

where

$$F(t) = \int_0^t \exp[Ex - \frac{G}{\beta}e^{-\beta x}]dx \tag{39}$$

with

$$E = \frac{4k_w}{D_H} \tag{40}$$

$$G = \frac{N_0 4\pi a^2 k_p}{\rho} \tag{41}$$

The detail of the development of equation (38) is given in ref. 25. The amount of fission product deposited on the surface of a particle, $Y(t)$, is governed by the equation

$$\frac{dY}{dt} = 4\pi a^2 k_p(m_x - m_{x,p}) \tag{42}$$

Equation (42) can be integrated to yield

$$Y(t) = Y_0 + 4\pi a^2 k_p \int_0^t (m_x(x) - m_{x,p})dx \tag{43}$$

The total amount of fission product leaving the pipe per unit time is thus given by

$$M_{x,L} = AU[\rho m_x(L/U) + N(L/U)Y(L/U)]$$ (44)

As it is shown in ref. 25, the parameter β is quite small under typical reactor conditions expected in a TMLB' accident (β is estimated to be about 10^{-6} sec^{-1} for both the upper plenum and the pressurizer). For cases with $m_{x,p} = m_{x,w} = 0$ and initially non-radioactive particles ($Y_0 = 0$), an analytical expression for $M_{x,L}$ can be readily generated by expanding equation (44) as a Taylor series in β and keeping only the first order term. One obtains

$$M_{x,L} = R\, m_{x,i}\, \rho AU$$ (45)

with

$$R = \frac{G}{E+G} + \left(\frac{E}{E+G}\right)e^{-(E+G)L/U}$$ (46)

$$+ \beta\left[\frac{EG(L/U)^2 e^{-(E+G)L/U}}{2(E+G)} - \frac{L}{U}\left(\frac{G}{E+G}\right)\right.$$

$$\left. + \frac{L}{U}\frac{EG}{(E+G)^2} e^{-(E+G)L/U} + \frac{G^2}{(E+G)^3}(1 - e^{-(E+G)L/U})\right]$$

The dimensionless parameter R has the obvious physical interpretation of being the fraction of incoming fission product which is released from the considered control volume.

Equation (46) indicates that the important physical parameters affecting the release of fission product in a flow of steam/nuclide vapor mixture and particles are E, G, β and L/U. Physically, 1/E, 1/G, 1/β and L/U can be interpreted as characteristic times for the various physical processes. 1/E is for the deposition of fission product vapor onto the wall of the control volume; 1/G is for the deposition of fission product vapor onto the surface of particles; 1/β is for the deposition of particles onto the surface of the control volume, while L/U is the characteristic time for convective mass transfer due to the bulk flow.

Values of R for some typical values of β, G and E predicted by equation (46) are tabulated and presented in Figures 3 and 4. L/U in these calculations is assumed to be 300 sec. which is the approximate value for the flow within the pressurizer at the early part of a TMLB' accident (see Table A1 for the relevant thermal-hydraulic data). As expected, R increases with increasing G when β is small. Physically, particles provide additional surfaces for fission product mass transfer. When there is little or no particle deposition onto the surface of the control volume, all fission products deposited on the particle's surface will add to the inventory of released fission product. When β is increased, however, R decreases with increasing G when E is small. Physically, a large β means that a significant fraction of particles is retained within the control volume by deposition onto the wall. When E is small, a significant amount of the fission product is deposited on the surface of the particles and remained within the control volume, the amount of fission product present in the exiting stream

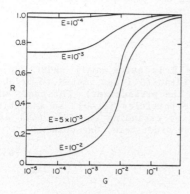

Fig. 3. Ratio of released fission product to incoming value (R) as predicted by equation (46) for different mass transfer characteristic times (E,G with $\beta = 10^{-5}$). Definitions of E, G and β are given in section 4.2.

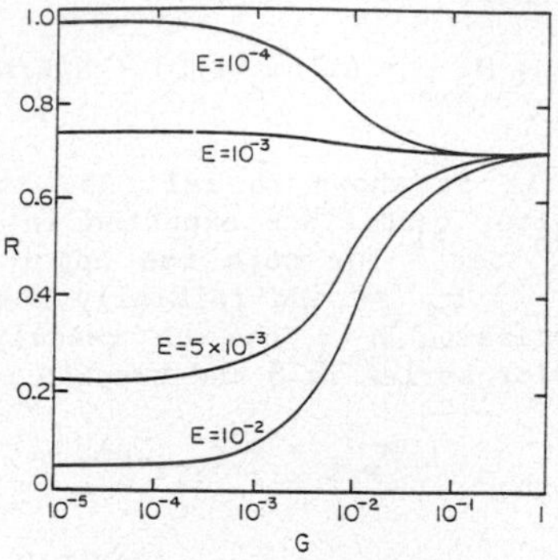

Fig. 4. Ratio of released fission product to the incoming value (R) as predicted by equation (46) for different mass transfer characteristic times (E, G with $\beta = 10^{-3}$). Definitions of E, G and β are given in section 4.2.

decreases. It is important to note that the behavior of R is clearly a sensitive function of β, E and G. A precise knowledge of the exact values of these various parameters at the condition of a hypothetical accident is thus important in generating accurate prediction of fission product retention in flows with particles.

The uncertainty in the evaluation of V_d and k_w, which are directly related to the parameters β and E, have been discussed in previous sections. For k_p, the existing approach [3] assumes that particles are spherical and at rest with respect to the flowing mixture. The mass transfer is assumed to be controlled entirely by molecular diffusion. k_p is given by

$$k_p = D/a \tag{47}$$

Theoretical expression of D for many binary gas systems are available from standard reference [14]. These expressions, however, can only be applied qualitativey to a steam/nuclide vapor mixture because the various atomic constants for such a mixture are not precisely known. Significant uncertainty thus exists in the evaluation of D and consequently the value of G.

4.3 Mass transfer from a steam/nuclide vapor bubble rising in a large pool with finite depth.

This problem has an obvious application for the determination of a scrubbing factor for the suppression pool. A number of important physical phenomena must be included in the analysis. Specifically, mass transfer of steam and nuclide vapor from the bubble to the pool are occuring simultaneously. Heat transfer from the hot bubble to the pool will lead, in addition to the condensation of steam, to the lowering of the bubble's temperature as it rises. As the bubble size decreases due to condensation, its rising velocity also changes. This change in velocity can have a significant effect on the overall rise time for the bubble and consequently the amount of nuclide vapor which will reach the top of the pool.

A simultaneous analytical consideration of all of the above phenomena is clearly quite extensive and beyond the scope of the present work. An excellent summary of the necessary conservation equations which include most of the above phenomena can be obtained from references [15] through [18]. To reduce the mathematical complexity, the following assumptions are utilized in the present analysis:

1. The bubble remains spherical and at saturated temperature and pressure conditions at all times. (Physically, the assumption of a spherical bubble is not too restrictive as this assumption has been used successfully in nearly all of the existing analysis on bubble dynamics. The assumption of saturation within the bubble is probably not accurate at the initial stages when bubbles are first injected into the pool. (Thermal-hydraulic conditions predicted by MARCH indicated that superheated steam/nuclide vapor mixture exists inside the reactor.) But rapid condensation and the breaking up of large bubbles into smaller ones should lead to a saturation condition within the bubble relatively quickly.

2. The bubble velocity as a function of radius is given by the terminal velocity correlation for a gas bubble in a liquid generated by Peebles and Garber [19]. (Realistically, a gas bubble which is initially at rest in a liquid pool will first rise very quickly, then oscillate about some average upward motion and finally settle into a constant terminal velocity. The present work thus assumes that this initial phase of bubble motion and the adjustment of bubble velocity due to the change in its radius all happen very quickly compared to the general motion of the bubble. It is important to note that since the correlation of Peebles and Garber was experimentally determined, it has already included the effect of bubble shape distortion on its motion.)

3. The heat transfer from the bubble to the pool is estimated by empirical correlation of turbulent flow around a solid sphere. (To predict heat transfer by any other method will involve a solution of two simultaneous equations which is beyond the scope of the present work.)

4. The mass transfer of nuclide vapor from the bubble to the pool is controlled by the diffusion of nuclide molecule in steam within the bubble, not by the turbulent convective flow outside of the bubble. (As noted in ref. 25, the concentration gradient of dissolved nuclide outside of the bubble should be quite small. The concentration gradient of nuclide vapor within the bubble, on the other hand, can be large because most fission products are quite soluable in water. The local concentration of nuclide vapor at the inside of the bubble surface, for example, should be zero because of the expected large partition coefficient (see p. 33-35 of reference 2).)

5. The bubble's kinetic energy due to its upward motion is negligibly small compared to its internal energy.

The physical model of the percent analysis is illustrated by Figure 5. Based on the above assumptions, the energy equation of a condensing bubble (based on the general expression given by equation 25 in reference [17]) can be written as

$$h(T_\infty - T)4\pi a^2$$

$$= \frac{d}{dt}[\frac{4}{3}\pi a^3 \rho]H + \frac{4}{3}\pi a^3 \rho \frac{du}{dt} + 4\pi a^2 \frac{da}{dt}(P-P_\infty) \qquad (48)$$

where

T_∞ = pool temperature (°K)

P_∞ = pool pressure (dyne/cm^2)

ρ = overall vapor density within the bubble (g/cm^3)

H = latent heat for the total mass transfer of steam and nuclide vapor (J/g)

u = specific internal energy of the vapor mixture (J/g)

P = pressure within the bubble (dyne/cm^2)

The latent heat of the condensation of steam is well known. The energy involved in dissolving the nuclide vapor in water, however, is unknown. The present work uses the following approximate expression for the total latent heat

$$\frac{d}{dt}[\frac{4}{3}\pi a^3 \rho]H = \frac{dM_s}{dt}(h_{sg}(T) - h_{s\ell}(T_\infty))$$

$$+ \frac{dM_i}{dt}C_{pi}(T - T_\infty) \qquad (49)$$

where

M_s = total mass of steam vapor within the bubble (g)

M_i = total mass of nuclide vapor within the bubble (g)

$h_{sg}(T)$ = specific enthalpy of saturated steam at temperature T (J/g)

$h_{s\ell}(T)$ = specific enthalpy of saturated water liquid at temperature T (J/g)

C_{pi} = specific heat at constant pressure for the nuclide vapor (J/g/°K)

Physically, equation (49) assumes that the energy released in dissolving nuclide vapor in water is small compared to the change of the enthalpy of the nuclide vapor and the latent heat of steam condensation. Since the rate of steam condensation is always much higher than the rate of mass transfer of nuclide vapor, equation (49) should be reasonably accurate.

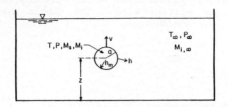

Fig. 5.   The physical model for the
pool scrubbing analysis.

The specific internal energy of the vapor mixture within the bubble is given by

$$u = \frac{M_s u_s + M_i C_{vi}(T - T_\infty)}{M_s + M_i} \tag{50}$$

where

u_s = specific internal energy of steam vapor (J/g)

C_{vi} = specific heat at constant volume for the nuclide vapor (j/g/°K)

The pressure within the bubble is related to the bubble radius and the pool pressure P_∞ by the following relation

$$P = P_\infty + \frac{2\sigma}{a} \tag{51}$$

where

σ = surface tension of the bubble in water (dyne/cm)

The assumption of saturation within the bubble requires that

$$P\left(\frac{M_s/A_s}{M_s/A_s + M_i/A_i}\right) = P_{sat}(T) \tag{52}$$

where

A_s = atomic weight of steam

A_i = atomic weight of nuclide vapor

The mass transfer of nuclide vapor from the bubble's interior to its surface is

$$\frac{dM_i}{dt} = -4\pi a^2 h_m M_i/(M_s + M_i) \tag{53}$$

To complete the mathematical description of the problem, the thermodyanmic equation of state of the mixture within the bubble must be specified. Utilizing the ideal gas law, one obtains

$$\frac{4}{3} \pi a^3 P = (\frac{M_s}{A_s} + \frac{M_i}{A_i}) RT \tag{54}$$

with

$$R = \text{universal gas constant } (8.31 \text{ J/}°\text{K/mole})$$

Equations (48), (51), (52), (53) and (54) constitute a set of five equations based on which the unknowns T, P, a, M_s and M_i can be determined numerically as a function of time. Once a is obtained, the bubble velocity can be generated from assumption 2. The bubble rise time and the scrubbing factor as a function of pool depth can also be readily determined.

Numerical results for the evolution of steam/HI vapor bubbles with different initial conditions (initial radius = 1 cm, 10 cm, initial temperature = 400 K) and different values of h_m rising in an infinite pool of temperature 300 K and 1 atm pressure are tabulated and presented in Figures 6 through 10. (Note that with the ideal gas and saturated bubble assumptions, initial values of M_s and M_i can be determined from the initial bubble radius and temperature.) Some interesting conclusions can be readily generated from these solutions. First, results in Figures 6, 8 and 10 show that the bubble's temperature and radius change continuously as it rises. The mass transfer coefficient for nuclide vapor has a significant effect not only on the mass transfer of nuclide vapor, but also on the condensation of steam and the reduction of the bubble size because of the saturated bubble assumption. The bubble's temperature, however, is insensitive to h_m and depends strongly on the initial bubble radius, indicating that condensation is the dominating mechanism of heat and mass transfer. In both the 1 cm and 10 cm bubbles, the bubble's temperature reaches the pool temperature very quickly.

It is also interesting to note that since the velocity of a 10 cm bubble is generally greater than that of a 1 cm bubble, for a fixed pool depth, the present result shows that the smaller bubble has a larger rise time than the smaller bubble. Qualitatively, the smaller bubble should thus have a larger scrubbing factor (which, as defined in reference [1], is the fraction of fission product scrubbed from the bubble). While results presented in Table E1 of reference [1] are consistent with the present conclusion, they were generated assuming bubble radius and rise time is independent parameters. The bubble velocity were also assumed to be approximately constant. Quantitatively, results presented in reference [1] are thus unreliable.

A great deal of work is needed before this phenomenon of pool scrubbing can be accurately understood. Nevertheless, results of the present calculation indicate conclusively that the existing approach to this problem as described in ref. 25 of reference [1] is inaccurate, even for qualitative application. The sensitivity of the result to the parameter h_m and the initial bubble's radius indicates that for accurate determination of the scrubbing factor, experimental and analytical works directed in determining these quantities accurately under conditions of a hypothetical accident are needed.

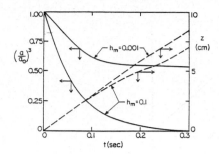

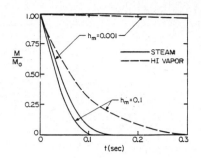

Fig. 6. Bubble radius (a) and rise distance (h) as a function of time
for different interior mass transfer coefficient h_m. (Initial
bubble radius $a_0 = 1$ cm, initial
bubble temperature 400°K.)

Fig. 7. Mass ratio inside the bubble
(m_0 = initial mass within the
bubble) as a function of time
for different interior mass
transfer coefficient h_m.
(Initial bubble conditions
are identical to those of
Fig. 6.)

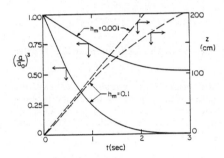

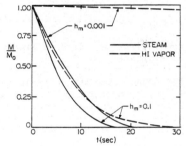

Fig. 8. Bubble radius (a) and rise distance (h) as a function of time
for different interior mass transfer coefficient h_m. (Initial
bubble radius = 10 cm, initial
bubble temperature 400°K.)

Fig. 9. Mass ratio inside the bubble
(M_0 = initial mass within the
bubble) as a function of time
for different interior mass
transfer coefficient h_m.
(Initial bubble conditions are
are identical to those of
Fig. 8.)

4. CONCLUSIONS

The current approaches used in predicting fission product retention by
reactor components during hypothetical accidents are reviewed. Based on the
expected thermal-hydraulic conditions existed within the component, a number
of deficiencies in the current modelling are identified and improvements are
proposed. Utilizing a one-dimensional model, dimensionless parameters which
are relevant for the different mass transfer mechanisms are identified.

Specifically, the present work concludes:
1. For reactor components such as the pressurizer and the upper plenum for
   which the length to diameter ratio is not too large, the current assumption of fully developed pipe flow existed within the component is an
   inadequate representation of its thermal-hydraulic state in a slowly developing accident such as TMLB'. A developing turbulent boundary layer
   is a more realistic representation based on which the mass transfer

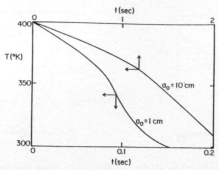

Fig. 10.   Bubble temperature as a function of time
for different initial bubble radius and
$h_m = 0.001$. (Cases with $h_m = 0.1$ yield
essentially identical results.)

coefficient for fission product deposition on the component wall should
be evaluated.

2.   For TMLB' type of accident in which the bulk flow through the various
reactor component is low, the current TRAP-MELT approach which assumes a
mixture of uniform concentration within the component is unrealistic.
It fails to predict the important delay of fission product release and
underestimates the overall amount of retention.

3.   For the case of fission product transport from a steam/nuclide vapor
mixture (without particulate), a one-dimensional analysis shows that the
dimensionless parameter relevant for the mass transfer process is
$k_w L/(UD_H)$ which can be interpreted physically as the ratio of the characteristic time for mass transfer by convection in the axial direction
(L/U) to that of mass diffusion in the radia direction (D_H/k_w).

4.   For the case of fission product transport from a steam/nuclide vapor and
particulate mixture, a one-dimensional analysis shows that the relevant
dimensionless parameters are D_H/k_w, $\rho/N_Q a^2 k_p$, D_H/V_d and L/U. These
parameters can be interpreted as characteristic times for the deposition
of nuclide vapor onto the wall of the reactor component, the deposition
of nuclide vapor onto the surface of particles, the deposition of particles onto the wall of the reactor component and convective mass transfer
due to the bulk flow respectively.

5.   In the scrubbing of fission product from a steam/nuclide vapor bubble by
a subcooled water pool, the size and rise velocity of the bubble are key
parameters for the determination of fission product retention. The two
parameters depend quite strongly upon each other can vary significantly
as the bubble rises in a pool, depending primarily on the steam condensation process. The approach utilized in reference [1], which assumed
a constant rise velocity and treated bubble radius and rise time as
independent parameters, is unreliable.

5. ACKNOWLEDGEMENT

One of the authors (W.W. Yuen) would like to thank the Electric Power Research Institute, Palo Alto for supporting this work. Discussions with J.A. Gieseke of Battelle Columbus Laboratories were very helpful.

6. NOTATION

a	bubble radius (cm)
A	cross sectional area of the control volume (cm^2)
A_s	molecular weight of steam
A_i	molecular weight of fission product vapor
A_w	surface area of a control volume (cm^2)
B	particle mobility, equation (34) (g/sec)
C_w	mass concentration of nuclide vapor (g/cm^3)
C_s	equilibrium mass concentration of nuclide vapor at the temperature of a surface (g/cm^3)
C_{vi}	specific heat at constant volume for nuclide vapor (J/(g-K))
C_{pi}	specific heat at constant pressure for nuclide vapor (J/(g-K))
D^p	diffusion coefficient of nuclide vapor in steam (cm^2/sec)
D_H	hydraulic diameter (cm)
D_p	Brownian diffusivity of particles (cm^2/sec)
E	parameter defined by equation (40), 1/E is the characteristic time for deposition of nuclide vapor onto the surface of a control volume (sec^{-1})
f	friction factor, equation (15)
F	function defined by equation (39)
g	gravitational constant (980 cm/sec)
G	parameter defined by equation (41), 1/G is the characteristic time for the deposition of fission product vapor onto the surface of particles (sec)
Gr	Grashof number
h	heat transfer coefficient (J/(cm-sec-K))
h_m	mass transfer coefficient for the transport of fission product vapor from the interior of the bubble to its surface (g/(cm-sec))
h_{sg}	specific enthalpy of steam (J/g)
h_{sl}	specific enthalpy of water liquid (J/g)
H	latent heat for the mass transfer of steam and nuclide vapor from the interior of a steam/nuclide vapor bubble to the surrounding liquid (J/g)
k	Boltzman constant (1.38×10^{-23} J/K)
k_2	exterior mass transfer coefficient of nuclide vapor from a steam/nuclide vapor bubble to the suppression pool (cm/sec)
k_w	deposition velocity for mass transfer of nuclide vapor from a bulk flow to the surface of a control volume (cm/sec)
k_p	deposition velocity for mass transfer of nuclide vapor from a nuclide vapor/steam mixture to the surface of a particle (cm/sec)
K^+	dimensionless parameter defined by equation (13)
ℓ	mean free path of gas molecules, equation (E1) (cm)
L	length of the control volume (cm)
m_x	mass fraction of nuclide vapor
$m_{x,i}$	mass fraction of nuclide vapor in the mixture flowing into a control volume
$m_{x,0}$	initial mass fraction of nuclide vapor in the mixture within a control volume
$m_{x,w}$	equilibrium value of m_x at the surface of a control volume
$m_{x,p}$	equilibrium value of m_x at the surface of a particle
M_s	total mass of steam vapor within a bubble (g)
M_i	total mass of nuclide vapor within a bubble (g)

$M^-_{x,L}$ mass flow rate of fission product out of a control volume (g/sec)

n concentration of gas molecules (cm^{-3})

N concentration of particles in a control volume (cm^{-3})

N_0 intial concentration of particles in a control volume (cm^{-3})

Nu_L average Nuselt number over a flat plate of length L

P pressure with a bubble ($dyne/cm^2$)

P_∞ ambient pressure of the suppression pool ($dyne/cm^2$)

q_m mass flux of nuclide vapor deposited on the wall of a control volume (g/cm^2)

Q dimensionless constant used in the definition of B, eqation (34)

R fraction of incoming fission product which is released from a control volume, equation (46)

$\bar{R}$ ratio of particle diameter to pipe diameter

R_e Reynolds Number, equation (5)

$S\bar{h}$ Sherwood Number, equations (3) and (10)

Sc Schmidt Number, equation (4)

SF scrubbing factor, equation (19)

t time (sec)

t_c time constant defined by equation (29) (sec)

T bulk fluid temperature (K)

T_w control volume's surface temperature (K)

T_∞ ambient temperature of the suppression pool (K)

u specific internal energy of a steam/nuclide vapor mixture (J/g)

u_s specific internal energy of steam (J/g)

U bulk mean velocity of the steam/nuclide vapor mixture (cm/sec)

U_f friction velocity (cm/sec)

V volume of the control volume (cm^3)

V_d deposition velocity for the transport of particles from a particle/vapor flow onto the surface of a control volume, equation (11) and (12) (cm/sec)

X dimensionless parameter used in the definition of B equation (34)

Y the amount of fission product deposited on the surface of a particle (g)

β parameter defined by equation (37), $1/\beta$ is the characteristic time for the deposition of particles onto the surface of a control volume (sec)

δ^w_L boundary layer thickness at the top of the control volume calculated from natural convection correlation (cm)

δ^F_L boundary layer thickness at the top of the control bolume calculated from forced convection correlation (cm)

ξ normalized mass fraction of fission product defined by equation (F2)

ξ_w normalized equilibrium mass fraction of fission product at the wall of the control volume defined by (F3)

ξ_i normalized initial mass fraction of fission product in the mixture within the control volume as defined by euation (F6)

η dimensionless length used in ref. 25

θ dimensionless time used in ref. 25

μ viscosity of the steam/nuclide vapor mixture (g/(sec-cm))

ν kinematic viscosity, μ/ρ (cm^2/sec)

ρ density (g/cm)

ρ_p particle density (g/cm)

σ surface tension (dyne/cm)

τ rise time of the bubble (sec)

ϕ dimensionless parameter defined by equation (17)

7. REFERENCES

1. Silberberg M. et al., "Technical Bases for Estimating Fission Product Behavior During LWR Accidents," USNRC Report, NUREG-0772, March 1981.

2. Gieseke, J.A., et al., "Analysis of Fission Product Transport Under Terminated LOCA Conditions," Battelle Columbus Laboratory Report, BMI-NUREG-1990, December 1977.

3. Jordon, H., Gieseke, J.A. and Baybutt, P., "TRAP-MELT Users Manual," Battelle Columbus Laboratories Report NUREG/CR-0632, BMI-2017, February 1979.

4. Wooton, R.O. and Avci, H.I., "MARCH (Meltdown Accident Response Characteristics) Code Description and User's Manual," Battelle Columbus Laboratories Report, NUREG/CR-1711, BMI-2017, October 1980.

5. Holman, J.P., Heat Transfer, 5th Edition, McGraw-Hill Book Company, New York, 1981.

6. Sehmel, G.A., "Particle Deposition from Turbulent Air Flow," J. Geophys. Res., Vol. 75, 1970, p. 1766.

7. Brock, J.R., "On the Theory of Thermal Forces Acting on Aerosol Particles," Journal of Colloid Science, Vol. 17, 1962, p. 768.

8. Soo, S.L., Fluid Dynamics of Multiphase Systems, Blaisdell Publishing Company, 1967.

9. Genco, J.M. et al., "Fission Product Deposition and Its Enhancement Under Reactor Accident Conditions: Deposition on Primary-System Surfaces," Battelle Columbus Laboratories Report, BMI-1863 (March 1969).

10. Weast, R.C. and Astle, M.J., Editor, CRC Handbook of Chemistry and Physics, CRC Press, Inc., Florida 1980.

11. Zebel, G., "Coagulation of Aerosols," Aerosol Science, Edited by C.N. Davies, Academic Press, New York, 1966, p. 31.

12. Knudsen, M. and Weber, S., Ann. der Phys., Vol. 36, 1911, p. 982.

13. Millikan, R.A., "The General Law of Fall of a Small Spherical Body Through a Gas, and its Bearing upon the Nature of Molecular Reflection from Surfaces", Physical Review, Vol. 22, 1923, p. 1.

14. Reid, R.C. and Sherwood, T.K., The Properties of Gases and Liquids, 2nd Edition, McGraw-Hill, New York, 1966.

15. Forster, H.K. and Zuker, N., "Growth of a Vapor Bubble in a Superheated Liquid," Journal of Applied Physics, Vol. 25, 1954, pp.474-478.

16. Plesset, M.S. and Zwick, S.A., "The Growth of Vapor Bubbles in Superheated Liquids," Journal of Applied Physics, Vol. 25, 1954, pp. 493-500.

17. Scriven, L.E., "On the Dynamics of Phase Growth," Chemical Engineering Science, Vol. 10, Nos. 1/2, 1959, pp. 1.

18. Florschuetz, L.W. and Chao, B.T., "On the Mechanics of Vapor Bubble Collapse," Trans. ASME. Journal of Heat Transfer, May, 1954, pp. 209-220.

19. Peebles, F.N. and H.J. Garber, "Studies on the Motion of Gas Bubbles in Liquids", Chem. Eng. Prog., Vol. 49, 1953, pp. 88-97.

318

20. Jackson, J.D. and J. Fewster, "Enhancement of Turbulent Heat Transfer due to Buoyancy for Downward Flow of Water in Vertical Tubes," Heat Transfer and Turbulent Buoyancy Convection, Edited by Spalding, D.B. and Afgan, N., Vol. 1, Hemisphere Publishing Corporation, McGraw-Hill, New York, 1977, P. 759.

21. Davies, C.N., "Deposition from Moving Aerosols," Aerosol Science, Edited by C.N. Davies, Academic Press, New York, 1966, P. 393.

22. Wallis, G.B., One Dimensional Two-Phase Flow, McGraw Hill, 1969.

23. Linke, W.F., Solubilities of Inorganic and Metal Organic Compounds, 4th Edition, American Chemical Society, Washington, D.C. (1965).

24. Reif, F., Fundamental of Statistical and Thermal Physics, McGraw-Hill Book Company, New York, 1965.

25. Yuen, W.W. and K.H. Sun, "A Study on the Retention of Fission Products by Reactor Components During Hypothetical Accidents," UCSB Report ME-82-3, 1982.

Multi-Phase Flow and Heat Transfer III. Part B: Applications
edited by T.N. Veziroğlu and A.E. Bergles
Elsevier Science Publishers B.V., Amsterdam, 1984 — Printed in The Netherlands   319

A FUEL ROD DEBRIS PACKING MODEL

Jeffrey A. Moore
Department of Physics and Engineering
Fort Lewis College
Durango, Colorado, 81301, U.S.A.

ABSTRACT

The nonrandom packing of fuel rod debris around and above the surviving fuel rod segments in a degraded core is analyzed with the spacer grids being modeled as a porous floor. The irregular shape of the debris is simulated by assuming that all of the spherical particles terminate their migration within the debris bed at their first two-point contact. The analytical approach is verified by comparing the computational results with experimental data for nonrandom packing. Specific calculations for the TMI-2 geometry reveal an average (horizontally integrated) nonrandom packing density between the fuel rods of approximately 0.30. If simulated vibrations are imposed, this value increases to 0.50. If the debris bed builds up above the fuel rod stubs, the average (horizontally integrated) packing density above these rods achieves a value of approximately 0.38 without vibrations; loosely packed gravel has an average random packing density of 0.45.

1. INTRODUCTION

One of the major phenomena often associated with a core melt or near core melt accident in a nuclear reactor core is the fracture of fuel pellets, oxidized cladding and solidified uranium dioxide/zircalloy eutectic when the overheated core is quenched. The resulting fragments settle in the core with the surviving fuel rod spacer grids acting as sieves to catch the larger particles. The remaining filtered debris comes to rest on the core lower support plate, or passes into the lower plenum of the reactor vessel.

The particular manner in which these particles pack to form debris beds within the core dictates the void fraction distribution in each packed bed, and thus has a major influence on the subsequent cooling history of the degraded core. This paper describes a method for predicting the (horizontally integrated) packing density function within a debris bed that rests on a porous floor and is composed of an arbitrary size distribution of solid particles that surround or bury the surviving segments of the fuel rods.

2. THE ANALYTICAL MODEL OF THE PACKING PROCESS

The packing of granular materials is important in a variety of disciplines, including powder metallurgy, soil mechanics, mining and heat exchanger design.

In addition to such obvious factors as particle shape and size distribution, parameters that influence the final packed state include the dynamics of particle deposition, externally applied compaction loads, vibrations within the packed bed, container wall effects and floor porosity.

Although a technique has recently been developed for modeling the dynamic packing of arbitrarily shaped blocks of geological materials (1), it would be impractical and unrealistic to attempt such a detailed modeling for the reactor meltdown problem. Rather, the debris is modeled as an arbitrary distribution of hard spheres. The irregular shapes of real debris are approximated by requiring that each particle terminate its migration within the packed bed when it realizes a two-point contact.

The random packing of smooth spheres on an infinite nonporous floor has been studied theoretically for single (2), binary (3), and log-normal size distributions (4). In our analytical model, the physical process of depositing particles on the debris bed is simulated on the computer in a manner similar to that of Visscher and Bolsterli (3). That is, spheres from an arbitrary initial distribution of particle sizes are dropped sequentially from a random point above the floor with an exclusion zone extending one particle radius outward from the surface of the cylinder. Each particle then moves under the vertical force of gravity until it either hits another particle on the debris bed or strikes the floor. If it hits another sphere, it rolls with its center moving on a vertical plane until it contacts either the cylinder, a second sphere or the floor. When it is in contact with two spheres, or one sphere and the cylinder, the position of the rolling sphere is considered to be stable. Particles that strike the porous floor pass through if their diameter is less than the prescribed floor porosity; otherwise, they stick at their point of impact. The particles are assumed to be deposited on the bed with a negligible velocity, and thus dynamic effects are eliminated. No external compaction loads are imposed, but the method of Reference (3) for simulating vibration effects is incorporated into this analysis.

For the reactor case, a rectangular floor of arbitrary porosity is introduced and a circular cylinder of arbitrary diameter and height is placed in the center of the rectangular floor. Periodic boundary conditions are imposed at the floor grid edges. The cylinder wall introduces a packing order into the debris bed. This packing order can be expected to extend approximately five maximum particle diameters from the cylinder wall (5). For the geometries of interest, the packing is nonrandom within the debris bed if a significant fraction of the fuel rod survives.

3. COMPARISON BETWEEN THEORY AND EXPERIMENT

A variety of experiments have been conducted to study the random packing of equal and unequal size spheres, and also irregular particles. Many of these experimental observations are summarized in Reference (6). However, these observations must be examined with care because a variety of particle deposition dynamics and compaction methods were utilized. For example, gently settled uniform glass spheres have resulted in a random packing density of 0.58 (7), whereas random packing densities of 0.64 have been achieved with uniform ball bearings (8). On the other hand, smooth spheres in a binary distribution can be packed to densities of 0.825 (3). Particle shape irregularities introduce a greater uncertainty in the data, even for random packing. For example, flat laminar particles and needle-shaped particles have produced packing densities as low as 0.10, and loosely packed crushed stone has yielded a value of 0.45 (9).

Perhaps the only experimental observations useful in assessing the analytical results obtained in this study for nonrandom packing are those of Benenati and Brosilow (5). These authors studied the void fraction distribution within a bed of uniform spheres that were placed in a cylindrical container but were not compacted. The ordered packing influence of the container wall on the (horizontally integrated) void fraction was obtained as a function of the distance from the container wall for a variety of container-to-sphere diameter ratios. The void fractions were evaluated at a sufficient height in the bed to eliminate the ordering influence of the floor. For completeness, these authors also conducted one experiment to study the nonrandom packing about a post; they concluded that the resulting void fraction distribution exhibits the same radial dependence as that which was observed for concave surfaces.

Of particular interest is the experiment employing a container-to-sphere diameter ratio of 5.60 because this gave a well-defined asymptotic nonrandom (horizontally integrated) packing density of 0.55 at approximately 3 sphere diameters from the wall. This experiment with smooth spheres was simulated for the convoluted problem of packing about a fuel rod by requiring that all of the particles roll until they achieve a stable three-point contact, and also imposing a fuel-rod-to-sphere diameter ratio of 5.60. Moreover, the floor edge was taken to be a minimum of three sphere diameters from the fuel rod wall. The corresponding input data and computational results are presented in Table 1. These results give an average (horizontally integrated) packing density of 0.56 above the region of floor effects. This result is well within the scatter of the experimental data of Reference (5).

Next, we considered the same physical problem but with irregularly shaped particles, stipulating that these particles must stop their packing movement in the debris bed at their first two-point contact. The computer output for this case is presented in Table 2. Note that the average packing density has decreased to 0.36, a density well below the value expected for random packing; however, strong nonrandom effects have already been demonstrated experimentally for this geometry.

4. DEBRIS PACKING FOR TMI-2

The TMI-2 fuel rod assembly contains 204 fuel rods and 8 Inconel "egg crate" spacer grids along the length of these fuel rods. Also included in each assembly are 16 guide tubes and 1 instrument tube. These spacer grids serve as the porous floor for debris bed packing. For computational purposes, only the horizontally repetitive single grid cell geometry depicted in Figure 1 is analyzed. The actual breakup of the fuel rods and the subsequent creation of a debris bed on each surviving spacer grid is idealized in Figure 2.

A lack of fragmentation data for the fuel rods required the use of in-pile experimental results (10) to obtain a particle size distribution for the fractured cladding and fuel pellets; these results are summarized in Figure 3. Because particles of less than 300 microns in diameter account for less than 3% by weight fraction of the debris, they were ignored in all of the TMI-2 calculations.

Table 1. Smooth Sphere Packing About a Cylinder

Diameter of particles = .1930(cm).
x-dimension of floor grid = 2.240(cm).
y-dimension of floor grid = 2.240(cm).
Diameter of cylinder in center of floor grid = 1.080(cm).
Height of cylinder in center of floor grid = 5.994(cm).

Packing Zone Boundary	Packing Zone Density	Number of Particles in Each Zone
.0(cm)		
	.489	93
.193(cm)		
	.548	121
.386(cm)		
	.551	122
.579(cm)		
	.565	115
.772(cm)		
	.568	126
.965(cm)		
	.564	115
1.16(cm)		
	.553	113
1.35(cm)		
	.555	116
1.54(cm)		
	.555	121
1.74(cm)		
	.545	115
1.93(cm)		
	.508	104
2.12(cm)		
	.437	97
2.32(cm)		
	.347	69
2.51(cm)		
	.232	49
2.70(cm)		
	.976e-01	21
2.90(cm)		
	.192e-01	3
3.09(cm)		
	.0	0
3.28(cm)		

Average packing density is .560 between .579(cm) and 1.737(cm) with a standard deviation of .00564.

Table 2.   Irregular Particle Packing About a Cylinder

Packing Zone Boundary	Packing Zone Density	Number of Particles in Each Zone
.0(cm)		
	.338	68
.193(cm)		
	.365	78
.386(cm)		
	.384	82
.579(cm)		
	.370	79
.772(cm)		
	.367	76
.965(cm)		
	.386	84
1.16(cm)		
	.362	75
1.35(cm)		
	.353	71
1.54(cm)		
	.330	72
1.74(cm)		
	.368	72
1.93(cm)		
	.397	90
2.12(cm)		
	.352	70
2.32(cm)		
	.324	70
2.51(cm)		
	.366	77
2.70(cm)		
	.362	75
2.90(cm)		
	.332	72
3.09(cm)		
	.320	68
3.28(cm)		
	.320	68
3.47(cm)		
	.298	62
3.67(cm)		
	.239	49
3.86(cm)		
	.142	30
4.05(cm)		
	.452e-01	10
4.25(cm)		
	.137e-01	2
4.44(cm)		
	.0	0
4.63(cm)		

Average packing density is .363 between .193(cm) and 2.895(cm) with a standard deviation of .01894.

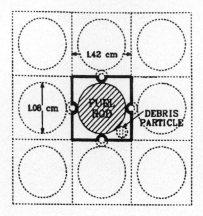

Figure 1. Fuel Rod Spacer Grid Unit Cell
(Maximum Debris Particle Pass-Thru Size = 0.17 cm.)

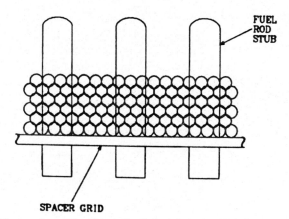

Figure 2. Idealized Fuel Rod Debris Model

For illustrative purposes, two TMI-2 cases were completed. To conserve computer time, we considered only the 48.5-cm segment of the fuel rod that is above the second spacer grid from the top. (A reasonable approximation would be to linearly scale the results that follow to any greater rod fracture length.) Case 1 corresponds to the upper 13.4% of the fuel rod fragmenting during reflood. Based on the average values for the data presented in Figure 3, this fragmentation resulted in the particle size distribution and surviving fuel rod length given in Table 3. Because the fuel rod surface was only 0.57 large particle diameters from the grid cell boundary, severe nonrandom effects were expected for this geometry. Indeed, the output summarized in Table 3 reveals that the strong ordering influence of the fuel rod wall reduces the average (horizontally integrated) packing density to 0.30, with the oscillations in this variable producing a relatively large standard deviation of 0.04. When the intact fuel rod was included, the maximum effective packing density was 0.66. Because of the elected distribution in particle sizes, a void large enough to pass the 0.10-cm-diam particle was relatively infrequent. Moreover, all particles had to cease their packing migration within the bed at their first two-point contact. Thus, the only particles that passed through the porous floor were essentially those 0.10-cm-diam particles that first hit the bare floor.

For completeness, Case 1 was rerun with simulated vibration effects. This calculation was accomplished by successively dropping each particle 10 times from different random points above the debris pile and then defining its final resting place as that which was closest to the floor. The enhanced packing density is displayed for this calculation in Table 4. Note that the average nonrandom packing density between the fuel rods increased to 0.50, whereas the maximum effective packing density including the fuel rods increased to 0.74. In addition, the number of small particles that passed through the porous floor also increased.

Case 2 corresponds to the upper 50% of the fuel rod fragmenting during reflood and has the same relative debris particle size distribution as in Case 1. The concomitant input data and resulting computer output are summarized in Table 5. The average (horizontally integrated) packing density over the 24.25-cm length of fuel rod stub dropped to 0.29. However, the debris bed above the rod began to approach the structure of a randomly packed bed: the (horizontally integrated) packing density achieved an average value of 0.38 and a maximum value of 0.45. As in the previous case, few of the small particles passed through the porous floor.

A complete presentation of the computer output for Cases 1 and 2 is given in Reference (11).

5. SUMMARY AND CONCLUSIONS

Employing available in-pile experimental observations for the fuel rod debris particle size, we achieved a relatively low nonrandom packing density of approximately 0.30 between the surviving segments of the fuel rods. Simulated vibration effects can be imposed to increase the between-rod packing density to 0.50.

Moreover, less than 3% of the particles fell through the porous spacer grid floor. As the fraction of the surviving fuel rod length is decreased, the debris bed builds up above the fuel rod stubs and the maximum

326

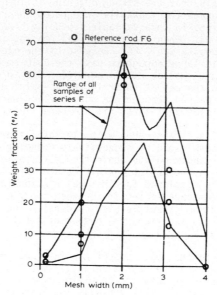

Figure 3. Debris Particle Size Distribution
(KFK In-Pile Tests, Series F)

packing density achieves a value of 0.45. This approximates the average random packing density observed experimentally for loosely packed gravel.

Preliminary data from TMI-2 indicate that approximately the upper 40% of the reactor core is totally void and about 40-cm of loosely packed debris covers an unknown solid structure. If we assume that this solid structure is the top of the unfractured fuel rods, then the TMI-2 degraded core corresponds to approximately 50% of the fuel rod fragmenting during reflood. Case 2 produced a debris bed height that is equal to 88% of the original fuel rod height, not the observed 60%. Thus, one plausible but highly speculative explanation of the TMI-2 final configuration is that only the lower spacer grid remained submerged in water and that all of the others melted before reflood. Then, during reflood, at least the upper half of the fuel rods fractured to produce the final debris bed and a significant mass of debris was flushed out of the core.

This analysis can be coupled to a hydrodynamics code to predict the influence of the particle settling rates on the time-dependence of the particle size distribution that appears above the debris bed during the packing process.

Table 3. Debris Packing with 13.4% TMI-2 Fuel Rod Fracture

--Input Data--

Number of Particles of Size 1 = 100
Number of Particles of Size 2 = 872
Number of Particles of Size 3 = 1575

Diameter of Particles of Size 1 = 0.3000(cm)
Diameter of Particles of Size 2 = 0.2000(cm)
Diameter of Particles of Size 3 = 0.1000(cm)

x-dimension of floor grid = 1.420(cm)
y-dimension of floor grid = 1.420(cm)
Hole diameter of floor grid = 0.17(cm)

Diameter of cylinder in center of floor grid = 1.080(cm)
Height of cylinder in center of floor grid = 42.012(cm)

--Output Data--

	Size 1	Size 2	Size 3
Number of particles in packed bed	100	872	1536
Number of particles that fall through floor grid	0	0	39

All particles that hit the floor but do not fall through stick at point of impact.

All other particles stop at first two-point contact.

The maximum packing density is 0.380 at 3.900(cm).

The maximum packing density including the cylinder is 0.662.

Average packing density is 0.304 between 0.000(cm) and 17.400(cm) with a standard deviation of 0.03985.

Table 4. Debris Packing with 13.4% TMI-2 Fuel Rod Fracture
(with Vibration Effects)

--Input Data--

Number of Particles of Size 1 = 100
Number of Particles of Size 2 = 872
Number of Particles of Size 3 = 1575

Diameter of Particles of Size 1 = 0.3000(cm)
Diameter of Particles of Size 2 = 0.2000(cm)
Diameter of Particles of Size 3 = 0.1000(cm)

x-dimension of floor grid = 1.420(cm)
y-dimension of floor grid = 1.420(cm)
Hole diameter of floor grid = 0.17(cm)

Diameter of cylinder in center of floor grid = 1.080(cm)
Height of cylinder in center of floor grid = 42.012(cm)

--Output Data--

	Size 1	Size 2	Size 3
Number of particles in packed bed	100	872	1512
Number of particles that fall through floor grid	0	0	63

All particles that hit the floor but do not fall through stick at
point of impact.

All other particles stop at first two-point contact.

Simulated vibration of Debris Bed

The maximum packing density is 0.521 at 4.200(cm).

The maximum packing density including the cylinder is 0.738.

Average packing density is 0.499 between .300(cm) and 10.500(cm) with
a standard deviation of 0.01497.

Table 5. Debris Packing with 50% TMI-2 Fuel Rod Fracture

--Input Data--

Number of Particles of Size 1 = 373
Number of Particles of Size 2 = 3253
Number of Particles of Size 3 = 5876

Diameter of Particles of Size 1 = 0.3000(cm)
Diameter of Particles of Size 2 = 0.2000(cm)
Diameter of Particles of Size 3 = 0.1000(cm)

x-dimension of floor grid = 1.420(cm)
y-dimension of floor grid = 1.420(cm)
Hole diameter of floor grid = 0.17(cm)

Diameter of cylinder in center of floor grid = 1.080(cm)
Height of cylinder in center of floor grid = 24.246(cm)

--Output Data--

	Size 1	Size 2	Size 3
Number of particles in packed bed	373	3253	5842
Number of particles that fall through floor grid	0	0	34

All particles that hit the floor but do not fall through stick at
point of impact.

All other particles stop at first two-point contact.

--Averages for Packing About the Cylinder--

Average packing density is 0.292 between 0.000(cm) and 24.300(cm) with
a standard deviation of 0.04695.

The average packing density including the cylinder is .613.

--Averages for Packing Above the Cylinder--

Average packing density is 0.379 between 24.300(cm) and 42.900(cm) with
a standard deviation of 0.03461.

ACKNOWLEDGEMENTS

The research reported here was completed at the Los Alamos National Laboratory and funded by the United States Nuclear Regulatory Commission.

REFERENCES

1. Walton, O.R., "Particle-Dynamics Modeling of Geological Materials," UCRL Report No. 52915 (March 25, 1980).

2. Jodrey, W.S. and Tory, E.M., "Simulation of Random Packing of Spheres," Simulation, 32, No. 1, 1-12 (January, 1979).

3. Visscher, W.M. and Bolsterli, M., "Random Packing of Equal and Unequal Spheres in Two and Three Dimensions," Nature, 239, No. 5374, 504-507 (October 27, 1972).

4. Powell, M.J., "Computer-Simulated Random Packing of Spheres," Powder Technology, 25, No. 1, 45-52 (January/February, 1980).

5. Benenati, R.F. and Brosilow, C.B., "Void Fraction Distribution in Beds of Spheres," A. I. Ch. E. Journal, 8, No. 3, 359-361 (July, 1962).

6. Fowkes, R.S. and Fritz, J.F., "Theoretical and Experimental Studies on the Packing of Solid Particles: A Survey," Bureau of Mines Information Circular No. 8623 (1974).

7. Alder, B.J., "Mixtures of Rigid Molecules," J. Chem. Physics, 23, No. 2, 263-271 (February, 1955).

8. Scott, G.D. and Kilgour, D.M., "The Density of Random Close Packing of Spheres," Brit. J. Appl. Phys. (J. Phys. D.), Series 2, 2, 863-866 (June, 1969).

9. Carman, P.C., "Fluid Flow Through Granular Beds," Trans. Inst. Chem. Eng., 15, 150-166 (1937).

10. Karb, E.H., Sepold, L., Hofmann, P., Peterson, C., Schanz, G. and Zimmerman, H., "KFK In-Pile Tests on LWR Fuel Rod Behavior During the Heatup Phase of a LOCA," Kernforschungszentrum Karlsruhe, Germany, Report No. 3028 (October, 1980).

11. Moore, J.A., "A Mechanistic Debris Bed Packing Model," Los Alamos National Laboratory Report No. LA-9619-MS (1983).

Multi-Phase Flow and Heat Transfer III. Part B: Applications
edited by T.N. Veziroğlu and A.E. Bergles
Elsevier Science Publishers B.V., Amsterdam, 1984 — Printed in The Netherlands

HYDRODYNAMIC AND THERMAL MODELING OF SOLID PARTICLES IN A MULTI-PHASE, MULTI-COMPONENT FLOW

A. M. Tentner and H. U. Wider
Reactor Analysis and Safety Division
Argonne National Laboratory
Argonne, Illinois 60439, U.S.A.

ABSTRACT

This paper presents the new thermal hydraulic models describing the hydrodynamics of the solid fuel/steel chunks during an LMFBR hypothetical core disruptive accident. These models, which account for two-way coupling between the solid and fluid phases, describe the mass, momentum and energy exchanges which occur when the chunks are present at any axial location. They have been incorporated in LEVITATE [1], a code for the analysis of fuel and cladding dynamics under Loss-of-Flow (LOF) conditions. Their influence on fuel motion is presented in the context of the L6 TREAT [2] experiment analysis. It is shown that the overall hydrodynamic behavior of the molten fuel and solid fuel chunks is dependent on both the size of the chunks and the power level. At low and intermediate power levels the fuel motion is more dispersive when small chunks, rather than large ones, are present. At high power levels the situation is reversed. These effects are explained in detail.

1.   INTRODUCTION

Unprotected LOF accidents in LMFBRs with a moderate positive void worth lead to overpower situations due to sodium boiling and voiding. This overpower, which can be enhanced by molten cladding relocation, leads to fuel melting and fuel pin breakup. The subsequent motion of both liquid and solid fuel components strongly influences the net reactivity, and thus, the accident sequence. Earlier safety analysis codes which describe these events have generally concentrated on modeling the hydrodynamic behavior of the liquid and vapor components, and assumed that the solid fuel is in the form of small fuel particles, moving with the same velocity as the liquid fuel component [3] [4]. However, it is now recognized that in addition to other components large solid fuel/steel chunks are likely to be present in the voided channels during the initiating phase of a LOF accident [5]. Pellet-size solid fuel chunks have been observed in various experiments, such as P3A [6] and L5 [2]. Under certain conditions, they can play a significant role in the early fuel relocation, due to jamming at the abrupt contractions characteristic for LOF geometries [7]. A typical LOF geometry, as modeled by LEVITATE, is shown in Fig. 1.

The presence of large fuel chunks requires a two-way coupling between the fuel chunks and the surrounding components in order to describe correctly the fuel dynamics in the presence of solid fuel chunks. Such a two-way coupling model has been recently incorporated into the LEVITATE channel treatment.

Returning to the equation (5) which applies to the geometry in Fig. 3, we can see that at an abrupt contraction the chunk convective flux differs from the

regular convective flux for a fluid component by the factor $\dfrac{A_i}{A_{i-1}}$. This is due to the fact that some chunks will hit the surface normal to the flow and will be prevented from leaving the cell. The continuity equation does not apply to the chunk field, which has an obviously discontinuous nature. It can be seen that the larger the difference between A_{i-1} and A_i, the more different the behavior of the chunk will be from the behavior of the fluid components. Another factor to consider is the size of the solid chunks. Thus, if large chunks are present and the cross sectional area of an individual chunk in cell i-1 becomes larger than A_i, the convective flux $(\rho'u)_{i-1/2}$ will be reduced to zero. This is an additional constraint, not reflected in eq. (5). The residual or complementary part of the convective flux in eq. 5 is for $u_i > 0$:

$$(\rho'u)_{i-1/2}^{residual} = \rho'_{i-1} \cdot u_i \cdot (1 - \frac{A_i}{A_{i-1}}) \tag{7}$$

This term, when integrated over time, measures the amount of chunks which have hit the wall normal to the flow during the current time step. Because these chunks will loose their momentum and come to a temporary rest, the abrupt contraction acts like a significant momentum sink, which has been separated from the other sources/sinks in eq. (2). While this momentum sink could be compared to the local pressure drop which appears at the abrupt area changes equation for the fluid components, its nature is very different and related to the discontinuous character of the chunk field. The chunk jamming at the abrupt contraction is illustrated in Fig. 4.

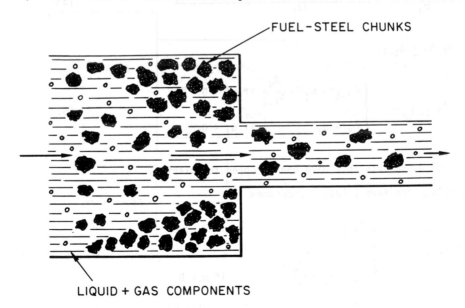

FUEL-STEEL CHUNKS

LIQUID + GAS COMPONENTS

Fig. 4. Chunk "Jamming" at an Abrupt Contraction.

The momentum loss at the abrupt contractions, due to the chunk field can play a significant role in the hydrodynamic behavior of the overall system, as will be shown in the next section.

Another aspect which will be shown to be significant is related to the heat transfer between the solid chunks and the surrounding components. As the radius of the chunks increases, their volume increases proportional to R^3, while the surface area will only increase as R^2. Thus, larger chunks will exhibit a lower heat transfer per unit mass and, during fast temperature transients, will exchange little energy with the surrounding components.

3.   DISCUSSION AND RESULTS

The presence of the solid fuel chunks can influence significantly the behavior of the dynamics of the multi-phase multi-component flow and thus the early fuel relocation. Jamming at the abrupt contractions occurs preferentially because of the larger chunks. The large chunks lose more momentum at the contractions and their coupling with the fluid components is weaker than that of smaller chunks. Thus the presence of larger chunks will tend to prevent the fuel from leaving the central disrupted region and thus will reduce the dispersive character of the early fuel motion.

This effect is illustrated in Fig. 5, which compares the fuel dispersal calculated by LEVITATE with data obtained in the L6-TREAT experiment analysis. In this experiment three thermally irradiated pins were subjected to a flow coastdown and a power pulse with a 10 times nominal power peak, simulating a loss of flow situation. The fuel dispersal is measured, in fig. 2, by the relative fuel worth. This integral indicator of fuel motion is obtained by calculating the total fuel worth at any given time and dividing it by the total fuel worth at time step zero, i.e. before the onset of fuel motion. A decrease in the relative fuel worth indicates a dispersive motion, i.e. away from the central regions of high fuel worth, while an increase in the relative fuel worth is indicative of a compactive trend.

Three LEVITATE runs were performed, varying the chunk radii in order to observe the jamming effect. It can be observed that the results obtained with constant chunk radius $R=10^{-3}$m indicate significantly less dispersal than these obtained with $R=10^{-4}$m. The third run was made with the more realistic assumptions that the size of the chunks generated via the pin disruption is large ($R=2.5 \, 10^{-3}$m) while all other chunks are fairly small ($R-10^{-4}$m). The results of this run are in fairly close agreement with the experimental data.

Another consequence of the chunk formation is that the separation of the moving fuel into solid fuel chunks and molten fuel allows the molten fuel temperature to respond faster to high power ramps. This leads to an earlier onset of high vapor pressures which enhances the dispersive character of the fuel motion. This effect becomes dominant at high power levels, as illustrated in fig. 6. For these runs, a steep power ramp was superimposed over the L6 power curve, so that the peak power was about 1000 times nominal as compared to the 10 times nominal in the experiment. The runs in which the larger chunk radii were used show a faster fuel dispersal, confirming the fact that at high power levels the energy exchange aspects have a larger influence on the overall fuel behavior than the momentum loss aspects, which were shown before to dominate at lower power levels.

where subscripts h and c refer to the hot and cold liquids respectively, $\alpha = \sqrt{k\rho c}$ and k is the thermal conductivity, ρ is the density and c is the specific heat. The temperature of spontaneous nucleation is approximately

$$T_s \approx \frac{16\pi\sigma^3}{3P_v^2 k} f(\theta) \tag{2}$$

where σ is the surface tension, P_v is the vapor pressure, k is the Boltzmann constant and $f(\theta)$ is a functional relationship involving the contact angle θ^* and is included in Eq. (2) to account for the possibility of spontaneous nucleation at the liquid-liquid interface rather than in the bulk of the volatile fluid (homogeneous nucleation). At this temperature, the exponential is a very strong function of temperature, causing the nucleation rate, i.e. number of bubbles formed per unit volume and time, to increase many orders of magnitude per °C. The large scale vapor explosion criterion therefore, becomes

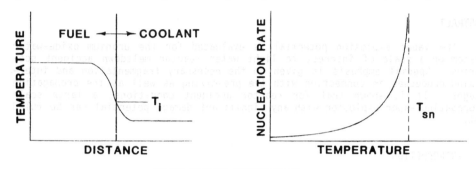

EXPLOSION CRITERION

$$T_i > T_{sn}$$

(a)   **(b)**

Fig. 1   Illustration of spontaneous nucleation model.

$T_i > T_s$ explosions are possible   (3)

$T_i < T_s$ explosions are impossible*   (4)

The evidence for this requirement is very persuasive, running from tests with single drops through a large number of pouring and mixing experiments in the kg range with many different liquids exploding at a precise temperature (see Fig. 2). None have yet been observed experimentally that violates the criterion [3]†.

Application of the vapor explosion criterion to the molten uranium oxide-water system clearly shows that the potential for such an explosion exist, i.e. T_i is substantially above the spontaneous nucleation temperature (approximately 305°C) as well as the thermodynamic critical temperature for water. However,

---

*θ depends on the wetting characteristics of a given system. As shown in Ref. [1], the spontaneous nucleation threshold is dependent on both contact mode and contact time, and these effects can be represented by specifying the contact angle α. In case of perfect wetting (θ = 0 which corresponds to f(θ) = 1), the limit of superheat (homogeneous nucleation) is attained.

it should be emphasized that this criterion only supplies necessary and not sufficient conditions for a large scale explosion. As such it is necessary to mechanistically address the fragmentation and intermixing process. Generally speaking this process has been postulated to take place in two stages including a relatively slow <u>premixing</u> phase involving fragmentation and intermixing on a coarse scale (of the order of 1 cm) in the film boiling mode, and a subsequent very rapid high pressure liquid-liquid fine-scale fragmentation and intermixing (sub mm scale) in connection with the <u>propagation</u> stage [4], (see Fig. 3). Some basic considerations are provided below relative to the intermixing potential for the uranium oxide-water system on a scale of interest to light water reactor accident scenarios.

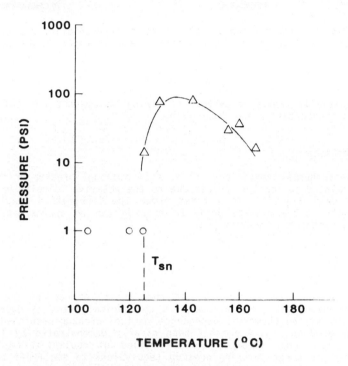

Fig. 2 Oil to Freon-22 peak observed pressures.

---

*It should be noted that small scale superheat limiting explosions are possible for $T_i < T_s$ as long as the hot liquid bulk temperature exceeds or equals T_s, i.e. following liquid-liquid contact and a brief waiting period a thin layer of the more volatile material can be heated to T_s. In fact, explosive vapor growth which results upon exceeding T_s locally can also serve as a trigger for a propagating vapor explosion where the bulk of the system has experienced premixing as a result of $T_i > T_s$.

†For the Liquid Metal Fast Breeder Reactor UO_2/Na case, T_i is well below T_s, and large scale vapor explosions are rigorously excluded [1]. In well over a hundred tests involving molten UO_2 and liquid sodium under a variety of contact modes including numerous in-pile tests, no large scale vapor explosions or energetic fuel-coolant interactions have ever been observed.

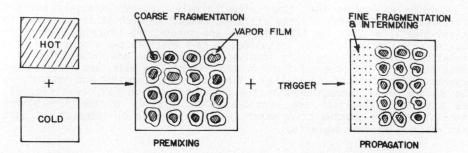

Fig. 3   Illustration of the two-stage fragmentation and
intermixing process postulated for vapor explosion.

3.   PREMIXING

In an initially separated system, premixing of the hot and cold liquids
involves both fragmentation and intermixing.

3.1.   Fragmentation

For non-isothermal conditions, (i.e. film boiling) breakup or fragmentation can be related to inertial forces due to the relative velocity between the
hot and the cold liquids. To a first order the penetration length (L) of
liquid mass of arbitrary shape prior to breakup can be approximated by the
corresponding isothermal behavior [5].

$$L \simeq 3d \sqrt{\frac{\rho_F}{\rho_c}} \tag{5}$$

where ρ_F is the density of fuel, ρ_c is the density of coolant and d is the
characteristic dimension (in the case of a cylindrical jet d is equal to the
diameter). The strong geometric dependence on fuel breakup predicted from Eq.
(5) is illustrated in Fig. 4 and has been clearly demonstrated by laboratory
experiments. For large simulated volume ratios of coolant-to-fuel (of the
order of 1000) and hence favoring breakup (equivalent to small jet behavior),
fragmentation sizes (4-6 mm) have been observed in good agreement with capillary breakup ($\sim \sqrt{\sigma/g\Delta\rho}$). This is in complete contrast to small simulated
volume ratio tests (of the order of 1) which is prototypic of a postulated
catastrophic large coherent fuel pour in the accident case, where the hot fuel
or liquid is observed to collect at the bottom of the test vessel remaining
essentially intact. Eq. (5) indicates that a volume ratio of coolant to fuel
of at least 10 is required to initiate breakup, while this ratio is 2-3 in the
reactor case. These data reflect both real materials [6] and simulant materials [7] as indicated in Table 1.

3.2.   Intermixing

To a first order, intermixing would appear possible if the vapor flux in
the nucleation limited film boiling regime ($T_i > T_s$) is well below the hydrodynamically limited critical heat flux value [8] (see Fig. 5).

Table 1

AVAILABLE EXPERIMENTS RELEVANT TO LWR FUEL FRAGMENTATION AND STEAM EXPLOSION POTENTIAL

Source	Material	Temperature	Mass	Coolant-to-Fuel Volume Ratio	Release Mode		Fragment Size	Trigger	Comments
					Pouring	Drop$^\Gamma$			
ISPRA	UO_2-H_2O	~ 3100°K^Y	~ 4 kg	> 1000	X		~ 4-6 mm$^+$	No	No Explosions
	UO_2-H_2O	~ 3100°K^Y	~ 4 kg	~ 2-4*		X	Large Lump^o	No	No Explosions
PURDUE	H_2O-N_2	~ 300°K	~ 3-1000 g	Large	X		~ 10 mm$^+$ (Ice Particles)	No	No Explosions
	H_2O-N_2	~ 300°K	~ 5000 g	Small*		X	Approximately No Fragmentation (Remain Liquid)	No	No Explosions

* - Prototypic ratio for LWR.

Y - Prototypic fuel temperature.

Γ - Corresponds to catastrophic fuel collapse into the lower vessel plenum.

\+ - Corresponds to capillary breakup

$$\sim \sqrt{\frac{\sigma}{g\Delta\rho}}$$

o - Insufficient time for breakup -- scale too large.

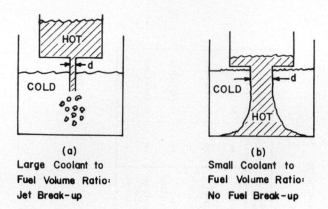

(a)

Large Coolant to
Fuel Volume Ratio:
Jet Break-up

(b)

Small Coolant to
Fuel Volume Ratio:
No Fuel Break-up

Fig. 4 Illustration of possible premixing
geometries in the film boiling regime.

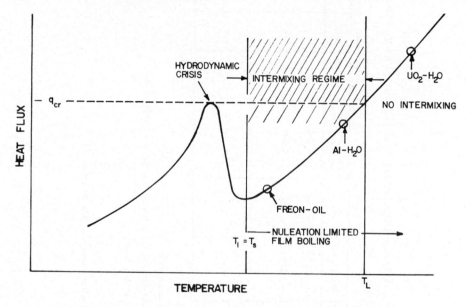

Fig. 5 Illustration of the premixing regime and premixing
potentials for different systems (P = 1 atm).

The critical heat flux can be expressed by the well-known Kutateladze
equation [9]:

$$q_{crit} = 0.14 \; h_{fg} \sqrt{\rho_g} \; \sqrt[4]{g\sigma(\rho_\ell - \rho_g)} \tag{6}$$

where h_{fg} is the latent heat of vaporization, σ is the liquid surface tension
and ρ_ℓ and ρ_g are the liquid and vapor densities, respectively. It is of key
interest to compare the critical heat flux value with the film boiling heat
flux given by

$$q_{FB} = \sigma\epsilon(T_h^4 - T_s) + h(T - T_s) \tag{7}$$

where σ is the Stefan-Boltzmann constant, ϵ is the emissivity which is usually set equal to unity, T_h is the temperature of the hot surface, T_s is the saturation temperature of the volatile liquid and h is the free convection heat transfer coefficient.

For some systems (such as Freon-oil) the film boiling heat flux is very low relative to the critical heat flux (see Table 2), hence favoring inter-mixing which is consistent with the experimentally demonstrated high explo-sivity for such low temperature systems [10]. In contrast, for the molten UO_2-saturated H_2O system, the film boiling vapor flux due to radiation domi-nated heat transfer is of the same order as the hydrodynamic crisis limit*. Since the premixing volume is proportional to $(q_{crit}/q_{FB})^3$, the potential of a

Table 2

COMPARISON OF FILM BOILING HEAT FLUX AND MAXIMUM NUCLEATE
BOILING HEAT FLUX FOR THREE DIFFERENT SYSTEMS (P = 1 atm)

System	Peak Nucleate Boiling Heat Flux, w/cm^2	Film Boiling, w/cm^2
Oil-Freon	~ 30	~ 1
Al-H_2O	~ 150	~ 30
UO_2-H_2O	~ 150	~ 300

propagating vapor explosion for the Freon-oil system is approximately 10^4 times larger than for the UO_2-saturated H_2O system. Propagating vapor explosions have never been observed with the latter system.

Based upon the above discussion, subcooled coolant conditions would appear to be required in order to allow premixing and produce a propagating vapor explosion which is consistent with experimental observations with the UO_2-H_2O system [11]. As long as significant subcooling prevails, the net vapor flux will remain well below that corresponding to the hydrodynamic crisis condition. However, as the coolant becomes saturated, the net vapor flux increases rapid-ly, and if a timely trigger is not provided, premature and benign separation of the fuel and coolant will result rather than an explosion.

An order of magnitude estimate of the time to reach saturated coolant conditions following initiation of premixing can be obtained by considering equal volume mixing, room temperature water and a film boiling heat flux of approximately 300 w/m^2, and results in a time of approximately 0.1 s. It follows that large quantities (many tonnes) of fuel must break up and intermix on a similar time scale, in order to have a significant damage potential. The breakup time can be approximated by:

---

*At 2500°C, the film boiling heat flux is approximately 3000 kw/m^2 as compared to the critical heat flux value of 1500 kw/m^2 at 1 bar pressure. The latter value increases with increasing pressure as illustrated by Eq. (6), and reaches a maximum of 4000 kw/m^2 at 100 bars.

friction factor for the suspension in terms of that for the gas alone, modifying it either additively or multiplicatively. The most widely used correlations, namely those of Rose and Duckworth [4], Rose and Barnacle [5], Dogin and Lebedev [6], and Pfeffer [7] were chosen for the initial stage of this investigation. One other correlation, which was discovered along with an extensive data set during the literature search, that of Konchesky [8], was also checked, but only against its' own data set. This correlation is given in dimensional form and contains parameters which made it difficult to apply in other situations.

During the literature search a special effort was made to collect the data sets on which the above authors based their correlations. This was possible only to a limited extent, because much of the data was not published along with the original reports or in some cases, because the publication was not available. Nonetheless the data sets of twelve different authors were obtained containing approximately 1000 experimental data among them. After examining the reports a few of those sets were judged to be inapplicable, generally because the data reported allowed inadequate acceleration length. The common practice of presenting data in graphical form led to the elimination of some other data. In other cases the graphically presented data was usable only with difficulty. Wherever graphs were used, much effort and care was taken to correctly interpret the information.

2. METHOD OF COMPARISONS

Each correlation was investigated by using it to calculate the pressure drop per unit length for a certain set of conditions and comparing it with the pressure drop that was actually measured. Since the end use of the correlation is the pressure drop in most cases it is thought advantageous to obtain this quantity rather than the derived friction factor. This avoided the difficulty of comparing friction factors which are defined differently. All computations were carried out on the University of Delaware's DEC-10 computer.

As a first step all data collected were converted to the units of the S. I. system. The quantities that were required by the correlations were:

1. The mass ratio of solids to gas flowing (loading).

2. The ratio of particle diameter to pipe diameter.

3. The ratio of solid density to gas density.

4. The Reynolds number for the gas flow in the pipe.

5. The Froude number for the gas flow in the pipe.

6. A kinetic energy term, $\dfrac{\rho_a V_a^2}{2D}$, where ρ_a and V_a are the density and superficial velocity of the gas and D the pipe diameter.

These quantities were calculated from the data expressed in S. I. units. Wherever the friction factor for air flowing alone in a pipe was required it was calculated from the Blasius equation,

$$f = 0.046 \; Re^{-0.2} \tag{1}$$

This function is a good approximation of the Fanning friction factor for turbulent flow in smooth pipes. In the cases where the larger D'Arcy factor C_f, was used, the Fanning factor, f, was simply multiplied by 4 so $C_f = 4f$. In one case, [4], a function based on the coefficient of restitution was provided. The scarcity of information regarding the coefficient of restitution for the materials which were often used in experiments limited the usefulness of this function. In addition the function provided values very close to one over its' range and appeared to be based only upon four data points. For these reasons the factor was eliminated from the analysis.

It is a common practice in the literature to report functions in graphical form. Functions given in this manner were represented for computational purposes by either cubic spline interpolation of a significant number of data points taken from the given curve or by piecewise linear interpolation of the data given.

The predicted pressure drop was calculated for each individual set of conditions by each of five methods.

1. Rose and Duckworth [4]

2. Rose and Barnacle [5]

3. Dogin and Lebedev [6]

4. Dogin and Lebedev modified as per Soo [2]

5. Pfeffer [7]

In addition the Konchesky's data sets [8,9] were used to check his own correlation.

The accuracy of each correlation was checked by determining how well it predicted the pressure drop for the various conditions encountered in the experimental data sets. The relative error for the ith data e, was calculated as:

$$e_i = \frac{y_i - \hat{y}_i}{y_i} \tag{2}$$

where y_i is the measured value and $\hat{y}_i$ the one predicted from the correlation. The average relative error, $\bar{e}$, and standard deviation of the error, σ, was computed for each data set in combination with each correlation:

$$\bar{e} = \frac{1}{N} \sum_{i=1}^{N} e_i \; , \tag{3}$$

$$f_T = f_a (1 + \dot{m}*)^{0.3} \qquad (8)$$

3.5 Konchesky [8]: The Konchesky correlation is strictly a regression of pressure drop on pipe diameter, mass flow of solid (coal), and specific gravity. Their regression was done in dimensional form so the relation is difficult to apply to other situations. They did not in any case include a particle size factor since they were working with coal as it would normally be conveyed, i.e., mixed particle sizes. Their equation is:

$$\frac{\Delta P}{L} = 0.00454 \ W^{0.688} \ \gamma^{0.410} \ e^{(7.833/D)} \qquad (9)$$

where W is the flow rate of coal in tons per hour, γ the specific gravity in pounds per cubic foot, D the pressure drop in inches of water, D the pipe inside diameter in inches and L the pipe length in feet.

3.6 Data: The data sets range in size from approximately thirty points to as many as five hundred. The materials used by the investigators include various grains, coal, sand, and a few synthetic materials. Pipe diameters varied from 3.2 mm to 203 mm. Loadings vary from less than one to twenty.

4. RESULTS AND CONCLUSIONS

Table 1 shows the average relative errors and standard deviation for each correlation as it was applied to each of the data sets. Figures 1 through 6 show the pressure drops measured by each

Table 1. Relative Error Analysis $\bar{e}(\sigma)$

Author Data	Rose and Duckworth	Rose and Barnacle	Dogin and Lebedev A=1.65x10^{-6}	A=2x10^{-7}	Pfeffer	Konchesky
Hariu	0.56(0.15)	0.32(0.15)	0.45(0.15)	0.62(0.15)	0.37(0.15)	
Hinkle	0.15(0.12	0.23(0.15)	0.06(0.15)	0.22(0.14)	0.02(0.15)	
Hitchcock	-0.02(0.43)	0.12(0.38)	-1.4(0.87)	0.02(0.41)	-0.28(0.51)	
Konchesky	0.12(0.43)	0.22(0.42)	-1.8(1.2)	0.01(0.46)	-0.09(0.50)	-0.07(0.34
Rose	0.08(0.15)	0.15(0.17)	-0.03(0.14)	0.17(0.18)	-0.15(0.17)	
Uematsu	0.27(0.15)	0.29(0.16)	0.23(0.13)	0.3(0.17)	0.15(0.10)	
Vogt	-0.16(0.64)	-0.27(0.58)	-0.33(0.56)	-0.12(0.66)	-0.59(0.61)	
Welshof	0.25(0.24)	0.32(0.25)	-0.27(0.16)	0.30(0.24)	0.05(0.24)	

368

1949.

13. Hinkle, B. L.,"Acceleration of Particles and Pressure Drops Encountered in Pneumatic Conveying", Ph.D. Thesis, Georgia Institute of Technology, 1953.

14. Hitchcock, J. A. and Jones, C.,"The Pneumatic Conveying of Spheres Through Straight Pipes, British Journal of Applied Physics 9, June 1958.

15. Uematsu, T. and Morikawa, Y., "Pressure Losses in Pneumatic Conveying Systems of Granular Materials", 3, #12, 1960.

16. Welshof, G., "Pneumatic Conveying at High Particle Concentration", VDI Forschungsheft 492 b 28, 1962.

17. Bethea, R. M., Duran, B. A. and Boullin, T. L., Statistical Methods for Engineers and Scientists, Marcel Dekker, Inc., New York, 1975.

Multi-Phase Flow and Heat Transfer III. Part B: Applications 369
edited by T.N. Veziroğlu and A.E. Bergles
Elsevier Science Publishers B.V., Amsterdam, 1984 — Printed in The Netherlands

MEASUREMENT OF PRESSURE FLUCTUATIONS IN PNEUMATIC CONVEYING

Alexander R. Peters and James E. Boucher
Mechanical Engineering Department
University of Nebraska-Lincoln
Lincoln, Nebraska 68588-0525, U.S.A.

ABSTRACT

     This paper reports transient pneumatic conveying pressure responses during
flow instabilities. Increased solids loading in a conveying system causes transition from homogeneous dilute phase conveying to a non-homogeneous state.
Pressure fluctuations occur when flow becomes non-homogeneous in the line. This
phenomenon is usually referred to as saltation and choking in horizontal and
vertical flows respectively. An experimental conveying loop, approximately
109 feet in length with both horizontal and vertical lengths, was made from 1.5"
Schedule 40 PVC pipe. Pressure taps were placed at seven points in the line.
Crushed black walnut shells were conveyed in this experiment. Two Kavlico Model
P609-10GP pressure transducers were used to monitor instantaneous pressure responses at different tap combinations. In addition to individual pressure
measurements, instantaneous pressure drop records were also obtained. By using
this system, it was possible to detect onset of non-homogeneous flow plus obtain
good quantitative pressure response measurements.

1. INTRODUCTION

     A pneumatic conveying system utilizes the flow of air or gas to impart the
necessary movement energy directly to particles of material in a pipeline. Of
all the solids-moving alternatives, pneumatic conveying is probably the most
suitable method for continuous transporting of small-sized solids (1). A major
advantage is its extreme flexibility with regard to space design. A major disadvantage has been the relatively high cost of energy compared with other bulk-
solids moving systems.

     Visual observation of the motion of solids in a glass pipe reveals that
the flow patterns are rather complex and are affected, among other factors, by
solids-gas ratio, Reynolds number of the flow, and specific properties of the
solids. At very low solid-gas ratios, the solid particles are quite uniformly
distributed throughout the pipe. This region, called "homogeneous," is characterized by the flow behavior in which radial and axial solids density variations are so small that clusters of solids cannot be identified. As the solids-
gas ratio is increased, solids begin to settle out along the bottom of a horizontal pipeline, partially obstructing the flow area. In vertical lengths a
sharp choking transition can occur with increased loading.

     The objective of this paper is to report transient pneumatic pressure
responses during flow instabilities. Increased solids loading in a conveying
system causes transition from a homogeneous state to a non-homogeneous state

which were detected using two pressure transducers.

2. PROBLEM DESCRIPTION

Despite many years of research in the study of pneumatic conveying systems, this method of handling solids is still an art (2). Much that has been written on the theory of solids conveying is applicable to only a few selected materials of certain particle sizes transported in a particular conveying system. One reason for the diversity of data is the variety of materials that can be conveyed. Each material can be conveyed by a broad range of air velocities at various solids-air loadings. The conveying air velocities are themselves a function of the particle-size spectrum and of the density, shape, and physical characteristics of the material, as well as whether the flow is horizontal or vertical.

Probably the most important factor in the transport of solid particles is the conveying velocity. It is difficult to determine the actual velocity of the particles of flowing material, but the air velocity of the conveying stream is easily measured and calculated. Reference to a conveying velocity usually refers to the superficial air velocity, which is calculated from the measured air flow and the cross-sectional area of the conveying line. A minimum superficial air velocity that will convey a given solids feed rate is called the "critical velocity." Since most systems have a combination of vertical and horizontal runs of piping, the only velocity of interest is the lowest one that will clean all materials out of the line. For a vertical lift, this may be called "dropout velocity;" for horizontal conveying, it is referred to as the "settling" or "saltation" velocity.

It is desirable to reduce the transport air velocity as low as possible to minimize power consumption, solids breakage and pipe abrasion. High velocities also promote electrostatic hazards which can lead to charge buildup, particle deposition, flow stoppage and high voltage arcs.

Pneumatic conveying systems are often classified as dilute phase and dense phase. Dilute or lean phase systems are normally material-into-air systems and are characterized by low solids loading ratios, low pressure differentials and high solids velocities. Dense phase systems are normally air-into-material systems with high solids loading ratio, high pressure differential and low solids velocities. Dense phase, or slug flow is often to be avoided because of the erratic nature of the flow, pressure fluctuations, high pressure drops, and pipeline vibrations.

The pipeline pressure drop for transporting a homogeneous two-phase mixture can be found by summing several terms. Factors which make up the total pressure drop include: 1.) acceleration of air to carrying velocity, 2.) acceleration of solid particles to conveying velocity, 3.) friction loss between gas and pipe wall, 4.) friction between particles, particles and air, and particles and pipe wall, 5.) pressure drop due to vertical lift of gas, and 6.) pressure drop due to vertical lift of solid particles. Inspection of each of these pressure drop expressions leads to the observation that the air mass velocity G (mass/time-area) is directly proportional to pressure drop. Figure 1 shows how pressure drop is related to air flow in the pipe. Zero loading, air flow only, is shown as line A. When solids, W_s, are added, this curve is displaced upward, such as shown by line B.

2.1. Saltation

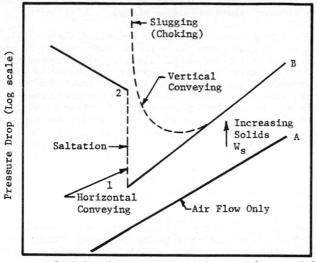

Figure 1. Pressure Drop Characteristics

In horizontal pneumatic conveying, saltation velocity is an important practical problem from a design standpoint. This velocity is defined as the mean air velocity at which almost all the particles slide on the bottom of a pipe. In a homogeneous two-phase state, pressure drop increases linearly with mass flowrate. As the gas velocity is reduced, the pressure gradient decreases until a point is reached where the pressure increases dramatically and the pipe operates in a dense phase regime. In the neighborhood of point 1 (Figure 1) particles begin to slide and settle out on the bottom of the pipe; that is, saltation occurs. This point is defined as the "saltation point" and the corresponding gas velocity is the "saltation velocity." Particles are observed to drop out of suspension and remain in a stationary layer on the bottom of the pipe, or stop rolling and sliding along the bottom. At point 2 (Figure 1) the deposit rate will equal the erosion rate, which has increased progressively due to the higher gas velocity in the restricted area of the pipe. In the region between points 1 and 2, the dispersion begins to collapse. Any further reduction in air mass flowrate will cause further increase in pressure drop as further deposition occurs (3–5).

2.2. Choking

In vertical pneumatic transport, particles are often carried up the column as an apparently evenly dispersed suspension with low volumetric concentration (generally less than 5%). If the air velocity is gradually reduced at the same mass flowrate of solids, the in-line solids concentration increases. A point will be reached where the air velocity is insufficient to support the particles as a uniform suspension. The entire suspension collapses and is then transmitted up the column in slug flow. The point of choking is the transition point from upflow of solids as a thin suspension to slugging flow. The gas velocity at this point is called the "choking velocity" and represents the saturated carrying capacity of the gas stream. It is analogous to saltation velocity in horizontal transport, since both represent saturation conditions for solids carrying capacity between dilute and dense phases.

Some gas-solid-tube systems exhibit the so-called sharp choking transition from a lean phase to slugging dense phase while other systems do not. As the gas velocity is reduced at a fixed solid flowrate, solid concentration in the tube increases. Two different types of behavior are possible. In one type of system, a sharp transition point will eventually be reached at which the uniform suspension collapses and the solids are then conveyed upwards in dense phase slugging flow with solids carried upwards mainly in wakes of rising slugs. In the second type of system no sharp transition is discernible when gas velocity is gradually reduced at fixed solid flowrate in lean phase flow. As gas velocity is reduced solid concentration increases and the suspension becomes progressively less uniform. Clusters and streams of particles appear and solids are conveyed upward with considerable internal recirculation. This mode of flow has been termed dense phase conveying without slugging and is analogous to a recirculating fluidized bed. Further reduction in gas flowrate will result in transition to slugging (6-8).

As may be seen in Figure 1, the pressure drop tends toward a minimum as air velocity and, simultaneously, solids velocity decreases. This is caused chiefly by a drop in air friction loss. As air velocities decrease below the point of minimum pressure drop, the amount of solids per unit length of line increases and the pressure drop increases, but at a more rapid rate, due to friction of material in the line, up to the point where slugging flow starts.

According to Zeng and Othmer (9) the choking velocity is approximately the same as the saltation velocity; but for mixed-sized solids, the saltation velocity is found to be several times greater than the choking velocity. For pneumatic conveying, it is generally recognized that the critical air flowrate at a given solid rate for mixed-size solids is higher than that for a uniform-sized material of the same mean diameter.

Little information is available in the literature for predicting the choking velocity and choking voidage. It has been pointed out that choking is not a clear-but phenomenon but involves a whole range of instabilities. It is important when designing vertical transport systems to predict conditions at choking.

3. EXPERIMENTAL APPARATUS AND PROCEDURES

An experimental pneumatic conveying apparatus was constructed to study pressure fluctuations that occur with the onset of saltation and choking.

3.1. Description of Apparatus

A conveying loop was constructed using opaque 1.5" Schedule 40 PVC pipe. Figure 2 shows dimensions and configuration, which consisted of both horizontal and vertical lengths. The initial section extended approximately 38 feet along the horizontal followed by a long-radius elbow and turn to the vertical. Elbows in the system had a bend radius equal to 12 diameters. The rest of the loop consisted of 14 feet vertical (upward), 12 feet horizontal, 14 feet vertical (downward) and 31 feet horizontal. The pipe terminated at a receiver and bag filter that had been constructed from two 40-gallon fiber drums. The bag filter had six 14-inch long by 4-inch diameter felted cotton bags. Approximately 7 square feet of filtering area was available in the bags. Most of the conveyed

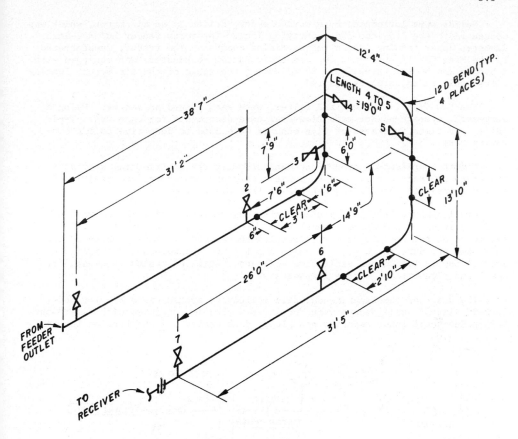

Figure 2. Conveying Loop

solid material was collected in the receiver drum by cyclone separation, with some carryover to the filter drum. For additional control of fine dust, an industrial-size vacuum was attached to the outlet from the filter drum.

Sections of opaque pipe were replaced with 1.5" I.D. clear acrylic tube in order to observe flow patterns. Clear sections were installed in the horizontal run ahead of the first vertical riser, in both the vertical riser and downcomer and in the horizontal run just after the downcomer. These locations are shown in Figure 2.

Seven 1/4-inch pressure taps were installed in the conveyor pipe, at locations noted as 1 through 7. Each tap had an isolation 1/4-inch root valve, so that two transducers could be moved to different combinations of these seven positions for pressure measurements. A braided copper cable was wrapped around the entire piping system and grounded.

Plant air, dry and oil free, was used as the conveying medium. Air flow was measured by a sharp-edged orifice plate meter with 0.650 inch diameter orifice plate. Air pressure was reduced in two steps of regulation, resulting in up to 100 scfm of 2 to 5 psi air available.

Solids were introduced by a screw feeder, driven by an air motor, which was connected to the 1.5-inch I.D. conveying line. The screw feeder had a 2-inch diameter auger feeding into a valved mixing chamber. The feeder, manufactured by the Mine Safety Appliance Co. for their Bantam Rockduster, was equipped with a hopper and hopper extension. By adjusting the speed of the air motor, loading rates could be adjusted as desired.

The two drums, receiver and filter, were each placed on scales. Using a stopwatch, average powder mass flowrates were determined for each run. Typically, run times in excess of five minutes were used to determine solids conveying rate.

Figure 3 presents a system schematic showing the experimental apparatus. Two pressure transducers, used to monitor pressure responses, are positioned at locations noted as PT1 through PT7 in this drawing.

Kavlico Model P609-10GP pressure transducers were used. These produce a linear voltage output for pressures from 0 to 10 psig. The output is 2.0 to 6.0 volts dc for 0 to 10 psig with 9 volts input, and 2.5 to 7.5 volts dc for this same pressure range with 12 volts input. Two separate power supplies were used to supply transducer excitation voltages. Shielded cables were used for power and signal leads to the transducers.

The heart of the experimental data collection system was a Nicolet Model 2092-3C digital oscilloscope with Model 206-2 plug-in, equipped with two-channel 4096-point total input capacity and floppy disk memory. A subtraction function,

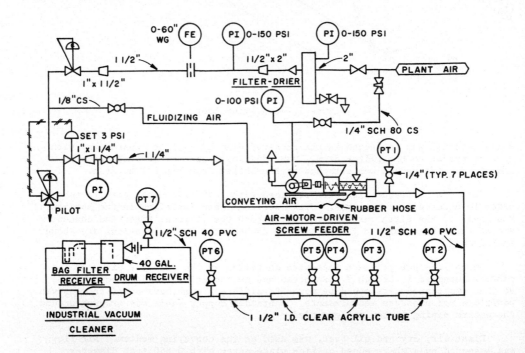

Figure 3. System Schematic Diagram

one of several operations available on this model, allows instantaneous generation of voltage difference curves as a function of time. Variable sweep speeds are available, from microseconds upward. Using a sweep of 2 or 5 milliseconds per point permits easy observation of events of fractional seconds duration.

An X-Y plotter connected to the oscilloscope is then used to produce a permanent record, or hard copy, of the pressure readings. The display, recorded on graph paper, can be taken from disk storage or from live storage. Figure 4 shows an electrical instrumentation schematic for this data gathering system. A Hewlett-Packard Model 7402 two-channel strip chart recorder was also used to continuously monitor pressure responses from both transducers. This instrument was connected in parallel with the Nicolet oscilloscope, permitting readings on both to be taken simultaneously.

3.2. Test Conditions

Ground black walnut shells, marketed by Hammon Products Company of Stockton,

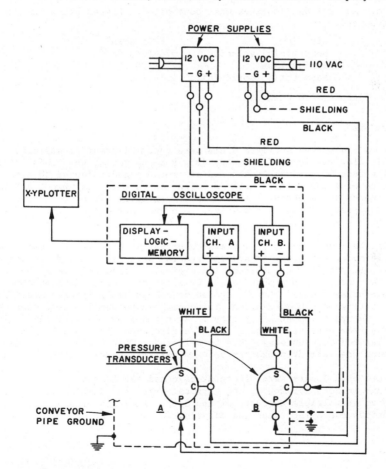

Figure 4. Electrical Instrumentation Schematic

Missouri, were selected as the material conveyed. Table I summarizes properties of the walnut shells.

Table 1. Ground Walnut Shells

Particle density	87.4 lb/ft^3
Bulk density	41.67 lb/ft^3

Size Distribution – U.S. Std. Sieve	
Retained on 40 Mesh	13.61%
60 Mesh	56.96%
100 Mesh	26.58%
200 Mesh	1.27%
Passes through	1.58%
	100%

Calculated Average Particle diameter 1.0×10^{-3} ft

Both air and solids flowrates were controlled. Listed below are ranges of flowrates used:

Air flow	7-31 scfm*
Superficial velocity	570-2525 fpm*
Solids flow	73-390 lb/hr
Mass ratio (solids/air)	1.7-11.2

*Standard Conditions

3.3. Procedure

Tare weights of the receiver drums were recorded prior to starting a run. Next the air flow was set and pressure drop reading across the fluid meter recorded. The feed hopper was loaded and air motor speed adjusted to give the desired value of auger speed. After a suitable run time, between 5 and 15 minutes, the feeder was stopped, timer stopped, conveying air shut off and the collected material weighed. From weight gain and lapsed time, the conveying rate in "lbs/hr" was calculated.

During the conveying period, visual observation of flow in clear sections of pipe were made. Pressure transducer records were taken using the two-channel strip chart recorder. When desired, the digital oscilloscope could be triggered to record fast response pressure data. This latter record, displayed on the oscilloscope screen, could then be stored on floppy disk, labeled and indexed for future reference. When desired, the oscilloscope results were converted to a hard copy record using an X-Y plotter. As noted previously, individual pressure records from both transducers could be displayed, or the difference (pressure drop between transducers) could be selected. This latter record gives an instantaneous pressure drop history between two transducer locations.

Pressure transducers were placed at selected tap locations to monitor pressure response in both horizontal and vertical legs, in addition to across the entire system or multiple segments of the loop. Numerous pressure records were made at various conveying rates and air flows.

4. RESULTS

Data, in addition to the air and solids conveying rates, consisted of graphs that depict instantaneous pressure responses between selected transducer locations for each set of operating parameters. The two-channel strip chart record shows individual pressure fluctuations. An example of this information appears in Figure 5. In this particular case, transducers were mounted at locations 1 and 2, in the first horizontal leg of the system. At an air flowrate of 12 cfm and mass ratio of 6.53, the flow is homogeneous throughout the system. This is indicated by the nearly constant pressure record at both locations 1 and 2. On the other hand, when the air flowrate was reduced to 7 cfm and the mass ratio increased to 9.54, the flow becomes non-homogeneous, as is indicated in the second part of the figure. The fact that the two pressure records are similar in shape (P-1 and P-2) suggests that a disturbance further downstream is transmitting signals upstream to these two transducers. It is probable that choking is occurring in the vertical leg. Pressure fluctuations on the order of 1 1/2 psi are noted in this second case.

An example of a fast response Nicolet oscilloscope record is given in Figures 6 and 7. In Figure 6, pressure records at locations 1 and 4 are given (8.5 cfm, 355 lb/hr, Mass ratio = 9.26). When the signals are electrically subtracted, their difference, which represents the pressure drop between locations 1 and 4, appears. Figure 7 shows this instantaneous Δp as a function of time. The entire time record shown by Figures 6 and 7 takes place in approximately 0.8 seconds. As may be noted in Figure 6, the flow is non-homogeneous. Maximum pressure variations on the order of 1.7 psi are observed. The flow between locations 1 and 4 is fairly homogeneous, except during the early part of the time interval (see Figure 7). The maximum Δp shown is 1.36 psi.

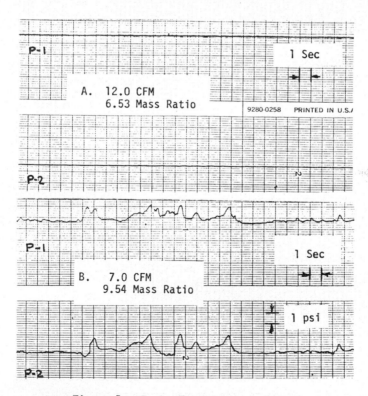

Figure 5. Strip Chart Pressure Record

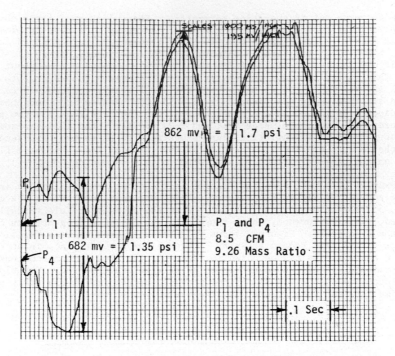

Figure 6. Nicolet Oscilloscope Pressure Record

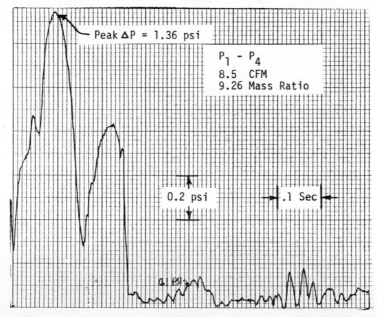

Figure 7. Pressure drop between pressure tap locations

These results, while preliminary in nature, permit one to observe and detect the transition from homogeneous to non-homogeneous flows in a pneumatic conveying system. In an attempt to quantify results into a single figure, a plot shown as Figure 8 was made. Here the maximum transient pressure excursion was divided by the steady state pressure drop between the same two locations for air flow. Plotted along the horizontal is a measure of material transported, described as (Mass ratio/cfm) and called Material Transport Time. Using information from a number of Nicolet graphs, data points were calculated and plotted. An approximate regression, shown as a dashed straight line, permits one to estimate the magnitude of pressure fluctuations expected when conveying walnut shells at different rates in this system.

5. CONCLUSIONS

The conclusions of this study are the following:

1. A pressure measuring methodology has been developed that permits detection of transition from homogeneous to non-homogeneous conveying.

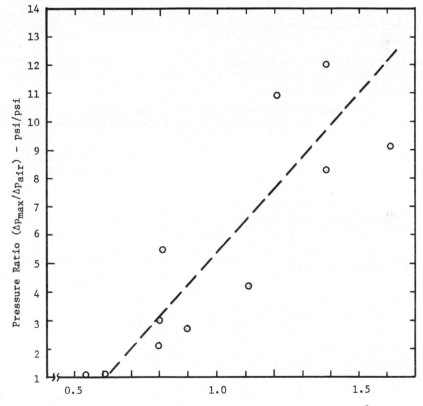

Figure 8. Flow Stability

        2. Instantaneous pressure drop records between different locations are easily measured using the Nicolet oscilloscope.

        3. The methodology developed can be used to study unsteady-state phenomena in two-phase flow, thus better understanding the onset of saltation and choking.

        4. Using this technique, it is possible to determine when the flow is choking, when there is saltation, and when both are occurring at the same time.

REFERENCES

1. Marchello, J. M., and Gomezplata, A., Gas-Solids Handling in the Process Industries, Marcel Dekker, Inc., New York (1976).

2. Kraus, M. N., Pneumatic Conveying of Bulk Materials, Second Edition, Chemical Engineering, McGraw-Hill Publications Co., New York (1980).

3. Matsumoto, S., Harada, S., Saito, S., and Maeda, S., "Saltation Velocity for Horizontal Pneumatic Conveying," J. of Chem. Eng. of Japan, Vol. 8, No. 4, 331-333 (1975).

4. Jones, J. P., and Leung, L. S., "A Comparison of Correlations for Saltation Velocity in Horizontal Pneumatic Conveying," Ind. Eng. Chem., Process Des. Dev., Vol. 17, No. 4, 571-575 (1978).

5. Tomita, Y., Jotaki, T., and Hayashi, H., "Wavelike Motion of Particulate Slugs in a Horizontal Pneumatic Pipeline," Int. J. Multiphase Flow, Vol. 7, No. 2, 151-166 (1981).

6. Yang, W.-C., "A Mathematical Definition of Choking Phenomenon and a Mathematical Model for Predicting Choking Velocity and Choking Voidage," AIChE Journal, Vol. 21, No. 5, 1013-1015 (1975).

7. Leung, L. S., Wiles, R. J., and Nicklin, D. J., "Correlation for Predicting Choking Flowrates in Vertical Pneumatic Conveying," Ind. Eng. Chem., Process Des. Dev., Vol. 10, No. 2, 183-189 (1971).

8. Leung, L. S., "A Quantitative Flow Region Diagram for Vertical Pneumatic Conveying of Granular Solids," Pneumotransport 5, Fifth Int. Conf. on Pneumatic Transport of Solids in Pipes, 35-46 (1980).

9. Zeng, F. A., and Othmer, D. F., Fluidization and Fluid Particle Systems, Reinhold, New York (1960).

Multi-Phase Flow and Heat Transfer III. Part B: Applications
edited by T.N. Veziroğlu and A.E. Bergles
Elsevier Science Publishers B.V., Amsterdam, 1984 — Printed in The Netherlands

381

NUMERICAL INVESTIGATION OF TWO-PHASE POLYDISPERSE FLOWS IN AXISYMMETRIC LAVAL NOZZLES TAKING INTO ACCOUNT THE PROCESSES OF FRAGMENTATION

E.G. Zaulichny, B.M. Melamed, A.D. Rytchkov
Institute of Theoretical and Applied Mechanics
Siberian Branch of the USSR Academy of Sciences
Novosibirsk 630090, USSR

ABSTRACT

Two-phase polydisperse flows in Laval nozzles are considered with an account being taken of coagulation and fragmentation processes of different size particles interacting with each other. Numerical parametric investigations of such flows are carried out with purpose of elucidating the influence of features of the accepted model of the medium on the integral characteristics of nozzles.

Notations

$\vec{U}\{u,v\}$ velocity vector and its projections on the x - and y - axes of the cylindrical coordinate system

ρ density

p pressure

T temperature

k, k_e isentropic exponents for gas (the "frozen" one) and for mixture (the "equilibrium" one)

c_p, c_δ specific heats of gas and particles

C_{R_i}, C_{α_i} drag coefficient and heat transfer coefficient, respectively

Φ_{ij} collisional efficiency coefficient

K_{ij} coagulation constant

H_0 total gas enthalpy

E_i total particle energy

n_i the number of particles per unit volume

e_{ij} coefficient of particle entrainment

Subscripts i, j refer to the i th and j th species of par-

The vector $A^{(i)}$ is expanded into Taylor series

$$A^{(i)} = A^{(i)}_0 + \left(\frac{\partial A^{(i)}}{\partial x}\right)_0 \bar{x} + \dots$$

where the values of components obtained at $\bar{x} = \text{depth}$ for $\bar{y}$ are used. Then,

The stability condition of the scheme (5)–(6) is

$$\dots$$

where $\dots$ is the maximum positive eigenvalue at the given point.

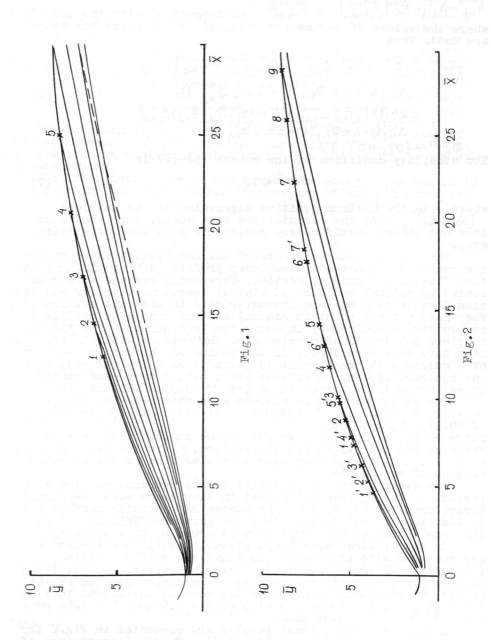

Fig.1

Fig.2

sults in downstream transfer of the location of their sedimentation on the nozzle contour. As the throat contour straightens for
contours with equal expansion degree the particles falling out
on the profile with break point (Fig.2) do not fall out on the
profile with smoothed throat (Fig.1). In accordance with this the
sedimentation onset of the first species I considered in computational model for the above contours is transferred downstream
by a factor of about two along the x -coordinate from the nozzle
throat for the profile with a smoothed throat as compared to the
nozzle with $z^* =1.0$. The near-wall region in the channel which
is bounded by its wall and by the limit trajectory of the smallest considered species is called the pure gas zone. The computational model of the two-phase flow assumes that in the pure gas
zone the particles may be present in an insignificant number that
are less than the size of the first species falling out on the
nozzle contour.

In Fig.2 by figures with primes the sedimentations locations
are denoted for particles species at $d_{43 6 x} =2.2$ μm. It can be se-
en that the increase in the amount of small particles results in
an earlier sedimentation of these particles on the profile surface under the same rest of conditions. In this case the smallest
species I is 0.28 μ and the rest of species enumerated in an increasing sequence are as follows: 2 $-0.41 \mu m$, 3 -0.6 μm ,
4 -0.9 μm , 5 -1.37 μm, 6 -2.26 μm, 7 -3.71 μm. As it was expected, the diminution of the mean mass particles size d_{43} at
this constant mass fraction z does not result in significant difference in the sedimentation location of equally sized particles on the nozzle contour. The velocity non-equilibrium of particles (their lag in speed from gas) is affected by their drag
in gas, the increase in the amount of small species increases in-
significantly the velocity of gas particles. The equal size particles fall out on the nozzle surface practically in the same
section $\overline{x}$.

Let us demonstrate the abilities of the model by other Figures. In Fig.3 we present a comparison of the computed pressure
values along the nozzle contour (solid line) with measured values
in corresponding sections (points). A satisfactory agreement between compared values can be asserted both in the pure gas zone
and in the zone with particles sedimentation on the channel surface. A satisfactory agreement between calculated and experimental pressure values on the nozzle surface in the zone of sedimentation of particles of the K th species corroborates the correct-
ness of the estimate that at the relation of the particles density ρ_p to gas density ρ_g being higher than 10^3 and at z indicated above the particles volume is less than 1% [10-12] and it
does not affect significantly the change in the pressure exerted
by the gas on the wall. The measurement of the pressure on the
wall by its draining in the particles sedimentation zone is possible by virtue of small angles of particles approach to the surface flowed around.

Fig.4 illustrates the magnitude of the deviation of computed values of the flow rate of the products of a mixture, G_{cal} ,
from one obtained in experiment, G_{exp} . A higher values of G_{cal}
as compared to G_{exp} are related to the fact that the increase in
the boundary layer thickness is not taken into account in the
computational model.

In Fig.5 the distribution of Mach numbers and gas temperature on the wall T_{gw} along the nozzle contour is presented for the

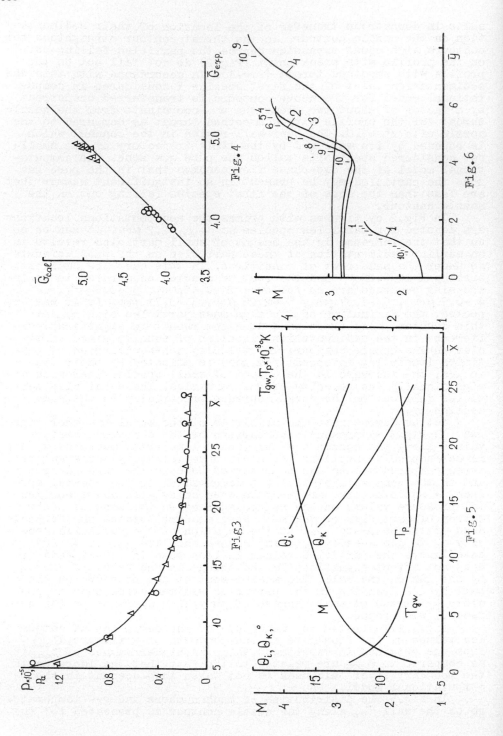

Fig.4

Fig.6

Fig.3

Fig.5

contour with M_0 =4.56 and z^* =0.65. In the same Figure also the changes in the particles temperature T_p , in the inclination angle of the contour to the nozzle axis Θ_K and in the angle Θ_i between the trajectory of corresponding species and the nozzle axis are shown in the interval of particles sedimentation. The curves of T_p and Θ_i are drawn for visual convenience after discrete values obtained in the place of sedimentation of different particles species. It is seen that the particles of larger sizes have a larger angle of approach to the contour and a larger difference $T_{gw} - T_p$ here.

The distribution of Mach numbers M in a number of cross-sections of the nozzle is shown in Fig.6 for M_0=4.56, z^*=1.0. The curve I corresponds to the section $\bar{x}$ =1.5, 2 - $\bar{x}$=11, 3 - $\bar{x}$=12.3, 4 - $\bar{x}$ =25.5. Vertical bars show the locations of limit trajectories of particles of different species. An abrupt increase in M is related to the presence of a zone of stratification of limit particles trajectories in which their concentration rapidly decreases.

The above presented computational technique enables one to determine the location of the particles sedimentation on the nozzle surface, the sedimentation intensity, the angles of particles approach to the wall which may serve as input data for the computation of quantities characterizing the nozzle wall ablation by some of the models describing this phenomenon [13] . The above listed parameters of the two-phase flow as well as mechanical characteristics of the original wall material or of its coke residue for the case of application of composite materials affect mainly the wall erosion by particles of the flow [13-15] .

Fig.7 shows the computed distribution of the specific mass of particles falling out on the wall for channel profiles with M_0=4.56 and for different z^* . The curve 1 refers to the contour z^*=1.0, 2 - z^* =0.95, 3 - z^*=0.85, 4 - z^* =0.65 for d_{436x}=4.75 μm; the curve 5 refers to the contour z^* =1.0 and to d_{436x}=2.2 μm . With reduction of d_{436x} the particles sedimentation on the wall begins at the point located much more closer to the critical section under the same rest of conditions. Conversely, the straightening of the nozzle throat moves the onset of particles sedimentation to the exit.

The experimentally determined magnitude of the mass flow rate of eroding the composite material by particles for conditions comparable with computation is shown in Fig.8 by approximating curves. A satisfactory qualitative agreement between the computed specific mass of particles M_p falling out on the wall and the mass ablation j_w of the wall material should be noted that confirms the assumption on proportionality of erosion magnitude to the particles flow. The experimental cross-sections of the ablation onset do not coincide with calculated cross-sections of small particles sedimentation. For the realization of the mechanical destruction of the wall the kinetic energy of falling particles should exceed the yield strength of the wall material[14] . In addition, the specific flow rate of small species should be sufficient to remove from the wall material an amount being sufficiently noticeable for measurement during the engine functioning. As the calculated value of the ablation onset a definite level of the mass M_p of particles falling out on the wall may be taken. By using the estimations carried out with [13] in view and comparing with experimental data a threshhold value $M_p \simeq 0.01 kg/m^2 s$ has been obtained. The calculated values of the ablation onset $\bar{x}_\ell$ for different nozzle contours are shown in Fig.9 by the curves $1 - z^* = 1.0$, $2 - z^* = 0.93$, $3 - z^* = 0.86$, $4 - z^* = 0.65$. Experimentally determined values of $\overline{w}_\delta$ for corresponding conditions are shown by points in the Figure. For contours with a break point the computation yields somewhat higher values of $\overline{w}_\delta$. In the rest of cases the agreement is satisfactory. Below we illustrate the influence of particles sedimentation and of the angle of their approach to the contour on the nozzle impulse.

In Fig.10 we present the distribution of the value of specific impulse I versus expansion degree $\overline{y}$ for different nozzles. The notations are explained in the Figure. It can be seen that on the nozzle contour with $M_0 = 4.24$ the particles of the th species intensively fall out on the wall at the least expansion degrees $\overline{y} \approx 6$ as compared to other profiles. This results in an abrupt reduction of the impulse. At the same time for profiles with greater M_0 there is practically no particles sedimentation on the wall because of a larger expansion of profiles , but in this case we have a loss in impulse due to scattering. The optimal contour of a nozzle is determined depending on the necessary expansion degree.

REFERENCES

1. Rytchkov, A.D., Chislennoie Issledovanie Dvukhfaznykh Techenii v Osesimmetricheskikh Soplakh Lowalia s Uchetom Processov Koagulatsii i Drobleniia Chastits Kondensata. Izv. AN SSSR, MZh G, 1980, No 1.

2. Sternin, L.E., Maslov, B.N., Schreiber, A.A., Podvysotskii, A.M., Dvukhfaznyie Mono- i Polidispersnyie Techeniia Gaza s Chastitsami. Moscow, "Mashinostroienie", 1980.

3. Sternin, L.E., Osnovy Gazodinamiki Dvukhfaznykh Techenii v Soplakh. Moscow, "Mashinostroienie", 1974.

4. Henderson, C.B., Drag Coefficients of Spheres in Continuum and Rarefied Flows. AIAA Journal, 1976, 14, No 6.

5. Alemasov, V.E., Dregalin, A.F., Tishin, A.P., Hudiakov, V.A., Termodinamicheskiie i Teplofizicheskiie Svoistva Productov Sgoraniia. Tom 1. Moscow, 1971.

6. MacCormack, R.W., The Effect of Viscosity in Hypervelocity Impact Cratering. AIAA Paper, 1969, No 354.

7. Rytchkov, A.D., Ob Odnoi Raznostnoi Skheme, Ispol'zuemoi pri Raschiotakh Neravnovesnykh i Dvukhfaznykh Techenii. Sb. "Gazovaia Dinamika", izd-vo TGU, Tomsk, 1977.

8. Fein, H.L., A Theoretical Model for Predicting Aluminium Oxide Particle Size Distributions in Rocket Exhausts. AIAA Journal, 1966, 4, No 1.

9. Dobbins, R.A., and Strand, L.D., A Comparison of Two Me - thods of Measuring Particle Size of Al_2O_3 Produced by a Small Rocket Motor. AIAA Journal, 1970, 8, No 9, 1544-1555.

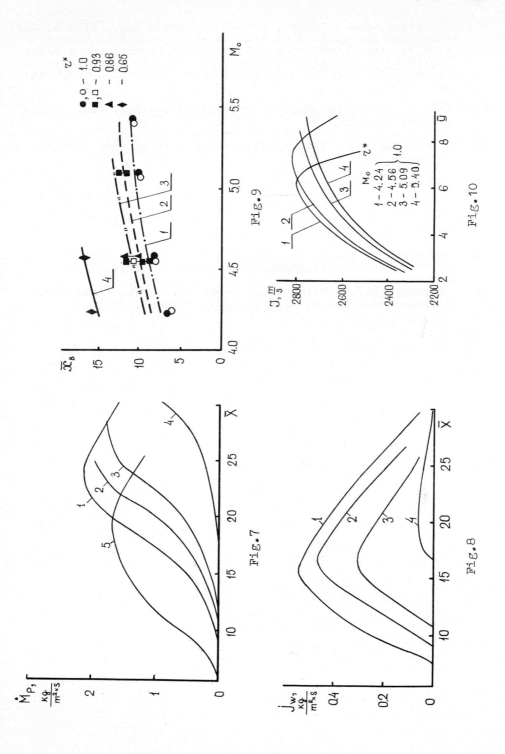

Fig.7

Fig.8

Fig.9

Fig.10

10. Zaulichny, E.G., Ivanov, V.Ya., Rytchkov, A.D., Sivykh, G.F., Two-Phase Nonequilibrium Flow in an Axisymmetric Channel with Viscous Interaction and Heat-Mass Transfer on a Wall. Multiphase Transport: Fundamentals, Reactor Safety, Applications. Vol. 1, Hemisphere Publishing Corporation, Washington, 1980.

11. Soo, S., Fluid Dynamics of Multiphase Systems. Russ. Trans. Moscow, Mir, 1971.

12. Boothroyd, R.G., Flowing Gas-Solids Suspensions. London , 1971.

13. Rafikov, R.V., Zaulichny, E.G., Rytchkov, A.D., Tenenev , V.A., Ivanov, V.Ya., Nefedov, I.V., Chislennoie Issledovaniie Dvukhfaznogo Techeniia v Osesimmetricheskom Kanale s Uchetom Real'nykh Mekhanizmov Razrusheniia Iego Stenok. Izvestiia SO AN SSSR, Ser. Tekhn. Nauk, 1981, No 3, vyp. 1.

14. Dvukhfaznyie Techeniia v Soplakh Raketnykh Dvigatelei (Obzor). Voprosy Raketnoi Tekhniki. Sb. perevodov, 1974, No 7.

15. Laderman, A.J., Lewis, C.H., Byron, S.R., Two-Phase Plume Impingement Effects, AIAA Journal, 1970, 8, No 10.

Multi-Phase Flow and Heat Transfer III. Part B: Applications
edited by T.N. Veziroğlu and A.E. Bergles
Elsevier Science Publishers B.V., Amsterdam, 1984 — Printed in The Netherlands

PROBLEMS ASSOCIATED WITH PARTICULATE FLOW IN TURBOMACHINERY

W. Tabakoff and A. Hamed
Department of Aerospace Engineering and Applied Mechanics
University of Cincinnati
Cincinnati, Ohio 45221, U.S.A.

ABSTRACT

A computational model for calculating the flow field, in the presence of solid particles, through a two-dimensional compressor cascade is presented. The method treats the particle phase in the Lagrangian system and the fluid phase in the Eulerian system. The equations of momentum of the fluid phase are modified to account for the momentum exchange between the two phases. The resulting modified momentum and continuity equations are reduced to the conventional stream function-vorticity formulation.

The analysis yields the change in the blade surface pressure distribution, the total pressure and velocity in the flow field. Based on this, it is possible to predict the decrease in turbomachinery performances under the presence of solid particles.

NOMENCLATURE

b	blade height in cascade, m
C_D	coefficient of drag
Ch	chord, m
d	diameter, m
D	drag force, N
F	force, N
i,j	grid point indices
J	transformation parameter
k	constant defined as $W/b\rho_g$
$\dot{m}$	mass flow rate of the particle along a trajectory, kg/sec
p	pressure static, N/m^2
P	pressure total, N/m^2
q	dynamic head, N/m^2
S	pressure coefficient
Δt	residence time, s
u,v	velocity in x,y directions, respectively, m/s
V	volume, m^3
W	mass flow rate of gas through one cascade passage, kg/sec
x,y	Cartesian coordinates
α	transformation parameter or particle concentration
$\bar{\alpha}$	acceleration, m/s^2
β	transformation parameter
$\beta_{(\ )}$	angle, radians

γ   transformed parameter
ξ,η   transformed coordinates
ζ   total pressure loss coefficient
μ   viscosity
ω   vorticity
ψ   stream function

Subscripts

e   exit
g   gas
i   inlet
p   particle
s   surface
x,y   x and y components
ξ,η   partial derivatives with reference to ξ,η

1. INTRODUCTION

The problem of gas flow mixed with solid particles in turbo-machinery has great importance in industrial, naval and aero-nautical applications. The effect of particle laden flow on the turbomachinery performance can be of considerable magnitude since the momentum exchange between the particles and the fluid can alter significantly the flow field.

Experimental work on turbine performance deterioration with presence of particulate flow is reported in reference [1] and a theoretical model in reference [2]. In this reference, the theoretical analysis is for one dimensional flow and assumes that the particles follow strictly the fluid streamlines. This approximation is not valid in all cases, therefore a new theoretical analysis for particulate flow in a two dimensional axial flow compressor cascade is initiated. The new analysis includes the inelastic rebound of the particles at the blade surfaces. Consistent with the discrete nature of the particles, the particle phase equations are solved in the Lagrangian system and the fluid phase equations are solved in the Eulerian system. In most cases of axial flow turbomachinery, the flow Reynolds number is very high, consequently the viscous effects are limited to the boundary layer region. Hence, the fluid is assumed as inviscid. Also, the fluid is treated as incompressible in the following analysis.

2. EULERIAN FORMULATION FOR THE FLUID PHASE

The continuity and momentum equations of the fluid phase are as follows:

$$\frac{\partial u_g}{\partial x} + \frac{\partial v_g}{\partial y} = 0 \tag{1}$$

$$u_g \frac{\partial u_g}{\partial x} + v_g \frac{\partial u_g}{\partial y} = -\frac{1}{\rho_g} \frac{\partial p}{\partial x} - \frac{F_x}{\rho_g} \tag{2}$$

$$u_g \frac{\partial v_g}{\partial x} + v_g \frac{\partial v_g}{\partial y} = -\frac{1}{\rho_g} \frac{\partial p}{\partial y} - \frac{F_y}{\rho_g} \tag{3}$$

where F_x and F_y are the interphase force components representing the momentum exchange between the fluid and particles.

The above equations can be reduced to the conventional stream function-vorticity formulations given below:

$$\frac{\partial \psi}{\partial y} \frac{\partial \omega}{\partial x} + \frac{\partial \psi}{\partial x} \frac{\partial \omega}{\partial y} = -\frac{1}{\rho_g k} \frac{\partial F_x}{\partial y} + \frac{1}{\rho_g k} \frac{\partial F_y}{\partial x} \tag{4}$$

$$\frac{\partial^2 \psi}{\partial x^2} + \frac{\partial^2 \psi}{\partial y^2} = \frac{\omega}{k} \tag{5}$$

where $k = \dfrac{W}{b\rho_g}$, $\omega = \dfrac{\partial u_g}{\partial y} - \dfrac{\partial v_g}{\partial x}$, $u_g = k\dfrac{\partial \psi}{\partial y}$, $v_g = -k\dfrac{\partial \psi}{\partial x}$,

W is the mass flow through the blade passage, and b is the blade height.

However it is to be noted that the vorticity production mechanism is due to the momentum exchange between the fluid and the particle. The above equations can be transformed into an arbitrary (ξ, η) coordinate system. The transformed equations are as follows:

$$\xi = f_1(x,y) , \qquad \eta = f_2(x,y)$$

and equations (4) and (5) become

$$\omega_\xi \psi_\eta - \omega_\eta \psi_\xi = \frac{1}{k\rho_g} [-x_\xi F_{x_\eta} + x_\eta F_{x_\xi} + y_\eta F_{y_\xi} - y_\xi F_{y_\eta}] \tag{6}$$

and

$$\left(\frac{\partial \left(\frac{\alpha}{J}\right)}{\partial \xi} - \frac{\partial \left(\frac{\beta}{J}\right)}{\partial \eta} \right) \frac{\partial \psi}{\partial \xi} + \left(\frac{\partial \left(\frac{\gamma}{J}\right)}{\partial \eta} - \frac{\partial \left(\frac{\beta}{J}\right)}{\partial \eta} \right) \frac{\partial \psi}{\partial \eta} + \frac{2}{J} \frac{\partial^2 \psi}{\partial \xi^2}$$

$$+ \frac{\gamma}{J} \frac{\partial^2 \psi}{\partial \eta^2} - 2\frac{\beta}{J} \frac{\partial^2 \psi}{\partial \xi \partial \eta} = \frac{J\omega}{k} \tag{7}$$

where $\alpha = x_\eta^2 + y_\eta^2$, $\gamma = x_\xi^2 + y_\xi^2$

$$\beta = x_\xi x_\eta + y_\eta y_\xi , \qquad J = x_\xi y_\eta - x_\eta y_\xi$$

Boundary Conditions

Referring to Fig. 1, the boundary conditions for the stream function equation (7) are:

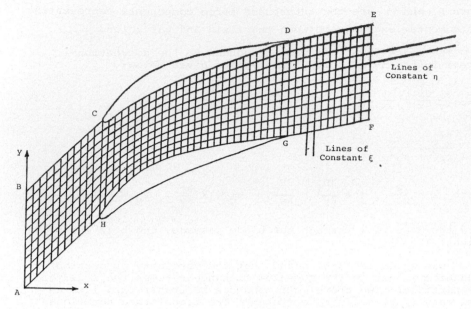

Fig. 1. Grid System.

For the upstream and downstream boundaries (AB) and (EF), the fluid inlet and exit angles β_1 and β_2 are specified as follows:

$$\tan \beta_1 = (\frac{v_g}{u_g})_{inlet} \quad , \quad \tan \beta_2 = (\frac{v_g}{u_g})_{exit}$$

For the periodic boundaries (AH), (BC), (GF) and (DE)

$$\psi_{(BC)} = \psi_{(AH)} + 1.0$$

$$\psi_{(DE)} = \psi_{(GF)} + 1.0$$

Along the airfoil surfaces

$$\psi_{HG} = 0 \ , \quad \psi_{CD} = 1.0$$

With these boundary conditions, the equation for the stream function may be solved in any arbitrary coordinate system.

3. LAGRANGIAN FORMULATION OF THE PARTICLE PHASE

It is assumed that only the fluid drag is the major force acting on the particles. The equation of motion of the single particle in Cartesian coordinates is given by:

$$m_p \frac{du_p}{dt} = D_x \quad ; \qquad m_p \frac{dv_p}{dt} = D_y \tag{8}$$

where D_x and D_y are the drag forces experienced by the particle with their mass, m_p, and u_p, v_p are particle velocities in the x and y directions. It was assumed that the particles are spherical and the drag forces are given by modified Stokes law.

The above equations for the particle trajectories constitute an initial value problem. However, at the point of impact of the solid particles with the body surface, the rebounding particle velocities and angles must be specified as functions of the incoming angles and velocities for inelastic collision. This data is provided in the form of experimental correlation[4] for various materials and particles combinations. In this study, the blade material is assumed to be 6064 T-6 aluminum and the particle material, quartz sand. For this material and particle combination, the restitution coefficients are given by the equations:

$$\frac{V_{n_2}}{V_{n_1}} = 0.993 - 1.76 \beta_1 + 1.56 \beta_1^2 - 0.49 \beta_1^3 \tag{9}$$

$$\frac{V_{t_2}}{V_{t_1}} = 0.998 - 1.66 \beta_1 + 2.11 \beta_1^2 - 0.67 \beta_1^3 \tag{10}$$

where V_{n_1} and V_{t_1} are the impinging normal and tangential velocities and V_{n_2} and V_{t_2} are the rebounding normal and tangential velocities, and β_1 is the angle between the impacting particle and tangent to the local surface.

4. THE INTERPHASE FORCE TERM

The interphase force term represents the mean values of the momentum exchange between the fluid and the particles. The computation of these forces at any given point is carried out by the method employed by Crowe [5]. If one considers a cell ABCD about point i,j as shown in Fig. 2, then the average interphase forces per unit volume at this point are given by:

$$F_x = \dot{m}_p \, \Delta\tau \, \bar{\alpha}_{xp}/V_{cell} \quad , \qquad F_y = \dot{m}_p \, \Delta\tau \, \bar{\alpha}_{yp}/V_{cell} \tag{11}$$

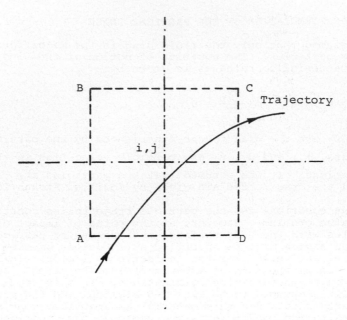

Fig. 2. Control Volume for Computing Force Terms.

where $\dot{m}_p$ is the mass flow rate of particles associated with this trajectory and $\Delta\tau$ is the residence time of the particles inside of the cell ABCD. V_{cell} is the cell volume and $\bar{\alpha}_{xy}$ and $\bar{\alpha}_{yp}$ are the average particle acceleration components through this cell along this trajectory in the x,y direction.

These force terms are calculated based on the solution of the particle equation of motion, and later used in the computation of the vorticity in the flow field.

5. NUMERICAL TECHNIQUE

The method of solution of the stream function-vorticity equations depend mainly on the choice of the coordinate system. In this case, the ξ coordinates were chosen as lines of constant x and the η coordinates as any body fitted coordinate lines.

In this particular problem, the vorticity generation mechanism is due to the interphase force gradients. In order to get fairly smooth force gradients at all points, the grid system has to be fairly coarse. Since the error in the computed vorticity at the body surface can be large in a coarse mesh, it was decided not to use the vorticity at the body surface during the computation. The vorticity equation can be solved easily by marching along lines of constant η if the term $\omega_\eta\psi_\xi$ in equation (6) is neglected. In the case that the lines of constant η and the streamlines are the same, there is no error in the calculated vorticity, since the term ψ_ξ is zero. However, the error in the computed vorticity can be minimized if the η coordinates are updated to the new streamline pattern at every iteration.

The advantages of this method is that the only information needed is the vorticity values at the inlet boundary. The solution does not use the vorticity at the blade surface. However, the marching solution of the vorticity equation needs the information on the vorticity at the trailing edge. Since the fluid total pressure is different at the trailing edge on the airfoil suction and pressure sides, the vorticity at the trailing edge is not zero. Consequently, the vorticity at the trailing edge is computed as the sum of the vorticities of the pressure and suction surfaces at the trailing edge [6].

The finite difference scheme of the stream function equation can be solved by a banded matrix solver. The solution was started with an assumed pattern of streamlines satisfying the inlet and outlet flow angle requirements. The vorticity was assumed to be zero and the irrotational flow streamlines were found for 21 streamlines. This was followed by the computation of the trajectories and the forces.

Once the force terms are computed, the vorticity is computed by marching along the η coordinates. This is followed by the stream function solution. Then the coordinates are updated and the process is repeated until the streamlines and the vorticity converge.

The pressure solution was obtained by integration of the momentum equation along η and ξ directions.

6. RESULTS AND DISCUSSION FOR CASCADES

The method was applied to a diffusing cascade of NACA 65(10)10 airfoils. The inlet gas velocity was 130 m/s and the ratio of the particle to gas velocities at the cascade inlet was assumed to be 0.80. The cascade had an air inlet angle of 45° and an exit angle of 18°. The numerical solution for particulate flow was obtained for 21 streamlines in the η direction. The upstream boundary was chosen at 0.5 Ch ahead of the leading edge of the cascade and the downstream boundary was chosen at 2.0 Ch behind the trailing edge. The $\Delta\eta$ spacing was 0.05 and the $\Delta\xi$ spacing was 0.05 Ch. The particle mass concentration, α, defined as the mass flow of particles per unit mass flow of air was maintained at 0.10 throughout the analysis.

Figure 3 shows the trajectories of 165 micron size particles through the cascade. Figure 4 shows the change of the streamline pattern for airflow with and without particles. It can be observed that the streamlines near the pressure surfaces move away from the surfaces and the streamlines near the suction surface move closer to the surface. However, the changes were only of the order of two percent or less of the blade spacing. The streamlines in general bend in the direction of the blade suction surface. The effect of the particles on the pressure distribution is shown in Fig. 5. The blade surface pressure distribution is

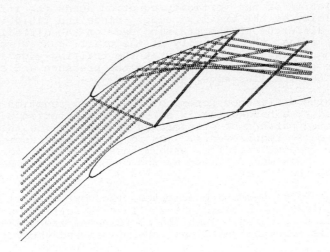

Fig. 3.   Particle Trajectories Through the Cascade,
          d_p = 165 Microns.

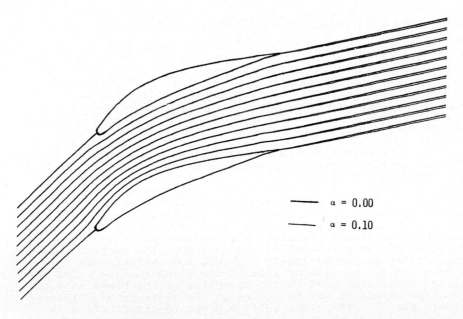

Fig. 4.   Streamline Pattern With and Without Particles,
          d_p = 165 Microns.

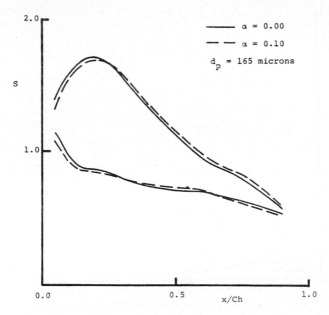

Fig. 5. Effect of Particles in the Surface Pressure Distribution.

plotted in terms of pressure coefficient, S, defined as:

$$S = (P_i - P_s)/q_i \tag{12}$$

There is a decrease of the surface static pressures on the blade suction surface. However, the pressure distribution on the blade pressure surface is more sensitive to the particle presence and does not follow a set pattern.

In addition, the mean exit total pressure is always less than the mean inlet total pressure. Hence one can define a loss coefficient based on the decrease in the total pressure due to particle. This loss coefficient is defined as,

$$\zeta_{particle} = (\bar{P}_i - \bar{P}_e)/q_i \tag{13}$$

Figure 6 shows the effect of particle diameter on the total pressure loss. The total pressure loss decreases rapidly as the particle size increases. Under identical conditions, the fluid drag is proportional to the total surface area of the particles. For the same particle concentration the total surface area is proportional to $1/d_p$. For this reason, the small particles give rise to a large decrease in total pressure.

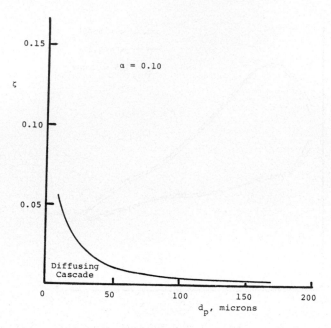

Fig. 6. Effect of Particle Size on the Total Pressure Loss Coefficient at Cascade Exit for α = 0.10.

7. RESULTS AND DISCUSSION FOR A RADIAL INFLOW TURBINE

These example calculations are for viscous fluid particulate flow in a radial inflow turbine. The analysis for viscous flow interaction is very similar to the analysis described above for inviscid flow. Further details for particulate viscous flow interaction is presented in reference [7].

The calculated fluid streamline pattern, in the absence of particles, for given conditions is shown in Fig. 7. The streamlines are plotted for the region between a pair of blades, represented by the heavy thick lines. The streamlines are designated by a stream function ratio ψ/ψ_{total} such that the distance between any two streamlines indicates the amount of the flow between them. Thus, for the given channel configuration, the streamline spacing is indicative of the velocity relative to the rotor, with close spacing indicating high velocities and with wide spacing indicating low velocities. From the inspection of Fig. 7, it can be observed that there is a recirculating zone near the blade pressure surface. The next step was to calculate the particle trajectory for particle size of 10 microns in diameter.

A time step of 0.0001 sec. is used in all the trajectory calculations, except preceding the solid surface impact. Lower time steps require higher computer time. The optimum time step is obtained after a few trials, in such a way that any further reduction in the time step does not change the magnitude of the interphase forces. The results for 10 micron particles are shown in Fig. 8.

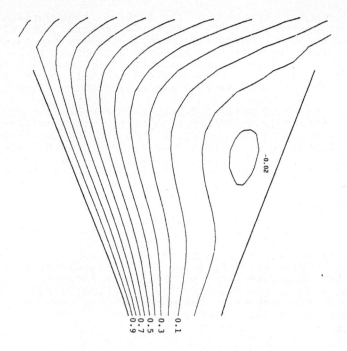

Fig. 7.   Relative Streamlines for Flow Through Radial
           Inflow Turbine (Without Particles).

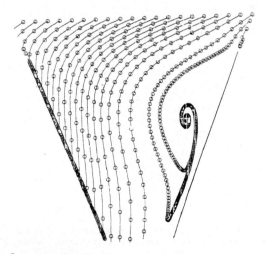

Fig. 8.   Trajectories of 10 Micron Particles.

The particles impacting on the suction blade surface are
slowed down considerably after repeated impacts. The flow pattern
present in the rotor channel is such that there are not many
particle impacts on the blade pressure surface. Figure 9 shows
the effect of particle concentration on the fluid flow pattern
inside the radial inflow turbine blade channel, for a mean
particle diameter of 10 microns and mass fraction 0.10.

After the solid particles impact on the blade surfaces, their
velocities differ considerably from that of the gas flow, both in
magnitude and direction, consequently affecting the fluid flow
properties. Figure 10 is obtained by superimposing streamlines
of Fig. 9 on Fig. 7. Comparing Figs. 7 and 9, it can be observed
that the fluid streamlines are shifted with the presence of the
particles. Near the blade exit, the streamlines are shifted in
the direction of the blade pressure surface. This shift will
become more dominant at higher particle concentrations. In the
vicinity of the blade suction surface, towards the exit, the
streamlines are less closely spaced, indicating that the fluid
has slowed down considerably due to the particles momentum loss.
Since ten micron diameter particles follow the fluid streamlines
more closely (Fig. 8), they enter into the recirculation zone and
cause further flow disturbances. All these flow changes will
affect the turbine performance. The numerical solution is obtained
on an AMDAHL 470 computer. The computer code requires about 500 k
in storage and 6 minutes of CPU time.

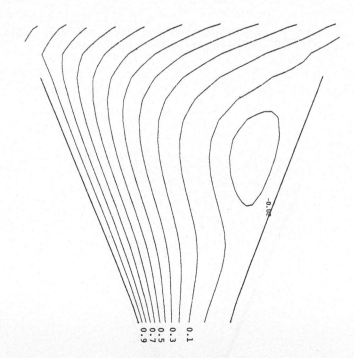

Fig. 9.   Effect of Particle Concentration on Fluid Streamline
          Pattern. (Particle Size = 10 Microns, Mass Fraction = 0.10).

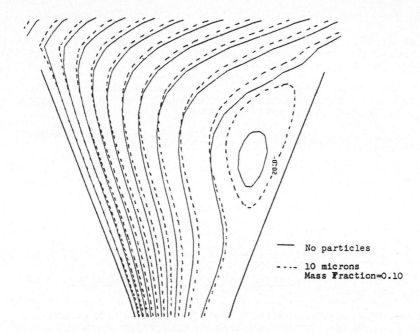

Fig. 10.   Effect of Particles on Fluid Streamline Pattern.

7.   CONCLUSIONS

It was found that the effect of solid particles on the
cascade flow field contributed to the bending of the streamlines
toward the blade suction surface. The computed difference in the
pressure coefficient for particulate flow, with 165 micron
particles diameter, is of the order of two percent over the
air flow only.   The most important finding was that the total
pressure loss associated with the particulate flow is very high
for very small particles as compared to large particles. Consequently, significant reduction in performance can occur in a
real multi-stage machine due to the changes in the pressure
ratios, particularly if the suspended particles are small.
In addition, the behavior of solid particles entrained by the
viscous gas flow in a inward radial inflow turbine rotor blade
channel is determined. The difference between using viscous or
inviscid flows is very much pronounced in certain blade regions.

ACKNOWLEDGEMENT

This research was supported by U.S. Army Research Office
under Contract No. DAAG29-82-K-0029.

REFERENCES

1. Tabakoff, W., Hosny, W. and Hamed, A., "Effect of Solid Particles on Turbine Performance," Journal of Engineering for Power, January 1976, pp. 47-52.

2. Tabakoff, W. and Hussein, M.F., "Effect of Suspended Solid Particles on the Properties in Cascade Flow," AIAA Journal, Vol. 8, August 1971, pp. 1514-1519.

3. Soo, S.L., Fluid Dynamics of Multiphase Systems, Blaisdell Publishing Co., 1967.

4. Wakeman, T. and Tabakoff, W., "Measured Particle Rebound Characteristics Useful for Erosion Prediction," ASME Paper 82-GT-170, 27th International Gas Turbine Conference, April 18-22, 1982, London, England.

5. Crowe, C.T. and Stock, D.E., "A Computer Solution for Two-Dimensional Fluid Particle Flows," International Journal for Numerical Methods in Engineering, Vol. 10, 1976, pp. 185-196.

6. Vavra, M.H., Aerodynamics and Flow in Turbomachines, John Wiley and Sons, Inc., New York, London, 1960.

7. Vittal, B.V.R. and Tabakoff, W., "Solution of Particulate Flow in a Radial Inflow Turbine," ASME Paper 83-GT-75, 28th International Gas Turbine Conference, March 26-31, 1983, Phoenix, Arizona.

Multi-Phase Flow and Heat Transfer III. Part B: Applications   407
edited by T.N. Veziroğlu and A.E. Bergles
Elsevier Science Publishers B.V., Amsterdam, 1984 — Printed in The Netherlands

THE EFFECT OF DIFFERENT PARAMETERS ON THE TRAJECTORIES OF PARTICULATES IN A
STATIONARY TURBINE CASCADE

A. F. (Abdel Azim) El-Sayed* and W. T. Rouleau
Department of Mechanical Engineering
Carnegie-Mellon University
Pittsburgh, Pennsylvania   15213, U.S.A.

ABSTRACT

    The contribution of different particulate parameters on particle traject-
ories in a stationary turbine cascade is examined here as a part of detailed
erosion studies in turbines. The particle trajectory depends on the gas flow
field, the physical characteristics of the particle, and the interaction
between the particles. To determine the contribution of each of the above
parameters, they were systematically varied and the particle trajectories were
traced. A laser-Doppler anemometer (LDA) was used to measure the 3-D gas flow
field within the cascade. Two cases were compared for the three-dimensional
viscous flow with and without secondary flow components. The effect of secondary flow on the particle trajectories was clearly emphasized. The contribution
of different particulate parameters (shape, material, diameter, initial location, and initial velocity) was examined. The effect of particle shape was
examined by considering five different shapes; namely sphere, oblate and prolate
spheroids, disk, and cylinder. These are shapes which are likely to be found
in field applications. The trajectories of particles other than spheres showed
considerable deviation from the trajectories of spherical shaped particles.
The trajectories of four different materials--silicon dioxide, aluminum oxide,
magnesium oxide, and iron oxide showed the importance of particle material.
Furthermore, three particle initial velocity ratios (Vp/Vg = 0%, 50% and 100%)
illustrated that velocity ratio is important.

    Finally, for investigating the effect of concentration on particle trajec-
tories, the particle-particle interaction was included. This was accomplished
by modifying the drag force to account for the particulate volume fraction.
The particle trajectories for a particle cloud having a particulate volume
fraction equal to 5% were traced and compared with the single particle trajectories.

1.   INTRODUCTION

    Gas-particle, two-phase (or particulate-laden) flow in turbomachines re-
sembles the operating media in numerous industrial applications. Examples may
be found in gas turbines using pulverized coal as an alternate energy source,
turboshaft engines likely to be found in helicopters [1,2] and engines of some
conventional airplanes, like the BOEING 737-200 which routinely operate from
unpaved runways of small cities [3,4]. The thorough and correct understanding
of such a particulate flow is a key factor in designing durable and reliable
turbomachines. Structural deterioration and performance degradation may be

*Presently with Mechanical Engineering Dept., Zagazig University, Egypt.

and f(Re) is defined in (1) and ρ_p is the density of the particle material.

The initial conditions are as follows:

$$x_p(0) = x_0$$
$$y_p(0) = y_0$$
$$z_p(0) = z_0$$

(5)

$$\dot{x}_p(0) = \lambda U_g$$
$$\dot{y}_p(0) = \lambda V_g$$
$$\dot{z}_p(0) = \lambda W_g$$

where x_0, y_0, z_0, are the particle's location at the turbine passage inlet; λ is the particle's velocity lag at inlet. The magnitude of λ varies from one case study to another in the range $0 \leq \lambda \leq 1$, as will be explained in the next section. U_g, V_g and W_g are, respectively, the x, y, z components of the gas velocity at the particle's inlet location (x_0, y_0, z_0).

A suitable time increment is selected and equations (3) are numerically integrated through the 3-D grid network, formed previously in the gas flow section with the locations of the LDA measurements as its nodes (i.e. 14x11x11 nodes). At the end of each time step two processes are performed:

1. Interpolation to define the gas velocity at the new particle location using an 8-point interpolation scheme.

2. Iteration to obtain a better location of the particle using the new gas velocity obtained in the previous step to obtain the particle's slip velocity and the associated drag coefficient (refer to [6] for detailed procedure).

The process is continued until the particle either leaves the turbine or impacts on a blade surface or an endwall.

3. INVESTIGATION OF THE EFFECT OF DIFFERENT PARTICULATE PARAMETERS ON THE PARTICLE'S TRAJECTORIES

The flight path of a particle suspended in a flowing gas depends on the gas flow field, the physical characteristics of the particle and the interaction between particles. In this section, the effect of one variable is examined at a time while the other variables remain unchanged.

3.1. Effect of Particulate Free Gas Flow Field

To examine the contribution of the gas flow field on particle trajectories, two cases were considered. These were the three-dimensional viscous gas flow with and without secondary flow. Secondary flow has a predominant effect on the particle's trajectories in a turbomachine. An approximate analysis of the effect of secondary flows on the motion of particulates in an axial flow gas turbine was clearly investigated in [22]. In that study [22], an analytical procedure was used in defining the gas flow. In this study, an experimental technique using the LDA was used in [9] to investigate the secondary flow in the axial cascade test section previously described. These measurements were utilized to trace the particle trajectories in the two cases of particulate-free gas flow with and without secondary flows. A thorough explanation of the difference between the particle trajectories and the associated erosion damage in the two cases where secondary flow is included and excluded is given in [23].

The geometry of the test section, cross section locations and the general pattern of secondary flows likely to be found in cross sections numbers 4 and beyond are illustrated in Figure 1. Nine particle initial locations were selected as shown in Figure 1. For a simplified description of the particle trajectories, each was given a two-digit label: the first is its diameter and the second defines its position. For example, particle 56 has a five micron diameter (16.4×10^{-6} ft.) and initial position number 6. The initial velocity of the particle V_p was assumed to be one-half of the corresponding gas velocity V_g at the same inlet location. Two particle materials were examined; these were silicon dioxide and aluminum oxide. The detailed comparison of the trajectories of particles with and without secondary flows of different materials, initial velocities and initial locations are presented in [23], and only few examples will be given here.

Figure 2 illustrates the trajectories of two silicon dioxide particles, namely P21 and P29, having the same diameter (2 microns) and initial locations 1 and 9. The particle (P21) is close to the blade pressure surface and lower endwall and has initial velocity components (0, 0, 8.37)ft/sec in the x, y, z directions. Particle (P29) starts its motion from the diagonally opposing location. It is close to the blade suction surface and upper endwall and has

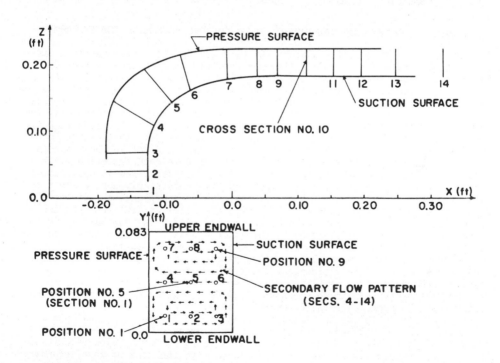

FIGURE 1. CASCADE GEOMETRY AND SECONDARY FLOW PATTERN

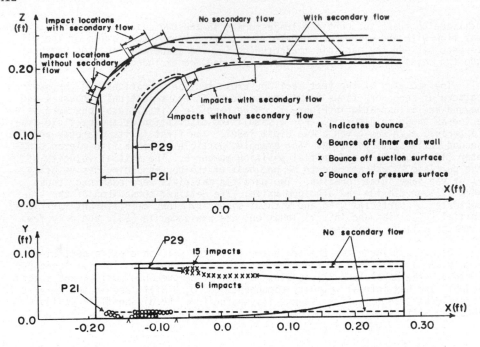

FIGURE 2. EFFECT OF SECONDARY FLOW ON THE PARTICLE TRAJECTORIES

initial velocity components equal to (0, 0, 9.15) ft/sec. Each particle has two trajectories, where secondary flow was included in one case and excluded in the other. Regarding the particle P21, it moved nearly in a straight path to impact the blade pressure surface in the initial portion of the turning section. Moreover, the two cases with and without secondary flow showed similar trends. These features are due to two facts: the first is the absence of the drag force normal to their original direction, as both air and particle had zero velocity components in the x and y directions. The second reason is that either there was no secondary flow at all (since the primary flow had not yet been turned) or else the secondary flow components had not yet developed appreciably. The first impact on the pressure surface with zero secondary flow is upstream of that with non-zero secondary flow components. This is simply due to the secondary flow velocity which has two components; one in the negative y-direction, or downward towards the lower endwall, while the other is directed from the blade pressure surface towards the blade suction side. The downward velocity component moved the particle towards the lower endwall. In both cases, the particles after their first impacts on the blade pressure surface were trapped in the boundary layer, where both air and particle attained very small velocities. Thus the particle incidence velocities and incidence angles were successively decreased from nearly 15 ft/sec to 6 ft/sec and from 15° to $1-3^\circ$. However, the particle with no secondary flow recovered from the impact series faster than the other case (with secondary flow). This is due to its two-dimensional motion (blade-to-blade) compared to the three-dimensional one when secondary flow is included. Consequently, the incidence velocity component in the x-z (blade-to-blade) surface is higher for zero secondary flow velocities. Thus, higher rebound velocities are associated with the case of zero secondary flows which assist the particle to move outside the boundary layer.

After its last impact with the blade pressure surface, the particle associated with the secondary flow field moved very slowly close to this surface and towards the lower endwall to impact it. When it rebounded, it was influenced by the secondary flow field which directed it to impact the pressure surface repeatedly (70 impacts). These impacts occurred with continuously decreasing velocities and angles. The particle was then influenced by the appreciable secondary flow and moved very close to the lower endwall and away from the pressure surface, towards the suction surface. This resulted in a single impact with the lower endwall and continuous motion toward the suction surface. This motion put the particle in locations where the secondary flow has an upward (towards upper endwall) velocity component. Thus, the particle responding to this upward motion left the passage at a y coordinate equal to four times its inlet value.

The particle with no secondary flow after the first set of impacts with the pressure surface, moved outside the blade boundary layer but still with small velocities insufficient to acquire drag force capable of pulling it away from the wall. Thus, again, it encountered the pressure surface with another set of successive impacts. After the last impact (downstream of the last impact of the case with secondary flow), it moved nearly in a straight line parallel to the straight portion of the blade.

In the x-y plane it is significant to note that the case with no secondary flow showed no motion in the y-direction as both gas and particle had zero velocity components in this direction and consequently no drag force to pull the particle from its original y location. Thus, we have only a blade-to-blade motion similar to all the previous investigations carried out by Tabakoff and his group and summarized in [24].

The motion of the particle P29 subject to the secondary flow was quite different from that of the particle P21 with included secondary flow effects. At first, it was accelerated and moved in a straight path. Within the turning section it was turned at a larger radius of curvature (compared to the associated streamline), due to its inertia. The secondary flow had components downward (towards the lower endwall) and another pointed toward the suction surface, which resulted in a series of successive impacts with the blade suction surface close to the upper endwall. The impact velocities increased from 12.5 to 15 ft/sec. This is due to the presence of the particle close to the suction surface as well as the acceleration of air flow outside the boundary layer owing to the decrease in cross-sectional area. The first impact occurred at a relatively high incidence velocity and angle (12.5 ft/sec and 18°); the particle rebounded to a location outside the boundary layer. The positive gas velocity gradient accelerated the particle while the secondary flow directed it again towards the suction surface where it once more impacted the suction surface at a higher incidence velocity and equal incidence angle. These events were repeated with a slight increase in the incidence velocity (to reach 15 ft/sec) and a slight decrease in the incidence angle (to reach 10°). The negative y-component of the secondary flow drove the particle close to the mean blade-to-blade surface or plane of symmetry. The secondary flow in this new location had a negligible negative y-component while its appreciable component was towards the blade pressure surface. This resulted in particle motion away from the suction surface and towards the pressure one. The particle finally left the passage very close to the mean blade-to-blade surface and the meridional streamline in the x-z plane.

If we examine the trajectory of the same particle without secondary flow, we find that its initial portion coincides with the trajectory just described with secondary flow effect. Next, it was turned in the turning section at higher radius of curvature than the case with secondary flow. Thus, it impacted the blade pressure surface downstream of the first impact with included secondary flow. This particle, being influenced only by the primary flow, was pulled away from the suction surface after a smaller number of impacts (15) and moved very close to the suction surface but outside the boundary layer. The trajectory of the particle is nearly parallel to the straight portion of the suction surface. The trajectory of the particle in the x-y plane is also parallel to both endwalls; that is, a blade-to-blade motion is found for the case with no secondary flow.

In conclusion, the secondary flow assists in trapping the particle in the boundary layer and increases the number of impacts in most cases. In some cases particles influenced by the secondary flow will be pulled away from the surface but will still move in the boundary layer. Secondary flow results in a particle motion outside its initial blade-to-blade surface. Thus, particle impact with either endwall is solely a contribution of the secondary flow field. Consequently, secondary flow is responsible for the successive impacts with the internal surfaces of the cascade, and thus increases its erosion damage.

3.2. Effect of Different Particulate Characteristics

In order to assess how the characteristics of the particles affect their trajectories, five different parameters associated with particles were examined. These were shape, material, diameter, initial velocity and inlet location.
Particle shape. Most solid particles of practical interest are of irregular shape. Nevertheless, the particles generally are assumed to be spherical. Some authors treated the motion of non-spherical particles as reviewed in [25-27]. The mathematical difficulties of analyzing the motion of non-spherical particles are obviously much greater than for spheres. The drag force is still related to the product of cross-sectional area normal to the flow, the fluid dynamic pressure, and a drag coefficient C_D.

The drag coefficients for non-spherical particles moving in a viscous fluid in the creeping regime have been theoretically expressed by many authors, as described in [28]. Consider an oblate spheroid having an equatorial radius a and polar radius b; its equation in the (x, y, z) frame is given by

$$\frac{x^2 + y^2}{a^2} + \frac{z^2}{b^2} = 1 \tag{6}$$

with b = a(1-e) and e is so small that squares and higher powers of it may be neglected. The drag force on such a spheroid when it moves parallel to its axis of revolution with velocity U (in the z-direction) is given by

$$D = 6 \pi \mu a U \left(1 - \frac{e}{5}\right). \tag{7}$$

For the general case of an oblate spheroid (e is not necessarily a small quantity, or in other words the spheroid shape may deviate considerably from the spherical shape), its equation in terms of the orthogonal coordinates (ξ, η, ϕ), is expressed as

$$z + i \rho = c \sinh (\xi + i\eta)$$
$$z = c \sinh \xi \cos \eta \tag{8}$$
$$\rho = c \cosh \xi \sin \eta,$$

where ξ = constant represents the surface of a spheroid and 2c is the confocal distance. The drag coefficient is given by

$$D = \frac{8 \pi \mu c U}{\lambda_0 - (\lambda_0^2 - 1) \cot^{-1} \lambda_0}, \tag{9}$$

where $\lambda_0 = \frac{b}{c} = [(\frac{a}{b})^2 - 1]^{-1/2}$.

To compare the resistance of the oblate spheroid with a sphere of radius a, equation (9) may be rewritten as

$$D = 6 \pi \mu a \ UK,$$

where

$$K = \frac{1}{\frac{3}{4} \sqrt{\lambda_0^2 + 1} \ [\lambda_0 - (\lambda_0^2 - 1) \cot^{-1} \lambda_0]} . \tag{9'}$$

For the special case of small e (large λ_0) (spheroid departs little in shape from a sphere),

$$D = 6 \pi\mu a \ U (1 - \frac{e}{5} - \frac{33}{175} e^2 - \ldots) \tag{10}$$

The value of K is a function of the ratio b/a and may be tabulated from the spectrum b/a $\longrightarrow$ 0 (disk) and b/a $\rightarrow$ 1 (sphere). Thus, the drag exerted on a sphere of radius a which is equal to the equatorial radius of the spheroid is higher than the spheroid.

A more appropriate comparison may be made between equation (7) and the resistance of a sphere of equal volume or surface area. The volume of a spheroid is $\frac{4}{3} \pi a^3 (1 - e)$. Hence, a sphere of equal volume will have a radius of

a $(1 - \frac{e}{3})$. Thus, the drag factor defined as the ratio between drag forces on a spheroid and a sphere of equal volume (DGFC) is

$$DGFC = \frac{Drag \ force \ on \ a \ spheroid}{(Drag \ force \ on \ a \ sphere \ of \ equal \ volume)} = \frac{1 - (e/5)}{1 - (e/3)} . \tag{11}$$

It follows that a sphere of equivalent volume has a smaller resistance than the oblate spheroid. Similarly, the surface area of a spheroid is $4 \pi a^2 (1 - \frac{2}{3}e)$.

Thus, the sphere of equal surface area will have a radius a $(1 - \frac{2}{3}e)$. It follows that the smaller resistance of a sphere holds on the basis of equal surface area.

The motion of a prolate spheroid [which is described by the same equation (6), with b = a (1 - e) but here e is negative] parallel to its axis of revolution may be calculated in a similar way. The drag coefficient is expressed also as

$$D = 6 \pi \mu b \ U \ K,$$

$$K = \frac{1}{\frac{3}{4} \sqrt{\tau_0^2 - 1} \ [(\tau_0^2 + 1) \coth^{-1} \tau_0 - \tau_0]} \tag{12}$$

$$\tau_0 = \frac{a}{c} = [1 - (\frac{b}{a})^2]^{-1/2} .$$

b is the equatorial radius and a is the polar radius. The factor K is still a function of the ratio b/a where a is the longest of the two semiaxes. When b/a = 1, we have an undeformed sphere, thus K=1; while the case of b/a = 0 if b is finite corresponds to a, K $\longrightarrow$ ∞. Expanding the factor K for small e, we have the same equation (10) with negative e. Thus, comparing the drag coefficients of a prolate and a sphere of equal volume will give the same equation

418

a flight path different from that of a sphere having the same material (SiO₂), same initial velocity and initial location. Both trajectories showed impacts on the blade suction surface. At most of its trajectory, the sphere moved away from the suction surface and lower endwall as compared to the spheroid. It is important to mention here that though a spheroid has two planes of symmetry, it doesn't have the unique point symmetry of a sphere. For such a spheroid it is hard to say that it will maintain a constant orientation throughout its trajectory, from inlet to exit. However, one may assume that the trajectory of this spheroid will attain a certain value between two flight paths representing the cases of a spheroid moving parallel or perpendicular to its axis of rotation.

Figure 3 illustrates also the flight path of a prolate spheroid having a length-to-diameter ratio of five. The drag coefficient of such a prolate spheroid will be 0.68 times the drag coefficient of a sphere having the same volume. The trajectories of both spheroid and sphere having the same material (SiO₂), same diameter (5 microns), same initial velocity (0, 0, 11.57)ft/sec, and same inlet position (P55) are shown in Figure 3. The oblate spheroid moved closer to the pressure surface and upper endwall than the spherical particle.

Figure 4 illustrates the trajectories of disk and cylindrical shaped particles. The drag coefficient for a cylinder having b/a = 0.7 and moving parallel to its axis is 2.5 times that of a sphere having the same material (SiO₂), same volume with radius of 2 microns, same initial velocity (0, 0, 11.4)ft/sec and same initial position (P26).

In the same figure, the trajectory of a circular disk moving broadside-on is drawn and compared to the spherical particle trajectory. Both particles have the same diameter (4 microns), same inlet location (P42) and same initial velocity (0, 0, 8.12)ft/sec. The drag coefficient of the disk is 0.85 times that of a sphere having the same diameter. The deviation of the trajectory of the disk from that of a sphere indicates the importance of the particle shape.

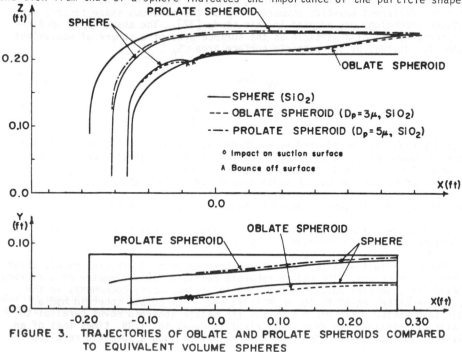

FIGURE 3. TRAJECTORIES OF OBLATE AND PROLATE SPHEROIDS COMPARED TO EQUIVALENT VOLUME SPHERES

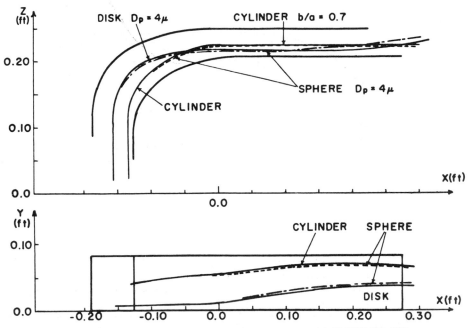

FIGURE 4. TRAJECTORIES OF CYLINDER AND DISK COMPARED TO SELECTED SPHERICAL PARTICLES

For the prolate spheroid, disk and cylinder the same argument concerning the particle orientation with respect to the flowing air is still correct. The possible location of any of them will lie between the trajectories for the particle moving parallel to and perpendicular to its axis.

    Particle material. To examine the effect of particle material on trajectories, four materials were chosen. These were iron oxide (Fe_2O_3) and magnesium oxide (MgO), in addition to the aluminum oxide (Al_2O_3) and silicon dioxide. Figure 5 illustrates these trajectories. Particle P23 shows the trajectories of three materials; namely iron oxide, aluminum oxide and silicon dioxide. The iron particles are the heaviest while the silicon dioxide ones are the lightest. All have the same diameter (2 microns), same position (#3) and same initial velocity component ((0, 0, 9.16)ft/sec). The three trajectories had an initial straight line portion, then turned to impact on the curved camber of the suction surface. The iron and aluminum particles impacted the surface three times compared to the seven impacts of the silicon dioxide particles. Next, the iron particle moved in a straight line before it was moved by the secondary flow away from the suction surface. Both alumina and silica particles were quickly influenced by the secondary flow to trace different trajectories from the iron particle, being moved away from the suction surface and towards the pressure surface. The deviation between the three particle trajectories in the x-y plane is slight and showed itself in the central portion of the trajectory.

    On the same figure, another comparison between the trajectories of the magnesium and silicon particles can be made. Both were labled P44. They had the same initial position #4 and initial velocity components (0, 0, 11.83)ft/sec. Both impacted the pressure surface on its turning section. The light SiO_2 particles impacted the pressure surface upstream the first impact of the magnesium particle. Furthermore, the SiO_2 particle impacted the pressure surface thirty-

432

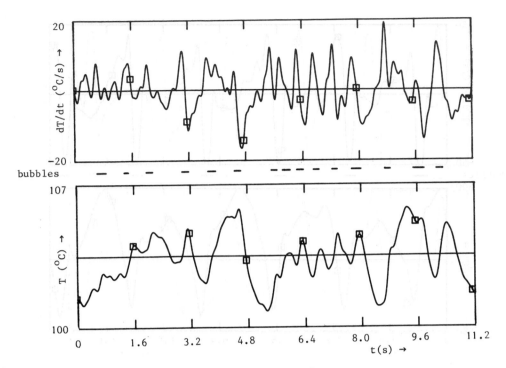

Fig. 5 Detector Temperature (T) and Signal (dT/dt) and Observed Bubbles in the Cold 2D-Model (Fluidized)

cross section and that their diameter is independent upon the horizontal position in the bed. The calculation procedure is described below.

The average chord length of the bubble areas ($<\ell>$) can be calculated from the average time interval of the signal peaks ($<t>$) and the bubble velocity (U_b, derived as indicated in 3.2):

$$<\ell> = U_b \cdot <t> \qquad (1)$$

The relation between the average chord length and the diameter of the bubble areas (D_{ba}) is given by:

$$<\ell> = \frac{4 \cdot \text{volume}}{\text{surface}} = \frac{2}{3} D_{ba} \qquad (2)$$

The relation between the point frequency and the total frequency of the bubble area is given by the probability (P) that a bubble area is detected by the detector:

$$F_{ba} = P \cdot f_{ba} \qquad (3)$$

This probability is equal to the bubble area cross section divided by the horizontal bed cross section (A) and is limited to unity

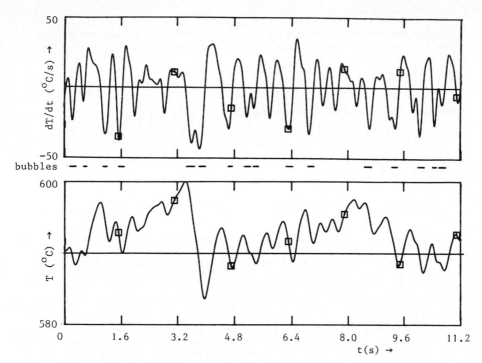

Fig. 6   Detector Temperature (T) and Signal (dT/dt) and Observed Bubbles in the
Hot 2D-Model (Fluidized)

$$P = \frac{\pi}{4} D_{ba}^2 / A \qquad\qquad \text{when} \quad A \geq \frac{\pi}{4} D_{ba}^2 \qquad\qquad (4)$$

$$P = 1.0 \qquad\qquad\qquad \text{when} \quad A < \frac{\pi}{4} D_{ba}^2 \qquad\qquad (5)$$

Using equation (1) through (5) the total bubble area frequency can be calculated. Because each bubble area contains a bubble void, this value is also the total bubble void frequency:

$$F_{ba} = F_{bv} \qquad\qquad\qquad\qquad\qquad\qquad\qquad\qquad (6)$$

Using the average bubble void diameter (see 3.4) the point frequency of the bubble voids can be calculated. For this purpose the equations (3), (4) and (5) should be used with the index "ba" replaced by "bv".

3.4. Bubble Void Diameter

In 3.3 a description has been given of the derivation of the bubble area diameter. The detector signal does not give information on the bubble void diameter. However, by use of the dynamical bed height (H) the bubble void diameter, averaged over any horizontal bed cross section, can be calculated. A way of determining the dynamical bed height is described in 4. The calculation procedure is as follows:

Examine an arbitrary horizontal slab of the bed with a height h during fluidization and h_{mf} during minimum fluidization. The total bubble void volume (V_t) of this slab is:

$$V_t = A \cdot (h - h_{mf}) \tag{7}$$

The number of bubble voids (n_{bv}) in the slab is:

$$n_{bv} = \frac{h}{U_b} \cdot F_{bv} \tag{8}$$

The average bubble void volume (V_{bv}) in the slab is:

$$V_{bv} = V_t / n_{bv} = \frac{h - h_{mf}}{h} \cdot U_b \cdot A / F_{bv} \tag{9}$$

When one assumes that changes in bubble void diameter is only caused by coalescence or splitting during which no bubble void volume is created or lost, the total bubble void volume flowing through the bed is for any horizontal slab:

$$\phi_{bv} = V_{bv} \cdot F_{bv} = \frac{h - h_{mf}}{h} \cdot U_b \cdot A \tag{10}$$

Because this is valid irrespective of the choice of the horizontal slab, ϕ_{bv} can also be obtained using the total bed height during fluidization (H) and during minimum fluidization (H_{mf}), and the bubble rise velocity, averaged over the entire bed height ($<U_b>$):

$$\phi_{bv} = \frac{H - H_{mf}}{H} \cdot <U_b> \cdot A \tag{11}$$

Combining equations (9), (10), and (11) one arrives at the equation by which the average bubble void volume can be calculated for any horizontal slab:

$$V_{bv} = \frac{H - H_{mf}}{H} \cdot <U_b> \cdot A / F_{bv} \tag{12}$$

The equivalent bubble void diameter (D_{bv}) is:

$$D_{bv} = (V_{bv} \times \frac{6}{\pi})^{1/3} \tag{13}$$

When this value is larger than the bed diameter, it can be concluded that slugging appears. In that case the slug height (H_s) can be obtained using the following equation:

$$H_s = \frac{H - H_{mf}}{H} \cdot <U_b> / F_{bv} \tag{14}$$

4. PRACTICAL EXPERIENCE

In order to test the detection system (hardware and software) an experiment

has been carried out in a coal fired fluidized bed combustor with an inner diameter of 20 cm and a thermal capacity of 60 kW. The bed material was sand with
an average particle diameter of 1 mm. The bed temperature was kept at a constant
$850^{\circ}C$, the statical bed height was 34 cm and various fluidization velocities
have been chosen.

As follows from the foregoing (3.4), it is not only necessary to equip the
combustor with fluidization detectors, but also with instrumentation to measure
the dynamical bed height. For this reason an instrumentation has been applied
which can serve both purposes.

4.1. Experimental Procedure

The in-bed instrumentation consists of a vertical cooling tube, partially
above and partially in the fluidized bed, on which nine thermocouples have been
mounted at a regular mutual distance of 5 cm (see Fig. 7). The lowest thermocouple was placed at 38 cm above the distributor plate. Because the thermocouples are cooled by the cooling tube they can serve as fluidization detectors.
Furthermore, it was expected that the temperature of the thermocouples above the
bed would have a distinctly lower temperature than the thermocouples in the bed
as an indication of the dynamical bed height.
The thermocouples ($\emptyset$ 1.6 mm) were connected to the cooling tube by steel strips
combined with high temperature soldering. The tube was cooled with water having
a temperature below $30^{\circ}C$ along the entire tube length. Outside the fluidized
bed the nine thermocouples have been connected to a data acquisition system for

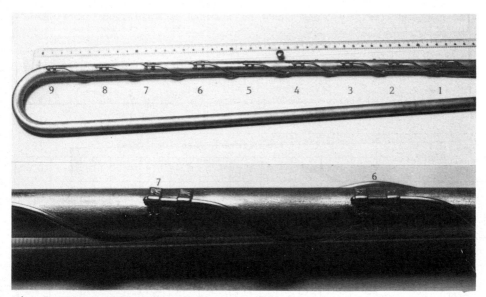

Fig. 7   Photograph of the Cooling Tube with Thermocouples
(Left is Lower Part, Right is Upper Part)

obtaining the temperatures (scan rate 32 Hz). Furthermore, the thermocouple voltages have been amplified by AC-amplifiers in order to obtain the temperature variations. The output voltages of the amplifiers were conducted to the data acquisition system too (scan rate 256 Hz).
At fluidization velocities of 0.3 m/s and lower the detector signals were zero which means that the bed was not fluidized. Data have been registrated on a magnetic tape during measurements with fluidization velocities of 0.4, 0.5, 0.6 --- 1.5 m/s.

4.2. Results

The recorded temperatures of the various thermocouples are shown in Table 1.
These temperatures have been used to obtain the dynamical bed height as a function of the fluidization velocity. When the temperature is high, the thermocouple finds itself in the bed and when the temperature is low, it is above the bed. Because a fluidized bed has no sharply defined top level, but changes over to a splashing zone and then to the free-board, an arbitrary choice had to be made concerning the distinction between low and high temperature as bed height indication. A thermocouple is said to find itself above the fluidized bed when its temperature differs significantly (20 per cent or more) from its temperature when it would have been in the bed.

During the experiment with the highest fluidization velocity (1.5 m/s) most or all thermocouples have been in the bed. The temperatures recorded during this experiment have been regarded as the temperatures of the thermocouples inside the bed. Subsequently for each thermocouple the fluidization velocity has been determined at which the thermocouple temperature has a value of 80 per cent of the temperature during this reference experiment. This has been done by linear interpolation between the values of Table 1. The results are shown in Fig. 8. From this figure it can be seen that the results for thermocouples 1 and 2 do not fit with those of the others. Probably the experiment with a fluidization velocity of 1.5 m/s cannot be regarded as a reference experiment for these two thermocouples because they were too close to the top level of the bed during this experiment.

From Table 1 it can be seen that the thermocouples which are in the bed do

Table 1. Measured Thermocouple Temperatures (in ^{o}C) at Various Fluidization Velocities

Thermo-couple	Distance to Distributor (cm)	Fluidization Velocity (m/s)											
		1.5	1.4	1.3	1.2	1.1	1.0	0.9	0.8	0.7	0.6	0.5	0.4
1	78	65	56	57	48	48	44	39	35	35	29	28	22
2	73	52	48	47	40	40	35	36	28	25	21	21	16
3	68	62	56	56	47	45	43	39	32	28	25	24	19
4	63	65	60	62	54	52	54	43	38	32	27	26	21
5	58	70	65	66	60	60	65	52	48	37	30	29	23
6	53	71	65	67	63	68	71	62	59	45	35	34	24
7	48	78	69	72	69	82	79	74	77	67	54	41	25
8	43	136	125	129	125	145	140	139	144	146	137	105	52
9	38	101	102	101	104	114	107	114	109	115	111	116	84

not have exactly the same temperatures. This is probably caused by slight differences in the thermal contact between the thermocouples and the cooling tube. Thermocouple 8 shows the strongest deviations and it can also be seen from Fig. 7 that this thermocouple is fixed to the cooling tube in a distinctly different way from the others.

The experiments with fluidization velocities of 1.5, 1.0 and 0.7 m/s have been selected for the determination of the various bubble quantities. During the experiments at 1.5 and 1.0 m/s the thermocouples 6, 7, 8 and 9 have been in the bed and at 0.7 m/s thermocouple 6 was just above the bed. These four thermocouples have been used as fluidization detectors and their detector signals have been determined. The signal of thermocouple 9 has been correlated with the signal of thermocouples 8, 7 and 6 respectively. As an example the signals of thermocouples 6 and 9 and their covariance are given in Fig. 9 for the experiment with a fluidization velocity of 1.5 m/s.
The time at which the first (τ_1) and the second (τ_2) maximum of the covariance of the various combinations occur, are shown in Table 2.

From the values of τ_1 and the vertical distance between the thermocouples, the bubble velocities have been calculated. From the difference between τ_1 and τ_2 the point frequencies of the bubble areas have been calculated ($f_{ba}=1/(\tau_2-\tau_1)$). The velocities and frequencies are given in Table 3.

Hereafter the average bubble area diameter has been calculated as described above in 3.3. It appeared that in all cases the bubble area filled the whole bed diameter from which a detection probability has been obtained of 1.

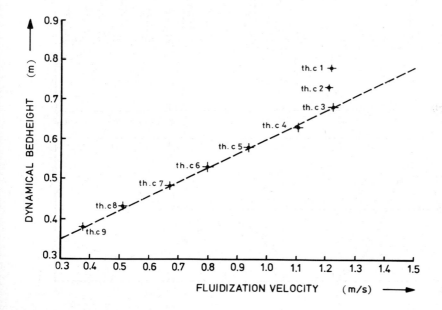

Fig. 8   Dynamical Bed Height as a Function of the Fluidization Velocity

438

Table 2.   Time (in ms)   at Which the First (τ_1) and the
           Second (τ_2) Maximum in the Covariance Occurs

Thermocouple Combination	U_f = 1.5 m/s		U_f = 1.0 m/s		U_f = 0.7 m/s	
	τ_1	τ_2	τ_1	τ_2	τ_1	τ_2
6.9	135	795	165	840	190	820
7.9	80	730	95	750	130	780
8.9	45	695	50	705	70	745

This means that the measured point frequency of the bubble areas is equal to the total frequency of the bubble areas and consequently equal to the total frequency of the bubble voids. By assuming that the bed height during minimum fluidization is equal to the statical bed height, all quantities needed to calculate the bubble void diameter are available. The results of these calculations are given in Table 4.

In order to obtain an idea of the span of life of the thermocouples in the fluidized bed, a duration experiment has been performed with the same kind of thermocouples, connected to the cooling tube with steel strips only. All thermocouples were situated in the fluidized bed. They started to show damage after 140 hours exposure to the bed. After 400 hours the upper five thermocouples still functioned properly but the lower four were broken down.

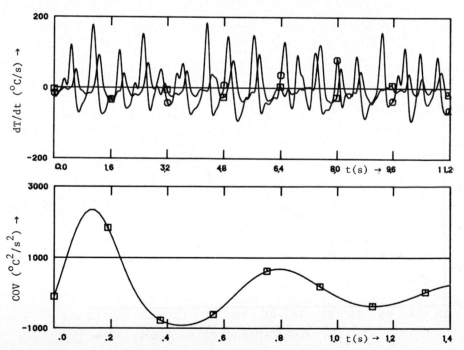

Fig. 9   Signals of Thermocouples 6 and 9 and Their Covariance

Table 3. Measured Values for the Bubble Rise Velocity (m/s)
and Point Bubble Area Frequency (Hz)

Thermocouple Combination	$U_f = 1.5$ m/s		$U_f = 1.0$ m/s		$U_f = 0.7$ m/s	
	U_b	f_{ba}	U_b	f_{ba}	U_b	f_{ba}
6.9	1.11	1.52	0.91	1.48	0.79	1.59
7.9	1.25	1.54	1.05	1.53	0.77	1.54
8.9	1.11	1.54	1.00	1.53	0.71	1.48

5. DISCUSSION

For this work the necessary definition of dynamical bed height was arbitrarily chosen and was based on the assumption that the temperature of a
fluidization detector above the bed is at least 20 per cent lower than the temperature of this fluidization detector while it is in the bed. This criterion
should be checked and if necessary changed with the aid of an experiment in a
fluidized bed with visible bed height.

During the study all data treatment has been done in a numerical way.
Nevertheless, it is expected that all information required can be obtained using
hardware instruments (AC-amplifier, differentiator, recorder and correlator).

In view of the expected span of life of the in-bed instrumentation
(140 - 400 hours) the instrumentation described can be applied for monitoring
an industrial fluidized bed combustor only after improving the construction.
The method as developed may be used during the first start-up of an industrial
fluidized bed combustor. Such an application in the semi-industrial facility
of Netherlands Energy Research Foundation ECN (2 MW$_{th}$) turned out to be satisfactory. Furthermore, the method can supply relevant data for research on i.e.
physical models for heat transfer in fluidized beds.

Table 4. Results of the Calculation of the Bubble Properties

U_f (m/s)	H [1] (m)	H_{mf} [2] (m)	$<U_b>$ [3] (m/s)	A (m^2)	F_{bv} [4] (Hz)	V_{bv} (m^3)	D_{bv} (m)	H_s (m)
1.5	0.775	0.340	1.11	0.0314	1.53	0.0128	0.29	0.41
1.0	0.600	0.340	0.91	0.0314	1.51	0.0082	0.25	0.26
0.7	0.480	0.340	0.79	0.0314	1.54	0.0047	0.21	0.15

1) from figure 8

2) by assuming that H_{mf} is equal to the statical bed height

3) for the traject between thermocouples 6 and 9, from Table 3

4) average values from Table 3

6. CONCLUSION

With use of cooled thermocouples as fluidization detectors in a fluidized bed combustor and the data treatment procedures as described, it is possible to detect local fluidization, to determine bubble properties (velocity, point- and total frequency and volume) and to measure the dynamical bed height. The latter has not yet been examined sufficiently.

REFERENCES

1. Werther, J. & Molerus, O., "The Local Structure of Gas Fluidized Beds - I. A Statistically Based Measuring System", Int. J. Multiphase Flow, 1, (1973), pp. 103 - 122.

2. Werther, J. & Molerus, O., "The Local Structure of Gas Fluidized Beds - II. The Spatial Distribution of Bubbles", Int. J. Multiphase Flow, 1, (1973), pp. 123 - 138.

3. Rowe, P.N. & Everett, D.J., "Fluidised Bed Bubbles Viewed by X-Rays", Part I - III, Trans. Instn. Chem. Engrs, 50, (1972), pp. 42 - 60.

4. Yasui, G. & Johanson, L.N., "Characteristics of Gas Pockets in Fluidized Beds", A.I.Ch.E.J., 4, (1958), pp. 445 - 460.

5. Davidson, J.F., Harrison, D., "Fluidization", Academic Press London and New York, (1977), p. 126.

6. Collins, R., University College London, Private Communication.

LIST OF SYMBOLS

A	Horizontal bed cross section	(m^2)
ba	Index indicating bubble area	
bv	Index indicating bubble void	
D	Diameter	(m)
F	Total frequency	(s^{-1})
f	Point frequency	(s^{-1})
H	Total bed height	(m)
h	Height of imaginary horizontal slab in the bed	(m)
H_s	Slug height	(m)
$<\ell>$	Average chord length of bubble area	(m)
mf	Index indicating the condition of minimum fluidization	
n	Number	$(-)$
P	Detection probability	$(-)$
T	Temperature	(^{o}C)
t	Time	(s)
U_b	Rise velocity of bubble voids and bubble areas	(ms^{-1})
U_f	Fluidization velocity	(ms^{-1})
V	Volume	(m^3)
V_t	Total bubble void volume	(m^3)
ϕ	Volume flow	$(m^3 s^{-1})$
τ_1	Time at which the first maximum occurs in a covariance series	(s)
τ_2	Time at which the second maximum occurs in a covariance series	(s)

Multi-Phase Flow and Heat Transfer III. Part B: Applications
edited by T.N. Veziroğlu and A.E. Bergles
Elsevier Science Publishers B.V., Amsterdam, 1984 — Printed in The Netherlands

HEAT TRANSFER IN VIBRATED FLUID BEDS - BASIC PRINCIPLES AND DESIGN CONSIDERATIONS

A. S. Mujumdar and Z. Pakowski
McGill University
Montreal, Canada

ABSTRACT

Vibrated aerated beds of solid particles are commonly found in industrial thermal processing of granular materials which are difficult to fluidize in conventional manner without mechanical assist. Heating, cooling, drying, granulation, instantizing of certain foodstuffs, coating etc are among the major applications of the vibrated bed technique. The objective of this paper is to present some recent data on heat transfer studies and discuss certain design considerations in the light of the data as well as visual observations.

1. INTRODUCTION

Vibrated beds are widely used for performing a series of technological processes (1, 14). Among them is thermal treatment of granular solids which includes drying, heating or cooling, agglomeration etc is the most important one. Thermal treatment basically consists of heat transfer processes between a gaseous heat carrier and solid-particle bed (convection), heated surfaces and the bed (mostly conduction) or an IR source and the bed (radiation). Among more exotic but less important heating methods are micro-wave heating of wet solids or induction heating of metallic particles.

A vibrated thermal processing unit typically consists of a rigid trough (straight or spiral) which is driven by a vibrator. The deck of the trough is generally perforated so that a heat carrier can be blown through the perforations. According to the heat supply mode, appropriate heaters are mounted in the installation, air heaters in the ducting or contact or IR heaters within the chamber itself. The trough is hooded if fumes, vapors or entrained fine particles are emitted during the process. The solid is fed at one end of the trough and thermally treated as it is conveyed by vibrational, gravitational or aerodynamic forces towards the exit.

The main objective of this paper is to identify the basic design problems of thermal processing in vibrated beds (VB) and describe some recent experimental results obtained with a view to providing relevant design information.

2. GENERAL DESCRIPTION OF THE DESIGN PROCESS

The main objective of the design process is to find the trough length, width and particle bed depth as well as the vibration parameters, gas flow velocity, temperature of the heating media to control material temperature

along the trough and so on. Taking as the basis for heat transfer calculations
the area of contact between both phases (bed-gas in convective (direct) heat
transfer or bed-heaters in conductive (indirect) heating) and available
temperature driving forces, the volume of the bed can be calculated. The
heat transferred minus heat loses will determine the accumulation rate of the
solid. The local temperature of the bed will then depend on the solid flowrate
and its heat capacity.

Due to a multitude of possible types of arrangements (Figure 1) of flow
and heat transfer no general procedure can be given. e.g. the case of drying
with contact heat transfer from a heated deck and gas flow above bed in
countercurrent with solid cannot be treated by the same procedure as heating
by hot gas flow through the bed in cross-flow to the solid. In most cases,
however, the process of heating of solid in a continuous VB processor can be
described by the following equation:

$$\frac{1}{s} \frac{d(W_s i_s)}{dl} = (qa)_{conv} + (qa)_{cond} + (qa)_{rad} \qquad (1)$$

where W_s - solid flowrate, kg/s;

i_s - $c_s t_s$ specific enthalpy of solid, kJ/kg K;

s - bed cross section area, m^2;

1 - length; m

$(qa)_{conv}$ - convective heat flux, kW/m^3

$(qa)_{cond}$ - conductive heat flux, kW/m^3 and

$(qa)_{rad}$ - radiative heat flux, kW/m^3.

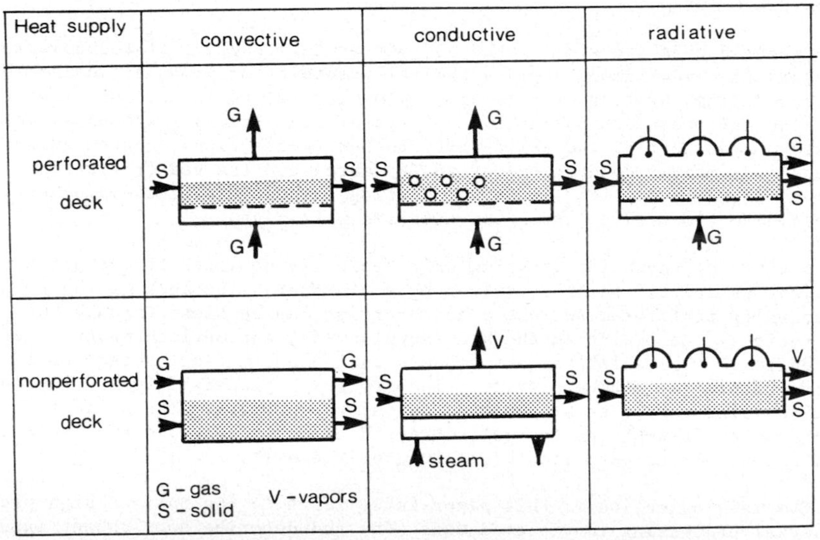

Heat supply	convective	conductive	radiative
perforated deck			
nonperforated deck			

G - gas   V - vapors
S - solid

Fig. 1 Typical arrangements of VB thermal processors for
particulate solids

In some cases where solid flow axial dispersion cannot be neglected one can use the following equation

$$\frac{1}{s} \frac{d(W_s i_s)}{dl} = E_1 \frac{d^2 i_s}{dl^2} \rho_s (1-\varepsilon) + \sum_i (qa)_i \qquad (2)$$

where E_1 – coefficient of axial dispersion, m^2/s;

ρ_s – solid density, kg/m^3;

ε – bed voidage

A basic problem in the thermal design of VB processes is to determine the values of all the Σqa components. In the following sections attention will be paid to the individual components of the total heat flux to the solid.

3. GAS-BED HEAT TRANSFER

In a typical VB solid processor gas flows through the bed in a cross-flow manner relative to the solid. This situation is similar to that encountered in plug-flow fluid bed dryers with the sole difference that the gas velocity in a VB can be set at any level below the entrainment velocity of particles i.e. even well below the minimum fluidization velocity u_{mf}.

Gas particle heat transfer is closely related to the gas velocity. Two major areas of application should be considered here: (1) VB with gas velocity below u_{mf}, (2) VB with gas velocity above u_{mf}.

3.1. Low Velocity Case

In the low velocity case the bed is kept in fluidized state by vibration. It is observed that if the vertical component of vibrational acceleration is larger than gravitational acceleration the bed becomes "vibrofluidized" i.e. starts moving and induces particle circulation. The transition occurs theoretically at $A\omega^2 \sin \alpha/g = 1$ but actually higher values are necessary (1.4 and above). Higher accelerations induce the so-called "flight period" or "throw period" in which the bed separates from the supporting grid as explained in Fig. 2.

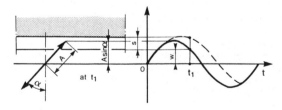

Fig. 2 Vibration cycle of a VB with a flight period

The gap formed below the bed in such situations does not influence gas-bed heat transfer but can seriously deteriorate contact heat transfer between the heated deck and the bed. The full history of the air gap formation can be obtained from an appropriate model of bed mechanics.

The assumption that the bed behaves as a stiff body is very often made which, in the case of gas flow through the gas distributor-bed system (Fig. 3), leads to the following equation:

$$m_A \frac{d^2(w + s)}{dt^2} + m_A g + p_A - P_A = 0 \tag{3}$$

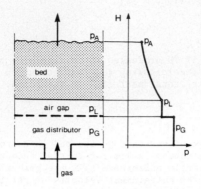

Fig. 3 Scheme of an aerated VB – gas distributor systems and related pressure profile.

On the other hand, assuming that the gas flowrate through the gas distributor and the bed is linearly dependent on the appropriate pressure drop as well as assuming gas incompressibility one obtains

$$\frac{ds}{dt} = k_d(p_G - p_L) - k_b(p_L - p_A) \tag{4}$$

In these equations in addition to the nomenclature given in Fig. 3, the additional symbols mean

g – gravitational acceleration, m/s^2;

m_A – unit load of the gas distributor grid kg/m^2;

k_d, k_b – permeability constants of gas distributor and bed respectively, $m^2 s/kg$.

Eqs. (3) and (4) have the following solution (1) for a gap thickness s

$$s = U_m t - K\{p_m t - \frac{1}{p} c_2 \exp(-pt) + \frac{Q_m}{\omega} \cos(\omega(t + t_0 - \gamma))\} + c_3 \tag{5}$$

where

$K = k_d + k_b$;

$P_m = P_A + m_A g$;

$Q_m = \dfrac{A\omega^2}{K\sqrt{p^2 + \omega^2}}$;

$p = \dfrac{1}{m_A K}$;

$\gamma = \tan^{-1}\omega/p$;

$$c_2 = (P_{LO} - P_m + O_m \sin(2\omega t_0 - \gamma)) \exp(p t_0);$$
$$U_m = k_d P_G + k_b P_A;$$
$$P_G = P_A + u/k_d + u/k_b;$$
$$c_3 = (p_m K - U_m) t_0 + \frac{KQ_m}{\omega} \cos (2\omega t_0 - \gamma) + \frac{K}{p} c_2 \exp (-p t_0)$$
$$P_{LO} = u/k_b;$$
$$t_0 = \frac{1}{\omega} \sin^{-1} ((g - \frac{P_{LO} - P_A}{m_A}) \frac{1}{A\omega^2})$$

The model is written for vibration vertical to the deck but can easily be modified for any angle of inclination. The vibration vector in practice may be inclined to allow solids conveying from the feed to the discharge end.

In the vibrofluidized state solid particles are generally well mixed except for the cases when they have certain surface stickiness as is the case in drying, due to surface moisture or in sintering when the particle surface may plasticize. Wet particles behave very differently from dry ones; even small amounts of moisture can degrade the fluid bed performance.

Assumption of perfect mixing of the bed is often made. In low velocity operations, the gas flow through the bed is close to a plug flow. No bubbles, channels or other disturbances are present in the gas phase.

Under these conditions gas-bed heat transfer is close to that in porous solids and can be calculated using well-known correlations e.g. the Yoshida correlations:

$$j_H = 0.61 \, Re_m^{-0.41} \, \phi_s; \qquad\qquad Re_m > 50$$

$$\qquad\qquad\qquad\qquad\qquad\qquad\qquad\qquad\qquad\qquad\qquad (6)$$

$$j_H = 0.91 \, Re_m^{-0.51} \, \phi_s; \qquad\qquad Re_m > 50$$

where $\quad Re_m = 6 \, G/a \, \mu \, \phi_s;$

$\quad\quad G \quad$ = gas mass velocity, $kg/m^2 s;$
$\quad\quad \mu \quad$ = gas viscosity, pa s;
$\quad\quad \phi_s \quad$ = particle sphericity; and $\quad j_H$ = heat transfer factor.

The specific gas-bed contact area, a, can be calculated as

$$a = \frac{6 \, (1 - \epsilon)}{d_p \, \phi_s} \qquad\qquad\qquad\qquad\qquad\qquad\qquad (7)$$

where $\quad d_p$ - particle diameter, m.

Experimental observations indicate that using empirical correlations for heat transfer in porous solids for VB's overestimates the heat transfer coefficient particularly for Re < ~ 20 (Re = ud_p/ν). This is especially true in the presence of surface moisture or other stickiness-producing agents. Apparently due to a finite degree of agglomeration the effective heat exchange area decreases resulting in lower effective heat transfer coefficients. It is advisable to use correlations recommended for the appropriate operating conditions e.g. the correlation given by Pakowski (2):

$$Nu = 0.827 \, Re_p^{1.04} \, (d_p/H)^{1.17} (A\omega^2/g)^{0.483} \tag{8}$$

where   H – bed height, m;
A – amplitude of vibration, m;
ω – angular frequency of vibration, 1/s.

is valid for shallow beds (< 6 cm) of small particles (d_p < 2 mm) for Re_p= 1-16 and $A\omega^2/g$ = 1 - 6, in presence of surface moisture.

It must be remembered that in most beds of particles below 2-3 mm, the volumetric heat transfer coefficients are very high (of the order of 5 - 25 kW/m^{3}s) which leads to the total consumption of the initial driving force t_G - t_b (gas temp. - solid temp.) within a distance of only a few centimeters from the distributor grid.

This leads to the conclusion that in deeper beds the convective heat transfer flux (qa)$_{conv}$ is in fact given by the conditions of equilibrium of bed and gas leaving the bed. Under these conditions:

$$q_{conv} = W_G c_G \, (t_{G1} - t_{G2}) \tag{9}$$

where t_{G2} = t_b and c_G – specific heat of gas, kJ/kg K

Only in case of large particles or very shallow beds are heat transfer coefficient values really needed.

As a valuable conclusion of this analysis one can assume that shallow beds are much more favorable for convective heat treatment of solids in VB.

As the total heat transferred depends only on the gas flowrate and its initial temperature, the former is not dependent on bed height. Therefore, the amount of heat transferred per kg of bed material is much higher in shallow beds than in deep ones. Bed height should then be selected from the conditions of equilibrium. e.g. assuming that equilibrium means 99% of the initial driving force consumed:

$$H = \frac{u_G \rho_G c_G}{(ha)_{conv}} \, \ell n(1-0.99) \tag{10}$$

where u_G – gas velocity, m/s;
(ha)$_{conv}$– volumetric heat transfer coefficient, W/m^{3}K

3.2. High Velocity case

Above u_{mf}, the gas flow in the bed is considered to flow in two distinct phases: emulsion phase and bubble phase. There is gas-solid heat transfer in each of these phases as well as between both phases. The total heat transferred depends on the fraction of the bubble phase and bubble size. In general, the larger the bubbles and the higher the ratio of gas in bubble phase to that in the emulsion phase, the lower the ratio of heat transferred to the total heat entering the bed with gas. In fact, the gas in bubble phase leaves the bed without sufficient contact with it which is tantamount to bypassing of the bed by a part of the gas.

Although, intuitively vibration of the bed should result in breaking up bubbles thus improving the overall performance of the bed as a heat exchanger, however, there is not enough experimental evidence to support or reject this hypothesis.

Heat transfer calculations in bubbling flows can be performed according to the Bubbling Bed Models a number of which are available in the literature (4).

If one can evaluate approximately the volume of gas bypassing then the recently published bypass model of Schlünder (5) can be used to estimate the resulting heat transfer coefficient. It can also be used for fitting curves to experimental data with the bypass ratio as a parameter. Work in this direction needs to be carried out.

4. HEATER-BED HEAT TRANSFER

In many cases, especially in drying when the heat carrier temperatures cannot be excessively high and gas velocities low, heating surfaces will provide auxilliary heat supply to the bed. Initially, heated decks were employed in vibrating conveyors to heat up solids during their transport. Usually the heating medium was condensing steam or a heat transfer liquid. In such cases, the decks are nonperforated and beds are mixed only by vibrational forces. This case is basically different from the one wherein vertical or horizontal heaters are immersed in the bed.

In the first case beds must be shallow otherwise there will not be intensive mixing and lower bed areas may roast excessively. In the second case, especially when horizontal heaters in the form of tube bundles are used, deeper beds can be fluidized as vibration is transmitted to the bed from all the immersed surfaces as well. In both cases effective heat transfer depends both on the intensity of mixing which provides particle renewal on the heating surfaces as well as on the time of contact of the heater and bed which can be limited due to the air gap formation.

Both of the above cases will be discussed separately below.

4.1. Heat Transfer from the bottom plate

Heat transfer in such a case can be described by a model due to Muchowski (6). The average (over a whole vibration cycle) heat transfer coefficient across the air gap is given by:

$$\bar{h}_g = \frac{1}{2\pi} \int_0^{2\pi} h_c \, d(\omega t) + h_r + h_k (1 - \beta)$$

where h_c – heat transfer coefficient by conduction in gaseous gaps

h_r – heat transfer by radiation

h_k – heat transfer by contact

β – flight time to vibration period ratio.

In this model h_c is a function of the air gap thickness. According to Muchowski

$$h_c = h_{c\,max} (\frac{d_p}{2} (\sigma + s))$$

where $h_{c\ max}$ is a certain constant that can be calculated according to appropriate theory;

σ – modified mean free path of gas molecules;

s – air gap thickness that can be calculated according to formula (5) (taking $k_d \to 0$ and $u = 0$ for impermeable plate).

Heat transferred through the gap is then carried away by circulation of the particles. Heat transfer coefficient for that stage of the process is given by

$$h_{circ} = \sqrt{\frac{\lambda_b\ c_b\ \rho_b}{\pi t}}$$

where t – local contact time of particles and surface.

The effective heat transfer coefficient (using bulk bed – heater surface temperature difference as the driving force) was found to be equal to

$$h_{total} = \frac{2\bar{h}_g}{\sqrt{\pi\tau_f}}\ (1 + \frac{1}{\sqrt{\pi\tau_f}}\ \ln\ \frac{1}{1 + \sqrt{\pi\tau_f}})$$

where $\tau_f = \dfrac{\bar{h}_g^2\ \beta/f}{\lambda_b\ c_b\ \rho_b}$;

λ_b – heat conductivity of bed, W/m K

c_b – solid heat capacity, kJ/kg K

The basic problem in application of this model is to find the circulation rate which controls the contact time t. In continuous troughlike VB processors one can assume it to be equal to the linear transportation velocity of solids which is given by:

$$u_s = \frac{W_s}{BH\ (1 - \varepsilon)_s}$$

where B – bed width

H – bed depth.

This formula is obtained with the assumption that bed depth is constant which is true in a properly designed conveyor.

4.2. Heat Transfer from Immersed Heaters

Vertical heaters in form of a flat panel or heated chamber walls virtually do not present any obstacle to the flowing and vibrating solid (13).

On the way past the heated surface vibrating solid particles "scour" the gaseous film covering the heater. This according to Gutman (7) is responsible for lowering the heat transfer resistance and increasing heat transfer coefficient as compared to nonvibrated bed. Heat transfer coefficient under these conditions initially increases with vibrational acceleration then remains constant or even decreases. In the area of increasing heat transfer coefficient it is dependent on the maximum throw height s_{max} according to the following formula by Gutman

$$h_{cv} = h_{c\ stat} \frac{1}{1 - k\ s_{max}}$$

where $h_{c\ stat}$ is heat transfer coefficient in a static bed, and k is a constant.
Typical behavior of this kind is presented in Fig. 4. The vertical heater in this case was 25 mm wide, 2 mm thick and 50 mm high and was immersed in a bed of glass ballotini d_p = 0.454 mm, 100 mm deep vibrated with amplitude 4.25 mm.

Heat transfer coefficients from vertical heaters can reach higher values than those from horizontal heaters of comparable area as depicted in Fig. 4. The curve for horizontal heater was measured by the authors with a ϕ 19 mm, 58 mm long cylindrical heater in 80 mm deep bed. Other conditions are the same.

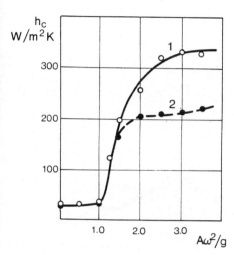

Fig. 4   Comparison of heat transfer coefficients for a vertical (1) and horizontal cylindrical (2) heater to a bed of glass ballotini d_p = 0.667 mm

However, the horizontal portion of the heater can be a great asset in fluidization of deeper beds and therefore is studied (e.g. (8)) as an alternative to the heated decks or vertical heaters. For shallow beds vertical panels may not provide adequate heat transfer surface. The reason for the lower values of heat transfer coefficient in the plateau area is the air gap formation on the heater. Contrary to the gap formation on the bottom of the bed the gap on the heater forms at the bottom of the heater when the bed ascends and on the top when the bed descends as schematically shown in Fig. 5. This was earlier reported in (9) and was also observed by Kossenko et al (10). (The air gap does not form until certain vibrational acceleration level is reached.) This fact proves that deeper beds do not behave as stiff bodies. If this were the case the air gap would be simultaneously formed at the bottom of the bed and at the top of the heater surface. This phenomenon requires further studies to describe the air gap formation process mathematically.

As shown in Figs. 6 and 7 the heat transfer coefficient curves display maxima. As noted earlier these probably correspond to the vibrational level at which air gap is formed. In the decreasing portion of the curves the gap is almost permanently formed which is especially pronounced if the beds contain any moisture that increases its stickiness (9).

450

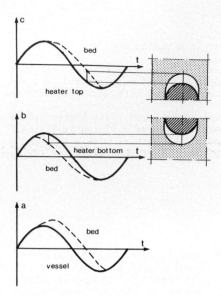

Fig. 5 Air gap formation in a deep
VB (a) on the gas distributor
grid (b) on the heater bottom
(c) on the heater top

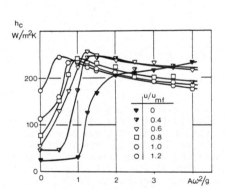

Fig. 6 Heat transfer coefficient
from a horizontal cylinder,
dia. 19 mm, to a bed of dry
glass ballotini d_p = 0.667 mm

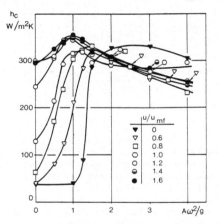

Fig. 7 Heat transfer coefficient
from a horizontal cylinder,
dia. 19 mm, to bed of dry
glass ballotini d_p = 0.454 mm

Air gap formation on selected shapes of heater are shown in Fig. 8 (12). The
gap is usually more pronounced for smaller particles (e.g. glass ballotini
d_p = 0.454 mm) as compared with larger particles (polyethylene resin d_p ~ 3 mm)

The research in this area is being continued, the following are immediate
directions:

- investigation of the heater size, shape and location in the bed on the heat transfer rate
- influence of particle surface stickiness on heat transfer
- mathematical description of the heat transfer process for a vertical heater
- influence of drying (evaporation of moisture) on immersed surface heat transfer.

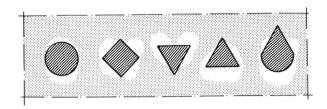

Fig. 8 Typical time averaged pictures of air gaps formed on various shape heaters.

4.3. Radiative Heat Transfer

This case is of limited importance as compared to convective or conductive heating. The area of its application is limited to rather shallow beds of not too small and sticky particles. In small beds mixing is very intensive so there is no danger of overheating the solids on the surface of the bed. The method was studiedfor its application to vacuum drying pharmaceuticals (11). It, however, can be used widely as an auxilliary source of heat in combined convective / radiative or convective / conductive / radiative heating of the beds.

If perfect mixing of the bed can be assumed then the radiative heat transfer to VB is basically similar to the design of any other radiative heat transfer system.

5. CONCLUDING REMARKS

Design of the vibrated bed thermal processors requires prior knowledge of heat transfer coefficients for convective, conductive and radiative heat transfer processes employed for thermal treatment of the particulate solids.

Convective heat transfer coefficients in VB operated below u_{mf} are necessary only in the area of large particle size and/or shallow beds because only then thermal equilibrium between the gas and solid is not reached. In heavily aerated VB Bubbling Bed Models have to be employed for heat transfer calculations. Much remains to be studied in this area; it is not clear if this is a desirable operating regime for vibrated beds.

Contact heat transfer coefficients are separately studied for the case when the trough deck is heated and in case of immersed heaters. A model by Muchowski can assist in the first case. In the second case the heat transfer mechanism is different for vertical heaters and for tubular horizontal heaters.

Both cases require further studies. Horizontal heaters which seem to be more helpful in the case of deeper beds are unfortunately less efficient at

bag filter; solids leave the loop through the cyclone bottom. Pressure ports are provided approximately every 1/3 of a meter around the loop for pressure monitoring with water manometers.

5.1 Void Fraction Measurement

One of the problems of studying a two-phase flow process is the inherent difficulties associated with measuring solid flux, superficial gas velocity, and void fraction. After reviewing various techniques used to obtain these quantities, the Auburn Monitor was selected for measuring void fraction in this study. The monitor operates in such a manner that the solid-gas flow is not obstructed during the process.

The Auburn monitor has been recently developed by Auburn International, Inc. Danvers, MA. It determines the voidage in a spool sensor piece by measuring the dielectric constants of the two phase flow without obstructing the flow itself. The monitor employs a patterned sequentially switched rotating field in a multi-electrode sensor. This permits the accurate determination of the voidage and prevents the measurement from being effected by the distribution of the discrete phase. The instrument consists of two major components: the sensing spool and the electronic unit interconnected with coaxial cables. The sensing spool contains an array of six electrodes, equi-spaced about the inner circumference. The electronics provide three functions: excitation of the sensor, measurement of the average dielectric constant within the sensing volume, and providing zero offset such that the background noise can be nulled. The monitor does not interfere with the flow and gives an output proportional to the void fraction.

The principle of operation of this instrument is based on the simple relation that the dielectric constant of a two component system is the sum of the individual dielectric constants multiplied by their respective volume fraction. Since the dielectric constant of the continuous phase is in general lower than that of the discrete phase the instrument can be zeroed when the voidage is unity. Similarly, the instrument can be spanned where the voidage is zero or some known value, say the voidage in a packed bed of the same material. This results in an output proportional to the solid fraction. The electric field is sequentially rotated around the spool by switching the excitation from one electrode to another at a rate in excess of 200 revolutions per second, providing an accurate value of the average void fraction.

The Auburn monitor was calibrated for voidage and solid mass flux.

6. RESULTS AND DISCUSSION

Experiments were conducted to study the effect of air flow rates through the nozzles N_1 through N_4 on the flow characteristics of sand particles through the loop shown in Figure 2. Operation of nozzle, N_1 established a superficial air velocity of 6.62 ms^{-1} through the riser. Fluidization was stable with a void fraction of 0.985 in the riser measured at about 0.75 m above the eductor section. The pressure drop in this region was about 6 cm of water per meter length of pipe. The bed exhibited a high degree of back mixing, characteristic of fast fluidization and a very desirable feature for coal combustion. There was some saltation in the eductor section of the loop. A sand height of 0.75 m was maintained in the stand pipe. In this study no restriction of any kind was provided in the stand pipe as done by some workers. A decrease in flow rate through the nozzle caused air to short circuit up the stand pipe and decreased the void fraction in the riser. At a velocity of 214 ms^{-1} (6.70 x $10^{-3}m^3s^{-1}$) through the nozzle, slugging was observed in the stand pipe with no fluidization in the riser.

Operation of nozzle, N_2 alone provided stable fluidization with a superficial velocity and voidage of 3.89 m/s and 0.980 in the riser, respectively. However, this low velocity caused heavy saltation in the eductor section filling half the pipe with sand particles. Decreasing the flow rate caused slugging in the stand pipe. An increase in the flow rate increased the voidage in the riser and decreased saltation in the eductor section.

The use of nozzles N_3 and N_4 independently at about the same flow rates as in the earlier experiments resulted in no solid flow through the loop due to heavy saltation in the eductor section.

The use of a combination of nozzles N_1 and N_2 with air flow rates through N_1 varying from 1.48 to 214 ms^{-1} (4.17 x 10^{-3} to 7.50 x 10^{-3} m^3s^{-1}) and through N_2 in the range of 5 to 45 ms^{-1} (0.1 x 10^{-3} to 1.57 x $10^{-3}m^3s^{-1}$) resulted in on flux through the loop. However, with the use of N_1 and N_3 combination stable operation could be achieved with good back mixing in the riser section.

Efforts were made to operate the LFB at minimal air flow rate to decrease the void fraction or increase the solid content. It was found that LFB could be operated at lower air flow rates using a combination of nozzles N_2 and N_3 rather than N_1 and N_3, decreasing the void fraction from 0.98 to 0.97. A void fraction of 0.97 corresponds to 6% solids with a mass flux of 50 Kg m^{-2}s^{-1}. Efforts to decrease the void fraction and increase solid mass flux are in progress.

7. CONCLUSIONS

Fluidization and fast fluidization have been the subject of numerous investigations. On the other hand studies on the LFB using the fast fluidization concept are rather scarce. In this investigation an LFB unit has been designed, fabricated, and successfully operated to study the effect of air injection through nozzles at various locations in the loop on the air-sand flow behavior.

ACKNOWLEDGEMENT

The authors are thankful to the U.S. Department of Energy (Morgantown Energy Technology Center, Morgantown, W.V.) for funding this project. Assistance of Wayne T. Robinson in the calibration of the Auburn Monitor is also gratefully acknowledged.

REFERENCES

1. Yerushalmi, J., M.J. Gluckman, R.A. Graff, S. Dobner, and A.M. Squires, "Production of Gaseous Fuels from Coal in the Fast Fluid Bed", Fluidization Technology, ed. D.L. Kearns, Hemisphere Publishing Corp., Washington, D.C., Vol. II, 1975, 437.

Ray, Berkowitz and Sumaria [10] , were among the first ones, who studied dynamics, using a very complex, non-linear plant model to examine local stability of a multicell fluidized-bed steam generator at different load levels. Another interesting dynamical analysis was presented by FAN and CHANG [11] , who investigated a shallow fluidized-bed combustor. The numerical results indicate that a critical bubble size and carbon feed rate exist, above which "concentration runaway" occurs.

In this paper a general and systematic, but not complete analysis of a non-catalytic solid-gas reaction in continuous fluidized-bed is presented. Beginning with a general dynamical PMA-type model, first its steady-state version is studied. Then a simplified, dynamical PMU-type model is deduced from this general one, and used for further analysis. The time dependence of the chemical reaction as well as that of the coupled heat and mass transfer are examined but the dynamics of the flow pattern in bed is not taken into consideration [12] . The simplified model represents, under certain circumstances, almost all significant properties of the original one, and at the same time it can be handled by standard methods well-known from the CSTR analysis [14-16].

2. PHYSICAL MODEL

Let us consider the following irreversible, simple, solid-gas reaction:

$$A + bB \rightarrow Product \qquad (1)$$
$$(s) \quad (g)$$

which takes place in a fluidized-bed reactor with continuous supply and removal of solid particles containing component A. The inlet gas stream fluidizes the bed as well as provides the reactant gas component B. The reactor is presented schematically on Fig. 1.

The so-called "zone-model" is supposed to be valid [13,17,18] for solid particles. They are considered to have the same size and porosity. The concentration of the reactant gas and that of the solid component A, and the temperature in the pores, and in the solid part of the particles are uniform.

It is assumed that the fluidized-bed can be represented by a Kunii-Levenspiel type three-phase model, where the solid phase and the interstitial gas are perfectly mixed, while the dilute phase consisting of bubbles of uniform size is in plug flow without axial dispersion [19]. The bubbles do not contain solid particles, therefore no chemical reaction takes place in this phase. Although the solid particles are perfectly mixed, they spend different lengths of time in the reactor. In steady state their residence time distribution can be expressed as:

$$f(\theta^*) = \frac{1}{\theta_p^*} \exp(- \frac{\theta^*}{\theta_p^*}) \qquad (2)$$

In addition, heat exchange between the interstitial gas and the reactor wall is supposed to be possible. The temperature of the reactor wall is uniform and the wall heat capacity is neglected.

These are the general hypotheses of the physical model and they, together with the conservation and transfer laws for mass and energy, and with the kinetic of the chemical reaction provide the basis for the mathematical description of the process. The details of the formulation of the model equations are not presented here. These, moreover the theoretical and empirical expressions giving

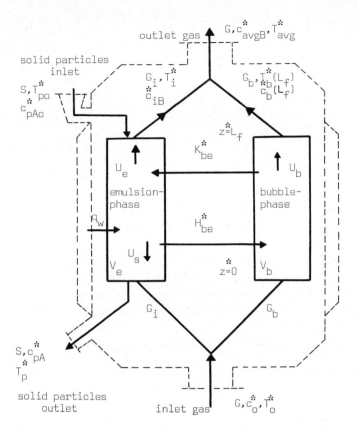

Fig. 1. Schematical layout of the reactor

the physical parameters used in the conservation equations can be found in [13] where a similar, but steady-state model was developed. The dimensionless variables and terms are given in Table 1.

3.   MATHEMATICAL MODEL

Let us consider a case where the particles are supplied at constant rate, and their initial and feed distributions are uniform in concentration of components A and B as well as in temperature. Then the concentrations and temperature associated with each particle will depend upon time and age, but not upon position [2] . Noting that $\Delta\tau=\Delta\theta$, mass and energy balances for a single solid particle can be set up. The mass balance for component A :

$$\frac{\partial c_{pA}}{\partial\tau} + \frac{\partial c_{pA}}{\partial\theta} = - R_A \tag{3}$$

Assuming a convective mass transport between the gas in pores of solid particles and the interstitial gas phase, we get an other mass balance for the reactant gas component B :

Table 1. Dimensionless variables and terms*

$$c_{(\ )} = \frac{c^*_{(\ )}}{c^*_{pAo}} \ , \quad T_{(\ )} = \frac{T^*_{(\ )}}{T^*_{po}} \ , \quad \tau = \frac{\tau^*}{t_o} \ , \quad \theta = \frac{\theta^*}{t_o} \ , \quad z = \frac{z^*}{L_f} \ , \quad N = \frac{E}{RT^*_{po}} \ ,$$

$$K_p = k_{ap} t_o \ , \quad H_p = \frac{h_{ap} t_o}{\rho_s c_s} \ , \quad \beta = \frac{T^*_{po} c_s \rho_s}{c^*_{pAo} M_A |-\Delta H|} \ , \quad F_o = \frac{k_o t_o c^{*2/3}_{pAo} |-\Delta H| M_A}{\rho_s c_s T^*_{po}} \ ,$$

$$f_b = \frac{L_f}{t_o U_b} \ , \quad f_e = \frac{L_f}{t_o U_e} \ , \quad \mu = \frac{\varepsilon_{mf}}{1-\varepsilon_{mf}} \ , \quad \varepsilon_R = \frac{(-\Delta H)}{|-\Delta H|} \ , \quad B_w = \frac{h_w a_{ex} L_f}{\delta \gamma_e U_e \rho_g c_g} \ ,$$

$$(K_{be})_b = \frac{K^*_{be} L_f}{U_b} \ , \quad (K_{be})_e = \frac{K^*_{be} L_f}{U_e \gamma_e} \ , \quad (H_{be})_b = \frac{H^*_{be} L_f}{\rho_g c_g U_b} \ , \quad (H_{be})_e = \frac{H^*_{be} L_f}{\rho_g c_g U_e \gamma_e} \ ,$$

$$A_e = \frac{k_{ap} L_f}{U_e} \ , \quad B_e = \frac{h_{ap} L_f}{U_e c_g \rho_g} \ , \quad P = \frac{\rho_s c_s}{c_g \rho_g}$$

* for computation of physical parameters see APPENDIX

$$\varepsilon_p \frac{\partial c_{pB}}{\partial \tau} + \frac{\partial c_{pB}}{\partial \theta} = bR_A + K_p(c_{iB} - c_{pB}) \tag{4}$$

The heat balance yields :

$$\frac{\partial T_p}{\partial \tau} + \frac{\partial T_p}{\partial \theta} = \frac{\varepsilon_R}{\beta} R_A + H_p(T_i - T_p) \tag{5}$$

According to the "zone-model", the reaction rate can be expressed as [13] :

$$R_A = \beta F_o c_{pA}^{2/3} c_{pB} \exp(-N/T_p) \tag{6}$$

Employing the assumptions made for the dilute phase, the mass and heat balances for the bubbles are :

$$f_b \frac{\partial c_b}{\partial \tau} + \frac{\partial c_b}{\partial z} = (K_{be})_b(c_{iB} - c_b) \tag{7}$$

$$f_b \frac{\partial T_b}{\partial \tau} + \frac{\partial T_b}{\partial z} = (H_{be})_b(T_i - T_b) \tag{8}$$

The convective terms on the right hand side represent the mass and heat exchange between the bubbles and interstitial gas.

The interstitial gas "exchanges" mass and heat with the solid and dilute phases. In addition, there is heat transfer between this phase and the reactor wall. That is why the balance equations for the interstitial gas are as follows:

$$f_e \frac{dc_{iB}}{d\tau} = c_o - c_{iB} + (K_{be})_e \mu(\tilde{c}_b - c_{iB}) + A_e \mu(\bar{c}_{pB} - c_{iB}) \qquad (9)$$

$$f_e \frac{dT_i}{d\tau} = T_o - T_i + (H_{be})_e \mu(\tilde{T}_b - T_i) + B_e \mu(\bar{T}_p - T_i) + B_w \mu(T_w - T_i) \qquad (10)$$

The interstitial gas is perfectly mixed,while the dilute phase is in plug flow, the space (bed height) average concentration and temperature of the bubbles are considered here as representative values.

$$\begin{bmatrix} \tilde{c}_b(\tau) \\ \tilde{T}_b(\tau) \end{bmatrix} = \int_0^1 \begin{bmatrix} c_b(\tau,z) \\ T_b(\tau,z) \end{bmatrix} dz \qquad (11)$$

Similarly,the age average state variables are used in Eqs.(9) and (10),because of the different concentrations and temperatures of the particles according to their different ages.These age average values can be computed on the basis of the density function of the particles RTD.

$$\begin{bmatrix} \bar{c}_{pA}(\tau) \\ \bar{c}_{pB}(\tau) \\ \bar{T}_p(\tau) \end{bmatrix} = \int_0^\infty f(\tau,\theta) \begin{bmatrix} c_{pA}(\tau,\theta) \\ c_{pB}(\tau,\theta) \\ T_p(\tau,\theta) \end{bmatrix} d\theta \qquad (12)$$

Considering the situation where all particles are initially at zero age and the solid stream is in steady-state,we get for the density function [2] :

$$f(\tau,\theta) = \begin{cases} \frac{1}{\theta_p} \exp(-\frac{\theta}{\theta_p})\psi(\tau) + \delta(\tau)\exp(-\frac{\theta}{\theta_p}) & \text{for } \theta \leqslant \tau \\ \\ 0 & \text{for } \theta > \tau \end{cases} \qquad (13)$$

where

$$\psi(\tau) = \begin{cases} 1 & \tau > 0 \\ 0 & \tau = 0 \end{cases} \qquad (14)$$

The boundary conditions for the dilute phase are :

$$c_b(\tau,0) = c_o$$
$$T_b(\tau,0) = T_o \qquad \text{for } \tau \geqslant 0 \text{ and } z=0 \qquad (15)$$

Assuming that the solid inlet does not contain reactant gas in the pores,the boundary conditions for the solid particle are :

$$c_{pA}(\tau,0) = 1$$
$$c_{pB}(\tau,0) = 0 \qquad \text{for } \tau \geqslant 0 \text{ and } \theta=0 \qquad (16)$$
$$T_p(\tau,0) = 1$$

The analytical,but even the numerical study of this model is very diffi-cult because of its high complexity.For steady-state case,when only two inde-pendent variables exist,the particle age and the position in the bed,a one di-mensional trial-error procedure has been developed to solve the appropriate integro-differential equations [13].This procedure is based on the method proposed by Hatfield and Amundson [2].As an illustration,Figs.2-4 show the concentrations and temperature of a solid particle as function of its age in the reactor as well as the reactant gas concentration and temperature distribution along the height of the bed.The values needed for computation are given in Table 2.

Table 2. Numerical values for calculation of steady-state
with the PMA model

$\theta_p^* = 360$ sec	$D_r = 1.2$ m	$c_s = 0.2$ cal/gr,K
$L_{mf} = 1$ m	$U_o = 30$ cm/sec	$T_o^* = 650$ K
$\Delta H = -1.10^2 \frac{cal}{gr}$	$\varepsilon_p = 0.35$	$\rho_g = 1.5 \; 10^{-3} \frac{gr}{cm^3}$
$d_b = 6$ cm	$T_w^* = 650$ K	$\rho_s = 1 \frac{gr}{cm^3}$
$T_p^* = 300$ K	$E/R = 1.4 \; 10^4$ K	$c_o^* = 1.5 \; 10^{-4} \frac{mol}{cm^3}$
$b = 2$	$\lambda_g = 8.10^{-5} \frac{cal}{cm,sec,K}$	$c_{pAo}^* = 1.10^{-2}$ mol/cm^3
$D_g = 0.2$ cm^2/sec	$\varepsilon_{mf} = 0.47$	$M_A = 50$ gr/mol
$\mu_g = 2.10^{-4} \frac{gr}{cm,sec}$	$d_p = 5.10^{-2}$ cm	$c_g = 0.25$ cal/gr,K
	$k_o = 4.10^{12} \frac{1}{sec}(\frac{mol}{cm^3})^{-2/3}$	

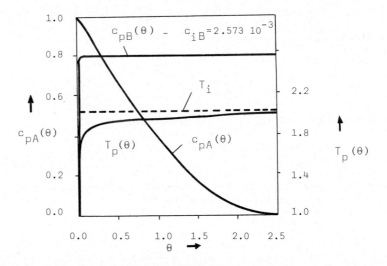

Fig. 2. The concentration of components A and B
in a solid particle and its temperature
as function of particle age.

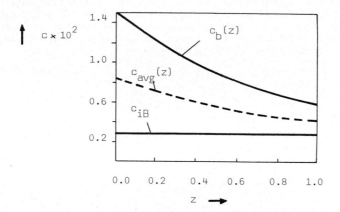

Fig. 3. Steady-state reactant gas concentration
distribution along the bed height.

The appropriate conversion X, can be computed as :

$$X = 1 - \bar{c}_{pA} = 1 - \frac{1}{\theta_p} \int_0^1 c_{pA}(\theta) \, \exp(-\frac{\theta}{\theta_p}) \, d\theta = 0.905$$

Simplifications being necessary for further steady-state, and especially dynamical analyses, can be carried out under certain circumstances. Let us see what kind of additional hypotheses are required to reduce the original model labelled as model Z_1.

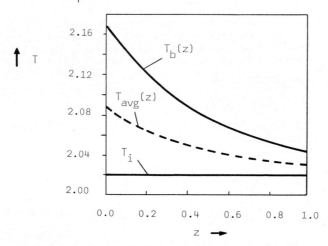

Fig. 4. Steady-state gas temperature distribution
along the bed height.

4. MODEL REDUCTION

As first additional hypothesis, H1, we suppose that the reaction rate is high and the solid residence time is long enough to consider the solid particles to have the same temperature and concentration in the reactor. In other words, these intensive variables can be "lumped" according to the particle age, . If we define the lumping by the following integral :

$$(\hat{\cdot}) = \frac{1}{\theta_p} \int_0^{\theta_p} (.(\theta)) \, d\theta \tag{17}$$

and using the boundary conditions (16), we get a PMU-type model for the process, model Z_2. Then the balance equations for the solid phase are :

$$\frac{d\hat{c}_{pA}}{d\tau} = \frac{1}{\theta_p} (1 - \hat{c}_{pA}) - \hat{R}_A \tag{18}$$

$$\varepsilon_p \frac{d\hat{c}_{pB}}{d\tau} = -b\hat{R}_A - \frac{\varepsilon_p}{\theta_p} \hat{c}_{pB} - K_p(\hat{c}_{pB} - c_{iB}) \tag{19}$$

$$\frac{d\hat{T}_p}{d\tau} = \frac{1}{\theta_p} (1 - \hat{T}_p) + \frac{\varepsilon_R}{\beta} \hat{R}_A - H_p(\hat{T}_p - T_i) \tag{20}$$

where the reaction rate is defined as :

$$\hat{R}_A = \beta F_o \hat{c}_{pA}^{2/3} \hat{c}_{pB} \exp(-N/\hat{T}_p) \tag{21}$$

In addition, the age average values, $\bar{c}_{pB}$ and $\bar{T}_p$ have to be replaced by $\hat{c}_{pB}$ and $\hat{T}_p$ respectively in the interstitial gas balance equations, Eqs.(9),(10).

The next reduction is possible, if the heat and mass transfer resistances between the solid phase and the interstitial gas are negligible, i.e. hypothesis H2. This can be valid at high mass and heat transfer coefficients, K_p and H_p. Let us express $(\hat{c}_{pB} - c_{iB})$ from Eq.(19) and substitute it into Eq.(9), taking into consideration that $\hat{c}_{pB} \equiv c_{iB} \equiv c_e$, and

$$f_e = A_e/K_p \tag{22}$$

then the component balance for the reactant gas B in the emulsion phase (interstitial gas + solid phase) yields

$$f_e(1 + \mu\varepsilon_p)\frac{dc_e}{d\tau} = c_o - c_e + (K_{be})_e \mu(\tilde{c}_b - c_e) - f_e\mu(b\hat{R}_A + \frac{\varepsilon_p}{\theta_p} c_e) \tag{23}$$

Likewise, expressing $(\hat{T}_p - T_i)$ from Eq.(20) and substituting it into Eq.(10), considering that $\hat{T}_p \equiv T_i \equiv T_e$, and

$$f_e P = B_e/H_p \tag{24}$$

the heat balance equation for the emulsion phase is the following :

$$f_e(1+ \mu P) \frac{dT_e}{d\tau} = T_o - T_e + (H_{be})_e \mu (\tilde{T}_b - T_e) + f_e P \mu (\frac{\varepsilon_R}{\beta} \hat{R}_A + \frac{1 - T_e}{\theta_p})$$

$$+ B_w \mu (T_w - T_e) \qquad (25)$$

Now, our model Z_3 consisting of the equations (18),(21),(23),(25) and Eqs. (7),(8) together with Eq.(11), is already much more simple than the original Z_1 model. However, further simplification can be carried out using a quasi-steady-state approximation (QSSA) for the bubble phase as hypothesis H3. This can be done if $f_b/(K_{be})_b$ and $f_b/(H_{be})_b$ are small values. The results of this reduction are the new balance equations for the bubbles :

$$\frac{dc_b}{dz} = (K_{be})_b (c_e - c_b) \qquad (26)$$

$$\frac{dT_b}{dz} = (H_{be})_b (T_e - T_b) \qquad (27)$$

These equations can be solved under boundary conditions (15). Then the quasi-steady reactant gas concentration and temperature profiles are in the dilute phase :

$$c_b(z) = c_e + (c_o - c_e) \exp(- (K_{be})_b z) \qquad (28)$$

$$T_b(z) = T_e + (T_o - T_e) \exp(-(H_{be})_b z) \qquad (29)$$

Computing the space average values according to Eq.(11) and substituting them into Eqs.(23) and (25), the heat and mass balances for the emulsion phase are :

$$f_e(1+ \mu \varepsilon_p) \frac{dc_e}{d\tau} = \tilde{K}(c_o - c_e) - f_e \mu (b\hat{R}_A + \frac{\varepsilon_p}{\theta_p} c_e) \qquad (30)$$

$$f_e(1+ \mu P) \frac{dT_e}{d\tau} = \tilde{H}(T_o - T_e) + f_e P \mu (\frac{\varepsilon_R}{\beta} \hat{R}_A - \frac{1 - T_e}{\theta_p})$$

$$+ B_w \mu (T_w - T_e) \qquad (31)$$

where

$$\tilde{K} = 1 + \mu \frac{(K_{be})_e}{(K_{be})_b} \left[1 - \exp(- (K_{be})_b) \right] \qquad (32)$$

$$\tilde{H} = 1 + \mu \frac{(H_{be})_e}{(H_{be})_b} \left[1 - \exp(- (H_{be})_b) \right] \qquad (33)$$

Equations (18) and (28)-(33) represent the model Z_4.

The last reduction leading to a pair of non-linear ordinary differential equations, can be applied to model Z_4, if we assume that $(1+\mu\varepsilon_p)/(1+\mu P)$ is a

472

very small value,hypothesis H4.This means that the concentration transient is significantly faster than that of the temperature.Therefore Eq.(30) has the form :

$$\tilde{K}(c_o - c_e) = f_e \mu (b\hat{R}_A + \frac{\varepsilon_p}{\theta_p} c_e) \tag{34}$$

Using this equation to eliminate c_e in the reaction rate equation (21),we get :

$$\hat{R}_A(\hat{c}_{pA}, T_e) = \frac{\beta F_o c_o \hat{c}_{pA}^{2/3} \exp(-N/T_e)}{1 + \frac{f_e \mu}{\tilde{K}} \left[\frac{\varepsilon_p}{\theta_p} + \beta b F_o \hat{c}_{pA}^{2/3} \exp(-N/T_e) \right]} \tag{35}$$

The other two remaining model equations can be written in the following form :

$$\frac{d\hat{c}_{pA}}{d\tau} = a_o - a_1 \hat{c}_{pA} - \hat{R}_A(\hat{c}_{pA}, T_e) \tag{36}$$

$$\frac{dT_e}{d\tau} = b_o - b_1 T_e + b_2 \hat{R}_A(\hat{c}_{pA}, T_e) \tag{37}$$

where the coefficients of these equations are :

$$a_o = a_1 = 1/\theta_p \tag{38}$$

$$b_o = \frac{\tilde{H}T_o + f_e \mu P/\theta_p + B_w \mu T_w}{f_e(1 + \mu P)} \tag{39}$$

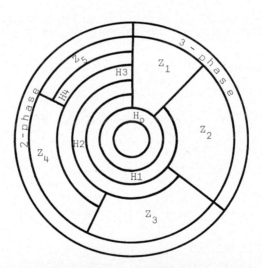

Fig. 5.   Types of model
Z_1-Basic Model,Z_i- Simplified Model,
H_o-General Hypotheses,Hj-Additional
Hypothesis

Greek Letters

α	volume ratio of cloud to bubble phase,	dimensionless
γ_e	volume ratio of solid to bubble phase,	dimensionless
δ	relative fraction of bubble phase,	dimensionless
$\delta(\tau)$	Dirac - delta function,	dimensionless
ε_{mf}	voidage of the emulsion phase at minimum fluidization velocity,	dimensionless
ε_p	particle porosity,	dimensionless
θ^*	age of particle,	sec
θ^*_p	mean residence time of particles,	sec
λ_g	heat conductivity of gas,	cal/(cm,sec,K)
μ_g	dynamic viscosity of gas,	gr/(cm,sec)
ρ	density,	gr/cm^3
Φ_s	sphericity of particles,	dimensionless
Φ_w	wall heat transfer parameter,	dimensionless
τ^*	time,	sec

Indeces

avg	average value
A	component A
b	bubble
B	component B
e	emulsion phase
g	gas
i	interstitial gas
o	inlet value
p	particle
s	solid phase

REFERENCES

1. Luss,D. and Amundson,N.R., AIChE J.,(1968), Vol.14, pp.211.

2. Hatfield,W.B. and Amundson,N.R., AIChE Symp.Series 67, (1971), pp.54.

3. Bokur,D.B. and Amundson,N.R., Chem.Eng.Sci., (1975), 30, pp.847.

4. Bokur,D.B. and Amundson,N.R., Chem.Eng.Sci., (1975), 30, pp.1159.

5. Bokur,D.B., Caram,H.S. and Amundson,N.R., Ch.11 in Wilhelm Memorial Volume
    Chemical Reactor Theory ,Ed.: Lapidus & Amundson,(1977),Prentice H;pp.686

6. Jaffrés,J.L. at al., ACS Symp.Series 168, (1981), pp.55.

7. Gordon,A.L. and Amundson,N.R., Chem.Eng.Sci., (1976), 31, pp.1163.

8. Gordon,A.L., Caram,H.S. and Amundson,N.R.,Chem.Eng.Sci.,(1978), 33, pp.713.

9. Rehmat,A. ,Saxena,S.C. and Land,R.H.,ACS Symp.Series 168,(1981), pp.117.

10. Ray,A., Berkowitz,D.A. and Sumaria,V.H., J. of Energy,(1978), Vol.2,pp.269

11. Fan,L.T. and Chang,C.C.,ACS Symp.Series 168, (1981), pp.95.

12. Didwania,A.K. and Homsy,G.M., J. of Fluid Mech., (1982), Vol.122,pp.433.

13. Paláncz,B., Chem.Ing.Techn. to appear

14. Varma,A. and Aris,R.,Ch.2. in Wilhelm Memorial Volume Chemical Reactor
    Theory ,Ed.: Lapidus & Amundson,(1977),Prentice H; pp.79.

15. Uppal,A., Ray,W.H. and Poore,A.B., Chem.Eng.Sci.,(1974), 29, pp.967.

16. Aris,R., Advances in Control and Dynamics of Systems ,(1979),Acad.P;pp.41

17. Beránek,J., Rose,K. and Winterstein,G., Grundlagen der Wirbelschicht Tech-
    nik , Krausskopf-Verlag,(1975), pp.146.

18. Székely,J., Ch.5. in Wilhelm Memorial Vol. Chemical Reactor Theory Ed.:
    Lapidus & Amundson, (1977) ,pp.269.

19. Kunii,D. and Levenspiel,O., Fluidization Engineering ,Wiley,(1969)

20. Zeeman,E.C., Sci.Am. 234.,No.4, (1976), pp.65.

21. Wen,C.Y., Proceedings of a Summer School on RTDT,Ed.:Pethő & Noble,
    (1982), Chemie-Verlag, pp.255.

22. Bailey,J.E.,Ch.12. in Wilhelm Memorial Volume Chemical Reactor Theory
    Ed.: Lapidus & Amundson, (1977), pp.758.

23. Aris,R. and Amundson,N.R., Chem.Eng.Sci.(1958), 7, pp.132.

24. Mansden,J.E. and McCracken,M.,"The Hopf Bifurcation and Its Application",
    Appl.Math.Sci., 19, Springer-Verlag,(1976)

25. Scheib,H.J., Ingenieur-Archiv, 51, (1981), pp.183.

APPENDIX

Parameter	Theoretical or empirical expression
Superficial velocity at minimum fluidizing condition	$U_{mf} = \dfrac{\phi_s^2 d_p^2}{150} \dfrac{\rho_s - \rho_g}{\mu_g} g \dfrac{\varepsilon_{mf}}{1 - \varepsilon_{mf}}$
Rising velocity of a single bubble relative to the emulsion phase	$U_{br} = 0.711 \, (gd_b)^{1/2}$
Absolute velocity of a crowd of bubbles in the bed	$U_b = (U_o - U_{mf}) + U_{br}$
Volume ratio of cloud to bubble phase	$\alpha = 0.6 \, d_b^{-1/2}$
Relative fraction of bubble phase	$\delta = \dfrac{U_o - U_{mf}}{U_b}$
Volume ratio of solid to bubble phase	$\gamma_e = \dfrac{1 - \delta - \alpha\delta}{\delta}$
Length of the fluidized bed	$L_f = \dfrac{L_{mf}}{1 - \delta}$
Surface area for external heat transfer	$S_{ex} = D_r \pi \, l_f$
Specific surface area for external heat transfer	$a_{ex} = \dfrac{4 S_{ex}}{L_f D_r^2 \pi}$
Mass transfer resistance between bubbles and interstitial gas	$1/K_{be}^* = 1/K_{bc}^* + 1/K_{ce}^*$
	$K_{bc}^* = 4.5 \dfrac{U_{mf}}{d_b} + 5.85 \dfrac{D_g^{1/2} g^{1/4}}{d_b^{5/4}}$
	$K_{ce}^* = 6.78 \, (\dfrac{\varepsilon_{mf} D_g U_b}{d_b^3})^{1/2}$
Heat transfer resistance between bubbles and interstitial gas	$1/H_{be}^* = 1/H_{bc}^* + 1/H_{ce}^*$

Parameter	Theoretical or empirical expression
	$$H^*_{bc} = 4.5 \frac{U_{mf}\rho_g c_g}{d_b} + 5.85 \frac{(\lambda_g \rho_g c_g)^{1/2}}{d_b^{5/4}} g^{1/4}$$
	$$H^*_{ce} = 6.78 (\rho_g c_g \lambda_g)^{1/2} (\frac{\epsilon_{mf} U_b}{d_b^3})^{1/2}$$
Mass transfer coefficient between interstitial gas and solid particles	$k_p = \dfrac{Sh_p D_g}{d_p}$, with $Sh_p = 2$
Heat transfer coefficient between interstitial gas and solid particles	$h_p = \dfrac{Nu_p \lambda_g}{d_p}$, with $Nu_p = 2$
Heat transfer with the reactor wall	$h_w = Nu_w \lambda_g / d_p$
	$$Nu_w = 0.16 \, Pr^{0.4} Re_p^{0.76} \frac{\rho_s c_s}{\rho_g c_g} Fr^{-0.2} (\frac{\phi_w L_{mf}}{L_f})^{0.36}$$
	$Pr = \mu_g c_g / \lambda_g$
	$Fr = U_o^2 / (g d_p)$
	$\phi_w = 0.6$
External surface area per unit volume of particles	$a_p = \dfrac{6}{d_p}$

Multi-Phase Flow and Heat Transfer III. Part B: Applications
edited by T.N. Veziroğlu and A.E. Bergles
Elsevier Science Publishers B.V., Amsterdam, 1984 — Printed in The Netherlands

487

DISPERSIVE STRESS IN SHEAR SLURRY FLOWS

M.C. Roco
Department of Mechanical Engineering
University of Kentucky
Lexington, Kentucky 40506, U.S.A.

C.A. Shook
Department of Chemical Engineering
University of Saskatchewan
Saskatoon, Sask. S7N 0W0, Canada

ABSTRACT

The dispersive stress between neighboring sheared layers of suspensions due to particle collisions has been introduced recently as a new term in the basic equations describing liquid-solid mixture flows. It has a strong effect on the concentration distribution and energy dissipation, especially at high concentrations and large particles.

In this paper we attempt to relate the coefficient of dynamic friction $\tan\theta$ (used in the expression for the dispersive stress) to a measure of the local ratio of inertial forces to viscous and gravitational forces. To select the significant dimensionless number we compare the model predictions to experimental concentration distributions and headlosses in a flat rectangular channel (ratio height/width = 1/4), as well as in circular pipes. The shear stress, concentrations and velocities are widely varied. A fine sand of $d = .15-.165$ mm was used in experiments, for which the Coulombic particle interaction may be neglected. The tests show that the coefficient $\tan\theta$ can be related to the local Froude number $Fr_* = \rho_L/\rho_M \cdot V_*^2/gd(S-1)$. The qualitative correlation between $\tan\theta$ and Fr_* is confirmed from another approach by using the supplementary headlosses caused by the dispersive stress at the pipe wall.

The semiempirical correlation suggested for the coefficient of dynamic friction increases the degree of generality of predictions of the dispersive stress, which is an essential element in numerical modeling of slurry flows.

INTRODUCTION

The flow behavior of concentrated suspensions is determined to a large extent by the interactions between solid particles. The interaction terms introduced in the governing equations are caused by collisions, Coulombic contacts and mixing eddies [1]. The transfer of momentum is mainly due to the the last two terms when the length scale of turbulence and local friction velocity are larger than particle dimension and settling velocity, respectively. The impingement energy between particles is then comparable to that of the dissipation rate. The corresponding repulsive normal stress acting between sheared layers of suspensions is denoted as "dispersive stress by particle impingement" (σ_{DS})(Figure 1). This stress has been introduced in the differential equations of motion to justify the experimental results [2]. It has an important role in turbulent slurry flow mechanism and its global evaluation is the object of this paper.

We initially employed for the flow field the relations suggested by Bagnold for the normal stress at the wall, $(P)_W$. Accordingly, the coefficient of the total dynamic friction $\tan\alpha$ varies with a dimensionless "grain flow number" between .32 and .75 (Figure 4). The dimensionless index may be converted to the following expression similar to a Reynolds number:

$$Re_* = \frac{\rho_S \cdot d^2 \cdot \sigma_S}{\lambda \cdot \mu^2} \tag{10}$$

where:

ρ_S - solid particle density

d - particle dimension

σ_S - total normal stress between solid particles

μ - dynamic viscosity of mixture

λ - average distance between particles

$$\lambda = [(C_M/C)^{.33} -1]^{-1}$$

The predictions for headlosses and concentration distributions using the Bagnold's correlation (see Bagnold line in Figure 4) do not agree with the experimental data obtained in the rectangular duct. Discrepancies are illustrated in Figures 4 and 5. The calculations made for both parameters show that the Bagnold's correlation $\tan\alpha = f(Re_*)$ is not general. In Figure 4 are compared $\tan\alpha = f(Re_*)$ resulted from the measured headlosses to the Bagnold line for three kinds of sand and one of nickel. In Figure 5 are illustrated the discrepancies between experimental results (Run A1) and calculated concentration with Bagnold's and two other correlations containing Re_*.

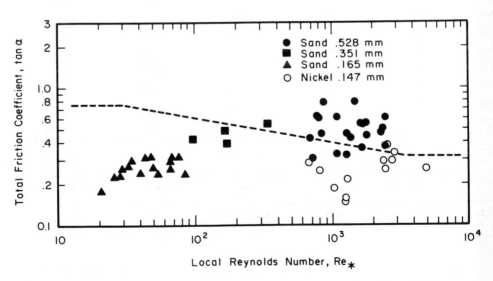

Fig. 4. The coefficient $\tan\alpha$ versus local Reynolds number (Re_*)

Investigations carried out in the sliding bed flow regime show that the coefficient of dynamic friction by Coulombic contacts may be assumed constant in particular conditions [5], [9].

Since $\tan\theta$ seems not to be correlated with Re_* by the Bagnold relationship, we first change the correlation $\tan\theta(Re_*)$ and if not satisfactory we look for other significant local parameters. The findings are presented in the following paragraphs.

3. EXPERIMENTAL EQUIPMENT

The loop system used for experiments with the rectangular pipe is composed of [4]:
    - horizontal pipes of approximate 30 m length, with a section of rectangular pipe (25 mm x 100 mm)
        - a rubber lined slurry pump Allis-Chalmers (76.2 mm x 76.2 mm)
        - a mixture tank mounted before the pump
        - rubber diaphragm valves for flow control (50.8 mm diameter)
        - glass section for observation
        - measuring equipment for flowrate, pressure and pressure gradient, sampling, temperature. A Caesium 137 gamma emitter provided a means to measure the mean value of concentration at any level in the cross-section.

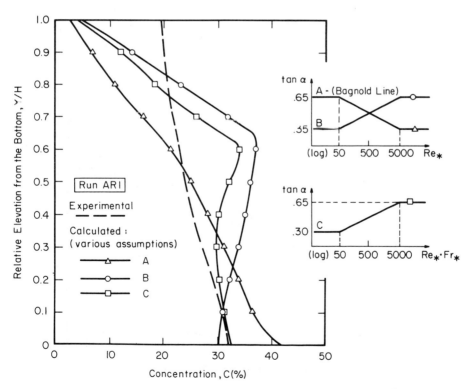

Fig. 5. Calculated concentration distribution using $\tan\alpha = f(Re_*)$/assumptions A and B/, and $\tan\alpha = f(Re_* \cdot Fr_*)$/assumption C/

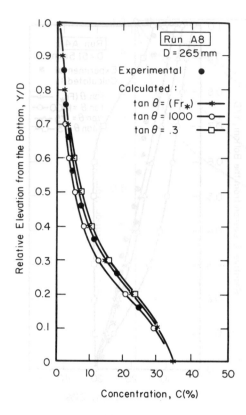

Fig. 10. Concentration distribution in the 263 mm circular pipe (Run A8)

α_M — mixing turbulence co-efficient

$\dfrac{\partial \sigma_{SL_y}}{\partial y}$ — partial derivative in the vertical direction of the supported load (acting normal to a plane y)

$\tan\beta$ — coefficient of Coulombic friction between solid particles

If there were no dispersive stress or Coulombic load (as approximately it is in very large pipes with fully turbulent flow and small particles) this would give:

$$\frac{dp}{dz} \simeq \frac{\rho_M}{\rho_L} \cdot \left(\frac{dp}{dz}\right)_L \tag{15}$$

where L denotes liquid alone, and M mixture.

The approach may be verified comparing the numerical predictions to the experimental distributions of the velocity, as presented in [1], [2]. In this paper we compare the head-losses using the boundary stress approach. The total solid content is divided into two fractions. Suspended load particles are considered to increase the boundary stresses because they increase the density and viscosity of the mixture. For sand of the size considered in this investigation, it is the density effect which is predominant in turbulent flows. 'Supported load' particles produces an additional stress proportional to their immersed weight. According to our analysis the coefficient of proportionality varies with the local flow conditions and the weight is transmitted between successive particle layers by the combined dispersive stress. At the pipe wall the pressure gradient can be expressed as follows:

$$R_h \cdot \frac{dp}{dz} = \frac{\rho_M}{\rho_L} \cdot (\tau_L)_w + (\tau_{DS})_w + (\tau_{SL})_w \tag{16}$$

where:

R_h — hydraulic radius of the cross-section

$(\tau_{DS})_w$, $(\tau_{SL})_w$ — the dispersive stress and supported load stress at the bottom wall

Rearranging the terms in equation (16)

$$\frac{i - (\rho_M/\rho_L) \cdot i_L}{C \cdot (S-1)} \simeq \frac{C_C}{C} \cdot \tan\alpha \qquad (17)$$

where:

C — average concentration in cross-section

C_C — concentration of particles whose weight is transmitted to the bottom by the combined dispersive stress

The friction coefficient $\tan\theta$ is determined by the ratio between the supported load and solid normal stress $k = \sigma_{SL}/\sigma_S$

$$\tan\alpha = k \cdot \tan\beta + (1-k) \cdot \tan\theta \qquad (18)$$

The experiments in the rectangular pipe chosen for test have large shear stresses, and then the ratio k tends to zero (i.e. $\tan\alpha \simeq \tan\theta$).

The left hand side in equation (17) can be computed directly from the data and plotted in terms of Fr_*. The conventional theory assumed $\tan\alpha$ is constant and C_C/C decreases with increasing velocity. However the data show that this is not the case and the product $(C_C/C) \cdot \tan\alpha$ increases with Fr_* while velocity increases (see Figure 11). This is in qualitative agreement to the proposed correlation $\tan\theta$ (Fr_*) (Figure 6). The fact that the variation continues

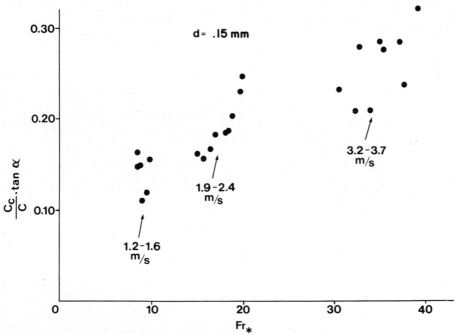

Fig. 11. Correlation $(C_C /C) \cdot \tan\alpha = f(Fr_*)$ from headlosses in the rectangular pipe

to higher Fr_* is due in part at least to the approximation which was used for ρ_M (ρ_M was calculated using total average concentration C).

The correlation $\tan\theta(Fr_*)$ was tested with good results for rectangular and circular pipe flow conditions (Table I). For moderate shear stresses and fine particle slurry flows $\tan\theta$ can be approximated to the average value .5. Comparisons made for fine sand .165 mm water mixture flows in pipes of diameter over 51.5 mm show a satisfactory agreement, as illustrated in Figure 12 (for concentration distributions) and Figure 13 (for headlosses).

6. CONCLUSIONS

1. The normal stress between two sheared layers of suspension has two components: one due to Coulombic contacts and other due to

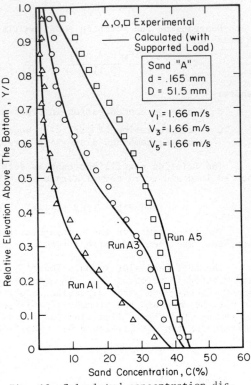

Fig. 12. Calculated concentration distribution ($\tan\theta$ = .5): Runs A1, A3, A5

turbulent collision interaction. Each component is related to a fraction of the solid shear stress by a different coefficient of dynamic friction ($\tan\beta$ and $\tan\theta$, respectively).

2. The coefficient of dynamic friction by particle impingement ($\tan\theta$) can be related to the local Froude number Fr_* (local ratio of the inertial force to gravitational force) as shown in Figure 6. Other terms in the momentum equation due to supported load and mixing effects were previously related to the same dimensionless number.

3. The dispersive stress σ_{DS} may be identified in the space/time average equation as a term containing the fluctuating part of the kinetic energy of solid particles. This fact would support a correlation between $\tan\theta$ and Fr_*.

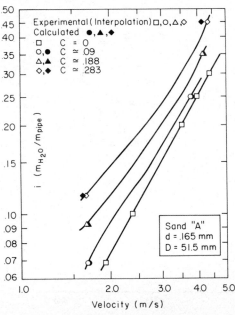

Fig.13. Calculated headlosses using $\tan\theta$ = .5: Runs A1-A6

4. The concentration distribution is determined by the dispersive stress σ_{DS} especially at high concentrations (C > 15%) and large shear stresses. The effects of the dispersive stress are smaller when the mixing influence (in large pipelines) or supported load (at velocities close to critical sediment values) are significant.

5. Headlosses should be computed in the boundary stress approach using the coefficient of dynamic friction by impingement $\tan\theta$ when the dispersive stress is predominant.

7. ACKNOWLEDGEMENTS

The studies reported here were supported by National Science Foundation Grant CPE 8205217 and Canadian NSERC Grant A1270.

REFERENCES

1. Roco, M.C. and Shook, C.A., "New Approach to Predict Concentration Distribution in Fine Particle Slurry Flow", AIChE 91st Annual Meeting, August, 1981, also Physico-Chemical Hydrodynamics (in press).

2. Roco, M.C. and Shook, C.A., "Modeling Slurry Flow for Different Particle Dimensions", Can. J. Chem. Engr. 1983 (in press).

3. Bagnold, R.A., Phil. Trans. Royal Soc., Series A, 1956, Vol. 249, No. 964.

4. Daniel, S.M., Ph.D. Dissertation, University of Saskatchewan, Saskatoon, September, 1965.

5. Wilson, K.C., "A Unified Physically-Based Analysis of Solid-Liquid Pipeline Flow", Proceedings Hydrotransport 4, paper A1, BHRA-Fluid Engineering, 1976.

6. Savage, S.B., "Gravitating Flow of Cohesionless Granular Materials in Chutes and Channels", J. of Fluid Mechanics, 1979, Vol. 92, p. 53.

7. Mandl, G., Luque, R.F., "Fully Developed Plastic Shear Flow of Granular Materials", Geotechnique, Vol. 20, 1970, p. 277.

8. Roco, M.C. and Balakrishnan, N., "Multidimensional Liquid-Solid Flow Analysis by Using an Integral Volume Approach", Annual Meeting of the Society of Rheology, paper D30, Chicago, November, 1982.

9. Televantos, Y., Shook, C.A., Carleton, A. and Streat, M., Can. J. Chem. Engr., 1979, Vol. 57, p. 255.

U_b = U_o $(\bar{z} = 1)$ = bottom velocity. Engelund (1974),
showed that the primary flow profile may be determined by an
equivalent sand roughness q for channels with plane bed covered with moving grains. His formula for U_b reads,

$$U_b = \sqrt{gmD} \; [1.9 + 2.5 \, Ln \, (Dq - 1)]$$

Engelund (1974), also established a useful alternate approximation to [7] as .

$$U_o = U_s \, \cos \, (\, \alpha \, \bar{z}) \, . \qquad\qquad [8]$$

where U_s = U_o $(\bar{Z} = o)$ = U_b + γ = surface velocity,

$$\alpha^2 = 14 \; (gmD)^{\frac{1}{2}} \; U_s^{-1}$$

The Secondary Flow :

Comparison of first order terms in [5] , gives eqns for secondary currents (u` , v` , w`) ,

$$\frac{\partial p'}{\partial x} = - U_o \, \frac{\partial u'}{\partial x} - w' \, \frac{d \, U_o}{dZ} + \nu \, \frac{\partial^2 u'}{\partial z^2} \; ,$$

$$\frac{\partial p'}{\partial y} = - U_o \, \frac{\partial v'}{\partial x} + \nu \, \frac{\partial^2 v'}{\partial z^2} \; , \qquad [9]$$

$$\frac{\partial p'}{\partial z} = - U_o \, \frac{\partial w'}{\partial x} \; .$$

$$\frac{\partial u'}{\partial x} + \frac{\partial v'}{\partial y} + \frac{\partial w'}{\partial z} = o$$

The above equations are subjected to boundary conditions at
the channel surface, bottom and side walls. The associated
boundary value problem is solved in the next section.

III. ANALYSIS OF THE SECONDARY FLOW :
----

1. The Governing Eqns :

The basic assumption is that the secondary flow velocity

components, and pressure vary sinusoidally in the longitudinal direction exactly the same manner as the side wall meandering . i.e.

$$P' = \rho U_b^2 \; P \; (\bar{y}, \bar{z}) \; e^{ikx}$$

$$U' = U_b \; U \; (\bar{y} , \bar{z}) \; e^{ikx}$$

$$V' = i \; U_b \; V \; (\bar{y} , \bar{z}) \; e^{ikx}$$

$$w' = i \; U_b \; W \; (\bar{y} , \bar{z}) \; e^{ikx}$$

[10 a]

$$d = D \; \bar{d} \; (\bar{y}) \; e^{ikx}$$

$$\eta = D \; \bar{\eta} \; (\bar{y}) \; e^{ikx}$$

where $\bar{y} = \dfrac{y}{D}$, $\bar{z} = \dfrac{z}{D}$. $P, U, V, W, \bar{d}, \bar{\eta}$ are unkown complex non-dimensional functions.

Inserting these expressions in [9] ,

$$\varepsilon \; \frac{\partial^4 W}{\partial \bar{z}^4} + \frac{\partial^2 W}{\partial \bar{y}^{-2}} + \frac{\partial^2 W}{\partial \bar{z}^2} = [(kD)^2 + \frac{U_{ozz}}{Uo}] \; W \qquad [10\ b]$$

$$\varepsilon \; \frac{\partial^2 V}{\partial \bar{z}^2} + V = \frac{1}{U_o} \int_0^{\bar{z}} U_o \; \frac{\partial w}{\partial \bar{y}} \; d\bar{z} + \beta \; \frac{dP}{d\bar{y}} \; (\bar{y}, o)$$

where $\varepsilon = i\nu \; (kD^2 U_o)^{-1} \qquad \beta = U_b \; (KDU_o)^{-1}$

and $\qquad U_{ozz} = \dfrac{d^2 U_o}{dz^2}$

2. The Boundary Conditions

[a] At the bottom $Z = D$ $(\bar{Z} = 1)$, the boundary conditions are :

(i) The river bed is impermeable

$$W \; (\bar{y}, 1) = - KD\bar{d} \; (\bar{y}) \qquad [11]$$

(ii) The horizontal velocity equals U_b i.e.
$(u' + U_o) = U_b$ at $\bar{Z} = 1$

or
$$\frac{\partial^2 W}{\partial z^2} + 2c\,\frac{\partial w}{\partial z} + c\,\frac{\partial V}{\partial y} = \frac{W}{U_b}\left(2d\,Uo + \frac{d^2 Uo}{dz^2}\right) \quad \text{at } \overline{z} = 1 \tag{12}$$

where $c = D\sqrt{gmD}\,\left(\nu\left(1.9 + 2.5\,\text{Ln}\,Dq^{-1}\right)\right)^{-1}$

(iii) The shear stress vector has the same direction
as the velocity vector

$$\frac{\partial V}{\partial z} = -cV \qquad \text{at } \overline{z} = 1 \tag{13}$$

[b] At the channel surface ($\overline{z} = o$). the boundary conditions are :

(i) The velocity vector is tangent to the surface

$$W\,(\overline{y}.o) = \frac{Us}{U_b}\,\overline{\eta}\,(\overline{y})\quad KD \tag{14}$$

(ii) The (dynamic) pressure vanish

$$P\,(\overline{y}.o) = -\,\overline{\eta}\,(\overline{y})\,F_r^{-2} \tag{15}$$

where $F_r^2 = U_b^2\,(gD)^{-1}$

(iii) The shear stress vanishes.

$$\overline{\eta}\,(\overline{y})\,\frac{d^2 Uo}{d z^2} = -\,U_b\,\frac{\partial W}{\partial z} \qquad, \overline{z} = o \tag{16}$$

and
$$\frac{\partial V}{\partial z} = O \qquad\qquad \overline{z} = o$$

[c] At the side walls $y = \pm B$. $(\overline{y} = \pm \frac{B}{D})$. walls are
impermeable,:

$$V\left(\frac{B}{D}, \overline{Z}\right) = V\left(-\frac{B}{D}, \overline{Z}\right) = O \tag{17}$$

3. The Solution

Using the modified Bernoulli method. the solution is expressed in series.

$$V\,(\overline{Y}.\overline{Z}) = \sum_{n=0}^{\infty}\,F_n\,(\overline{y})\,G_n\,(\overline{Z}) \tag{18}$$

$$W\,(\overline{y}.\overline{Z}) = \overline{y}\,f\,(\overline{z}) + \sum_{n=0}^{\infty}\,H_n\,(\overline{y})\,R_n\,(\overline{Z})$$

from [17] and [11] gives. $\dfrac{\partial W}{\partial \overline{y}} = f\,(\overline{z})$ at $\overline{y} = \pm\,\dfrac{B}{D}$ [19]

from [17] and [18] shows that

$$Fn(\overline{y}) = \frac{dH_n}{d\overline{y}} = 0 \qquad [20]$$

$$at \ \overline{y} = \pm \ B/D$$

Substituting [18] into [9] using [20] get

$$Fn(\overline{y}) = \cos(\overline{y} \ e_n D)$$

$$Hn(\overline{y}) = \sin(\overline{y} \ e_n D) \qquad [21]$$

where

$$e_n = \pi(n + \tfrac{1}{2})B^{-1}$$

Next expand all the non-homogeneous terms in [11] in series of $F_n(\overline{y})$ and $H_n(\overline{y})$. A set of non-homogeneous ordinary Differential Equation are obtained to determine $G_n(\overline{z})$ and $R_n(\overline{z})$. The following approximation are introduced to facilitate this process :

(i) based on eqn [8], $\dfrac{d^2Uo}{dz^2} \ Uo^{-1} = -\alpha^2 = $ constant $\qquad [22]$

(ii) Based on eqn [7], $\dfrac{d^2Uo}{dz^2} \ U_b^{-1} = - \overline{\alpha}^2 = $ constant $\qquad [23]$

hence $\overline{\epsilon}^1 = \dfrac{kD^2Uo}{i\nu} = \dfrac{kD^2}{i\nu} \ \dfrac{\overline{\alpha}^2}{\alpha^2} = $ constant

(iii) For natural meandering flows ,

$$kD \ll 1 \quad and \quad U_s^2 \ll gD$$

Now the boundary conditions [11]-[16] are expanded in series of $F_n(\overline{y})$ and $H_n(\overline{y})$. The coefficients of expansion are termed C_{1n} , C_{2n} , C_{3n} , C_{4n} , C_{5n} , C_{6n} respectively. The final solution is obtained in terms of these coefficients. The procedure is standard but tedious. Hence we suffice by writing the final solution.

$$V(\overline{y},\overline{z}) = \sum_{n=0}^{\infty} F_n(\overline{y}) [C_{1n} e^{s\overline{z}} + C_{2n} e^{-s\overline{z}}] - \frac{kDUo}{U_b} \cos(\frac{\pi D}{B} \overline{y})$$

$$- \frac{1}{s^2} \ \sum_{n=0}^{\infty} \ An \qquad [24]$$

where $S = (M\Theta)^{\frac{1}{2}}$,

$$M = ikD^2U_b \nu^{-1} \ , \qquad \Theta = \overline{\alpha}^2/\alpha^2$$

$$An = e_n kDM \ [\ \Theta^2 + \frac{\alpha^2}{M} \Theta^2 + \frac{\overline{\alpha}^2}{M} + (1+\Theta^2)(e_n D)^2 - (\frac{\pi D}{B})^2) tn]$$

$$+ e_n D^2 g \ Rn(o)/(i\nu kUs)$$

$$tn = ((k^2 + e_n 2)D^2 - \alpha^2)^{-1}$$

and

$$W(\bar{y},\bar{z}) = 2 \, k \, \bar{\alpha}^2 \, \frac{B}{\pi} \, \bar{z} \, \sin \left(\frac{\pi D}{B} \, \bar{y} \right) \qquad [25]$$

$$+ \sum_{n=0}^{\infty} Hn(\bar{y}) \, [C_{3n} \, e^{r_n \bar{z}} + C_{n4} \, e^{-r_n \bar{z}}$$

$$+ C_{n5} \, e^{h_n \bar{z}} + C_{n6} \, e^{-h_n \bar{z}} \,] - \bar{z} \sum_{n=0}^{\infty} \, Jn$$

where

$$r_n = [\frac{s^2}{2}(1+Ln)]^{\frac{1}{2}} \quad ,$$

$$h_n = [\frac{s^2}{2}(1-Ln)]^{\frac{1}{2}} \quad .$$

$$L_n = [1- 4 \, s^2 tn]^{\frac{1}{2}} \quad ,$$

and

$$J_n = \frac{16 \, (-1)^n \, kB \, \bar{\alpha}^2}{\pi^2(1-2n)(3+2n)}$$

Using the mass conservation eqn [9], U as defined [10], is determined by.

$$U = - \frac{1}{kD} (\frac{\partial v}{\partial \bar{y}} + \frac{\partial w}{\partial \bar{z}}) \qquad [26]$$

Hence. if the channel bed is fixed, the flow pattern is now completely defined as follows ;

(i)   The primary flow is determined by [7] or [8].
(ii)   The necessary flow is determined by [10],[24],[25],[26].

IV   Sediment Distribution and Transport

Consider a sediment laden flow in an open channel which meander in a manner similar to the one described so far.   Two types of sediment transports are considered : (i) transport caused by the gradient of sediment concentration and, (ii) the transport by the fluid flow.   Under steady conditions, the average concentration of suspended sediment, C, is constant. Sediment conservation equation is given by, chin and Mcsparran (1966)

$$\frac{\partial}{\partial y}(Cv) + \frac{\partial}{\partial z} \, [(w+v_s)C] = \frac{\partial}{\partial y} \, (\epsilon_y \, \frac{\partial C}{\partial y}) + \frac{\partial}{\partial z} \, (\epsilon_z \, \frac{\partial C}{\partial z}) \qquad [27]$$

where $\epsilon_y . \epsilon_z$ are the y and z components of the diffusion coefficient for sediment transport. and V_s = setting velocity of the representative particle under gravity.

Invoking von Karman's similitude theory and Prandtl mixing length theory.

$$f_2 = a\ constant \doteq - 1.73$$

Linearizing [40] using [36] ,

$$\Delta_y = \frac{f_2}{D}\ \frac{d\bar{d}}{dy} + \frac{v(\bar{y},1)}{U_b} \qquad\qquad [41]$$

Inserting [39] and [41] into [37] get,

$$\frac{d^2\bar{d}}{dy^2} = - \frac{1}{f_2}[\frac{d\bar{d}}{dy} + kDj\ \frac{U_b}{\bar{u}}\ \bar{U}\ ] \qquad\qquad [42]$$

eqn [42] determines the variation of the bed elevation $\bar{b}$ in the transverse direction. The boundary condition is that the transverse sediment transport should vanish along the side walls. Using the fact that the transverse velocity vanishes at the side walls,together with eqn [41],gives,

$$\frac{d\bar{d}}{dy} = 0 \qquad at\quad \bar{y} = \pm\ \frac{B}{D} \qquad\qquad [43]$$

solving [42] and [43]

$$\bar{d}(\bar{y}) = B_1 e^{-y/f_2} + B_2 + G\ (y) \qquad\qquad [44]$$

where B_1 , B_2 , $G(y)$ are to be determined according to the real situation at hand.

VI.   CONCLUSION :

       We have presented an analytical model of which simulates the manner by which strong secondary currents in meandering rivers affect both suspended sediment distribution and variation of bed topography. The flow is decomposed into depth-dependent longitudinal primary current and smaller,by an order of magnitude, secondary currents in the three directions. The boundary value problem for the secondary currents are solved in terms of infinite series. The variation in the longitudinal direction of all physical quantities is sinusoidal. The obtained currents are introduced into the conservation equation of the sediment concentration. Constant concentration curves in steady flow are obtained. Bed elevation variation due to secondary currents is cast into a second order differential-Equation whose solution gives the bed topography variation.

Multi-
edited
Elsevie

EXPEF
RECUF

Ibrah
Milit
Kobry

ABSTR

vario
have
consi
in th
as th
rated
liqui
tube
ments
gatio
iour

1.

sion,
tanks
reduc
impor

ing te
fixat:
dragg
Usual
requi
troub

overc
built
of the

2.

REFERENCES

[1]   Chiu.C and Mcsparran.J.(1966),"Effect of Secondary Flow
      on Sediment Transport", J.Hyd.Div.,ASCE.HY5.

[2]   Chiu.C.Hsiung.D.E.and R.C.,(1978),"Three-dimensional
      Open Channel Flow". J.Hydr.Div.,ASCE.HY8.

[3]   Chiu.C.,Hsiung,D.E. and Lin,R.C.(1978), "Secondary Cur-
      rents Under Turbulence in Open Channels of Various Geo-
      metrical Shapes", Proc. XVII th. Congress of IAHR.

[4]   Chiu.C.,and Hsiung D.E.,(1981),"Secondary Flow,Shear
      Stress and Sediment Transport",J.Hydr.Div., ASCE.HY7.

[5]   Engelund.F.A.,(1974),"Flow and Bed Topography in Chan-
      nel Bends",J.Hydr.Div.,ASCE.HY11.

[6]   Einstein H.A. and Li.H.,(1958),"Secondary Flow in Straight
      Open Channels", Trans.AGU. Vol. 39.

[7]   Einstein H.A.,and Chien.Ning."Effects of Heavy Sediment
      Concentration Near the Bed On Velocity and Sediment Dis-
      tribution", MRD Sediment Series No.8,Univ.of Calif.,1955.

[8]   Gottlieb.L.,(1974),"Flow in Alternate Bends",Prog.Report,
      Tech..Univ.of Denmark.

[9]   Gottlieb.L.(1976),"Three-dimensional Flow Pattern and
      Bed Topography in Meandering Channels",Series paper 11.
      Tech.Univ.of Denmark.

[10] Yen.B.C.(1972),"Spiral Motion of Developement Flow in
      Wide.Curved Open Channels", Sedimentation Sym.,Col.Sta-
      te Univ.,U.S.A.

APPENDIX

The following symbols are used in the paper :

x,y,z = the longitudinal.transverse and vertical coordi-
               nates respectively.

u,v,w = the local velocity components in the x,y,z direc-
               tions.

u',v',w' = the perturbation of secondary currents.

U_o = the depth-dependent primary flow.

U_b,U_s = the value of U_o at bottom and surface.

Po,P = the hydrostatic and the dynamic pressures.

518

sediments in water flumes. It consists of a down slotted tube
embedded in these sediments, (Fig. 1). The tube is divided to
main three parts:

- Entrance part, opens in the clear water, such that it
  ensures water flow inside the tube.
- Active slotted part,embedded in the settled sediments.
- Outlet part, transports out the recuperated solid-liquid
  mixtures.

As the water is allowed to flow through the tube, a pressure
difference arrises, between the inner and outer sides of the tube.
This difference creates seepage flow across the sedimentary mass.
The seepage flow, against the gravity forces,results in a liqui-
dized solid-liquid layer near the pipe slot, (3). This provokes
the solids entrainement by the water flow inside the tube. It
should be noticed that the entrained solids, increase the pressure drop in the tube, which accelerates the solids entrainement
process. The solid-liquid mixture is then, carried out through
the tube to the delivery side of the system.

The preliminary investigation of this system, had shown that
the slot has to be situated in the down side of the tube. This
position prevents the solids to fall inside the tube, and clog it,
if the water flow rate decreases or even stopped.

3. EXPERIMENT

To obtain the main parameters that influence the system

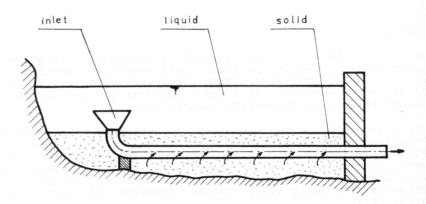

inlet   liquid   solid

Fig. 1. Proposed Recuperating System

EXPERIMENTAL FACILITIES

Figure (1) shows a schematic layout of the testing facility used for measuring the head, discharge and efficiency characteristics of irrigation pumps at different levels of mud concentrations.

In the figure, the pump 'P' which is under consideration, is driven by a variable speed DC motor–dynamometer 'D' which is used to control the pump speed at any desired value and at the same time, measure its input mechanical power.

The pump circulates a mud-water mixture in a closed-loop fashion by taking the pre-mixed mixture from the storage tank 'T', via a streamlined bell-mouthed suction pipe, and pumping it through the delivery pipe which delivers the mixture back into the tank. The delivery pipe incorporates a short venturi flow meter 'FM' to monitor the flow rate of the mixture and a flow control valve V_2 to regulate this flow at any desired rate. The outlet of the delivery pipe is directed towards the tank bottom to stir any settled mud and at the same time avoid any direct disturbance to the pump inlet. A screen 'S' is placed between the delivery and suction pipes in order to dampen the water oscillation and straighten the flow before entering into the pump. The pump suction and delivery heads are measured by the manometers M1 and M2 to determine the pump's total manometric head at any pump discharge.

The tank is provided by a fresh water supply pipe 'p_s', a drain pipe 'p_d' and a shut-off valve V_1 to allow for changing the tested pump without emptying the tank.

The described facility is used to test three different pumps whose inlet/outlet pipe diameters are 2/1½ in. and 6/6 in.

EXPERIMENTAL RESULTS

a. Head-Discharge Characteristics

Figure (2-a) shows the effect of varying the pump discharge and mud concentration level on the pump manometric head in meters of the mixture for a pump that has an inlet/outlet pipe diameter ratio of 2/1½ in. The pump normal operating head and discharge are 12.5 meter of water and 8 lit/sec, respectively, at a maximum efficiency of 68%. The figure indicates a significant drop in the pump manometric head, at any given discharge, as the mud concentration is increased. Such a result conforms with the trends reported in References [1 - 4] for other solids-water mixtures.

Similar results are also obtained for two larger size pumps that have inlet-outlet pipe diameter ratio of 6/6 inches but operate at optimal discharges of 55 and 35 lit/sec, respectively, while their manometric head is maintained at 5 meters of water. Figures (2-b) and (2-c) show the corresponding characteristics of these two pumps at different values of mud concentrations.

These two figures indicate that, for any given discharge, the pumping head drops with increasing values of mud concentrations but not as much as in the case of the small pump of figure (2-a).

The three pumps, which have been subjected to extensive testing in this study, are selected on purpose to cover the high head - low discharge range, the medium head - high discharge range as well as the low head - medium discharge range. This ensures a good understanding of the behavior of a wide spectrum of irrigation pumps in muddy waters.

The obtained head-discharge characteristics are replotted in the head ratio-concentration plane (H_R-C) for different values of pump discharge. The head ratio H_R, being equal to the ratio of the head developed on the mud slurry expressed in meters of mud to the head developed when pumping water at the same flow rate [1 - 4], serves as a measure of deterioration of pump performance with the mud concentration.

Figures (3-a), (3-b) and (3-c) show the resulting H_R-C characteristics for the three pumps under consideration. Such characteristics indicate that, for a particular pump, operating at a constant discharge, the head ratio H_R drops in almost a linear fashion with increasing values of mud concentrations. This result conforms with the general trends reported in references [1 - 4] for other solid-water mixtures.

However, the figures also show that the linear relationship between H_R and C depends also on the pump flow rate. Such dependence is manifested by the fact that for a particular value of C, the head ratio H_R drops considerably with increasing values of pump discharge. Similar finding has been reported by Cave [4] for tests carried out on a 12 in. gravel pump.

A more general correlation is obtained as shown in Figure (4) by plotting the slopes of the H_R-C linear characteristics against the dimensionless discharge ratio Q_R. The figure suggests that most of the obtained results fall on a single characteristic curve. With appropriate curve fitting, it is found that the equation of this curve can be adequately expressed by:

$$\frac{dH_R}{dC} = 0.02 \ (1 + Q_R^2) \tag{1}$$

where dH_R/dC - slope of the H_R-C characteristics. Accordingly, the head ratio H_R can be expressed as follows:

$$H_R = 1. - 0.02.C.(1 + Q_R) \tag{2}$$

This equation can be used to predict the performance of similar irrigation pumps when handling muddy waters of concentration up to 15% by volume.

b. Efficiency-Discharge Characteristics

Figures (5-a), (5-b) and (5-c) show the effect of varying the mud concentration on the overall efficiency of the three pumps under consideration at different values of flow rates. The figures indicate a significant drop in the efficiencies of the three pumps with increasing levels of mud concentration to the extent that the peak efficiencies drop to nearly half their values when the mud concentration is increased from 0% to 15%. Such a considerable loss in the pumping efficiency is definitely alarming and must be considered during the selection process of the irrigation pump-motor set.

The figures also show that the peak efficiencies at increasing values of mud concentrations occur at decreasing values of pump discharges. Such a phenomenon can be attributed to the fact that the high-mud concentration mixtures produce higher hydraulic friction losses at high discharges. Accordingly, shifting the peak efficiencies to lower discharges.

The loss in the pumping efficiency due to the presence of mud is measured by the efficiency ratio $\overline{\eta}$ which is set equal to the ratio of the efficiency when pumping a specific mud-water mixture to the corresponding efficiency when pumping water at the same flow rate.

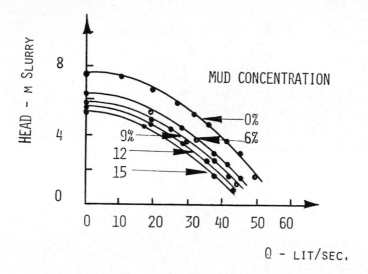

Figure(2-c)- Effect of mud concentration on head-discharge characteristics of a pump with Q_o = 35 lit/sec.

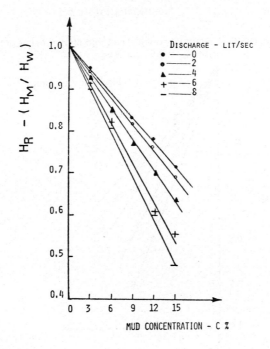

Figure(3-a) - The head ratio-mud concentration characteristics for a Pump with Q_o = 8 lit/sec.

534

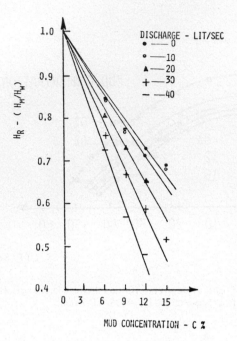

Figure(3-b)- The head ratio-mud concentration characteristics for a Pump
with Q_o = 55 lit/sec.

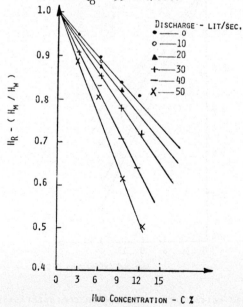

Figure(3-c)- The head ratio-mud concentration characteristics for a Pump
with Q_o = 35 lit/sec.

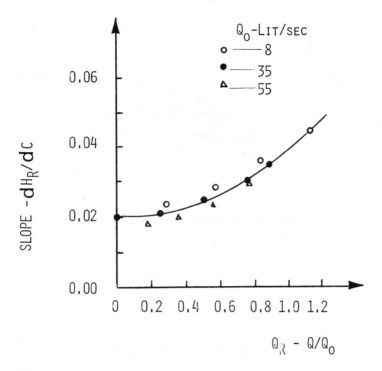

Figure(4) - Generalized dH_R/dC - Q_R Characteristics for the tested
irrigation pumps.

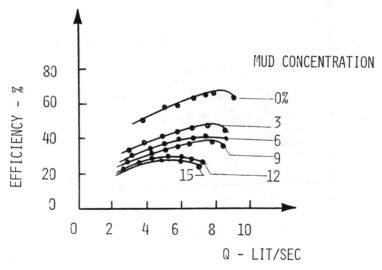

Figure(5-a)- Effect of mud concentration on Efficiency-discharge Characteristics for a pump with Q_O = 8 lit/sec.

mud have an effect which lies inbetween these two extremes.

CONCLUSIONS

This study has presented an experimental evaluation of the performance of irrigation pumps while handling muddy waters. The obtained results are extremely important for such a widely used class of pumps which have been considered for a long time as conventional pure water pumps.

The results show that the head and efficiency ratios, of a group of different dimension pumps, depend not only on the mud concentration but also on the pump discharge.

General correlations are developed from the experimental data which can be used to predict the performance of similar pumps when handling mud and other solid-water mixtures. Examples are given for coal, sand and ilmenite in order to provide comparisons with mud.

Deterioration of pump performance with running time is found to have a significant effect on the pump shut-off head and also at high discharges.

The presented results would provide quantitative means that are essential to the proper selection and operation of irrigation pumps in muddy waters.

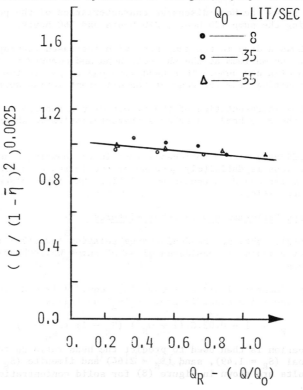

Figure(6) - Generalized $C/(1-\bar{\eta})^2$ -Q_R characteristics for the tested irrigation pumps.

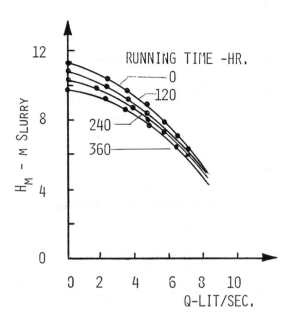

Figure(7-a)- Effect of running time on pump H-Q characteristics for a
pump with Q_o=8 lit/sec. and C = 15%.

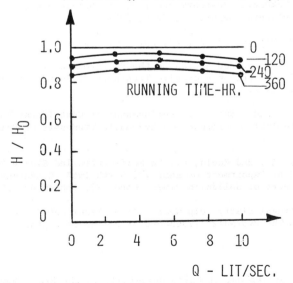

Figure(7-b)- Effect of running time on dimensionless H-Q characteristics
for a pump with Q_o = 8 lit/sec and C = 15%.

A viable latent heat storage arrangement which has been suggested is the containment of the phase change material within rectangular cavities [2]. In one such arrangement, the top or bottom surface of the cavity is heated (cooled) by a working fluid whose energy is to be stored (withdrawn). Charging the storage unit causes inward melting of the medium contained in the cavity; conversely, discharging the unit causes inward solidification of the material. The onset of thermally induced convection in the liquid can be caused by the presence of either a horizontal temperature or a negative (downward) vertical temperature gradient. Freezing in two-dimensional rectangular cavities has been studied both analytically [3-5] and experimentally [2,5,6]. Melting [7] and solidification [8] in the presence of natural convection has also been studied analytically. An extensive and current survey of literature is available [1].

In this paper, experiments are described which examine the fundamental heat transfer processes that accompany the freezing or melting of a phase-change material contained in an externally cooled or heated rectangular cavities. The work was motivated by current interest in latent heat-of-fusion energy storage devices in which a phase-change material is encapsulated within the container. The results obtained here may also be relevant to other applications such as the freezing of foodstuffs and biological materials and to the field of metallurgy in general. Heat transfer during freezing and melting was studied by imposing two different types of thermal boundary conditions on the phase-change material: 1) uniform temperatures at either the top or bottom of a two-dimensional rectangular cavity and 2) uniform temperatures at the vertical sides and bottom of a two-dimensional rectangular cavity.

In this investigation, experiments were performed with a phase-change material (99-% pure, n-octadecane) contained in the two types of cavities by imposing isothermal surface temperature boundary conditions. The solid-liquid interface position was determined photographically. The convective motion in the test cells during melting was visualized by using aluminum powder as a flow tracer. The local heat transfer coefficients along the heated walls during melting were determined by a shadowgraph technique. Numerous melting and freezing experiments for different externally imposed thermal boundary conditions were performed and are discussed. Instantaneous, heated-surface-averaged heat transfer coefficients were determined from a knowledge of the material melted. The experimental data are presented in dimensionless form in terms of relevant dimensionless groups. Empirical correlations are given for volume of material melted and the average heat transfer coefficient. Where possible, the predicted solid-liquid interface positions are compared with experimental data.

2. MELTING AND FREEZING IN A RECTANGULAR TEST CELL HEATED OR COOLED FROM THE TOP OR BOTTOM

2.1 Experiments

2.1.1 Test Cell and Instrumentation

The melting and freezing experiments were performed in a rectangular test cell whose inside dimensions were 89mm high, 76mm wide, and 38mm deep. The top and bottom of the cell served as the heat source/sink they were made of copper in which numerous channels were milled for the purpose of circulating the working fluid. All of the vertical sidewalls were made of Plexiglas. The two vertical sidewalls which supported the cell were made of 1.27 cm thick plate. For better insulation, the front and back walls were made of 0.318 cm and 0.635 cm thick Plexiglas plates and had a 0.318 cm air gap between them. The entire test cell was placed in a Lexan box in which the air temperature was maintained at a constant value close to the fusion temperature.

Five copper-constantan thermcouples were epoxyed into small holes close to the heat exchanger surface in both the top and bottom of the test cell. These thermocouples were used to measure and check the uniformity of the surface temperature of the heat exchanger. A total of eighteen copper-constantan thermocouples (0.127 mm in diameter placed in a 1.27 mm OD stainless steel tube) were located at the center of the test cell. They were spaced equally in order to measure the vertical temperature distribution. The thermocouple rack could be removed when it was not in use. A vertical slot 1.59 mm wide by 76 mm high was milled in one of the vertical sidewalls to allow for the insertion of a movable thermocouple probe in the liquid region. To prevent the PCM from freezing at the outside of the vertical slot, an electrical (small diameter cable) heater was immersed and heated.

2.1.2 Experimental Procedure

In both the freezing and melting experiments, research grade (99% pure) n-octadecane was used. The PCM has well-established thermophysical properties and a fusion temperature close to the ambient. In the experiments, great care was taken to degasify the liquid PCM. The degasification was accomplished by simultaneously heating and pulling a vacuum on the liquid for two hours. Provisions were also made to avoid entrapping any air bubbles during the filling process. A very small amount of aluminum powder was added to the liquid n-octadecane as a tracer for flow visualization. A uniform initial temperature (close to the fusion temperature with little superheating or supercooling) of the PCM in the test cell was established before each experiment, and only a single phase was allowed to exist. This was accomplished by circulating a constant temperature fluid through both heat exchangers at slightly above or below the fusion point for a sufficient period of time. Melting and solidification was initiated by switching one of the heat exchangers to another constant temperature bath preset at a different temperature.

For the two-dimensional observation of flow patterns (i.e., at a plane perpendicular to the z-axis), a slit (0.318 cm wide) was illuminated with a parallel beam from a mercury light source. The light beam external to the region was blocked with an aluminum foil which was fixed in a frame and which was movable in the direction of observation. This made flow visualization possible not only in the central region but also near the front or back walls of the cell.

The solid-liquid interface position and the flow patterns were photographed at predetermined time intervals. The selection of the shutter speed and of the aperture size of the camera was established by trial and error. The melted or frozen areas were measured with a planimeter.

2.3 Analysis of Solid-Liquid Interface Motion

2.3.1 Physical Model and Basic Equations

Even though great care was excersized to minimize heat losses (gains) from the test cell, they could not be eliminated completely. As a result of the temperature gradient in the horizontal plane, one expects natural convection to be established in the melt. The prediction of transient natural convection in a three—dimensional system with a moving phase—change boundary is a formidable problem which does not appear to have been solved. Therefore, the present analysis was restricted to predicting only the interface position through the use of a simple one—dimensional model.

If heat losses (gains) from the sides and edges can be neglected, heat transfer can be treated as one—dimensional, and the solid—liquid interface during stable conditions (i.e., melting from above and freezing from below) can be predicted from the classical Neumann model [1,9] for phase—change heat transfer.

For unstable configurations (i.e., melting from below and freezing from above) one expects, natural convection to be present in the melt after the critical Rayleigh number has been exceeded. As the phase change continues, the geometry (aspect ratio) of the liquid region will change. The solid—liquid interface is not expected to be planar because of the natural circulation patterns established in the melt. The rate of melting (freezing) will be controlled by the relative magnitude of heat transfer to each side of the solid—liquid interface and the latent heat—of—fusion absorbed (liberated) at the interface. We assumed that phase—change heat transfer is one—dimensional, and that all of the thermophysical properties, except the density in the buoyancy term, are constant. The motion of the solid—liquid interface owing to the expansion or contraction during phase transformation were accounted for. In the absence of better data, the natural convection heat transfer in the liquid at the interface was modeled by steady—state heat transfer coefficients in cavities in the absence of phase change [10]. This idealization may not be justifiable during rapid phase change, but for small interface velocities, the flow may be considered quasi—steady because it takes a much shorter time to establish natural convection in the liquid than to establish the temperature distribution in the solid.

The nonlinearity introduced by the motion of the solid—liquid interface and the natural convection in the liquid precluded closed—form analytical solution of the model equations. Among the different techniques available for solving the problem [1,9], the integral method was adopted because of its simplicity. Here, we consider only the case of freezing from above because it is identical to melting from below.

In dimensionless form, the energy balance at the interface ($\xi = \delta$) becomes

$$\frac{d\delta}{d\tau} = \frac{\partial \Theta}{\partial \xi}\Big|_\delta + (\frac{k_\ell}{k_s})\frac{\overline{Nu}_1}{(1-\delta)}(\Theta_{wb} - 1) \tag{1}$$

where

$$\overline{Nu}_1 = \overline{h}(1 - \delta)H/k_\ell \tag{2}$$

The energy equation in the solid can be expressed as

$$Ste\frac{\partial \Theta_s}{s\partial \tau} \doteq \frac{\partial^2 \Theta_s}{\partial \xi^2} \tag{3}$$

The initial, boundary, and interface conditions become, respectively,

$$\Theta_\ell = \Theta_o \quad , \quad \tau \leqslant 0 \tag{4}$$

$$\Theta_s = 0 \quad , \quad \xi = 0 \tag{5}$$

$$\Theta_\ell = \Theta_{wb} \quad , \quad \xi = 1 \tag{6}$$

$$\Theta_\ell = \Theta_s = 1, \quad \xi = \delta \tag{7}$$

Assuming a second-order polynomial in ξ for the temperature distribution,

$$\Theta = \mathcal{P}(\tau)(\xi/\delta) + [1 - \mathcal{P}(\tau)](\xi/\delta)^2 \tag{8}$$

which satisfies the temperature boundary conditions, the equation for the interface velocity can be solved. The unknown function $\mathcal{P}(\tau)$ can be found by the following procedure [11]. Differentiation of Eq. (7) with respect to time τ at $\xi = \delta$ yields

$$\frac{\partial \Theta_s}{\partial \xi} \frac{d\delta}{d\tau} + \frac{\partial \Theta_s}{\partial \tau} = 0 \tag{9}$$

We eliminate $d\delta/d\tau$ and $\partial\Theta_s/\partial\tau$ by combining it with Eqs. (1) and (3) to obtain the necessary condition to determine $\mathcal{P}(\tau)$,

$$\left[\mathrm{Ste}_s \left[\frac{\partial\Theta_s}{\partial\xi} \right] \left[\frac{\partial\Theta_s}{\partial\xi} \right] + \left[\frac{k_\ell}{k_s} \right] \frac{\overline{\mathrm{Nu}}_1}{(1-\delta)} [\Theta_{wb} - 1] + \frac{\partial^2\Theta_s}{\partial\xi^2} \right]_{\xi = \delta} = 0 \tag{10}$$

Substituting Eq. (8) into Eq. (10) and solving, we obtain

$$2 - \mathcal{P} = \left[-(b + c) + \sqrt{(b+c)^2 + 4c} \right]/2 \tag{11}$$

where

$$b = \left[\frac{\delta}{1-\delta} \right] \left[\frac{k_\ell}{k_s} \right] \overline{\mathrm{Nu}}_1 (\Theta_{wb} - 1) \quad ; \quad c = 2/\mathrm{Ste}_s$$

The final form of the differential equation for the solid-liquid interface position becomes

$$\frac{d\delta}{d\tau} = \frac{2 - \phi}{\delta} + \left[\frac{k_\ell}{k_s}\right] \frac{\overline{Nu}_1}{(1-\delta)}[\theta_{wb} - 1]$$

(12)

This equation can be solved numerically for the interface position δ.

2.4 Freezing From Below and Above

2.4.1 Flow Visualization

Flow visualization experiments clearly showed that natural convection was absent when freezing was from below. Since the fluid was stagnant, heat transfer in the liquid PCM was by conduction only. However, natural convection was established immediately after the start of solidification from above. Initially, a number of heated fluid parcels (thermals) individually rose from the heated bottom plate and moved toward the central region of the cell. A number of cooled fluid parcels descended from the cold top plate and moved toward each of the vertical side walls. After some time (Figure 1a), the heated thermals were observed to oscillate slowly from left to right like a plume in the central region of the cell. The cool thermals were observed to descend along the sidewalls. After this flow development period, a pair of two-dimensional convection rolls, with axes perpendicular to the longer dimension of the test cell, was established (Figure 1b). These convection rolls were symmetric and persisted for the duration of the freezing process. This flow pattern differed somewhat from the one observed for laminar natural convection in a cavity heated from below in the absence of a moving boundary caused by phase change [13]. The number of convection cells remained constant during the experiment and did not depend on the aspect ratio. This persistence may be attributed to the stabilizing effect of the freezing front. Near the front or the back wall, the velocity of the convection rolls was smaller than in the central region. This is attributed to the wall shear which retarded the fluid motion.

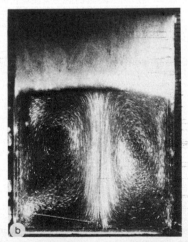

Fig.1 Flow visualization for freezing of n−octadecane from above, T_{wb}=30°C, T_{wt}=14.8 C: a) oscillation of a plume in central region, t=3 min 40 sec, b) well established natural convection cells t=327 min 25 sec.

   The freezing front was found to be flat initially and concave to the solid later.   During the initial stages of the freezing process, conduction is the dominant mode of heat transfer controlling the motion of the phase-change boundary.   The heat sink exerts a stabilizing influence on any local perturbation of the interface.   As the freezing progresses and the interface moves away from the sink, natural convection heat transfer gradually becomes of an order of magnitude comparable to conduction in the frozen layer because of its increased thermal resistance.   The interface becomes unstable, and natural convection eventually begins to play the dominant role in controlling the shape and motion of the interface.   It was found that later in the process, the interface became concave to the solid (Figure 2).   This occurred not only in the x-direction because of the convection rolls, but also in the z-direction because of a reduced intensity of natural convection heat transfer near the walls in comparison with that in the central region. Later, the motion of the interface was completely terminated.

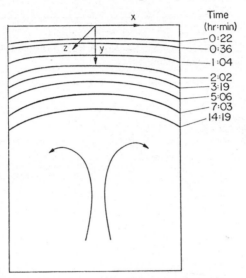

Fig.2 Solid-liquid interface position for freezing of n-octedecane from above, T_{wt}=19.8°C, T_{wb}=29.4°C.

2.4.2 Comparison of Measured and Predicted Interface Positions

   A comparison of measured and predicted solid-liquid interface positions during freezing from below and above is shown in Figures 3 and 4, respectively.   The agreement between the experimental data and the predictions is generally good.   For the freezing of a superheated liquid from below (Figure 3), the data from later in the process are above the predictions.   The measured growth rate of the solid crystal at these times is higher than the predicted one.   This may be attributed to the growth of dendrites overlying the freezing front.   The dendrites enlarge the total area of the solid-liquid contact, which is not taken into account in the analysis, leading to a higher rate of freezing.   This has also been observed by others [12].

548

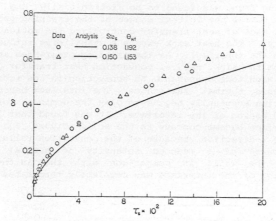

Fig. 3 Comparison of predicted and measured solid-liquid interface positions for freezing of n-octadecane from below.

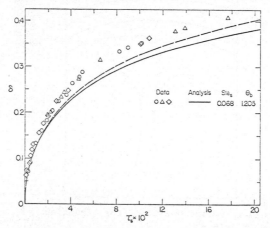

Fig. 4 Comparison of predicted and measured solid-liquid interface positions for freezing of n-octadecane from above. Dashed line denotes positions predicted using 10% lower Nusselt numbers.

For the freezing of a superheated liquid from above (Figure 4) the experimental data from later in the process (for three different runs) are also above the predictions based on the simple convection model described in Section 2.3.   In the model, the errors arise from neglecting the effect of the changing aspect ratio of the liquid-filled cavity and using the steady-state heat transfer coefficient in the absence of a moving boundary [10]. These errors may not be significant, provided that one-dimensional freezing is achieved. When some of the freezing experiments with identical superheat-ing in the liquid were repeated with either the same or different top plate temperatures, all the data points fell on the same line. This confirmed the results from the parametric study which showed that the prediction for the freezing front is not sensitive to the change in the Stefan number when the Stefan number is close to or less than 0.1. The discrepancy later in the process is due to the deviation of the freezing front from a flat surface.

At these freezing stages, the solid—liquid interface is concave to the solid, three—dimensional like a dome. The measured and averaged interface position is based on the freezing front at the front wall. This, of course, will overestimate the "average" interface position. However, both the experimental data and the predictions show that natural convection in the liquid can not only retard but can also eventually terminate the motion of the solid—liquid interface.

2.5 Melting from Below

2.5.1 Flow Visualization

In the experiments focusing on melting from below, after a critical Rayleigh number based on the melt layer depth was exceeded, a creeping motion was established in the fluid. Later, a number of small convection cells was observed. Several photographs taken at selected times are shown in Figure 5. Unfortunately, the convection cells could not be counted clearly from the photographs taken early in the process. The cells were small, and the fluid motion was not well enough organized to be clearly distinguishable with the flow visualization technique employed. The flow in the small convection cells (Figures 5a and 5b) was three—dimensional which produced a melting front distributed with hemispherical capped cells. The edges of the hemispherical cells were hexagonal in shape. This differed somewhat from the findings of others for the melting of ice [14] and stearic acid [15] from below. The hemispherical caps on the solid—liquid interface appeared shortly after natural (Benard) convection was initiated. Under the hemispherical cap,

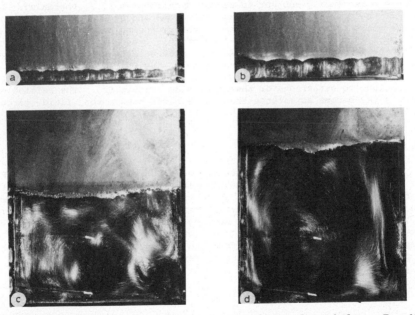

Fig.5 Flow visualization for melting of n—octadecane from below, $T_{wb}=46^{o}C$: a) Benard convection, t=8.25 min; b) growth of Benard cells, t=11 min., c) intense thermals produce horizontal mottion near the heated bottom and interface, t=87 min, d) counter clockwise unitary circulation, t=122 min.

the fluid was observed to rise in the central region of the cell and descend along the edges. Near the wall of the test cell, only one-half of the natural convection cell and one-half of the hemispherical capped cell were observed.

The total number of Benard cells (or the corresponding number of hemispherical caps) was found to depend on the temperature at the boundary (bottom plate) and the aspect ratio [(H−s)/W] of the liquid filled cavity formed on the top by the moving boundary. Unfortunately, the cells could not be counted clearly at every illuminated plane perpendicular to the direction of observation during the experiments because the cells were not always in the same plane. In general, it was found that the higher the temperature of the bottom plate and the smaller the aspect ratio, the larger was the number of the convection cells. For example, when the bottom temperature was 54°C, a maximum of 17 cells (determined by counting the hemispherical cells) was observed in the front side of the test cell immediately after natural convection was initiated. The size and survival of the convection cells depended on the depth of the melt layer. As a result of the increase in the depth of the melted layer, the number of the cells decreased as the neighboring cells merged (see Figures 5a and 5b). The merged cells were found to be distorted and elongated in the direction of the interface motion and did not appear to have the same structure as the original ones. The curvature of an individual cell was reduced as a result of the growing and combining of the convection cells in the melt layer.

As the melting progressed and the Rayleigh number increased, owing to the increase in the melt layer depth, the convection cells become unstable and were more readily subject to random disturbances (i.e. thermals). Later, some of the convection cells became distorted and broke down locally so that a large scale motion of flow driven by the ascending and descending of thermals may have taken over. The flow at this stage had no definite pattern. During the remaining part of the melting process, the heated (cooled) thermals which were released from the edge of the thermal boundary layer adjacent to the bottm plate (solid-liquid interface) showed a persistent horizontal motion to either one or the other side (Figure 5c). The release of the thermals may become so intense that at a given time, the fluid circulated counterclockwise in a unitary fashion (Figure 5d).

2.5.2 Temprature Distribution

Some typical temperature distribution measurements are shown in Figure 6. The initial temperature distribution in the solid was set close to the fusion temperature. Later in the phase transformation process the temperature in the liquid was uniform. This was due to the effective mixing by natural convection. A heated (or cooled) thermal boundary layer near the bottom plate (or the solid-liquid interface) where the temperature had a steep gradient could be observed from the temperature distributions later in the process. The experimental data confirmed the expectations.

2.5.3 Comparison of Data and Predictions

A comparison of the experimental solid-liquid interface position data with predictions based on the one-dimensinal convection model discussed in Section 2.3 is shown in Figure 7. The solid-liquid interface position predicted by the Neumann model is also included for the purpose of com-

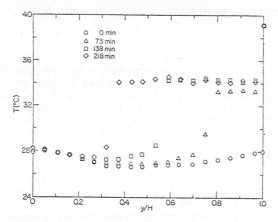

Fig.6 Temperature distribution at the centerline of the test cell during melting of n-octadecane from below, T_{wb}=37.4°C.

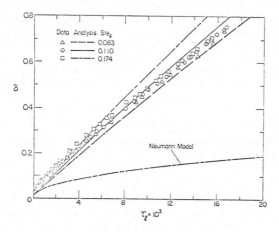

Fig.7 Comparison of measured and predicted solid-liquid interface positions during melting of n-octadecane from below.

parison. Only the data from experiments concerning the continuous melting of a flat or a near-flat interface were used. All the data points for different boundary temperatures tended to collapse into a single line and did not follow the line of the individual predictions, except for the case of Ste_l = 0.0827. This may have been due to the heat losses from the liquid PCM in the test cell to the ambient air in the Lexan box. The temperature in the box was set close to the fusion point of the PCM. The higher the Stefan number was, the higher was the operating temperature of the system, and the larger were the heat losses from the liquid in the cell. Hence, the discrepancy was greater between the data and the predictions than at lower source temperatures. The Neumann model [1] greatly underpredicted the rate of melting of the material.

3. MELTING AND FREEZING IN A RECTANGULAR, ISOTHERMAL WALL
   CAVITY

3.1 Test Cell

A rectangular test cell for the PCM was 150 mm high, 80 mm wide, and 50 mm deep. The upper surface of the material was bounded by an insulated air gap which accommodated the increase in volume caused by the melting and superheating of the melt. Melting or freezing occurred from two vertical sides and the bottom. The front and back faces of the test cell were made of Plexiglas to allow visualization, photography, and optical measurements. To reduce natural convection from the test cell to the ambient laboratory environment, a second Plexiglas plate was installed parallel to the first one. The air gap between the two vertical Plexiglas plates minimized heat transfer between the test cell and the ambient environment. The vertical walls and the bottom surface of the test region could be heated by circulating a working fluid from a constant temperature bath through multiple channels milled in the copper blocks which formed the heat exchanger. Different aspect ratio test cells could be made by filling the cell to different levels with the liquid and adjusting the height of the insulation so that there was a layer of air between the top of the upper surface and the insulation.

The surface temperatures of the cavity were measured by 16 copper-constantan thermocouples embedded along the inner surface of the heat exchanger. A shadowgraph system was employed to measure the local heat transfer coefficients along the heated surfaces of the cavity [16]. The shadowgraph system used was adapted from an earlier study [17], and the data reduction procedure was also identical to that described elsewhere [16,17].

3.2 Test Procedure

In the melting experiments, the test cell, free of paraffin and cleaned, was filled with degasified liquid n-octadecane (99% pure). It was given sufficient time to solidify and reach a uniform ambient temperature in a temperature-controlled laboratory environment. The initial temperature of the solid was maintained close to the melting temperature and was typically a maximum of only a couple of degrees Celsius below the fusion temperature. The insulation covering the faces of the test cell was removed to facilitate photography and was then immediately replaced. The maximum beam deflection in the shadowgraph system and the melt profiles were photorgraphed with a 35 mm camera.

In order to gain some qualitative insight into the development of convective motion in the melt zone during the course of melting, an extremely small amount of aluminum powder was used as a flow tracer. The powder was well mixed with the liquid PCM before the freezing of the liquid in the cavity was initiated. During the melting experiments, the test cell was illuminated from the front by two lamps located equiangularly with respect to the longitudinal axis of the test cell. The convective flow patterns were then observed and photographed. During flow visualization experiments, no heat transfer data were taken.

In the freezing experiments, the degasified liquid n-octadecane was either heated or cooled by circulating a working fluid through the passages in the test cell walls until the desired uniform temperature in the fluid was reached and all of the flow currents had died out. Freezing the PCM was ini-

tiated by circulating a cold working fluid through the walls of the cell.
The insulation covering the front and back of the test cell was removed for
the purpose of photographing the solid—liquid interface position and was then
immediately replaced.

3.3 Results and Discussion

3.3.1 Melting and Volume of Material Melted

The contours of the unmelted solid during a run were photographed, and
tracing the surface contour yielded the position and shape of the melt front
corresponding to the time of the run. Figure 8 illustrates the timewise
movement of the front position during inward melting in a rectangular cavity
having aspect ratios of 1.5. The contours displayed in the figure show that
early in the process (t = 11 min), the melt layer was practically uniform in
thickness over about 95 % of the cavity height. In the upper region, there
was greater displacement of the melt front from the wall. During the early
stages, the melting patterns were similar to those for melting from an isothermal vertical wall [18]. As time passed and heating continued, the front
moved inward in a nonuniform manner, and the unmelted solid shrank in height.
At the bottom of the cavity, the solid/liquid interface appeared to be irregular and remained so until about t = 90 min.

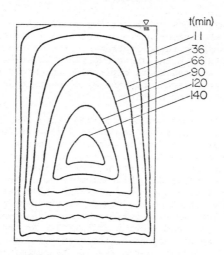

Fig.8 Solid—liquid interface contour variation during melting of n-
octadecane in a rectangular cavity, T_w=42°C, Ste_ℓ=0.133, AR = 1.5.

These findings indicate that early in the process, conduction is the
sole transport mechanism along the sidewalls and the bottom. The two factors
which contribute to the departure of the actual behaviour from that of pure
conduction are the density-change-induced motion and the natural convection
motion in the melt layer. The liquid n-octadecane is appproximately 10% less
dense than the solid, and this causes the liquid from the melt to expand and
overflow the solid into the space above it. It appears that this density-
induced flow and recirculation are responsible for the initial departures of
the melting front from that expected for pure conduction. As the heating
continues, natural convection motions develop in the melt layer. The motions

start first at the vertical side walls, with upflow near the cell walls and downflow adjacent to the solid/liquid interface. This recirculating flow brings relatively hot liquid to the upper part of the solid and accelerates the inward and downward progression of the interface. The transition from conduction to natural convection in the horizontal melt layer at the bottom occurs later, and for some time there is little or no interaction of flow between the horizontal and the vertical liquid layers. Eventually, as time passes, the recirculating flow in the two regions (i.e., adjacent to one sidewall and half of the bottom) combine to form a single recirculating flow pattern. The density-change-induced motions and natural convection are mutually aiding, but the latter is believed to be the dominant mechanism during most of the melting period.

The timewise variation of the instantaneous melt volume was determined by integrating the melt contours. Attempts were made to correlate the volume of material melted vs. $\tau_\ell Ra^{1/4}$ and other dimensionless parameters, but these were not successful in collapsing the data onto a single curve. The data were correlated using a least squares fit by the equation

$$V/V_o = 51.3 \tau_\ell^{0.802} \, Ste_\ell^{0.198} \, AR^{-0.264} \tag{13}$$

where V_o is the initial, total volume of the solid PCM in the cavity. The negative exponent of the aspect ratio in Eq. (13) indicates that the melting rate decreases as the aspect ratio increases. The melting time of a rectangular body determined from the above empirical equation is about one fifth or one sixth of that predicted by an expression which was based on the conduction model [19]. The presence of natural convection in the liquid is seen to greatly enhance the rate of melting.

Fluid motions in the melt can be inferred from the melt patterns and were visualized by using aluminum powder as a flow tracer. The first clear indication that the transition from conductive to convective heat transfer modes is very rapid can be inferred from the melting patterns at the bottom. After the transition to convection, cells started forming at the bottom of the solid. These cells appeared to be hemispherical. They had circular cross sections and were distributed uniformly over the whole interface. This, however, cannot be clearly seen from the traces of the photographs shown in Figure 8. Similar "cellular" melting patterns have been observed for the melting from below of ice [14] and stearic acid [15]. As the melting progressed and the melt layer thickness increased, the vortex cells grew in size and their number decreased. Similar decreases of the number of vortex cells with the increase in the Rayleigh number for a horizontal fluid layer heated from below has been observed [20,21]. As time passed, the thickness of the melt layer inreased to the point where the flow in the horizontal layer began to interact with the flow along the vertical sidewalls. Finally, the cellular flow pattern was completely destroyed and replaced by an "irregular" motion. This was evidenced by the melting patterns, because the flow visualization method used could not provide sufficiently clear photographs of the motion during the transition period between small, regular, cellular, and larger-scale recirculating flow.

The two photographs in Figure 9 provide evidence of the development of the vortex motion in the bottom melt region. This was accomplished, as described above, by using aluminum powder as the fluid flow tracer. Early in the process, the fluid motion in the layer is not sufficiently well organized to be clearly visualized with this technique. Figure 9a shows that a vortex pair exists which is symmetrical about the midplane of the cavity right below the bottom of the solid PCM core. Figure 9b reveals that the main vortex circulation has been completely suppressed and that the secondary vortex circulation now extends and dominates the entire bottom melt zone.

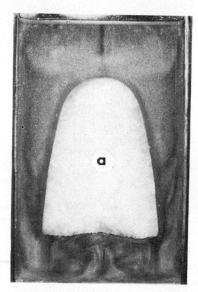 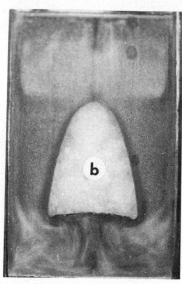

Fig.9 Photographs illustrating the development of the vortex circulation patterns in the bottom melt zone during melting of n-octadecane in a rectangular cavity: a) t = 90 min., and b) t = 120 min.

3.3.2 Heat Transfer

Local heat transfer coefficients along the heated vertical surface of the cavity were measured using a shadowgraph technique. The results are discussed in detail elsewhere [22], and only some typical results are included here. Figure 10 shows the local heat transfer coefficient variation along the vertical wall of the isothermal cavity. It is clear from the figure that conduction is the dominant mode of heat transfer in the early stages of melting (Fo = 0.07054), since the local Nusselt numbers are independent of the vertical position along the cavity wall. As the melting progressed and natural convection developed, the heat transfer became more non-uniform. The results show that the heat transfer coefficients decrease monotonically with the distance along the vertical walls. This is due to the fact that as the liquid is carried upwards along the vertical walls, it gains heat and the temperature gradient is reduced. Consequently, the heat transfer coefficient reaches a minimum at the free surface. The heat transfer coefficients also decrease with time as a result of the growth of the melt region, which decreases the temperature gradients in the liquid. This tends to impede the melt motion. As for the local heat transfer coefficient along the bottom heated surface of the cavity, the variation of the local Nusselt number with time and position shows rather complex trends because of changing vortex circulation patterns resulting from the altered shape and size of the melt zone [22].

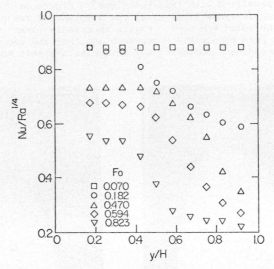

Fig.10 Variation of the local heat transfer coefficient along the vertical heated surface of the cavity, Ste_ℓ=0.130, S_c=0.03, Ra=3.75x10^6, and AR=1.5.

The instantaneous surface-averaged Nusselt number was determined on the basis of the variation of the melted volume of PCM with time and can be expressed as

$$\overline{Nu} = h\ell_{eq}/k = 22 \left[\frac{AR}{2AR+1}\right]^2 \left[\frac{V}{V_o}\right]^2 \frac{1}{Ste_\ell Fo}[1 + \frac{1}{2}(Ste_\ell - Sc)]/\tau_\ell \qquad (14)$$

The characteristic ℓ_{eq}, was chosen to be the instantaneous melt layer thickness equivalent to a corresponding unidirectional melting problem and was defined as

$$\ell_{eq} = V/A_h = [H/(2AR+1)](V/V_o) \qquad (15)$$

In order to establish the Nusselt-Rayleigh number relationship, the instantaneous Rayleigh number based on the characteristic length, ℓ_{eq}, was used. The data for various Stefan numbers and aspect ratios are plotted in Figure 11. An empirical correlation based on the least squares curve fit was obtained.

$$\overline{Nu} = 0.36 \ \overline{Ra}^{0.26} \qquad (16)$$

The correlation covers the Rayleigh number range $2.0 \times 10^3 \leqslant \overline{Ra} \leqslant 6.0 \times 10^6$. A poorer but almost equally good empirical correlation for the surface-averaged Nusselt number was found to be

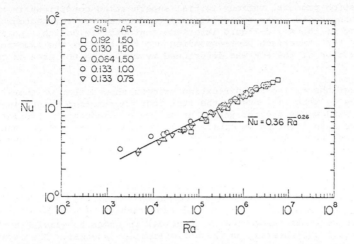

Fig.11 Nusselt number dependence on the average Rayleigh number.

$$\overline{Nu} = 0.43 \ \overline{Ra}^{1/4} \tag{17}$$

The equivalent length ℓ_{eq} was employed as a characteristic length for the process because the use of the height H or the equivalent height H_{eq} in the Nusselt and Rayleigh numbers did not collapse the experimental data into a single, unique curve. The equivalent length ℓ_{eq} can be calculated from the melted volume V/V_o given by Eq. (14) in terms of relevant, independent problem parameters. The average Nusselt number $\overline{Nu}$ can be evaluated from the empirical Eq.(16) or it can be determined from Eq.(17) by substituting Eq.(14) for V/V_o.

It should be stressed that a direct comparison of Eq. (16) with a similar correlation obtained by Marshall [2] cannot be made. His correlation was based on a rather scattered data set for five different thermal conditions. In the present work, the data were restricted to one of the five thermal conditions considered in his study. A close examination of the data [2] for the identical boundary conditions as in this work revealed that the appropriate best-fit curve through those data was a straight line with a slope of 1/4. This contrasts with the one with a slope of 1/3 given in the reference, which was proposed as a general correlation for all aspect ratios and different thermal conditions. It is concluded that the empirical correlations, Eq. (16) and (17), appear to be more appropriate for the specific thermal conditions at the cavity walls used in this investigation.

3.4 Heat Transfer During Solidification

Solidification experiments were performed by freezing n-octadecane in a two-dimensional, rectangular, isothermal wall cavity that was open at the top. Two aspect ratios of the cavity, 1 and 1/2, were used. To provide some insight into the effect of the initial superheating of the liquid on the solidification process, various initial superheating conditions in the liquid PCM were considered. The solid-liquid interface motion was photographed, and the tracing of the solid-liquid interface contours yielded the position and shape of the interface corresponding to the time of the experiment. The frozen fraction of the PCM was determined by measuring the area of the liquid region.

Photographs of the solidification process showed that whiskers formed on the window. This is due to the fact that the thermal conductivity of the Plexiglas window was greater than that of the n-octadecane. Similar observations have been reported by others [12]. Because of the volume change of the n-octadecane, the liquid level dropped continuously during the course of the solidification process. As a result, the shape of the interface at the free surface appeared to be concave.

The variation of the frozen fractions with time τ_s for different Ste_s (i.e., different wall temperatures imposed on the cavity) was determined from the contours (Figure 12). The initial temperatures of the liquid PCM in these experiments were maintained nearly identical in order to eliminate the effect of the initial superheating on the solidification process. The figure shows that the dimensionless time, τ_s, is an appropriate time scale for the solidification process because all the data appear to collapse into a unique, single curve. The figure also shows that the slope of the curve decreases continuously with time and indicates a gradual slowdown of the freezing rate during the course of the solidification. This is due to the increased thermal resistance of the solid PCM as a result of the increased thickness of the frozen layer.

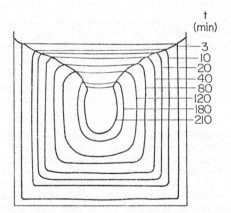

Fig.12 Solid-liquid interface contours during freezing of n-octadecane in a rectangular cavity, $Ste_s=0.148$, $S_h=0.010$ and AR=1.0.

Several experiments have been reported [1] which demonstrate that natural convection in the liquid is signficant during freezing. The process can be drastically slowed down and ultimately terminated by natural convection during the outward solidification of an initially superheated liquid PCM. However, there are some instances where the presence of temperature gradients in the liquid does not induce significant natural convection, e.g., in a stably stratified fluid in which liquid of lesser density is situated above liquid of greater density. This situation was more likely to occur for

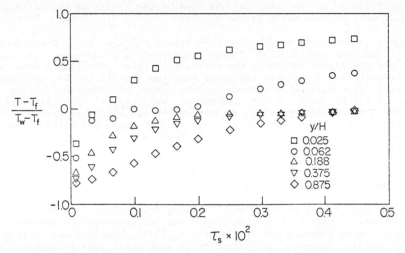

Fig.13 Temperature history during inward freezing of n-octadecane in a rectangular cavity, $Ste_s = 0.141$, $S_h = 1.10$, and AR = 1.0.

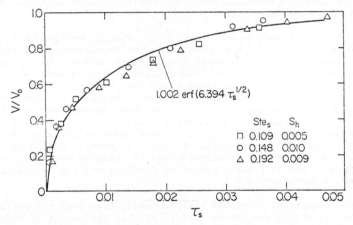

Fig.14 Variation of frozen volume fraction with dimensionless time during freezing of n-octadecane in a rectangular cavity, AR=1.0.

inward freezing of this study. In order to clarify the signficance of the effect of initial liquid superheating, several experiments were conducted with initial liquid temperatures that greatly exceeded the fusion temperature of the PCM. The temperature history in the liquid during the course of solidification, shown in Figure 13, was recorded by inserting a probe of five type-T thermocouples situated at different heights along the cavity's plane of symmetry. The figure shows that the superheated liquid was cooled down to nearly its fusion temperature ($\Theta = 0$) in a rather short period of time, $\tau_s = 0.005$ (approximately 20 minutes) in comparison with the time period that spanned complete solidification (about 4 hours). This indicates that the initial superheating of the liquid PCM was dissipated rather rapidly so that the significance of the effect of the superheated liquid on the solidification could only be seen early in the process.

Figure 14 shows the variation of the frozen fraction with dimensionless time τ_s for various initial liquid superheating conditions. The Stefan numbers and the aspect ratios of the cavities in these experiments were kept nearly identical. The figure indicates that the initial liquid superheating is of little significance in determining the rate of freezing during the internal solidification in an isothermal wall cavity. The effect of the aspect ratio of the cavity on the solidification was also investigated [21]. Owing to the fact that the initial liquid superheating has little effect on the solidification in the cavity, no significant effect of the aspect ratio of the cavity was found.

4. CONCLUSIONS

In the experiments concered with freezing from below and melting from above, natural convection in the liquid was not observed because the fluid was stable. When freezing was from above, natural convection was initiated, but only two main symmetric convection cells with their axes perpendicular to the longer dimension of the test section were observed to persist for the entire freezing process. The freezing front was found to be flat initially, but it later became concave into the solid. The natural convection in the melt not only gradually retarded but also completely terminated the motion of the solid—liquid interface.

Different flow structure in the liquid during melting and freezing experiments were observed. During melting, the buoyancy—induced flow in the bottom melt layer was found to change from cellular motion to large scale-motion. The enhanced melting rate, the retarded freezing rate, and the shape of the solid—liquid interface provided conclusive evidence of the important role played by natural convection in the liquid during phase transformation. However, there are situations, depending on the orientation of the heat sources and sinks, where natural convection heat transfer in the liquid need not be considered. For example, in the stable cases of freezing from below and melting from above, natural convection was absent. During the freezing of a superheated liquid in a rectangular cavity, natural convection was found to occur only in the very early period of freezing and can be neglected during the phase transformation.

A simple analytical model has satisfactorily described the motion of the solid—liquid interface and can be used to predict the effect of natural convection on the interface velocity at low Stefan numbers.

The experimentally determined instantaneous melted volume and surface averaged heat transfer coefficients were correlated in terms of relevant dimensionless parameters.

ACKNOWLEDGEMENTS

This work was supported by the Heat Transfer Program of the National Science Foundation under Grant CME-3014061.

NOMENCLATURE

AR — aspect ratio of the cavity, H/W

A_h — heated surface area of the cavity

c — specific heat

Fo — Fourier number, $\alpha t/H^2$, for cavity heated or cooled from top or bottom, and $\alpha t/w^2$ for isothermal wall cavity

g — gravitational acceleration

H — height of the cavity

Δh_f — latent heat of fusion of the phase change material

h — local heat transfer coefficient

$\bar{h}$ — average heat transfer coefficient

k — thermal conductivity

ℓ_{eq} — characteristic length, defined by Eq.(15)

Nu — local Nusselt number, hr_m/k_ℓ

$\overline{Nu}$ — Nusselt number defined in Eq.(2)

$\overline{Nu}^\ell$ — surface-averaged Nusselt number, defined by Eq.(14)

Pr — Prandtl number, ν/α

Ra — Rayleigh number, $g\beta(T_{wb}-T_f)(H-s)^3/\nu\alpha_\ell$, for cavity heated from top or bottom and $g\beta(T_w - T_f)r_h^3/\nu\alpha_\ell$ for isothermal wall cavity

$\overline{Ra}$ — instantaneous average Rayleigh number, $g\beta(T_w-T_m)\ell_{eq}^3/\nu\alpha_\ell$, in Eq.(16).

γ_h — $H/(2AR+1)$

S_c — subcooling parameter, $c_s(T_f-T_o)/\Delta h_f$

S_h — superheating parameter, $c_\ell(T_o-T_f)\Delta h_f$

Ste_ℓ — Stefan number, $c_\ell(T_w-T_f)\Delta h_f$

St_s — Stefan, number, $c_s(T_f-T_w)\Delta h_f$

s — solid-liquid interface position

T — temperature

T_m — film temperature of melt, $(T_w + T_f)/2$

V — melted volume of solid PCM

V_o — initial volume of solid PCM

W — width of the cavity

x — horizontal distance from the central symmetric line

y — vertical distance from the top of the cavity

z — horizontal distance normal to the x-y plane

α — thermal diffusivity

β — thermal expansion coefficient of liquid

ρ — density

ν — kinematic viscosity

τ — FoSte

Θ — dimensionless temperature $(T - R_w)/T_f-T_w)$

ξ — dimensionless distance, y/H

δ — dimensionless interface position, s/H

Subscripts

b — refer to bottom plate

f — refer to fusion point of PCM

ℓ — refer to liquid state

o — refer to initial state

s — refer to solid state

t — refer to top plate

w — refer to wall

REFERENCES

1. Viskanta, R., (in press). Phase—Change Heat Transfer in Solar Heat Storage: Latent Heat Materials, G.A. Lane, editor, CRC Press, Boca Raton, FL.

2. Marshall, R.H. 1978. Natural Convection Effects in Rectangular Enclosures Containing a Phase Change Material. Thermal Storage and Heat Transfer in Solar Energy Systems, F. Kreith, R. Boehm, J. Mitchell, R. Bannerot, editors, ASME, N.Y., pp. 61-69.

3. Shamsundar, N. and Sparrow, E.M. 1975. Analysis of Multidimensional Conduction Phase Change via the Enthalpy Model. J. Heat Transfer, Vol.97, pp. 333-340.

4. Shamsundar, N. and Sparrow, E.M. 1976. Effect of Density Change on Multidimensional Conduction Phase Change. J. Heat Transfer, Vol.98, pp. 550-557.

5. Saitoh, T. 1978. Numerical Method for Multidimensional Freezing Problems in Arbitrary Domains. J. Heat Transfer, Vol. 100, pp. 294-299.

6. Patel, G.S., Goodling, J.S. and Khader, M.S. 1978. Experimental Results of Two-Dimensional Inward Solidification, Heat Transfer—1978, Hemisphere Publishing Corp., Washington, DC, Vol. 3, pp. 313-316.

7. Sparrow, E.M., Patankar, S.V. and Ramadhyami, S. 1977. Analysis of Melting in the Presence of Natural Convection in the Melt Region, J. Heat Transfer, Vol. 9, pp.520-526.

8. Ramachandran, N., Gupta, J.P. and Jaburia, Y. 1982. Thermal and Fluid Flow During Solidification in a Rectangular Enclosure. Int. J. Heat Mass Transfer, Vol. 25, pp.187-194.

9. Lundardini, V.J. 1981. Heat Transfer in Cold Climates. Van Nostrand Reinhold Co., New York.

10. Hollands, K.G.T. and Raithyby, G.D. (in press). Natural Convection. in Handbook of Heat Transfer, Second edition, W.M. Rohsenow et al., eds., McGraw-Hill Book Co., New York.

11. Goodman, T.R. 1964. Application of Integral Methods to Transient Nonlinear Heat Transfer. Advances in Heat Transfer, Academic Press, New York, Volume 1, pp. 52-122.

12. Sparrow, E.MM., Ramsey, J.W., and Kemink, R.G. 1979. Freezing Controlled by Natural Convection. J. Heat Transfer, Vol.101, pp. 578-584.

13. Krishnamurti, R. 1973. Some Further Studies on the Transition to Turbulent Convection. J. Fluid Mech., Vol. 60, pp. 285-303.

14. Yen, Y.-C. 1980. Free Convection Heat Transfer Characteristics in a Melt Layer. J. Heat Transfer, Vol.102, pp. 550-556.

15. Hale, N.W., Jr. and Viskanta, R. 1980. Solid—Liquid Phase—Change Heat Transfer and Interface Motion in Materials Cooled or Heated from Above and Below. Int. J. Heat Mass Transfer, Vol. 23, pp. 283-292.

16. Hauf, W. and Grigull, U. 1970. Optical Method in Heat Transfer. *Advances in Heat Transfer*. Academic press, New York, Vol.6, pp. 133–366.

17. Bathelt, A.G. and Viskanta, R. 1981. Heat Transfer and Interface Motion During Melting and Solidification Around a Finned Heat Source/Sink. *J. Heat Transfer*, Vol.103, pp. 720–726.

18. Hale, N.W., Jr. and Viskanta, R. 1978. Photographic Observation of the Solid–Liquid Interface Motion During Melting of a Solid Heated from an Isothermal Vertical Wall. *Lett. Heat Mass Transfer*, Vol. 5, 329–337.

19. Solomon, A.D. 1980. An Expression for the Melting Time of a Rectangular Body. *Lett. Heat Mass Transfer*, Vol. 7, pp. 379–384.

20. Koschmieder, E.L. 1974. Bernard Convection. *Adv. Chem. Phys.*, Vol. 26, 177–211.

21. Ho, C. –J., Solid–Liquid Phase–Change Heat Transfer in Cavities. PhD Thesis, Purdue University (in preparation).

22. Oertel, J. Jr. and Buehler, K. 1978. A Special Differential Interferometer Used for Heat Convection Investigation. *Int. J. Heat Mass Transfer*, Vol. 22, pp. 1111–1115.

Multi-Phase Flow and Heat Transfer III. Part B: Applications
edited by T.N. Veziroğlu and A.E. Bergles
Elsevier Science Publishers B.V., Amsterdam, 1984 — Printed in The Netherlands

SOLIDIFICATION IN SPHERES - THEORETICAL AND EXPERIMENTAL INVESTIGATION

Luiz F. Milanez and Kamal A.R. Ismail
Department of Mechanical Engineering
Campinas State University
13100 - Campinas, S. Paulo, BRAZIL

ABSTRACT

The objective of the present study is to investigate theoreti cally and experimentally the solidification process in spherical geometry and its major parameters such as the instantaneous posi- tion of the solid liquid interface, temperature profile and heat transfer rate. In the theoretical part the integral method is applied to spherical geometry. An experimental testing rig to simulate the radial solidification process is designed, installed and instrumented. The theoretical calculations based upon the integral method are found to agree very well with the experimental results and with the results obtained by means of a numerical method.

1. INTRODUCTION

Transient heat transfer problems involving phase change are important and have many industrial applications such as in fabri- cation of ice, deep freezing of food, solidification or fusion of metals in foundry work and energy storage. The solution of this problem is difficult and analytical solutions are available only for the simple plane geometry. For problems of inward solidifica- tion of spheres the solutions obtained were: simplified [1, 2], graphical [3], numerical [4] and in terms of series [5, 6]. A solution using integral method [7] as suggested by Goodman [8] was not very successful. The first and apparently only experimental analysis of solidification in spherical geometry is dated back in 1945 [9].

2. FORMULATION OF THE PROBLEM

Consider a sphere of radius R containing phase change material in the liquid phase at its fusion temperature T_f, being cooled externally by a fluid at constant temperature T_∞ with a heat transfer coefficient h. The equations describing the process are

$$\frac{\partial T}{\partial t} = \frac{k}{\rho c} \left[\frac{2}{r} \frac{\partial T}{\partial r} + \frac{\partial^2 T}{\partial r^2} \right] \qquad r_f < r < R \qquad (1)$$

$$\frac{dr}{dt} = \frac{k}{\rho L} \frac{\partial T}{\partial r} \qquad\qquad r=r_f \qquad\qquad (2)$$

$$-\frac{\partial T}{\partial r} = \frac{h}{k} (T-T_\infty) \qquad\qquad r=R \qquad\qquad (3)$$

$$T = T_f \qquad\qquad r=r_f \qquad\qquad (4)$$

Introduce the following dimensionless variables:

$$X = 1 - \frac{r}{R} \qquad\qquad \varepsilon = 1 - \frac{r_f}{R}$$

$$\tau = \frac{\alpha t}{R^2} \qquad\qquad \theta = \frac{T-T_\infty}{T_f-T_\infty} \qquad\qquad (5)$$

and the dimensionless numbers:

$$\text{Ste} = \frac{c(T_f-T_\infty)}{L} \qquad \text{Stefan number} \qquad (6)$$

$$\text{Bi} = \frac{hR}{k} \qquad\qquad \text{Biot number} \qquad\qquad (7)$$

Rewriting equations (1) to (4) in terms of the new variables

$$\frac{\partial \theta}{\partial \tau} = \frac{2}{X-1} \frac{\partial \theta}{\partial X} + \frac{\partial^2 \theta}{\partial X^2} \qquad 0 \leqslant X \leqslant \varepsilon \qquad (8)$$

$$\text{Ste} \frac{\partial \theta}{\partial X} = \frac{dX}{d\tau} \qquad\qquad X=\varepsilon \qquad\qquad (9)$$

$$\frac{\partial \theta}{\partial X} = \text{Bi}\theta \qquad\qquad X=0 \qquad\qquad (10)$$

$$\theta = 1 \qquad\qquad X=\varepsilon \qquad\qquad (11)$$

Equation (8) can be written alternatively as

$$\frac{\partial}{\partial X} \left[(1-X)^2 \frac{\partial \theta}{\partial X} \right] = (1-X)^2 \frac{\partial \theta}{\partial \tau} \qquad 0 \leqslant X \leqslant \varepsilon \qquad (12)$$

The integral equation for the solid phase from X=0 to X=ε can be obtained integrating equation (12) and applying equation (9):

$$\frac{(1-\varepsilon)^2}{\text{Ste}} \frac{d\varepsilon}{d\tau} - \left(\frac{\partial \theta}{\partial X}\right)_{X=0} = \int_0^\varepsilon (1-X)^2 \frac{\partial \theta}{\partial \tau} dX \qquad (13)$$

3. SOLUTION OF THE PROBLEM

A simplified solution of the problem defined by equations (8) to (11) can be determined by assuming a profile of the form

$$\theta = \frac{1}{1-X} \left[A+B(X-\epsilon) \right] \tag{14}$$

where A and B are coefficients to be determined by applying the boundary conditions of the problem, equations (10) and (11), and the result is

$$A = 1 - \epsilon \tag{15}$$

$$B = \frac{(1-\epsilon)(Bi-1)}{(1-\epsilon)+Bi\epsilon} \tag{16}$$

Hence, the temperature profile in the solid is

$$\theta = \frac{1-\epsilon}{1-X} \left[\frac{BiX+1-X}{Bi\epsilon+1-\epsilon} \right] = \frac{\frac{X}{1-X} + \frac{1}{Bi}}{\frac{\epsilon}{1-\epsilon} + \frac{1}{Bi}} \tag{17}$$

Substituting this expression in equation (9) one obtains

$$\frac{d\epsilon}{d\tau} = \frac{Bi.Ste}{(1-\epsilon)(Bi\epsilon+1-\epsilon)} \tag{18}$$

and the integration of this equation allows the calculation of the solidification time as

$$\tau = \frac{1}{Ste} \left[\frac{\epsilon^3}{3} \left(\frac{1}{Bi} - 1\right) - \epsilon^2\left(\frac{1}{Bi} - \frac{1}{2}\right) + \frac{\epsilon}{Bi} \right] \tag{19}$$

which are the same results obtained by London and Seban [1] disregarding the variation of the internal energy of the solidified material.

A more precise solution can be obtained by assuming a second degree polynomial representation for the temperature profile:

$$\theta = \frac{1}{1-X} \left[A+B(X-\epsilon) + C(X-\epsilon)^2 \right] \tag{20}$$

In order to determine the coefficients A, B and C, three boundary conditions are needed. Equations (10) and (11) can be used for this purpose while equation (9) has to be modified to eliminate $d\epsilon/d\tau$. This can be readily done by differentiating equation (11) and eliminating the term $d\epsilon/d\tau$ between the resulting equation and equation (9). The resulting new condition is

$$\left(\frac{\partial\theta}{\partial X}\right)^2 = -\frac{1}{Ste}\frac{\partial\theta}{\partial\tau} \qquad X = \epsilon \tag{21}$$

Applying the boundary condition (11) in equation (20) results

$$A = 1 - \varepsilon \tag{22}$$

Now by eliminating $\partial\theta/\partial\tau$ between equations (12) and (21) one obtains

$$\left(\frac{\partial\theta}{\partial X}\right)^2 + \frac{1}{Ste(1-X)^2} \frac{\partial}{\partial X}\left[(1-X)^2 \frac{\partial\theta}{\partial X}\right] = 0 \qquad X = \varepsilon \tag{23}$$

which upon substituting the expression of the temperature profile given by (20) and its derivative results

$$(B+1)^2 + \frac{2C(1-\varepsilon)}{Ste} = 0 \qquad \text{for } \varepsilon \neq 1 \tag{24}$$

Applying the boundary condition (10) in the temperature profile (19) one has

$$C = \frac{B[1+\varepsilon(Bi-1)] - (1-\varepsilon)(Bi-1)}{\varepsilon^2(Bi-1) + 2\varepsilon} \tag{25}$$

Eliminating C between equations (24) and (25) one obtains an equation of the second order for B whose solution is

$$B = -\left[1 + \frac{1-\varepsilon}{Ste} \frac{\varepsilon(Bi-1)+1}{\varepsilon^2(Bi-1)+2\varepsilon}\right] \pm$$

$$\sqrt{\left[1 + \frac{1-\varepsilon}{Ste} \frac{\varepsilon(Bi-1)+1}{\varepsilon^2(Bi-1)+2\varepsilon}\right]^2 - \left[1 - \frac{2(1-\varepsilon)^2}{Ste} \frac{Bi-1}{\varepsilon^2(Bi-1)+2\varepsilon}\right]} \tag{26}$$

The positive sign is chosen due to the fact that

$$\left.\frac{\partial\theta}{\partial X}\right|_{X=\varepsilon} = \frac{B+1}{1-\varepsilon} > 0 \tag{27}$$

and hence the value of B is

$$B = -1 + \left[\frac{1-\varepsilon}{Ste} \frac{\varepsilon(Bi-1)+1}{\varepsilon^2(Bi-1)+2\varepsilon}\right]\left\{\sqrt{1 + \frac{2Ste \cdot Bi[\varepsilon^2(Bi-1)+2\varepsilon]}{(1-\varepsilon)[\varepsilon(Bi-1)+1]^2}} - 1\right\} \tag{28}$$

Therefore the temperature profile can be obtained by substituting equations (28), (22) and (25) in equation (20). Now, in order to determine the velocity of the solidification front one introduces in equation (13) the velocity profile and performs all the indicated operations. The result is

$$\frac{d\varepsilon}{d\tau} = - Bi\left[\frac{2(1-\varepsilon)-\varepsilon B}{\varepsilon(Bi-1)+2}\right] \cdot \left\{\frac{-(1-\varepsilon)^2}{Ste} + \frac{1}{[\varepsilon(Bi-1)+2]}\right. \cdot$$

$$\cdot (Bi-1)\left[\frac{4\varepsilon}{3} - \frac{13}{4}\varepsilon^2 + \frac{5}{3}\varepsilon^3\right] + (2-6\varepsilon+3\varepsilon^2) +$$

$$+ \left[\left(\frac{\varepsilon^4}{12} - \frac{\varepsilon^3}{6}\right)(Bi-1) + \left(\frac{\varepsilon^3}{4} - \frac{2}{3}\varepsilon^2\right)\right]\frac{dB}{d\varepsilon} +$$

$$+ B\left[(\frac{\varepsilon^3}{3} - \frac{\varepsilon^2}{2})(Bi-1) + (\frac{2\varepsilon^2}{4} - \frac{4\varepsilon}{3})\right] - \left\{(Bi-1)(\frac{2\varepsilon^2}{3} - \frac{13}{12}\varepsilon^3 + \frac{5}{12}\varepsilon^4) + \right.$$

$$+ (2\varepsilon - 3\varepsilon^2 + \varepsilon^3) + B\left[(\frac{\varepsilon^4}{12} - \frac{\varepsilon^3}{6})(Bi-1) + (\frac{\varepsilon^3}{4} - \frac{2}{3}\varepsilon^3)\right]\left. \right\} .$$

$$\cdot \frac{(Bi-1)}{[\varepsilon(Bi-1)+2]^2} - (1-\varepsilon)^2 \Bigg\}^{-1} \tag{29}$$

The relation between the solidification time and the position of the fusion front can be determined by integrating numerically (29) using adequate integrating scheme.

The determination of $d\varepsilon/d\tau$ when $\varepsilon=0$ can not be obtained from equation (29) because of the singularity in this equation when $\varepsilon=0$. This value can be obtained independently by combining equations (9), (10) and (11) resulting

$$\frac{d\varepsilon}{d\tau} = Ste.Bi \qquad \text{when} \qquad \varepsilon=0 \tag{30}$$

The heat transfered through the spherical surface is

$$q_o = - k(\frac{\partial T}{\partial r})_{r=R} = h(T_o - T_\infty)$$

or in a dimensionless form

$$\phi_o = \frac{q_o}{kL/cR} = \frac{Q_o}{4\pi RkL/c}$$

where $Q_o = 4\pi R^2 q_o$, and hence,

$$\phi_o = Ste(\frac{\partial\theta}{\partial X}) = Ste.Bi.\theta \qquad \text{for} \qquad X=0 \tag{31}$$

4. EXPERIMENTAL APPARATUS

The first and apparently the only experimental verification of the solidification process in spherical geometry is dated back to 1945 due to London & Seban [9]. In their experiments they used a tomato of 60mm diameter fitted with a thermocouple with the tip of the thermocouple as near as possible of the centre of the tomato. Due to the lack of experimental information on solidification and fusion in spherical geometry and the need of such information in order to compare and verify the theoretical and numerical results, an experimental analysis is initiated. An experimental rig is designed to simulate the process of radial solidification in such a way to permit measurements of the movement of the solid-liquid interface and the temperature profile in the solid phase. The apparatus uses water as well as air as refrigerating fluids. The phase change materials used were tin and lead. Measurements of the position of the solidification front used both dipstick and thermocouple techniques. Details of the experimental apparatus are shown in figures (1) and (2). Initially the mould is filled with the phase change material. The electric furnace is then turned on with a high power input to allow the complete melting of the material. After the material is melt, the power is gradually reduced while the material is continuously stirred in order to assure uniform temperature within the mould. When the temperature of the liquid metal is about 3K above the melting point the power is turned off and the refrigerating fluid line is open. As the mould is in thermal equilibrium with the furnace, the heat is not transferred through the side walls, being removed essentially through the bottom of the mould where the cooling effect is taking place. Therefore the heat flux occurs only in the radial direction. As the material is being radially solidified, the movement of the interface is measured by means of the dipstick or by means of a set of thermocouples evenly spaced.

5. RESULTS AND DISCUSSIONS

The movement of the solidification front is determined experimentally for a sphere of 100 mm radius from the reference point zero at the surface up to 600mm at intervals of 5 mm. From these measurements it is possible to plot the curves and determine the Biot number from equation (30). Figures (3), (4), (5) and (6) show the position of the interface for lead and tin with water and air as coolants as indicated in the figures. It can be noticed that the agreement between the theoretical and experimental results are good. Additional curves are put to compare with these results using a numerical method of the moving mesh type. Temperature profiles in the solidified phase for different positions of the solidification front are shown in figure (7) and the velocity of the interface is shown in figure (8). Agreement between the results from the integral method and the numerical calculations are found to be satisfactory. Curves showing the variation of the dimensionless heat flux rate at the surface of the sphere computed by the integral and numerical methods are shown in figure (9) for different Biot and Stefan numbers. The agreement is found to be very close. Variation of the temperature of the external surface of the sphere with time are shown in figures (10) and (11) which indicate close agreement between the theoretical and the experimental results.

As it can be seen from the above discussions, the theoretical calculations based upon the integral method are found to agree very well with the experimental results and with the calculations based upon a moving mesh numerical method.

NOMENCLATURE

    c   specific heat

    k   thermal conductivity

    L   latent heat of fusion

    r   radial position

    r_f radius of solidification front

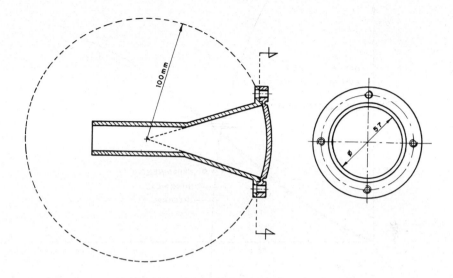

Fig. 1. Proposed Mould: Solid Angle of a Sphere.

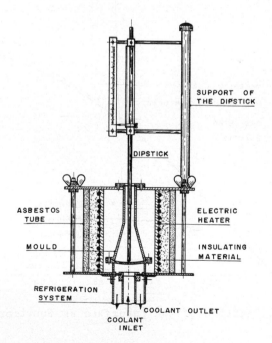

Fig. 2. Proposed Experimental Apparatus for Radial Solidification in Spheres.

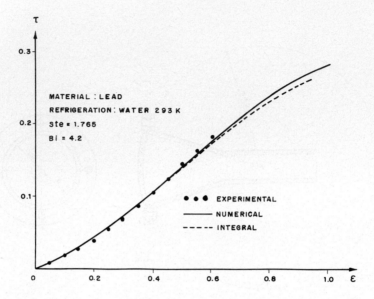

Fig. 3. Position of the Interface as Function of Time.

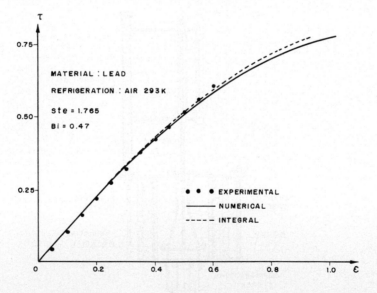

Fig. 4. Position of the Interface as Function of Time.

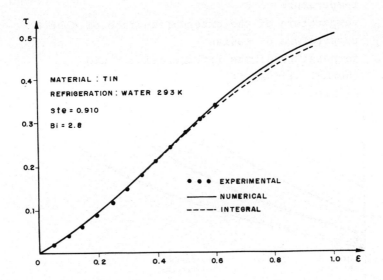

Fig. 5. Position of the Interface as Function of Time.

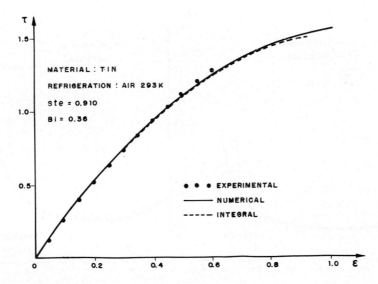

Fig. 6. Position of the Interface as Function of Time.

R   radius of sphere

t   time

T   temperature

T_0 temperature of the external surface of sphere

T_f temperature of fusion

T_∞ temperature of the refrigerating fluid

α thermal diffusivity

ρ density

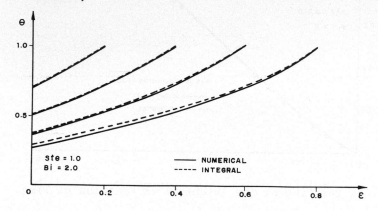

Fig. 7. Temperature Profile in the Solid Phase for Different Positions of the Interface.

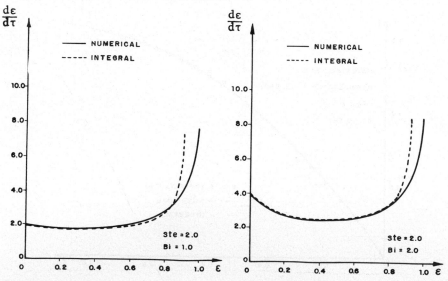

Fig. 8. Velocity of the Interface.

REFERENCES

1. London, A.L. and Seban, R.A. - "Rate of Ice Formation", Tran-
    sactions of the ASME, vol. 65, pp. 771-778, 1943.

2. Kern, J. and Wells, G.L. - "Simple Analysis and Working
    Equations for the Solidification of Cylinders and Spheres",
    Metallurgical Transactions, vol. 8B, pp. 99, 1977.

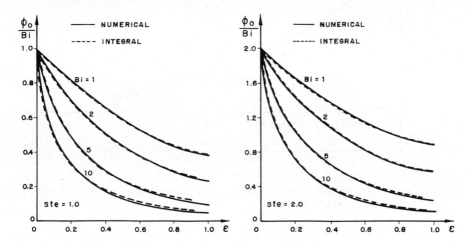

Fig. 9. Heat Flux Rate at the Surface of the Sphere as a Function
of the Interface Position.

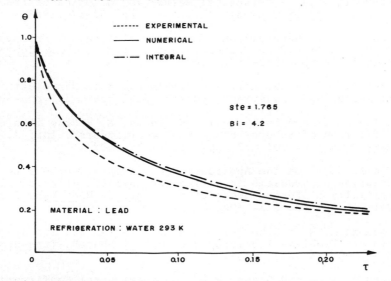

Fig. 10. Variation of the External Surface Temperature of the
Sphere with Time.

576

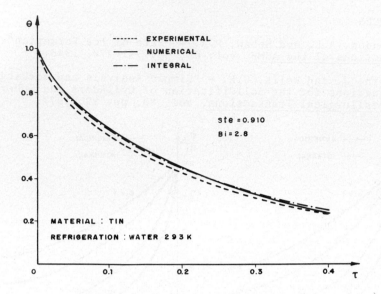

Fig. 11. Variation of the External Surface Temperature of the Sphere with Time.

3. Longwell, P.A. - "Graphical Method for Solution of Freezing Problems, AIChE Journal, vol. 4(1), pp. 53-57, 1958.

4. Tao, L.C. - "Generalized Numerical Solutions of Freezing a Saturated Liquid in Cylinders and Spheres", AIChE Journal, vol. 13(1), pp. 165-169, 1967.

5. Megerlin, F. - "Geometrisch Eindimensionale Wärmeleitung Beim Schmelzen und Erstarren", Forsch. Ing. Wes., 34, pp. 40, 1968.

6. Riley, D.S. and Smith, F.T. and Poots, G. - "The Inward Solidification of Spheres and circular Cylinders", Int. J. Heat Mass Transfer, vol. 17, pp. 1507-1516, 1974.

7. Poots, G. - "On the Application of Integral Methods to the Solution of Problems Involving the Solidification of Liquids Initially at Fusion Temperature", Int. J. Heat Mass Transfer, vol. 5, pp. 525-531, 1962.

8. Goodman, T.R. - "The Heat Balance Integral and its Application to Problems Involving a Change of Phase", Transactions ASME, pp. 335-342, 1958.

9. London, A.L. and Seban, R.A. - "Experimental Confirmation of Predicted Water Freezing Rates", Transactions of the ASME, vol. 67, pp. 39-44, 1945.

Multi-Phase Flow and Heat Transfer III. Part B: Applications   577
edited by T.N. Veziroğlu and A.E. Bergles
Elsevier Science Publishers B.V., Amsterdam, 1984 — Printed in The Netherlands

AN ANALYTICAL/EXPERIMENTAL APPROACH TO STUDY SOLIDIFICATION OF METALS IN COOLED
AND MASSIVE CYLINDRICAL MOULDS

Rezende Gomes dos Santos
Department of Mechanical Engineering
State University of Campinas (UNICAMP)
13100 - Campinas - SP, Brazil

ABSTRACT

A method to analyse the inward solidification of a circular cylinder
initially at melting temperature with coolant fluid outside the cylinder is
proposed and extended to study the solidification of metals initially at temperatures above the melting temperature and also to study the solidification
in massive moulds. The method is based on an analytical solution obtained for
slab shaped bodies, in which it was introduced a geometric modifying factor,
and led to equations that give the solidification time as function of solidified thickness. The theoretical results obtained by using the equations are
compared with experimental results obtained with the aid of a special device
developed by the author to simulate the solidification of cylindrical geometries and to permit the determination of solidified thickness as function of
solidification time by using the dip-stick technique. The comparisons made
in several different cases showed a good agreement among experimental results
and calculations furnished by the proposed technique.

1. INTRODUCTION

The mathematical analysis of the solidification of metal is quite complex
because it involves transient heat transfer with phase transformation and,
consequentely, differencial equations with no linear boundary conditions. For
this, in spite of its importance in many practical applications, exact solutions are available only for a few special cases, in general applied to slab
shaped geometries [1-10]. The difficult to study analitically the problem
increases with the complexity of the geometry and even for cylindrical and
spherical geometries exact solutions are not known [1-3].

In the case of cylindrical geometry only analytical solutions based on
mathematical approximations and numerical solutions are available. The
analytical solutions do not furnish good agreement with experimental data,
specially for the solidification of metals [11,12]. Some numerical solutions
show a good agreement with experimental data but suffer from lack of generality and/or simplicity [13,14].

In this work the inward solidification of circular cylinders is analysed
experimentally and theoretically, and a new analytical method is proposed to
study the solidification of metals in cooled or massive moulds. The equations
obtained permit the prediction of the solidified thickness as function of
solidification times, using the physical properties of the metal/mould system.

The predictions of the method were compared with experimental results obtained by the author and numerical results obtained by Tao showing a good agreement in all the cases studied. Comparison were made, also, among the method proposed, the experimental data and two other approximated analytical solutions.

2. EXPERIMENTS

2.1. Experimental Procedure

The experimental examination of the inward solidification of cylindrical geometry was made by freezing lead and tin in cooled and massive moulds.

The cooled mould was simulated by a special radial water cooled device designed and constructed by the author in order to simulate the solidification in cylindrical geometries, and to permit the determination of solidification times as function of solidified thickness of the metal by using the dip-stick technique. The experimental set-up is showed in the Figure 1.

The massive mould was built in low carbon steel with the following dimensions: internal radius 0,03 m, external radius 0,08 m and 0,14 m high. So the wall thickness is 0,05 m what is enough to consider the mould as semi infinite (the solidification process is completed before the temperature of the external surface of the mould has changed). In this case the solidification times as function of solidified thickness were obtained by using the thermocouple technique.

In order to study the influence of the superheat of the liquid metal in the solidification time, some experiences were carried out with the liquid metal initially at the melting temperature and at temperatures 30, 60 and 90 K above this one.

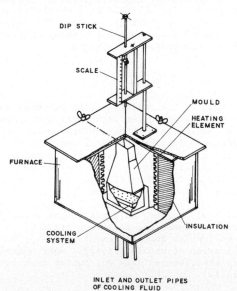

Fig.1 - Water cooled experimental device.

2.2. Experimental Results

In the results the solidification times were plotted as function of the ratio between the solidified volume of metal and the surface area of the metal/mould interface that is used to represent the solidified thickness because it was found that this ratio is more convenient to study the cylindrical geometry.

Figures 2 and 3 show the experimental results obtained during the solidification of lead and tin, initially at the melting temperature, in cooled cylindrical mould.

Figures 4 and 5 show the results obtained during the solidification of lead and tin, initially at temperatures above the melting temperature (ΔT = 30, 60 and 90 K), in cooled cylindrical mould.

The Figure 6 shows the results obtained during the solidification of tin, initially at the melting temperature, in massive semi infinite mould.

3. ANALYTICAL APPROACH

The method developed in this work is based on an analytical solution obtained by Garcia and Prates for slab shaped bodies [9,10]. The idea was to introduce in this solution a geometric modifying factor able to take into account the modifications introduced by the mould curvature [15].

A experimental analysis has shown that, comparing a slab and a cylinder, for the same values of the ratio between the solidified volume of metal and the surface area of metal/mould interface, the solidification times are almost the same in the begining of the process and the total solidification time of the cylinder is two times bigger than the corresponding time for the slab. So, the modifying factor must vary from 1 in the begining to 2 in the end of

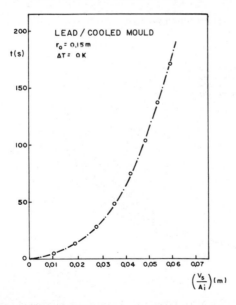

Fig.2 - Experimental results: lead in cooled mould.

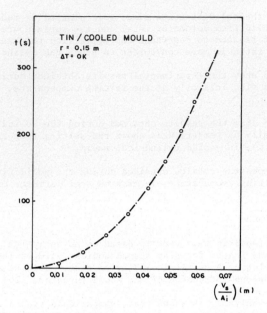

Fig.3 - Experimental results: tin in cooled mould.

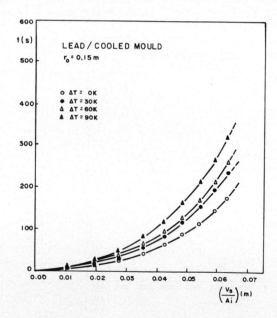

Fig.4 - Experimental results: lead superheated in cooled mould.

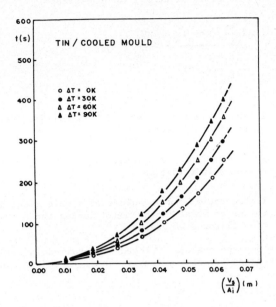

Fig.5 - Experimental results: tin superheated in cooled mould.

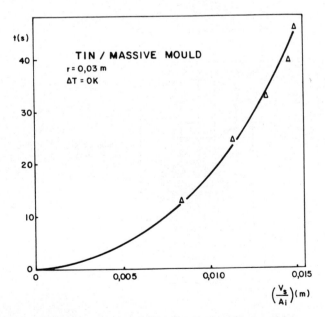

Fig.6 - Experimental results: tin in massive moulds.

582

the solidification process. The following factor satisfying these conditions was developed to be introduced in equations for the slab:

$$\Theta = \left[\frac{(\frac{V_s}{A_i})_{p\,(S=r_o)}}{(\frac{V_s}{A_i})_{c\,(r_f=0)}} - (\frac{V_\ell}{V_T})_c \right] \tag{1}$$

where:

$(\frac{V_s}{A_i})_{p\,(S=r_o)}$ is the ratio for a slab of thickness $S=r_o$ in the end of the solidification

$(\frac{V_s}{A_i})_{c\,(r_f=0)}$ is the ratio for a cylinder of radius r_o in the end of the solidification

$(\frac{V_\ell}{V_T})_c$ is the ratio between volume of liquid metal and total volume of metal in the mould during the solidification of a cylinder of radius r_o

The nomenclature for the symbols employed is detailed in the Appendix and the system of reference for slab and cylinder is showed in the Figure 7.

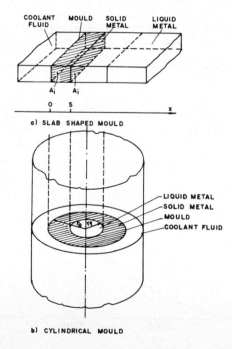

Fig.7 – Schematic representation of the solidification of slabs and cylinders.

Some simplifications can be introduced in the equation 1 that assumes the from:

$$\theta = \left[2 - \left(\frac{r_o}{r_f} \right)^2 \right] \tag{2}$$

Considering the following basic assumptions: heat flow geometrically unidimensional; constant heat transfer coefficient between the metal and the coolant; constant physical properties and melt and coolant with constant temperature, the relation between solidified thickness and solidification time for the slab is given by [9,10].

$$t_1^* = \left[\frac{1}{4\phi_1^2} \left(\frac{V_s}{A_i} \right)^{*2} + \frac{H}{c_s(T_m - T_o)} \left(\frac{V_s}{A_i} \right)^* \right] \tag{3}$$

for metals initially at the melting temperature solidified in cooled and massive moulds and

$$t_2^* = \left[\frac{1}{4\phi_2^2} \left(\frac{V_s}{A_i} \right)^{*2} + \frac{1}{\sqrt{\pi} \, \phi_2 \, \exp(\phi_2^2) \, \mathrm{erf}(\phi_2)} \left(\frac{V_s}{A_i} \right)^* \right] \tag{4}$$

for metals initially at a temperature above the melting temperature solidified in cooled moulds.

The dimensionless parameters are given by:

$$t^* = \frac{t \, h^2}{k_s \rho_s c_s} \tag{5}$$

$$\left(\frac{V_s}{A_i} \right)^* = \frac{h}{k_s} \left(\frac{V_s}{A_i} \right) \tag{6}$$

and ϕ_1 and ϕ_2 are solidification constants defined by:

- for massive moulds: $\quad \sqrt{\pi} \, \phi_1 \, \exp(\phi_1^2) \, \left[M + \mathrm{erf}(\phi_1) \right] = \dfrac{c_s(T_m - T_o)}{H} \tag{7}$

- for cooled moulds: $\quad \sqrt{\pi} \, \phi_1 \, \mathrm{erf}(\phi_1) \, \exp(\phi_1^2) = \dfrac{c_s(T_m - T_o)}{H} \tag{8}$

$$\frac{\exp(-\phi_2^2)}{M + \mathrm{erf}(\phi_2)} - \frac{m(T_v - T_m) \exp(-n^2 \phi_2^2)}{(T_m - T_o) \mathrm{erfc}(n \, \phi_2)} = \frac{\sqrt{\pi} \, H \, \phi_2}{c_s(T_m - T_o)} \tag{9}$$

where:

$$n = \sqrt{\frac{k_s \rho_\ell c_\ell}{k_\ell \rho_s c_s}} \quad , \quad m = \sqrt{\frac{k_\ell \rho_\ell c_\ell}{k_s \rho_s c_s}} \quad M = \sqrt{\frac{k_s \rho_s c_s}{k_m \rho_m c_m}} \tag{10}$$

Introducing the geometrical factor (equation 2) in the equations 3 and 4 we have the correspondent equations for cylinders:

$$t_1^* = \left[2 - \left(\frac{r_f}{r_o}\right)^2\right] \left[\frac{1}{4\phi_1^2}\left(\frac{V_s}{A_i}\right)^{*2} + \frac{H}{c_s(T_f-T_o)}\left(\frac{V_s}{A_i}\right)^*\right] \qquad (11)$$

and

$$t_2^* = \left[2 - \left(\frac{r_f}{r_o}\right)^2\right] \left[\frac{1}{4\phi_2^2}\left(\frac{V_s}{A_i}\right)^{*2} + \frac{1}{\sqrt{\pi}\,\phi_2\exp(\phi_2^2)\,\text{erf}(\phi_2)}\left(\frac{V_s}{A_i}\right)^*\right] \qquad (12)$$

4. APPLICATION OF THE METHOD AND COMPARISONS

The equations 11 and 12 were applied to study the dependence between solidified thickness and solidification time for lead and tin solidified in cooled and massive moulds, and the results were compared with experimental data. The experimental results were converted to the dimensionless form by employing the appropriate values of the heat transfer coefficient .

Figures 8 and 9 show the curves representing the predictions of the proposed technique compared with the experimental results obtained for the solidification of lead and tin, respectively, initially at the melting temperature, in cooled cylindrical molds (r_o = 0,15 m).

Figures 10 and 11 show the results for lead and tin initially at a temperature above the melting point (ΔT = 30, 60 and 90 K) in cooled cylindrical moulds (r_o = 0,15 m).

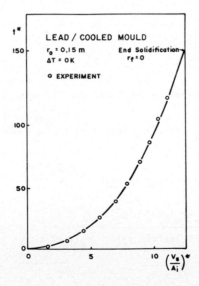

Fig.8 - Comparison between experimental results and theoretical curves.

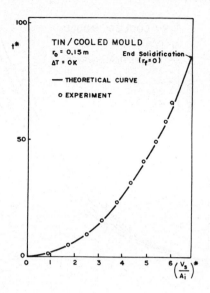

Fig.9 - Comparison between experimental results and theoretical curves.

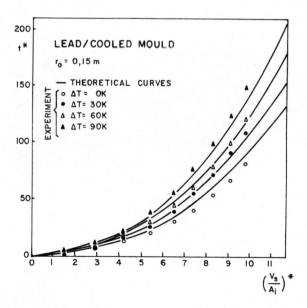

Fig.10 - Comparison between experimental results and theoretical curves.

586

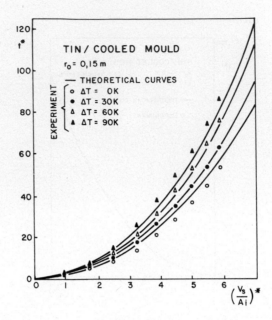

Fig.11 - Comparison between experimental results and theoretical curves.

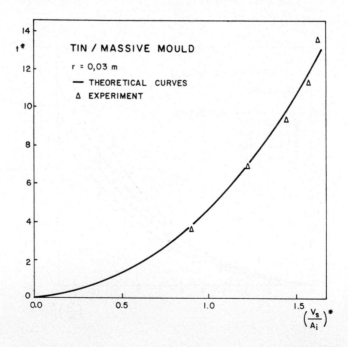

Fig.12 - Comparison between experimental results and theoretical curves.

Finally the Figure 12 the results for tin initially at the melting temperature in massive mould (r_o = 0,03 m).

It can be seen that the agreement between the proposed method and the experimental results is quite good in all the cases analysed.

To assess the performance of the proposed technique, it was compared with other existing in literature as the London and Seban and the Shih and Tsay solutions [11,12]. The results can be observed in the Figure 13. As it can be seen the proposed method shows a best agreement with experimental results than the other two. It was compared also with a numerical method (see Figure 14) showing a quite good agreement [14].

5. CONCLUSIONS

The method proposed was applied for several cases of solidification of metals and when compared with experimental data and a numerical solution showed a good agreement. This permit to conclude that, in spite of its simplicity and easy manipulation, this technique is a useful and confident tool to predict the solidification time as function of solidified thickness during the solidification of metals, initially at any temperature, in cooled and massive cylindrical moulds.

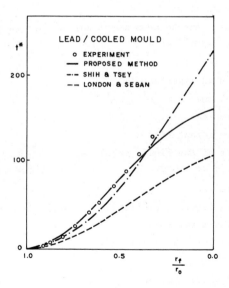

Fig.13 - Comparison between the proposed method and other techniques.

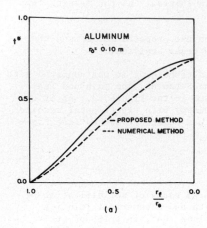 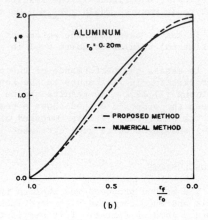

(a)　　　　　　　　　　　　　　　　　(b)

Fig.14 - Comparison between the proposed method and a numerical method.

APPENDIX

Nomenclature

A_i = area of the metal/mould interface (m^2)

c = specific heat (J kg^{-1} K^{-1})

H = latent heat of fusion (J kg^{-1})

h = heat transfer coefficient between metal and coolant (W m^{-2} K^{-1})

k = thermal condutivity (W m^{-1} K^{-1})

r_o = radius of cylinder (m)

r_f = radius of solid/liquid interface (m)

S = thickness of a slab (m)

T_m = melting temperature (K)

T_o = temperature of the coolant (K)

T_v = initial temperature of the liquid metal (K)

ΔT = supercooling ($\Delta T = T_v - T_m$)　(K)

t = solidification time (sec)

V_ℓ = volume of remaining liquid metal during the solidification (m^3)

V_s = volume of solidified metal (m^3)

V_T = total volume of metal in the mould (m^3)

ρ = density (kg m^{-3})

Θ = dimensionless geometrical factor

ϕ = dimensionless solidification constant

Subscripts:

c = cylinder

ℓ = liquid metal

m = mould

p = slab

s = solid metal

REFERENCES

1. Carslaw, H.S. and Jaeger, J.C., Conduction of Heat in Solids, pp.282-296, Oxford University Press, London, 1959.

2. Ruddle, R.W., The Solidification of Castings, pp. 72-120, The Institute of Metals, London, 1957.

3. Geiger, G.H. and Poirier, D.R., Transport Phenomena in Metallurgy, pp. 329-360, Addison Wesley, Massachussetts, 1973.

4. Lighfoot, N.M.H., Proceedings London Mathematical Society, 31(1930) 97.

5. Shwartz, C., Zeitschrift für Angewandte Mathematik und Mechanik, 13(1933) 202.

6. Lyubov, Y., Doklay Akad. Nauk S.S.S.R. 68 (1949) 847.

7. Stefan, J., Ann. Phys. u Chem. 42 (1891)139.

8. Chvorinov, N., Die Giesserei 27 (1940) 177.

9. Garcia, A. and Prates, M., Met. Trans. B9 (1978) 449.

10. Garcia, A., Clyne, T.W. and Prates, M., Met. Trans. B 10 (1979) 85.

11. London, A.L. and Seban, R.A., Trans. A.S.M.E. 65 (1943) 771.

12. Shih, Y.P. and Tsay, S.Y., Chem. Eng. Sci. 26 (1971) 809.

13. Sarjant, R.J. and Slack, M.R., J. Iron Steel Inst. 177 (1954) 428.

14. Tao, L.C., A.I.Ch.E.J., 13 (1967) 165.

15. Santos, R.G. and Prates, M., Proceedings of the Eight Canadian Congress of Applied Mechanics, Moncton, June, 1981.

Multi-Phase Flow and Heat Transfer III. Part B: Applications   591
edited by T.N. Veziroğlu and A.E. Bergles
Elsevier Science Publishers B.V., Amsterdam, 1984 — Printed in The Netherlands

MATHEMATICAL HEAT TRANSFER MODEL FOR THE ANALYSIS OF SOLIDIFICATION OF BINARY
ALLOYS

Amauri Garcia
Faculty of Engineering
State University of Campinas - UNICAMP
13100 Campinas, S.P., Brazil

ABSTRACT

This paper presents a heat transfer model describing the temperature
distribution and the movement of solidus and liquidus isotherms during solidification of binary alloys. The model is completely analytical and is based on
the principle of respresenting the Newtonian thermal resistance by virtual
extra-thicknesses of material. The analysis can be used to account for solidification behaviour over a wide range of conditions including any case of time
variable metal/mold heat transfer coefficients. A series of experiments were
designed to check the model predictions and a comparison was made with the
predictions furnished by a finite difference technique. In order to assess
the application of the model to industrial practice, data were examined from
measurements made in steel plants and presented in the literature. The model
is also shown to be an useful tool to predict casting structure parameters.
It can be used to calculate the rate of displacement of solidus and liquidus
isotherms and temperature gradients, which are the fundamental parameters of
structure control. The agreement observed among model predictions, experimental results and numerical techniques was acceptable in all cases examined.

1. INTRODUCTION

Mathematical modelling is a powerful tool for process control and optimization in a number of areas, and in particular for processes involving
liquid/solid transformation like static and continuous casting, welding, purification, and crystal growth. The progress of solidification influences
strongly the structure characteristics in normal casting operations and have
to be carefully controlled in order to make possible purification and growth
of perfect crystals. A large number of approaches have been developed to
treat heat flow during solidification, and these techniques can be classified
in three main groups: exact analytical [1], heat balance integral methods
[2], and numerical models [3-5].

Mathematically exact treatments available until recently were limited to
the condition of perfect metal/mold thermal contact, what is physically
unrealistic in most practical cases [6-7]. When mathematical approximations
are permited more physically realistic boundary conditions can be adopted.
Popular among these approaches has been to arbitrarily describe the metal
thermal profile by a mathematical function [8]. A heat balance is then
commonly applied under given boundary conditions and generally numerical integration is necessary to yield the final solution. Most of them are

restricted to the case of the mold acting as a perfect heat sink, and limited to the case of congruently-freezing metals, although one solution of this kind has been derived dealing with the solidification of binary alloys [9]. Finally, there has been extensive application of numerical methods to this class of problem, most of them variants of the basic finite difference technique [3-5]. This represents a very powerful tool and exhibits enormous versatility.

Recently, the present author developed an exact analytical model to treat the generalized plane front solidification problem, whose basic form has been previously outlined for the case of solidification in cooled and massive molds [10-12]. An assessment was also made of some practical implications, and application to continuous casting and for cases of variable interfacial thermal resistance and finite melt superheat has also been described [7,13-14]. In this paper, the fundamental assumption of the above mentioned model, which is based on the mathematical expedient of representing the Newtonian thermal resistance by virtual adjuncts of material, is extended to cover the important case of alloys freezing with a wide mushy zone. The most important industrial application of such a model is for the continuous casting of steel, in which heat flow is essentially unidimensional, partly as a result of the low thermal conductivity of ferrous material. The model predictions are compared with measurements referring to the continuous casting of a 0,62% carbon steel as well as with results obtained during directional solidification of an Al4.5Cu alloy in a water cooled mold. In both cases measurements of the secondary dendrite arm spacings at varying distances of the surface of the two castings mentioned, are compared with model predictions associated with theoretical and experimental relationships related to the dendritic growth during solidification.

2. ANALYSIS

The model is completely analytical and is derived under a set of physical suppositions similar to those frequently assumed in analytical treatments of the solidification problem [10-13]. They are:
(a) The conductive heat flow is unidimensional
(b) The Newtonian interface resistance is represented by a heat transfer coefficient h
(c) The material properties are invariant within the same phase
(d) In the liquid only conductive heat flow is considered
(e) The latent heat of fusion evolved during solidification is taken into account by adjusting the specific heat over the range of solidification ($c_2 = c_3 + H/\Delta T$)

The fundamental assumption of the model involves considering the Newtonian resistance as equivalent to a pre-existing adjunct of material, which is introduced into a virtual system. In this way, in the virtual system the basic Fourier field equation is exactly applicable, and the solution obtained in this system can be related to the real system by simple relationships. Under these assumptions the model is completely described by the following equations:

position of solidus isotherm

$$t = \frac{S_s^2}{4a_1\phi_1^2} + \frac{L_o S_s}{2a_1\phi_1^2} + \frac{(L_o^2 - S_o^2)}{4a_1\phi_1^2} \tag{1}$$

position of liquidus isotherm

$$t = \frac{S_L^2}{4a_2\phi_2^2} + \frac{L_o S_L}{2a_2\phi_2^2}$$ (2)

Temperature distribution in solid

$$T_1 = T_o + \frac{(T_s - T_o)}{erf(\phi_1)} erf(\phi_1 \frac{x + L_o}{S_s + L_o})$$ (3)

Temperature distribution in mushy

$$T_2 = T_L - \frac{(T_L - T_s)}{erf(\phi_2) - erf(n\phi_1)} \left[erf(\phi_2) - erf(\phi_2 \frac{x + L_o}{S_L + L_o}) \right]$$ (4)

Temperature distribution in liquid

$$T_3 = T_p - \frac{(T_p - T_L)}{1 - erf(m\phi_2)} \left[1 - erf(m\phi_2 \frac{x + L_o}{S_L + L_o}) \right]$$ (5)

ϕ_1 and ϕ_2 are the solidification constants, determined by:

$$\frac{(T_L - T_s)}{erf(\phi_2) - erf(n\phi_1)} = \frac{k_1 \exp\left[(n^2 - 1) \phi_1^2\right] (T_s - T_o)}{k_2 n \, erf(\phi_1)}$$ (6)

$$\frac{(T_L - T_s)}{erf(\phi_2) - erf(n\phi_1)} = \frac{k_3 \exp\left[(1 - m^2) \phi_2^2\right] (T_p - T_L)}{k_2 \left[1 - erf(m\phi_2)\right]}$$ (7)

and S_o and L_o are given respectively by:

$$S_o = \frac{2 k_1 \phi_1 (T_s - T_o)}{\sqrt{\pi} \, erf(\phi_1) \exp(\phi_1^2) (T_L - T_o) h}$$ (8)

$$L_o = \frac{\phi_2}{n\phi_1} S_o$$ (9)

where:

a   — thermal diffusivity ($m^2 s^{-1}$)
c   — specific heat (J/kg K)
h   — Newtonian heat transfer coefficient (W/m^{2}K)
H   — latent heat of fusion (J/kg)
L_o — thickness of total pre-existing adjunct to metal in virtual
        system (m)
k   — thermal conductivity (W/m K)

m — constant of metal = $(a_2/a_1)^{1/2}$ (dimensionless)
n — constant of metal = $(a_1/a_2)^{1/2}$ (dimensionless)
S — thickness of solid pre-existing adjunct to metal in virtual system (m)
t — time (s)
T — absolute temperature (K)
T_0 — temperature of mold cooling fluid (K)
ΔT — non-equilibrium solid/liquid temperature range (K)
x — distance from metal/mold interface (m)

Subscripts
g — global
L — liquidus
p — pouring
S — solidus
1,2 and 3 respectively solid, mushy and liquid

For the continuous casting of alloys, equation from (1) to (9) are applicable over the volume element shown in Figure 1, taken as representative of the ingot behaviour during the process. These equations can also be used to calculate the solidification parameters which control the structure characteristics. Cellular spacings and dendrite arm spacings depend on thermal gradient (G_L), growth rate (V_L) and local solidification time (t_{SL}) is defined as the time required for a given fixed location to go from the liquidus temperature to the solidus temperature. These spacings are related with G_L, V_L or

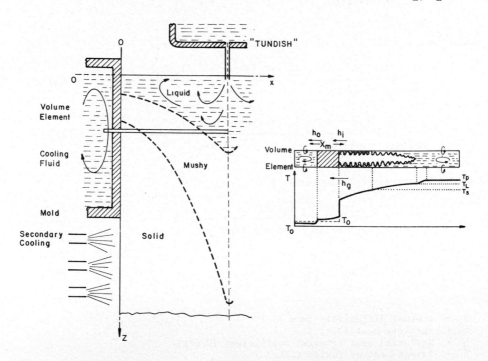

Fig. 1. Illustration of sectional geometry in the continuous casting mold.

t_{SL} via experimental and theoretical equations [15]. The theoretical analysis of Hunt permits the determination of cellular and primary dendrite arm spacings and is given by [16]:

$$\lambda_1 = 2,83 \ (\Theta \ D_3 \ \Delta T \ k')^{\frac{1}{4}} \ G_L^{-\frac{1}{2}} \ V_L^{-\frac{1}{4}} \tag{10}$$

where:

λ_1 = cellular or primary dendrite arm spacing [μm]
Θ - Gibbs-Thompson coefficient [K.m]
D - solute diffusivity in liquid [m^2/s]
k' - partition coefficient [dimensionless]

Thermal gradient and growth rate can be determined respectively by equations (5) and (2) of the proposed solidification model and are given by:

$$G_L = \left(\frac{\partial T_3}{\partial x} \right)_{x=S_L} = \frac{2 \ (T_p - T_L) \ m \ \phi_2}{\sqrt{\pi} \ [1 - \mathrm{erf}(m\phi_2)] \ \exp(m^2\phi_2^2) \ (S_L + L_o)} \tag{11}$$

$$V_L = \frac{2a_2\phi_2^2}{S_L + L_o} \tag{12}$$

For the case of secondary dendrite arm spacings (λ_2) correlation with thermal variables have the form [15]:

$$\lambda_2 = C_1 \ t_{SL}^{C_2} \tag{13}$$

where C_1 and C_2 are theoretical or experimental constants depending or the approach used to derive the relationship. In this case, equations (1) and (2) of the solidification model can be used to calculate t_{SL} ($t_{SL} = t_{S-LL}$). It is known that mechanical properties as well as the degree of homogeneity which can be obtained in a cast alloy with homogenization treatment depend on these segregate spacings.

3. RESULTS AND DISCUSSION

Monitoring the progress os solidification of alloys is generally more difficult than with pure metals because the growth front extends over an appreciable mushy zone. Information may be obtained via thermocouples embedded in the casting, and by using a dipstick arrangement. In the last case uncertainties will be generated as to the depth the dipstick will penetrate into mushy zone Figure 2 shows the progress of solidification during freezing of an Al 4.5 Cu alloy against a thin water-cooled mold. Dipstick and thermocouple measurements are compared with theoretical predictions furnished by equations (1) and (2) as well as with predictions given by a finite difference technique [17]. Figure 3 shows a similar comparison concerning the cooling of two points of the ingot during solidification. It can be seen, in both cases, that the model predictions give good correlation with experimental thermal data and with the finite difference technique, and are compatible with dipstick readings in terms of the slight expected penetration of the dipstick into the mushy zone.

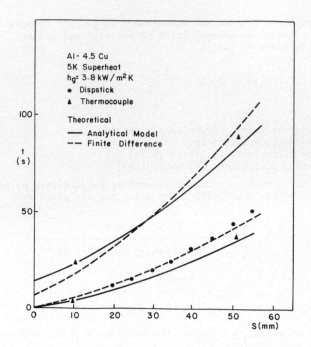

Fig. 2. Comparison between experimental results and calculated mushy zone progression for chill freezing of Al-4,5 Cu.

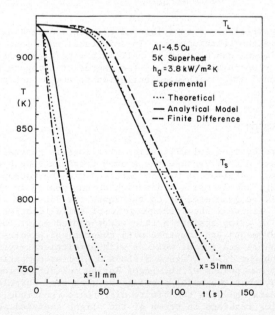

Fig. 3. Comparison between experimental and calculated thermal histories.

The solidification model can also be used to investigate the influence of thermal variables like heat transfer coefficient (h) and degree of super-heat, on the progress of solidification Figure 4 shows the influence of h on the displacement of solidus and liquidus isotherms. As expected, the solidus isotherm is much more sensitive on modification of h, and sensible differences on local solidification time can be attained by modifying the metal/mold inter facial thermal resitance.

For the case of Al-4,5 Cu, there are extensive experimental data in the literature concerning the secondary dendrite arm spacing, and these have been rationalized by Bower et al. into the equation [15]:

$$\lambda_2 = 7,5 \ t_{SL}^{0,39} \tag{14}$$

It is also possible to develop theoretical relationships, like the analysis of Feurer, which leads to [18]:

$$\lambda_2 = 12,1 \ t_{SL}^{0,33} \tag{15}$$

Metallographic examinations were carried out on the Al-4,5 Cu casting direct-ionally solidified in the water cooled mold already mentioned, at varying distances from the surface through which heat was extracted. Figure 5 shows the experimental points obtained and each one represents the average spacing between a number of secondary arms. These points are compared with theoreti-cal predictions based on the calculated local solidification time.

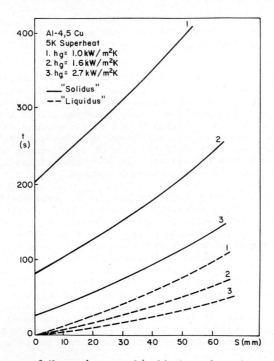

Fig. 4. Influence of Newtonian metal/mold thermal conductance (h).

598

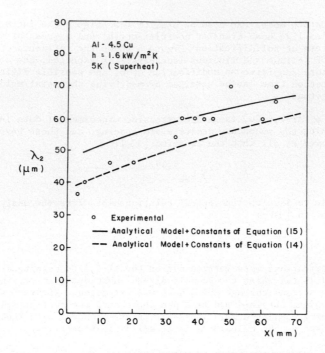

Fig. 5. Comparison between experimental and computed values of secondary
dendritic arm spacings (λ_2).

The above analysis can also be used to investigate the influence of
thermal variables like heat transfer coefficient (h) and superheat (ΔT_V) on
the secondary arm spacing. These influences are respectively analysed in
Figures 6 and 7, and it can be seen that these variables affect strongly the
values of λ_2. Figure 7 shows that higher values of ΔT_V produce smaller
secondary arm spacings. This is connected to the experimental evidence that
increasing progressively the superheat, the rate of displacement of the
liquidus isotherm is much more reduced then those of the solidus isotherm and
consequently the values of λ_2 are progressively reduced.

The analytical solidification model represented by equations from (1) to
(9) can also be used to analyse industrial situations like solidification
during continuous casting of steel. To simplify the application of the model
for this case, Figure 8 presents values of the solidification constants ϕ_1
and ϕ_2 in function of the steel carbon concentration. A comparison between
model predictions and experimental data (breakout) refering to solidification
of a 0,62% carbon steel in a continuous casting mold [13] is presented in
Figure 9. The experimental points lie near the dendrite tips, indicating that
liquid removal techniques do not give rise to significant dendrite remelt,
except a little low down in the mold. Another aspect of interest, is the
effect of carbon concentration in steel on the extent of the mushy zone. This
is analysed in Figure 10, where it can be seen that smaller concentrations
give rise to smaller values of local solidification times and consequently to
smaller values or of λ_2.

The secondary dendrite arm spacing can also be calculate for the solidi-

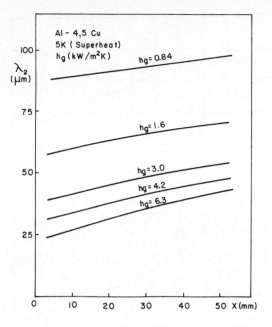

Fig. 6. Influence of Newtonian metal/mold thermal resistance on values of secondary dendrite arm spacings (λ_2).

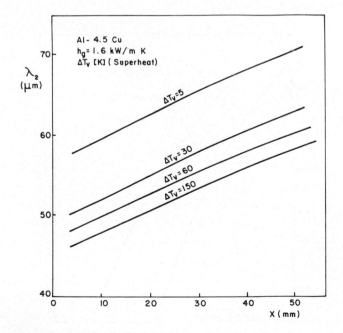

Fig. 7. Influence os superheat (ΔT_V) on secondary dendrite arm spacings (λ_2).

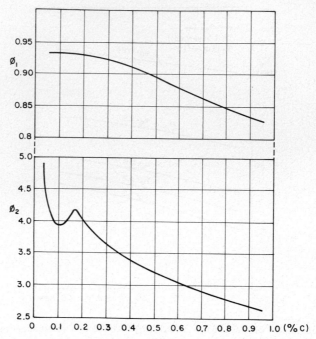

Fig. 8. Variation of solidification constants ϕ_1 and ϕ_2 in function of steel carbon content (% weight).

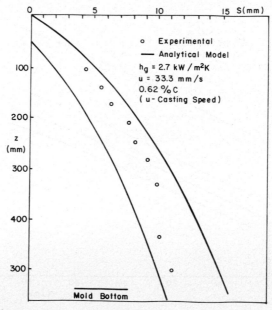

Fig. 9. Comparison between experimental and calculations referring to the progression of mushy of a 0,62% C steel in a continuous casting mold.

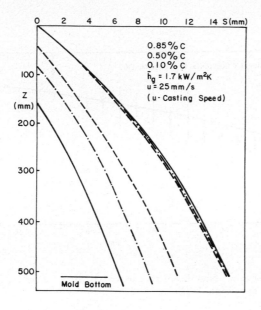

Fig. 10. Influence of steel carbon content on mushy zone progression in the continuous casting mold.

fication conditions observed in the continuous casting mold, by using equations (1) and (2) and the relationship obtained by Jacobi et al., for that steel composition, and given by [19]:

$$\lambda_2 = 15,8 \ t_{SL}^{0,44} \tag{16}$$

The comparison between theoretical predictions and experimental results is presented in Figure 11 and a good agreement is observed, expect near the casting surface. This is apparently due to the utilization of as mean value of global heat transfer coefficient along the continuous casting mold. It is well known that in the early begining os solidification in the continuous casting mold a much better metal/mold thermal contact is observed (higher values of h). If variation of the heat transfer coefficient along the mold was known, certainly a much better agreement would be observed near the casting surface.

4. CONCLUSIONS

It may be concluded that the analytical solidification model emerges as a useful tool to describe directional solidification of alloys, not only in certain static casting situations, but also in the important industrial application of the continuous casting of steels. The technique has the advantage of requiring considerable less computation, as well as easily allowing a complete description of the thermal characteristics. It can also be used to calculate rate of displacement of solidus and liquidus isotherms and thermal gradients, which are the fundamental parameters of structure control.

602

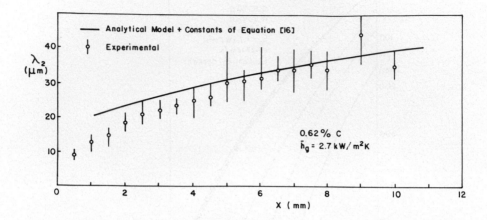

Fig. 11. Comparison between experimental and calculated values of secondary dendrite arm spacings (λ_2).

These together with cellular or dendritic growth models permit prediction of structure in any point from the casting surface.

ACKNOWLEDGEMENTS

The author would like to acknowledge financial support from FAPESP (the Scientific Research Foundation of the State of São Paulo, Brazil).

REFERENCES

1.   Mori, A. and Araki, K., _Int. Chem. Eng._, 1976, Vol. 16, pp. 734-743.

2.   Goodman, T.R., _Trans. ASME_, 1958, Vol. 80, pp. 335-342.

3.   Erickson, W.C., _AFS Int. Cast Metals_ J., 1980, Vol. 4, pp. 30-41.

4.   Ohnoka, I. and Fukusako, T., _Trans. Iron Steel Inst. Japan_, 1977, Vol. 17, pp. 410-418.

5.   Clyne, T.W., _Metal Sci._, 1982, Vol. 16, pp. 441-450.

6.   Clyne, T.W. and Garcia, A., _J. Mat. Sci._, 1981, Vol. 16, pp. 1643-1653.

7.   Clyne, T.W. and Garcia, A., _Int. J. Heat Mass Transfer_, 1980, Vol. 23, pp. 773-782.

8.   Hills, A.W.D., _Trans. TMS-AIME_, 1969, Vol. 245, pp. 1471-1479.

9. Miyazawa, K. and Muchi, I., <u>Trans. Iron Steel Inst. Japan</u>, 1975, Vol. 15, pp. 37-45.

10. Garcia, A. and Prates, M., <u>Metall. Trans.</u>, 1978, Vol. 9B, pp. 449-457.

11. Garcia, A., Clyne, T.W. and Prates, M., <u>Metall. Trans.</u>, 1979, Vol. 10B, pp. 85-92.

12. Garcia, A. and Prates, M., In: Veziroglu, T., <u>Multiphase Transport</u>, Hemisphere Publishing Co., N.Y., 1980, Vol. 1, pp. 487-513.

13. Clyne, T.W., Garcia, A., Ackermann, P. and Kurz, W., <u>J. Metals</u>, February 1982, pp. 34-39.

14. Garcia, A., Medeiros, M.D. and Prates, M., <u>Solidification Technology in the Foundry and Casthouse</u>, London, The Metals Society, to be published.

15. Bower, T.F., Brody, H.D. and Flemings, M.C., <u>TMS-AIME</u>, 1966, Vol. 236, pp. 624-634.

16. Hunt, J.D., <u>Solidification and Casting of Metals</u>, London, The Metals Society, 1979, pp. 3-9.

17. Garcia, A. and Clyne, T.W., <u>Solidification Technology in the Foundry and Casthouse</u>, London, The Metals Society, to be published.

18. Feurer, U., <u>Proc. Int. Symp. Eng. Alloys</u>, Delft, 1977, pp. 131-145.

19. Jacobi, H. and Schwerdtfeger, K., <u>Metall. Trans.</u>, 1976, Vol. 7A, pp. 811-820.

Multi-Phase Flow and Heat Transfer III. Part B: Applications 605
edited by T.N. Veziroğlu and A.E. Bergles
Elsevier Science Publishers B.V., Amsterdam, 1984 — Printed in The Netherlands

THE HEAT AND MASS TRANSFER OF A SMALL PARTICLE IN TURBULENT FLOW

Lang Wah Lee
Department of Mechanical Engineering
University of Wisconsin-Platteville
Platteville, Wisconsin 53818, U.S.A.

ABSTRACT

A mathematical model is presented for predicting the heat and mass transfer rate for a small spherical particle suspended in a turbulent fluid. The
root-mean-square particle-fluid relative velocity obtained from solving the
equation of motion is considered to be the equivalent convective velocity.
In accordance with a recent finding, the equation of motion used in this work
is without pressure gradient term. The resulting convective velocity is
found to be about an order of magnitude larger than that predicted by previous
works. The convective transfer rate evaluated from this convective velocity
agrees well with experimental data measured from the mass transfer process
in stirred tanks for particles ranging from 100 microns to 500 microns in
diameter.

1. INTRODUCTION

The heat and mass transfer for a small spherical particle suspended in
a turbulent fluid is an unsteady state process, and the instantaneous rate of
the transfer cannot be obtained analytically because the convective velocity
is a random function of space and time. The determination of the transfer
rate thus requires some forms of parameters to characterize the effect of
turbulence.

Friedlander [1] assumed without qualification that the heat and mass
transfer in dilute turbulent suspension of small particles was predominantly
influenced by the root-mean-square particle-fluid relative velocity (to be
called as rms relative velocity later in the text). This rms relative
velocity was found by solving the particle equation of motion derived by
Tchen [2] for a known turbulent flow field. Based on this approach,
Harriott [3] postulated a slip-velocity model in which the rms relative
velocity was used as an equivalent convective velocity to evaluate the
particle Reynolds number. This Reynolds number was then substituted into
the steady state forced convection equations by Ranz and Marshall [4] to
evaluate the Nusselt number Nu and Sherwood number Sh. The equations are
as follows:

$$Nu = 2 + 0.6 \ Re^{0.5}Pr^{0.33}$$
$$Sh = 2 + 0.6 \ Re^{0.5}Sc^{0.33}$$

(1)

The predicted transfer coefficients were then compared with the ones measured from the process in stirred tanks and the deviation between them were found to be as high as 250%. Harriott attributed such deviation to the unsteady state effect so he further suggested the transfer coefficients be estimated as a multiple of the transfer which would result if the particle is falling through a quiescent fluid at its terminal velocity. However, there are no reliable methods to predict this multiple. Other investigators have tried the slip-velocity theory and have met with little success.

New developments in the particle motion and heat and mass transfer provide the means to re-examine the slip-velocity theory. Such an endeavor seems worthwhile because the theory has the merit of simplicity and directness. In light of this thought, this paper will proceed to re-examine the theory. The major emphasis centers around two aspects. The first one is the particle-fluid relative motion (to be termed broadly as particle hydrodynamics) and the result shows that the slip-velocity found by previous investigations is incorrect. The second one is the way to predict the rate of heat and mass transfer (to be termed as particle heat and mass transfer) and the result indicates that equation (1) may not be the best correlation for predicting the transfer rate. The appropriateness of using the rms relative velocity as the convective velocity is also studied. The predicted results from the present method are then compared with the measured values and good agreement between them is achieved.

Finally, the limitation of the present method is discussed.

2. PARTICLE HYDRODYNAMICS

The rms values of the particle velocity and the particle-fluid relative velocity for a small spherical particle in turbulent suspension are determined by the equation of motion. While solving the equation of motion, a knowledge of the statistical description of the turbulent flow field is required. In what follows these two topics will subsequently be discussed.

2.1 Equation of Motion

The equation of motion used in previous works was derived by Tchen. In Tchen's equation, the motion of a spherical particle was affected by the Stokes drag, the pressure gradient in the fluid, the force to accelerate the added mass, the Basset history force and the force produced by a potential field. An inconsistency in Tchen's derivation of the equation was discovered by Soo [5] who proved that the pressure gradient in the fluid could not exert a net force on the particle and thus should not be included in the equation of motion. The equation of motion, after dropping the pressure gradient term from Tchen's equation and adding a correction factor to the viscous drag term to enlarge its range of applicability, takes the following form:

$$(4\pi/3)a^3\rho_p \, du_p/dt = 6\pi\mu aF(u-u_p) + (2\pi/3)a^3\rho\frac{d}{dt}(u-u_p) + 6a^2\sqrt{\pi\rho\mu}\int_{t_o}^{t}\frac{\frac{d}{d\tau}(u-u_p)}{\sqrt{t-\tau}}d\tau + f$$

$$(2)$$

where a is the radius of the particle

  u_p and u are the velocity component of the particle and the fluid, respectively

  μ is the viscosity of the fluid

ρ_p, ρ are the density of the particle and the fluid, respectively

   f is the gravitation force
   F is the correction factor to the Stokes drag

Equation (2) will be used in this model to evaluate the rms relative velocity.

   The assumptions made in deriving equation (2) were discussed in great detail by Hinze [6] and will not be repeated here. Of all the assumptions, the most stringent one is the no-overshooting condition which requires the particle to remain in the same eddy during its flight. To satisfy this relation, the particle must be many times smaller than the smallest eddy (about 50μm in diameter for the turbulent fluid in a stirred tank) and the density difference between the particle and the fluid should not be large. To satisfy the density requirement is not difficult in many cases, but the size requirement seems to place a severe restriction on the usefulness of equation (2). However, considering the fact that the energy content in the smallest eddies is low, it is conceivable that these eddies would have very little effect on the motion of the particle. Thus, from a practical point of view, equation (2) probably will hold for particles up to a few hundred microns in diameter, especially with inclusion of the correction factor for the Stokes drag coefficient. This correction factor F was given by Beard et al [7] as follows

$$F = 1 + cRe^a \tag{3}$$

where $c = 0.0806$, $a = 1$, for $0.1 < Re < 10$
         $c = 0.115$, $a = 0.802$, for $10 < Re < 20$
         $c = 0.189$, $a = 0.632$, for $20 < Re < 400$

   To evaluate the rms relative velocity, equation (2) will have to be recast in terms of the relative velocity u_R $(= u - u_p)$. Introducing u_R into equation (2) and rearranging the terms yields:

$$\dot{u}_R + Fa\beta u_R + \beta\left(\frac{3\alpha}{\pi}\right)^{\frac{1}{2}} \int_{t_o}^{t} \frac{\dot{u}_R}{t - \tau} + f - \left(1 - \frac{1}{3}\beta\right)\dot{u} = 0 \tag{4}$$

where the dot represents the time derivative following the particle motion,

   $\alpha = 3\nu/a^2$
   $\beta = 3\rho/(2\rho_p + \rho)$
   ν = kinematic viscosity of the fluid

If the pressure gradient term were included in the equation of motion, the coefficient in the last term of equation (4) would have been $(1-\beta)$ instead of the $(1-\frac{1}{3}\beta)$. The three components of u_R (in x, y, z direction) together with the free fall velocity are solved separately and then combined to yield the resultant relative velocity.

   Since the fluid velocity u is a random function of space and time, only the rms value of u_R can be obtained. In order to do this, a statistical description of the flow field, particularly the Lagrangian energy spectrum of the fluid flow, must be provided.

The analysis done in this paper is for the process in stirred tanks where a substantial body of experimental data on the hydrodynamics and heat and mass transfer was being published. This makes it possible to compare the predicted results with the measured ones. The hydrodynamics in a stirred tank is discussed in the following section.

2.2 The Lagrangian Energy Spectrum of the Flow

Hinze [6] proposed using an exponential function $\exp(-t/T_L)$ to describe the Lagrangian autocorrelation of the turbulent fluid. The corresponding energy spectrum $E_L(\omega)$ can then be obtained from the Fourier transformation of the exponential function and the result is

$$E_L(\omega) = (u_i'^2/\pi)\,[2T_L/(1+\omega^2 T_L^2)] \tag{5}$$

where T_L is the Lagrangian time scale
u_i' is the turbulent intensity
ω is the angular frequency

The Lagrangian time scale is related to u_i^2 and the dissipation rate ε. According to Hinze [6], the relation is

$$T_L = 0.2353\,\frac{u_i'^2}{\varepsilon} \tag{6}$$

The turbulent intensity and dissipation rate for the fluid in a stirred tank are related to geometrical configurations of the system and the operating condition. After summarizing previous investigations, Lee [8] proposed the following relations:

$$\varepsilon = \frac{4}{\pi}\,N_p \rho N^3 D^5/(g_c T^2 H) \tag{7}$$

$$u_i' = 0.676\,\sqrt[3]{N_p}\,\overline{W/D}\,\,ND^2/(T^2 H)^{1/3} \tag{8}$$

where N_p is the power number
N is the impeller speed
D is the diameter of the impeller
T is the diameter of the stirred tank
W is the impeller blade width

Thus the Lagrangian energy spectrum can be determined from a stirred tank operating at a given condition by combining equations (5), (6), (7) and (8).

2.3 The Calculated rms Relative Velocity

The relative velocity can be solved readily using the spectral analysis given by Chao [9] and the result is

$$u_R'^2 = \int_0^\infty \frac{\Omega_R}{\Omega}\,E_L(\omega)\,d\omega$$

where $\Omega_R = (1 - \frac{1}{3}\beta)^2\omega^2$

$\Omega_2 = [F\alpha\beta + \beta(\frac{3\alpha\omega}{2})^{0.5}]^2 + [\omega + \beta(\frac{3\alpha\omega}{2})^{0.5}]^2$

$E_L(\omega)$ is the Lagrangian energy spectrum specified in equation (5).

Since the drag correction factor F in equation (4) is not known apriori, an iteration procedure is required to obtain the solution. The detail of the procedure was given by Lee [8].

Calculation is carried out under the conditions parallel to those reported in Harriott's experiment [3]. The calculated results are presented in Fig. 1 alongside the effective slip-velocity and the terminal velocity obtained from Fig. 14 of Harriott's paper [3]. The effective slip-velocity is the convective velocity required to achieve the measured mass transfer rate and is calculated from measured mass transfer data via equation (1). The terminal velocity was considered by Harriott to be the lower limit of the slip-velocity. The big difference between the effective slip-velocity and the terminal velocity was pointed out by Harriott to show the inadequacy of the slip-velocity method. However, the slip-velocity (or the rms relative velocity) predicted by the present method is very close to the effective slip-velocity, indicating the rms relative velocity evaluated from the present method can be used as the convective velocity.

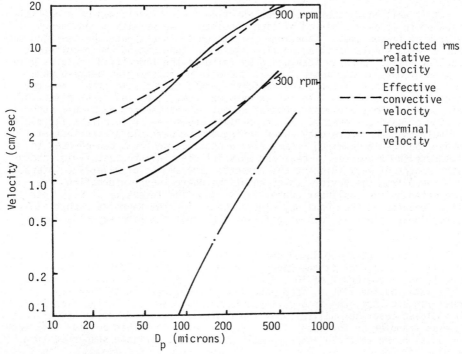

Fig. 1. Comparison between the Predicted rms relative Velocity, the Terminal Velocity and the Effective Convective Velocity based on Mass Transfer Coefficient System, NaOH-H_2O-Hr

3. THE PARTICLE HEAT AND MASS TRANSFER

For a particle in turbulent suspension, the relative velocity between the particle and the surrounding fluid constitutes the convective velocity. Since the relative velocity is a random function of space and time, the process is unsteady and the instantaneous heat and mass transfer rate cannot be obtained analytically. The transport rate can only be obtained in an average sense.

In this paper, the rms relative velocity obtained with the method mentioned in previous section is considered as the convective velocity. From this convective velocity, the transfer rate can be evaluated by the use of the steady state forced convection equation. Such a procedure needs justification because the unsteady state effect is not considered. Moreover, a realistic equation for predicting the heat and mass transfer rate has to be selected among the many existing equations. These two aspects will be discussed in the following sections.

3.1 The Unsteady State Effect

Experimental evidence from heat transfer in an oscillating flow field will be quoted to assess the unsteady state effect. It is therefore necessary to establish a relationship between the heat transfer in an oscillating field and that in a turbulent flow field.

It is well known that the turbulent flow field experienced by a particle suspended in it over a period of time consists of waves with a continuous spectrum of frequencies. Thus, one is tempted to investigate the heat and mass transfer in a simple oscillating flow field and then extend the result to a turbulent field. Such an idea would be valid if the flow field contains only one kind of wave at one moment, or if there is no interaction between waves when the particle is acted on by different waves at the same time. For the latter case, if there is interaction between waves, then the total effect of these waves on heat and mass transfer cannot be represented by simple super-position of wave components. For a particle suspended in turbulent fluid, it is likely that there are many waves acting on the particle at one instant. Whether the result obtained from a simple oscillating fluid can be used to illustrate the character of heat and mass transfer in turbulent fluid depends on the amount of wave interactions present. The amount of wave-wave interaction can be seen from the Navier-Stokes equations where these interactions are represented by the nonlinear convective terms. The intensity of such inter-actions depends on the local Reynolds number. For low Reynolds number flow, the nonlinear effect is less significant than that for high Reynolds number flow.

For a small particle suspended in a turbulent fluid, the nonlinear effect may not be significant if the Reynolds number calculated from the rms relative velocity is of the order of 10. In fact, the Navier-Stokes equation can be approximated by the linear Oseen equation at the lower end of this Reynolds number range. Thus, the information obtained from an oscillating flow field can at least describe qualitatively the character of low Reynolds number heat and mass transfer in turbulent flow. Approximation can then be made that each wave acts on the particle independently for a small particle suspended in a turbulent fluid.

In what follows some of the published heat and mass transfer data from oscillating flow field are discussed. It must be emphasized that this type of flow field is not used to replace the actual turbulent field. Instead, the character of the heat and mass transfer from an oscillating flow field is used to illustrate the character of the transfer process of a small particle in a turbulent field.

The convective heat and mass transfer in oscillating flow field is an unsteady state process. Due to the effect of thermal inertia, the temperature response curve for low frequency oscillations may be quite different from the one for high frequency oscillations. In other words, the transient effect needs to be considered.

Analytical solution to this problem is still unknown when the oscillating flow changes both in magnitude and direction (means the stagnation point is also changing). Since analytical results are not available, it is necessary to rely on experimental evidence to provide an estimation on the gross effect of vibration on heat transfer.

Experimental work on the heat transfer from a low frequency (4.25 cps) vibrating wire to originally still water was performed by Deaver et al. [10]. They showed that the empirical formula for steady state forced convection is still applicable provided that the average velocity of vibration over a half-cycle is taken as the convective velocity. Faircloth et al. [11] measured the heat transfer rate from a wire vibrating in a plane normal to an air stream (the Reynolds number calculated from the instantaneous velocity is below 15) and found the time average Nusselt number increased by only 10 percent above that found for a steady flow condition.

From the above review, it is postulated that heat transfer from a small object to an oscillating flow field can be approximated by the use of heat and mass transfer formula based on steady state condition with the appropriate choice of characteristic velocity as the equivalent convective velocity. In this model the rms relative velocity is taken as the equivalent convective velocity. Faircloth's experimental data is used to lend support to this hypothesis since the flow situation in Faircloth's experiment is rather similar to that of a particle suspended in turbulent flow.

Faircloth's data is presented in Fig. 2, the measured average Nu was given as 0.8. To check our hypothesis, the rms relative velocity is calculated from the data and is found to be 3.74 ft/sec. The Reynolds number calculated from this rms relative velocity is 1.25 (the mean fluid temperature is taken as the reference temperature to find the fluid properties). The steady state heat transfer formula from a cylinder to air was given by McAdams [12] as:

$$Nu = 0.32 + 0.43 \ Re^{0.52} \qquad\qquad 1 < Re < 1000$$

Substitute the Reynolds number into the above equation yields a Nusselt number of 0.804 which is very close to the measured average Nu of 0.8. Thus, the example provides some support to the hypothesis that the rms relative velocity can be used as the equivalent convective velocity for an unsteady state heat and mass transfer process.

3.2 The Prediction of Heat and Mass Transfer Rate

The rate of heat and mass transfer from a sphere is significantly altered by ventilation and has been the subject of numerous theoretical and experimental investigations. Dimensional analysis for forced convection indicates

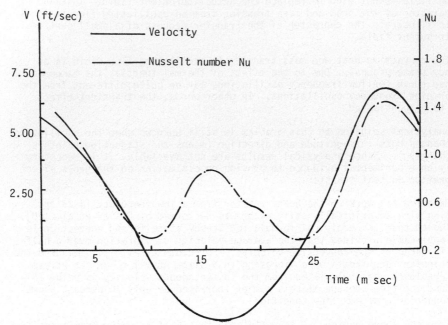

Fig. 2 Plot of Wire Velocity & Nu number vs. time (Data from Faircloth [11])

that Nu is a function of Re and Pr and Sh is a function of Re and Sc. In the case of convective heat transfer from a flat plate, laminar boundary layer theory shows that the Nu can be approximated by the expression

$$Nu = 0.664 \ Pr^{0.33}Re^{0.5}$$

for $Pr \simeq 1$. Many investigators have attempted to correlate data from other bodies including spheres by generalizing this expression to

$$Nu = A + BPr^{0.33}Re^{0.5}$$

When the Reynolds number is zero, heat is transferred from the sphere by pure conduction and the Nu is theoretically equal to 2. This limit is often used to define the constant A.

The diffusion of heat and mass is analogous for linear diffusion and, therefore, it should be possible to obtain the Sherwood number from the formula for Nusselt number by replacing the Prandtl number with the Schmidt number. Frössling [13] was the first to give a correlation of this form for heat and mass transfer from spheres and suggested the following forms

$$Nu = 2 + 0.552 \ Pr^{1/3}Re^{1/2}$$
$$Sh = 2 + 0.552 \ Sc^{1/3}Re^{1/2}$$

The correlation expressed in equation (1) which is widely used in many textbooks is another example along this line of thought.

For Reynolds number less than one and Peclet number Pe > 100, Friedlander [14] suggested the following equation

$$Sh = 0.99 \ Pe^{0.33} \tag{9}$$

Recent studies have shown that the exponent number (= 0.33) of the Prandtl and Schmidt number in equation (1) is inadequate when the Reynolds number is larger than one. In fact, the 0.33 value is obtained from the case of developing boundary layer which probably exists in the front half of the sphere (near the stagnation point). This would not apply to the wake region where the value of the exponent may change to 0.5 for unsteady state flow. In view of this, Whitaker [15] proposed the following equation for a sphere:

$$Nu = 2 + (0.4 \ Re^{0.5} + 0.06 \ Re^{0.667}) \ Pr^{0.4} \tag{10}$$

Considering the similarity between heat and mass transfer, equation (10) can be used for predicting mass transfer coefficient by replacing the Prandtl number with the Schmidt number, thus

$$Sh = 2 + (0.4 \ Re^{0.5} + 0.06 \ Re^{0.667}) \ Sc^{0.4} \tag{11}$$

To sum up, equation (9) will be used in the calculation for the flow where Re is less than one and Pe is larger than 100, while equations (10) and (11) are used when the Reynolds number is higher than one.

4. CALCULATED RESULTS

The rms relative velocities obtained with the method described in section 2 are used in equations (9), (10), (11) to predict the heat and mass transfer rate. The predicted mass transfer rates are then compared with Harriott's experimental data [3]. The detail description of the experimental conditions in Harriott's paper enables us to evaluate the Lagrangian energy spectrum with the method mentioned in section 2.3. Comparison for heat transfer is impossible because of the lack of heat transfer data.

The mass transfer data provided by Harriott were obtained from measuring the neutralization of ion exchange beads (Dowex 50-WX8) in stirred tanks. The physical properties of the solids and the solutions used in his experiment are listed in Table 1.

Table 1 Physical Properties at 20°C

Particle and Solution	Particle Density ρ_p (g/cc)	Viscosity μ (cp)	Diffusivity $Dv \times 10^5$ (cm/sec)	Schmidt No. Sc
NaOH + HR in water	1.24	1.0	1.93	518
NaOH + HR in .2% methocel solution	1.24	6.7	1.83	3670

The stirred tank used in the experiment is baffled and has a diameter of 4 inches. The depth of the solution is 5 inches. The impeller is a six-blade turbine type with a diameter of 2 inches and a width-diameter ratio of 0.4. The operating speed is 300 rpm and the power number for the system is estimated to be 5 from information provided by Bates et al. [16].

The predicted Sherwood number is then converted into the mass transfer coefficient k_c through the following relation

$$k_c = \frac{D_v}{d} \text{Sh}$$

where D_v is the diffusivity

d is the particle diameter

The values of k_c are then compared with the measured ones provided by Harriott. The results are presented in Tables 2 and 3.

Table 2 The predicted and measured k_c for NaOH + HR in 0.2% methocel solution

| Particle dia. (cm) | Predicted values | | Measured k_c (cm/sec) | % difference between predicted and measured values |
	Particle Reynolds number	k_c (cm/sec)		
0.007	0.029	0.0123*	0.0165	25%
0.01	0.07	0.0116*	0.0135	14%
0.02	0.34	0.00975*	0.009	8.3%
0.03	0.806	0.0079	0.0076	3.9%
0.04	2.125	0.0076	0.007	8.6%

"*" denotes that the value is calculated from Sh evaluated from equation (9)

Table 3 The predicted and measured k_c for NaOH + HR in H_2O

| Particle dia. (cm) | Predicted values | | Measured k_c (cm/sec) | % difference between predicted and measured values |
	Particle Reynolds number	k_c (cm/sec)		
0.005	0.332	0.0213*	0.028	24%
0.007	0.708	0.0195*	0.022	11%
0.01	1.50	0.0172	0.018	4.4%
0.02	5.64	0.0153	0.014	9.2%
0.03	11.37	0.0143	0.0125	14%
0.04	18.19	0.0134	0.0118	14%
0.05	25.9	0.0128	0.0112	14%

"*" denotes that the value is calculated from Sh evaluated from equation (9)

5. DISCUSSION

Comparison shown in Tables 2 and 3 indicates that the present method works well for particles ranging from 100 microns to 500 microns in diameter. When the particle is smaller than 100 micron, the predicted mass transfer coefficients are significantly lower than the measured ones. This may be due to the unsteady state effect caused by the smallest eddies. These eddies may be able to penetrate the boundary layer of a very small particle because the inertia of the particle is not large enough to retard the motion of the smallest eddies. Such penetration would enhance the transport process and the effect was discussed by Harriott [17]. Thus the heat and mass transfer equation used in this paper may not hold for particles smaller than 100 microns in turbulent flow.

As particle size increases to 200 micron, the predicted transfer rates become larger than the measured ones. This may be due to the over-prediction of rms relative velocity from equation (4). According to Odar et al. [18], the force to accelerate the added mass and the Basset force term needed to be modified when the particle Reynolds number is larger than one. Thus, the force term expressed in equation (4) may not accurately represent the true value acting on the particle. Moreover, it must be kept in mind that the equation of motion was derived under the assumption of no-overshooting, which can be observed only for particles many times smaller than the smallest eddies. However, the agreement between the predicted and measured results seems to indicate this restriction can be relaxed and the equation of motion can be used for particles up to a few hundred microns in diameter.

The slip-velocity model thus gives an adequate account of the unsteady state heat and mass transfer process for a small spherical particle suspended in turbulent flow. However, when the particle diameter is less than 100 microns, the motion of the smallest eddies would modify the heat and mass transfer mechanism and the equation for unsteady state heat and mass transfer probably would be more appropriate.

REFERENCES

1. Friedlander, S.K., "Mass and Heat Transfer to Single Spheres and Cylinders at Low Reynolds Numbers" AIChE Journal, Vol. 3, p. 47, 1957.

2. Tchen, C.M., "Mean Value and Correlation Problems Connected with Motion of Small Particles Suspended in a Turbulent Fluid", Ph.D thesis, Delft, 1947.

3. Harriott, P., "Mass Transfer to Particles: Part I", AIChE Journal, Vol. 8, p. 93, 1962.

4. Ranz, W.E. and Marshall, W.R. Jr., "Evaporation from Drops, I", Chem. Eng. Prog., Vol. 48, p. 141, 1952

5. Soo, S.L., "Net Effect of Pressure Gradient on a Sphere", The Physics of Fluids, Vol. 19, No. 5, 1976

6. Hinze, J., Turbulence, Second Edition, McGraw Hill Co., 1975

7. Beard, D.V. and Pruppacher, H.R., "A Determination of the Terminal Velocities and Drag of Small Water Drops by Means of a Wind Tunnel", J. Atmos. Sci., Vol. 26, p. 1066, 1969

8. Lee, L.W., "The Relative Fluid-Particle Motion in Agitated Vessels", AIChE Symposium Series, 208, Vol. 77, p. 162, 1981

9. Chao, B.T., "Turbulent Transport Behavior of Small Particles in Dilute Suspension", Osterrichisches Ingenicur - Archiv, Vol. 18, 7, 1964

10. Deaver, F.K., Penny, W.R. and Jefferson, T.B., "Heat Transfer from a Horizontal Wire to Water", Trans. ASME, J. Heat Transfer, 84, p. 251, 1962

11. Faircloth, J.M. Jr. and Schaetzle, W.J., "Effect of Vibration on Heat Transfer for Flow Normal to a Cylinder", Trans. ASME, J. Heat Transfer, 91, p. 140, 1969

616

12. McAdams, W.H., _Heat Transmission_, McGraw-Hill Co., 1942.

13. Frössling, N., "On the Evaporation of Falling Drops", _Beitr. Geophys_, Vol. 52, p. 170, 1938

14. Friedlander, S.K., _AIChE J._, Vol. 7, p. 347, 1961

15. Whitaker, S., "Forced Convection Heat Transfer Correlations for Flow in Physics, Past Flat Plates, Single Cylinders, Simple Spheres, and for Flow in Packed Bed and Tube Bundles", _AIChE Journal_, Vol. 18, p. 361, 1972

16. Bates, R.L., Fondy, P.L. and Corpstein, R.R., "Examination of Some Geometric Parameters of Impeller Power", _Ind. Eng. Chem. Proc. Design Develop._, 2, p. 310, 1962

17. Harriott, P., "A Random Eddy Modification of the Penetration Theory", _Chem. Eng. Sci._, Vol. 17, p. 149, 1962

18. Odar, F. and Hamilton, W. S., "Forces on a Sphere Accelerating in a Viscous Fluid", _J. Fluid Mech._, Vol. 18, p. 302, 1964

Multi-Phase Flow and Heat Transfer III. Part B: Applications   617
edited by T.N. Veziroğlu and A.E. Bergles
Elsevier Science Publishers B.V., Amsterdam, 1984 — Printed in The Netherlands

ANALYSIS OF THERMAL RADIATION EFFECTS ON VAPORIZATION OF
COAL-OIL MIXTURES

M. A. Colaluca,   J. K. Chou
Mechanical Engineering Department
Texas A & M University
College Station, Texas   77843, U.S.A.

ABSTRACT

    Calculations were made to determine the effect of absorp-
tion of continuum black body radiation on the solution to
theoretical single droplet vaporization. Differential equations
describing the process were derived assuming quasi-steady state
conditions in the vapor field surrounding the particle. The
energy equations for the droplet and coal particles included
absorption terms. The liquid fuel was taken to be kerosene
since absorptivity data were readily available for this substance. The equations were numerically integrated and the
results compared to solutions obtained for kerosene-ignoring
radiation.

    The results showed that a 25 micron Coal-Oil Mixture (COM)
droplet subjected to radiation from a 2300°K source vaporized
up to 30% more rapidly than the same droplet without radiation.
The vaporization time reduction increased with droplet size and
radiation source temperature, however, and optimum coal concentration was observed. Little difference in vaporization time
was observed for coal concentrations greater than 50%.

1.   INTRODUCTION

    The combustion of mixtures of coal particles and fuel oil
is an attractive method of extending dwindling petroleum supplies. Finely ground coal can be mixed with liquid fuel oil to
form a mixture that behaves as a liquid. The mixture can be
sprayed into conventional combustion chambers and burned without
having to significantly modify the combustion equipment. A basic
understanding of the combustion characteristics of these sprays
is essential to the design of efficient combustion systems.

    One approach to developing basic understanding of spray
combustion processes is to study vaporization and subsequent combustion of single droplets [1]. The results for single droplet
studies are then extended to predict the behavior of sprays consisting of collections of single droplets. The advantage of
this approach is that, ideally, the single droplet vaporization

and combustion process can be modelled theoretically and a
detailed analytical study made. In addition, careful experiments can be conducted on single droplets closely matching the
analytical model [2].

The theory of single droplet vaporization and combustion is
well established for one-component liquid fuel [e.g., 3-6].
These studies are carried out assuming the absorption of radiation by the droplet is negligible. Berlad and Hibbard [7]
concluded that fuels generally absorb at wavelengths different
from those of typical flame emission. Other studies [8,9] have
also concluded that radiation is unimportant for typical single
component droplet processes.

However, Hottel et al. [8] did demonstrate that radiation
might be significant in the case of furnace-fired residual oils.
Continuum radiation from soot and other particulates could provide radiation at wavelengths corresponding to fuel droplet
absorption characteristics. Furthermore, the addition of solids
can increase the overall absorption of the mixture droplet
through solid gray-body absorption [7,10].

The purpose of this study was to determine the effect of
absorption of continuum black body radiation on the solution to
theoretical Coal-Oil Mixture (COM) droplet vaporization. The
classical quasi-steady solution was assumed to describe the
vapor flow field surrounding the particle. The energy equations
including radiation terms were written for the fuel liquid and
solid coal phases assuming spatially uniform but time varying
temperatures for both phases. The liquid fuel was specified to
be kerosene since experimental absorption data was readily
available for this substance. The equations were numerically
integrated and the results compared to solutions obtained by
neglecting radiation absorption by the COM droplet.

The inclusion of radiation absorption did increase the
vaporization rate for COM when compared to non-absorbing droplets. A 30% increase in vaporization rate was observed for a
25 micron droplet containing a 10 micron coal particle when
subjected to radiation from a 2300°K black body source. In addition, the vaporization rate for absorbing COM droplets was
greater than the rate for absorbing fuel droplets without coal.
An optimum coal concentration was observed. That is, the
greatest increase in vaporization rate occurred for the 25
micron droplet containing a 10 micron coal particle. Droplets
with more of less coal experienced a lesser degree of increased
rate.

2. ANALYSIS

The model for vaporization process is shown schematically
in Figure 1. A spherically symmetric coal particle of radius c
is surrounded by a spherically symmetric fuel droplet of radius
b. Heat crosses the boundary of the fuel droplet by radiation
and conduction from the vapor field. Some of the radiation is
absorbed and some is transmitted to the coal particle. Heat

Table 1 Physical Properties

Diffusion Coefficient - D_V $(\frac{cm^2}{sec})$ .1

Vapor Specific Heat - C_{pv} $(\frac{cal}{gm\ °K})$

Vapor Thermal Conductivity - λ_v $(\frac{cal}{cm\ sec\ °K})$ .000025

Liquid Density - ρ_D $(\frac{gm}{cm^3})$ .75

Liquid Specific Heat - C_{PD} $(\frac{cal}{gm\ °K})$ .55

Heat of Vaporization - L $(\frac{cal}{gm})$ 78

Solid Density - ρ_C $(\frac{gm}{cm^3})$ 1.2

Solid Specific Heat - C_C $(\frac{cal}{gm\ °K})$ .25

Figure 1. Droplet Model

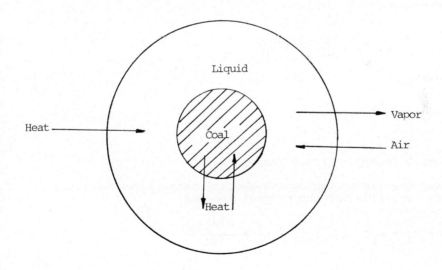

also crosses the coal particle boundary by radiation and conduction to the liquid phase of the droplet. The heat vaporizes
the fuel resulting in a concentration gradient of fuel vapor in
the field surrounding the droplet. The fuel vapor diffuses
toward the ambient while an inert (in this case air) diffuses
toward the droplet. It is assumed that no chemical reactions
occur.

The solution for the temperature and concentration profiles
have been developed for the vapor field around the droplet [e.g.,
11] and will not be repeated here. From these solutions, the
rate of mass flux and the rate of conduction heat flux at the
droplet surface are given as

$$G_b = \frac{\rho_v D_v}{b} \ln(1 + \frac{m_{v,o} - m_{v,\infty}}{1 - m_{v,o}}) \tag{1}$$

and

$$Q_o = \frac{C_{p,v} G_b (T_o - T_\infty)}{\exp(\frac{G_b (pv^b)}{\lambda_v}) - 1} \quad . \tag{2}$$

The mass balance on the fuel phase yields,

$$\dot{m}_{in} = \dot{m}_{out} + \frac{dm_D}{dt}$$

or

$$0 = 4\pi b^2 G_b + \frac{d}{dt} (\frac{4}{3}\pi b^3 \rho_D)$$

and finally,

$$\frac{db}{dt} = -\frac{G_b}{\rho_D} \quad . \tag{3}$$

2.1 Droplet Energy - Liquid Phase

Under the assumption of not spatial temperature variation,
the energy equation for the droplet becomes

$$A_s Q_R + A_s Q_o + A_c Q_c = \dot{m}_D h_D + \frac{d(m_D u_D)}{dt} \tag{4}$$

where

$$A_s = 4\pi b^2$$

$$A_c = 4\pi c^2 \quad .$$

The radiation contribution is given by

$$Q_R = f \, \sigma \, T_\infty^4$$

where f = the fraction of emitted black body radiation absorbed by the liquid phase of the fuel droplet. The conduction heat flux at the liquid-solid interface is given by

$$Q_c = \frac{\lambda v}{c} (T_D - T_c) \quad .$$

When these expressions are substituted into (4), the energy balance for the liquid phase becomes

$$4\pi b^2 f \sigma T_\infty^4 + 4\pi b^2 Q_o + 4\pi \frac{c^2 \lambda_c}{c} (T_D - T_c) =$$

$$4\pi b^2 G_b L + \frac{4}{3} b^3 \rho_D C_{PD} \frac{dT_D}{dt}$$

(5)

where

$$\dot{m}_D = 4\pi b^2 G_b = -\frac{dm_D}{dt} \quad .$$

2.2 Droplet-Energy Solid Phase

The energy balance on the coal particle can be written as

$$A_c Q_R^* - A_c Q_c = \frac{d(m_c u_c)}{dt} \quad .$$

Allowing the solid to emit and absorb as a gray body, the radiation term can be written as

$$Q_R^* = (1-f) \, \sigma \, (T_\infty^4 - T_c^4) \quad .$$

Substituting this and the expression for Q_c into (6) yields

$$\frac{4}{3}\pi c^3 \rho_c C_c \frac{dT_c}{dt} = 4\pi c^2 (1-f) (T_\infty^4 - T_c^4)$$

$$+ 4\lambda c^2 \frac{\lambda_c}{c} (T_D - T_c) \quad .$$

(7)

2.4 Radiation Fraction - f

By definition

$$f_b = \frac{\int_0^\infty E_{\lambda,T} \, \alpha_{\lambda,b} \, d\lambda}{\sigma \, T_\infty^{\,4}}$$

where

f_b = fraction of black body radiation absorbed by a fuel droplet liquid phase of radius b, and

$\alpha_{\lambda,b}$ = spectral absorptivity of droplet with radius b at wavelength λ.

The spectral absorptivity can be determined [9] as

$$\alpha_{\lambda,b} = \mathrm{erf}\,(1.75 \; \gamma_\lambda b)^{\frac{1}{2}}$$

where the "smoothed" coefficient is given by

$$\gamma_\lambda = \frac{1}{x} \, \mathrm{erf}^{-1} \, \bar{\alpha}_{\lambda,x}$$

and $\alpha_{\lambda,x}$ is the band absorption measured for a thin layer x for wavelength band $\lambda \pm \Delta\lambda$.

Also,

$$f_c = \frac{\int_0^\infty E_{\lambda,T} \, \alpha_{\lambda,c}}{\sigma \, T_\infty^{\,4}} \, d\lambda$$

where f_c = fraction of black body radiation that would be absorbed by a fuel droplet liquid phase of radius c so that

$$f = f_b - f_c$$

The radiation fraction f was calculated as outlined above at each step during the numerical integrations of the differential equations (5) and (7). The kerosene band absorption data of reference [7] was used to calculate the "smoothed" coefficient. The physical properties needed were assumed constant for the calculations and are listed in Table 1.

3.   RESULTS AND DISCUSSION

Figure 2 shows the variation of droplet radius with time for a 25 micron COM droplet. Also included for comparison is the variation of radius for a pure fuel droplet without any radiation absorption (curve D). The effect of pure liquid phase absorption can be seen by comparing curves A and D. Curve A represents the solution for a vanishing small amount of

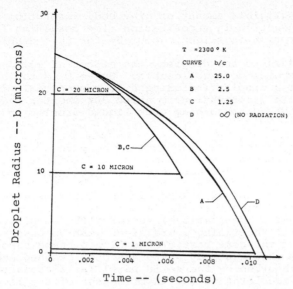

Figure 2. Effect of Coal Addition on Vaporization of 25 Micron Droplet.

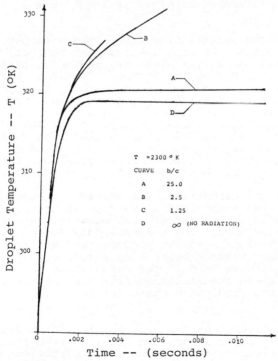

Figure 3. Effect of Coal Addition on Typical Droplet Temperature for 25 Micron Droplet

coal and thus a negligible amount of gray body absorption by
the solid. The reduction in vaporization time resulting from
liquid phase absorption is slight, about 5% for this droplet.

As coal is added to the droplet, the effect of solid gray
body absorption becomes more noticeable. Curve B is the variation of radius for a droplet including a 10 micron coal particle.
The time to complete vaporization is about 30% shorter than the
time to vaporize the non-absorbing pure liquid droplet to
10 microns.

As the percentage of coal increases, the percent difference between the absorbing and non-absorbing solutions decreases.
For example, only about a 10% decrease in vaporization time is
realized for the droplet with a 20 micron coal particle. Note
that the solutions for the 10 and 20 micron particles were
virtually identical during the lifetime of the 20 micron particle. At high coal concentrations, the droplets completely
vaporize while still undergoing a transient heat-up process.
The contribution of the coal particle to liquid phase vaporization is greatest once a steady state has been reached and all
of the radiation absorbed by the solid particle is conducted to
the liquid phase, accelerating the vaporization of the liquid.
During heat-up, the radiation is simply contributing to the
internal energy of the droplet and not the liquid phase vaporization and little difference is noted in liquid phase
vaporization time.

Figure 3 shows the variation of liquid phase temperature
for several COM droplets. For the 1 micron and no coal, no
radiation conditions, the liquid phase has reached a steady
state temperature early during the droplet lifetime. The
temperature for the 10 micron case is approaching a steady
state value as complete vaporization is approached. Some of
the gray body radiation is contributing to liquid vaporization
while still contributing to internal energy of the solid. On
the other hand, the highly loaded droplet (20 micron coal
particle) has barely emerged from the heat up process when complete vaporization occurs. Little time was available for the
absorbed radiation to contribute to droplet vaporization.

Figure 4 shows the effect of radiating temperature on
vaporization time. The higher radiative flux at the higher
temperature results in larger differences between the absorbing
and non-absorbing solutions. The spectral data for kerosene
indicates maximum absorption at 3.5 microns, which corresponds
to the wavelength near maximum black body emissive power for
the radiating temperatures used in the calculations. The temperature effect is due to the increased flux and not due to
better matching of the absorption characteristics for the liquid
phase.

The effect of particle size on the solution is shown in
Figure 5. The liquid phase absorption does contribute to a
difference in the solutions for the 50 micron particle. A 15%
reduction in vaporization time is noted for the absorbing 50
micron droplet with a vanishing coal particle while only a 5%

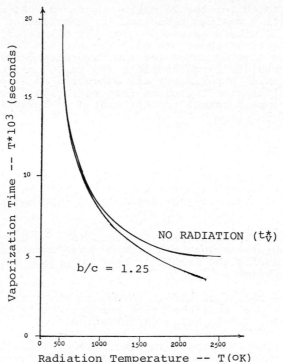

Figure 4. Effect of Temperature on Vaporization Time for 25 Micron Droplet.

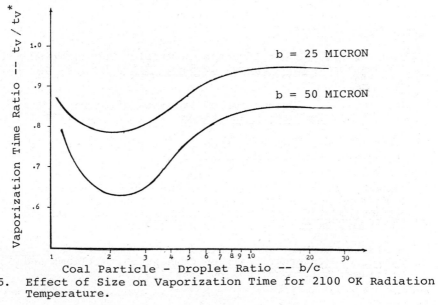

Figure 5. Effect of Size on Vaporization Time for 2100 °K Radiation Temperature.

reduction was noted for the 25 micron particle. As with the smaller droplet, a maximum reduction in vaporization time was noted for the 50 micron particle. For the 50 micron droplet, the maximum occurs at about b/c of 2 where a 40% reduction in vaporization time is noted for the absorbing droplet. It is important to consider large droplets (50 to 100 microns) since most single droplet experiments are conducted on droplets this size. This calculation indicates that radiation effects noticed in these experiments cannot be anticipated in "practical" droplet sizes.

4. CONCLUSIONS

1. For typical droplet size investigated (25 microns), the effect of radiation is greatest on the solution for vaporization times for COM droplets of intermediate coal loadings.

2. The effect of the radiation is negligible on highly loaded COM droplets because the liquid phase completely vaporizes during the transient heat up period of the droplet lifetime.

3. Most of the radiation effect is due to gray body absorption by the coal particle being conducted back to the liquid phase, accelerating vaporization.

4. For large "experimental" droplets, the contribution due to radiation absorption has a greater effect on vaporization time than it does for typical droplets. Also, liquid phase absorption as well as gray body solid absorption was important in the larger (50 micron) droplets.

REFERENCES

1. Law, C. K., "Recent Advances in Droplet Vaporization and Combustion", Prog. Energy Combust. Sci., Vol. 8, pp. 171-201 (1982).

2. Williams, A., "Combustion of Droplets of Liquid Fuels: A Review", Combustion and Flame, Vol. 21, pp. 1-31 (1973).

3. Faeth, G. M., "Current Status of Droplet and Liquid Combustion", Prog. Energy Combust. Sci., Vol. 3, pp. 191-224 (1977).

4. Godsave, G. A. E., "Studies of the Combustion of Drops in a Fuel Spray: The Burning of Single Drops of Fuel", Fourth Symposium (International) on Combustion, Williams and Williams, Baltimore, pp. 818-830 (1953).

5. Spalding, D. B., "The Combustion of Liquid Fuels", Fourth Symposium (International) on Combustion, Williams and Williams, Baltimore, pp. 847-864 (1953).

6. Goldsmith, M. and Penner, S. S., "On the Burning of Single Drops of Fuel in an Oxidizing Atmosphere", Jet Propulsion, Vol. 24, pp. 245-251 (1954).

7. Berlad, A. and Hibbard, R., "Effect of Radiant Energy on Vaporization and Combustion of Liquid Fuels", NACA RM E-52I09, Washington, D.C. (1952).

8. Hottel, H. C., Williams, G. C., and Simpson, H. C., "Combustion of Droplets of Heavy Liquid Fuels", Fifth Symposium (International) on Combustion, Reinhold, New York, pp. 101-129 (1955).

9. Friedman, M. H. and Churchill, W. S., "The Absorption of Thermal Radiation by Fuel Droplets", Chem Engng Prog Symp Ser, Vol. 61 (1965).

10. Braide, K. M., et al., "The Combustion of Single Droplets of Some Fuel Oils and Alternative Liquid Fuel Combinations", Journal of the Institute of Energy, Sept. 1979.

11. Spalding, D. B., Combustion and Mass Transfer, Pergamon Press, Oxford (1979).

Symbol List

A - surface area

b - instantaneous liquid phase droplet radius

c - coal particle radius

C_p - constant pressure specific heat

D - binary diffusion coefficient

$E_{\lambda,T}$ - spectral black body emissive power

G_b - vapor mass flux at droplet surface

h - enthalpy

L - heat of vaporization

m - mass (fraction with subscript)

$\dot{m}$ - mass flow rate

Q - heat flux

t - time

T - temperature

u - internal energy

λ_v - vapor thermal conductivity

λ - wavelength

ρ - density

σ - Stefan-Boltzman constant

Subscripts

C	- droplet solid phase
D	- droplet liquid phase
o	- gas-liquid interface
R	- radiation
v	- vapor
∞	- ambient

Superscripts

| * | - coal particle |

Multi-Phase Flow and Heat Transfer III. Part B: Applications
edited by T.N. Veziroğlu and A.E. Bergles
Elsevier Science Publishers B.V., Amsterdam, 1984 — Printed in The Netherlands

ANALYSIS OF TWO-PHASE FLOW IN A DIRECT COAL LIQUEFACTION PREHEATER

M. Perlmutter, R.M. Kornosky, and J.A. Ruether
U.S. Department of Energy
Pittsburgh Energy Technology Center
Pittsburgh, Pennsylvania 15236, U.S.A.

ABSTRACT

A theoretical analysis of two-phase flow and heat transfer is being performed to mathematically describe the phenomena occurring in a direct coal liquefaction preheater. The process being modeled features a high throughput recirculation loop that is capable of providing high mass velocities through an instrumented test section. Two 20-foot vertical tubes, 1.1-inch-i.d. and joined in a U-bend at the top, are included in the recycle loop and form the test section. The liquid phase consists of a slurry of coal-derived liquid and coal. The vapor phase consists primarily of hydrogen and coal-derived liquid vapors. Data collected from this apparatus, which is operable at conditions typical of pilot liquefaction plants, will be compared to the results of this analysis.

The analytical method includes separate continuity and momentum equations for each continuum phase, and a mixture energy equation. A mean flow approximation is used to represent fluid friction by velocity differences rather than by velocity gradients. The field equations are supplemented by friction factor and heat transfer correlations that are flow regime dependent. These correlations and flow regime maps have been developed. The thermophysical properties for the liquid and vapor phases will be determined using available process simulation programs.

1. INTRODUCTION

The Pittsburgh Energy Technology Center (PETC) of the U.S. Department of Energy is performing an experimental program that simulates a direct coal liquefaction preheater [1]. The function of the preheater is to heat hydrogen and a slurry of coal and coal-derived liquids to reaction temperatures ($\sim 800°F$) at elevated pressure (~ 2000 psi). Since the cost of the preheater represents a major capitol investment in the coal liquefaction process, improving our understanding can prevent expensive overdesigning.

A schematic diagram of the experimental preheater test unit being modeled is shown in Figure 1. This apparatus features a recirculation loop that is capable of providing high mass velocities through an electrically heated, instrumented test section. The test section consists of two 20-foot vertical tubes, 1.1-inch-i.d., joined with a U-bend at the top. The apparatus is operable at conditions of temperature, pressure, coal slurry concentration, and hydrogen and slurry rates typical of coal liquefaction

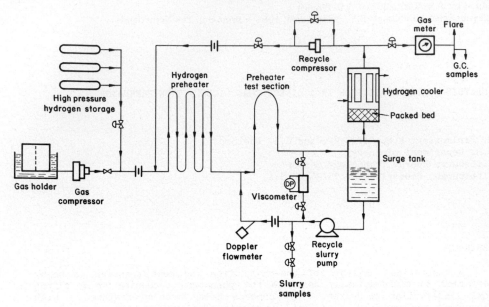

Figure I – Schematic diagram of the PETC preheater test unit

pilot plant preheaters. The liquid phase consists of a slurry of coal-derived liquid and coal, while the vapor phase consists primarily of hydrogen and coal-derived vapors. Data collected from this apparatus will be compared to the results of this analysis.

2. RECENT ANALYTICAL STUDIES OF MULTIPHASE FLOW

One of the simplest techniques for analyzing multiphase flow is the homogeneous equilibrium model (2). This model assumes that the multiphase fluid can be replaced by an equivalent pseudofluid that obeys the usual single fluid equations. The properties of the pseudofluid must then be defined, usually based on empirical correlations. Because of the complex physical and chemical processes and the high variability of conditions in the preheater, the simplified homogeneous model is unlikely to describe conditions in the preheater.

Through subsequent development the homogeneous mixture model was extended to the drift flux model. In this model only one mixture momentum equation and one energy equation are used; however, these equations contain terms that account for the effects of velocity differences between the phases. While the drift flux model predicts experimental data within its range of applicability, the model shows many of the shortcomings of the homogeneous mixture model because of its use of overall mixture correlations.

More recent analyses employ the multistream model (2). The multistream model assumes that the pressure of each stream is equal and the temperature of each stream either equal or unequal. Therefore, the various models are referred to as the UVET (unequal velocity, equal temperature) and UVUT (unequal velocity, unequal temperature). The multistream model allows direct incorporation of more complete descriptions of the physical processes occurring, including chemical reactions and solid particle streams.

2.1 Analysis of the PETC Preheater Test Unit

The analytical approach selected for use in mathematically describing the phenomena occurring in the PETC preheater test unit includes a two-stream model consisting of a liquid stream and a vapor stream. For each stream, continuity and momentum equations are written, while an overall energy equation is written for the combined stream. Wall friction and interstream friction are based on mean stream velocities rather than velocity gradients (2,3,4). The field equations are supplemented by friction factor and heat transfer correlations (2,4) that are flow regime dependent. The flow regime maps selected for use in this study are based on the results of Oshinowo and Charles (6). The simulation program "PROCESS" (5) was used to determine the liquid and vapor phase thermophysical properties.

The analytical technique employed in this analysis has been used extensively in studies dealing with water-steam systems characteristic of nuclear power reactors. This is believed to be the first attempt to apply this type of multistream model in the study of two-phase flow and heat transfer in direct coal liquefaction processes. Because of the wide variations in coal liquids, coal slurry properties, and flow regimes that occur in a coal liquefaction preheater, the more rigorous analytical procedure required by the UVET model is preferred over the homogeneous or the drift flux models. Although the existence of unequal stream temperatures may be significant, application of the UVUT model will not be considered as part of this study.

3. FLUID PROPERTIES

A knowledge of fluid properties as a function of temperature and pressure is required to analyze the preheater performance. An increasingly popular method of obtaining fluid properties is to use commercially available, comprehensive chemical process simulation programs such as "ASPEN Plus" (7) and "PROCESS" (5,8). As indicated previously, "PROCESS" was used to determine the liquid and vapor phase thermophysical properties required in the analysis.

As input variables to "PROCESS", hydrogen is specified as a library component, while the coal-derived liquid is specified by boiling point cuts having known normal boiling points and densities. Property parameters, such as the critical properties, molecular weight, and acentric factor, are available in the library for hydrogen. The property parameters for the coal-derived liquid cuts must be calculated. Several options that are available for this calculation include the Cavett, Lee-Kesler, and Twu options. The Twu option was used for the coal-derived liquid (5,8). Given the property parameters, "PROCESS" will then calculate the fluid properties, such as enthalpy, density, viscosity, and surface tension. Of the several options available for these calculations, the Soave-Redlich-Kwong method with the Lee-Kesler option for the density was used (5,8).

3.1 Fluid Properties Example

The example to be considered simulates hydrogen and coal-derived liquid being heated from 200°F to 800°F at 2000 psig. The stream flows are composed of 150 lb/hr of hydrogen and 2000 lb/hr of coal-derived liquid. The coal-derived liquid, referred to as CL1, has the characteristics of the process solvent from Wilsonville Run No. 164 containing 15 weight percent light solvent refined coal. The characteristics of the liquid are given in Table 1. The specific gravity is based on a Watson characterization factor of 10. For input into the program, the coal-derived liquid is divided into five pseudocomponents, with the normal boiling point and specific gravity entered for each pseudocomponent. The fluid properties as calculated by "PROCESS" are shown in Table 2. These fluid properties will subsequently be used to calculate the pressure drop and heat transfer, and to determine the flow regimes in two-phase flow.

4. FLOW REGIME MAPS

Prior to calculating the pressure drop and heat transfer, the flow regime must be determined. Of the many flow regime maps proposed, those based on the results of Oshinowo and Charles (6) are some of the few that consider both vertical upflow and vertical downflow. For this reason, their flow maps are presented in Figures 2 and 3. Also shown on these figures is the expected flow regimes that would be encountered if the preheater were operated at the process conditions identified in the previous fluid properties example given in Section 3.1 for a tube diameter of 1.1 inch. Based on these process conditions, Figure 2 shows that the frothy slug flow regime is expected in vertical upflow, while Figure 3 shows that for vertical downflow, the falling bubbly film or froth flow regime is expected.

4.1 Flow Patterns in Vertical Upflow

Flow patterns in vertical upflow as described by Hughes et al. (2), and Oshinowo and Charles (6) are shown in Figure 4. For increasing vapor flow rate, these are the following:

Bubble Flow. Bubbles are assumed to be uniformly distributed throughout the continuous liquid phase.

TABLE 1. - CHARACTERISTICS OF THE COAL-DERIVED LIQUID (CL1)

Pseudocomponent Number	Weight Percent	Normal Boiling Point, °F	Specific Gravity, 60°F/60°F
1	24.0	400	0.951
2	26.0	500	0.986
3	13.0	600	1.019
4	12.5	750	1.066
5	24.5	1000	1.134

TABLE 2. – PROPERTIES OF THE COAL-DERIVED LIQUID AND HYDROGEN AT A PRESSURE OF 2000 PSIG

LIQUID TEMPERATURE, °F	MASS FLOW, lb/hr	ENTHALPY, Btu/lb	DENSITY, lb/ft³	HEAT CAPACITY, Btu/lb °F	THERMAL CONDUCTIVITY, Btu/hr ft °F	VISCOSITY, cp	SURFACE TENSION, Dynes/cm
200	2000	61.76	62.193	0.4212	0.0644	0.8907	29.90
250	1997	83.51	60.777	0.4435	0.0619	0.6547	27.02
300	1990	106.40	59.290	0.4653	0.0594	0.5015	24.14
350	1977	130.30	57.775	0.4867	0.0567	0.3975	21.28
400	1954	155.30	56.278	0.5075	0.0539	0.3242	18.45
450	1915	181.30	54.830	0.5279	0.0614	0.3000	15.69
500	1857	208.20	53.458	0.5476	0.0536	0.2451	13.01
550	1773	236.00	52.221	0.5666	0.0565	0.2037	10.46
600	1660	264.50	51.181	0.5848	0.0571	0.1679	8.11
650	1516	293.50	50.415	0.6020	0.0580	0.1378	6.01
700	1347	322.90	50.009	0.6180	0.0500	0.1135	4.21
750	1162	352.70	50.006	0.6327	0.0636	0.0943	2.75
800	976	382.70	50.338	0.6462	0.0691	0.0788	1.57

VAPOR TEMPERATURE, °F	MASS FLOW, lb/hr	ENTHALPY, Btu/lb	DENSITY, lb/ft³	HEAT CAPACITY, Btu/lb °F	THERMAL CONDUCTIVITY, Btu/hr ft °F	VISCOSITY, cp
200	150	562.30	0.546	3.4800	0.0993	0.0092
250	153	726.10	0.520	3.4190	0.1035	0.0096
300	160	869.20	0.509	3.2880	0.1067	0.0102
350	173	974.50	0.518	3.0650	0.1082	0.0109
400	196	1027.00	0.555	2.7480	0.1079	0.0119
450	235	1024.00	0.627	2.3700	0.1061	0.0131
500	293	981.30	0.740	1.9960	0.1043	0.0146
550	377	920.20	0.899	1.6680	0.1036	0.0163
600	490	860.00	1.104	1.4080	0.1049	0.0180
650	634	811.60	1.346	1.2180	0.1082	0.0197
700	803	779.20	1.610	1.0870	0.1130	0.0213
750	988	763.00	1.869	0.9992	0.1188	0.0226
800	1174	760.60	2.101	0.9431	0.1255	0.0237

634

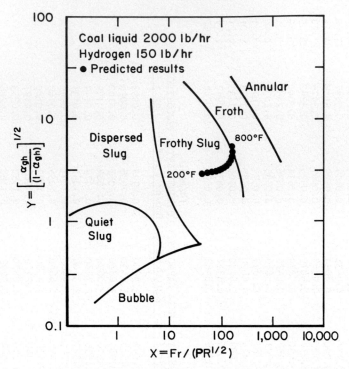

Figure 2- Flow map for two-phase vertical upflow.

Slug Flow. Wall is covered with liquid. The gas phase is large elongated slugs in a continuous liquid. This flow is approximated as annular flow for purposes of computation.

Froth or Churn Flow. Flow is turbulent, with the vapor bubbles having irregular shapes. This regime is approximated as bubble flow.

Annular Phase. The vapor phase flows upwards in the tube core with the liquid moving upward on the tube wall as an annular film.

4.2 Flow Patterns in Vertical Downflow

For vertical downflow, the observed flow patterns are also shown in Figure 4. These flow patterns, described in order of increasing vapor flow rate, are the following:

Coring-Bubble Flow. The gas phase is again dispersed in the form of individual bubbles in the downward flowing liquid. However, these bubbles migrate toward the axis of the tube to form a core of dispersed bubbles.

Bubbly Slug Flow. This flow pattern is characterized by the presence of large air bubbles of the Taylor type. The trailing upper end of the bubble or slug is fairly round due to its buoyancy relative to the liquid.

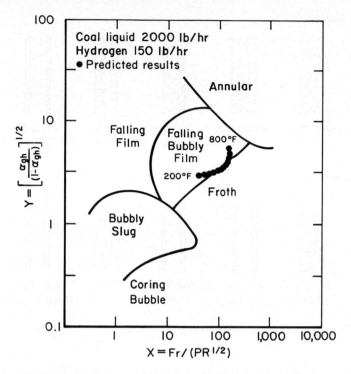

Figure 3-Flow map for two-phase vertical downflow.

Falling Film Flow. The liquid is flowing in the form of a thin film that, in general, contains no gas bubbles. The surface of the film is wavy, and the gas core contains very few or no liquid droplets.

Falling Bubbly Film Flow. This flow pattern is similar to the falling film flow except that the liquid film is thicker and contains small dispersed air bubbles.

Froth Flow. This flow pattern is similar to froth flow in vertical upflow.

Annular Flow. The description of the annular flow pattern in a downward flowing two-phase mixture is the same as that for vertically upward flow.

4.3 Mathematical Definition of Flow Regime Maps for Vertical Flow

As shown in Figures 2 and 3, the flow regime maps are plotted as functions of dimensionless variables X and Y. The ordinate Y in the flow regime maps is given by the following:

$$Y = \left(\alpha_{gh}/(1 - \alpha_{gh})\right)^{1/2} \qquad 4.1$$

636

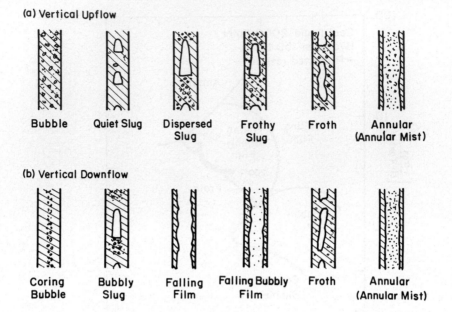

(a) Vertical Upflow

Bubble   Quiet Slug   Dispersed Slug   Frothy Slug   Froth   Annular (Annular Mist)

(b) Vertical Downflow

Coring Bubble   Bubbly Slug   Falling Film   Falling Bubbly Film   Froth   Annular (Annular Mist)

Figure 4 - Flow patterns observed in (a) vertical upflow and (b) vertical downflow.

where α_{gh} represents the ratio of the vapor volume to the total volume based on equal velocity streams. Mathematically, this can be expressed as

$$\alpha_{gh} = (w_g/\rho_g)/(\frac{w_g}{\rho_g} + \frac{w_\ell}{\rho_\ell}) \qquad 4.2$$

The abcissa X in the flow regime maps is given by

$$X = Fr/PR^{1/2} \qquad 4.3$$

where the Fr is the Froude number, and PR is the coefficient of physical properties. These dimensionless groups can be further expressed as

$$Fr = J^2/gD \qquad 4.4$$

where J is the velocity defined as

$$J = (\frac{w_g}{\rho_g} + \frac{w_\ell}{\rho_\ell})/A \qquad 4.5$$

and

$$PR = \frac{\mu_\ell}{\mu_w} (\frac{\rho_\ell}{\rho_w} (\frac{\sigma}{\sigma_w})^3)^{-1/4} \qquad 4.6$$

where the subscript w refers to water at 68°F and 1 atm. The water values are taken as μ_w of 1 cp, ρ_w of 62.3 lb/ft^3, and σ_w of 73 dynes/cm.

5. GOVERNING FIELD EQUATIONS

One dimensional steady-state field equations can be developed for unequal-velocity, equal-temperature, steady two-phase flow systems (UVET). As derived by Lyczkowski and Solbrig [3], the one-dimensional continuity equations include

Vapor Continuity

$$\frac{\partial}{\partial x} (\alpha_g \rho_g v_g) = m \tag{5.1}$$

where m is the mass transferred from the liquid to the gas phase per unit volume per unit time and

Liquid Continuity

$$\frac{\partial}{\partial x} (\alpha_\ell \rho_\ell v_\ell) = -m \tag{5.2}$$

with $\alpha_g + \alpha_\ell = 1$

Assuming the mass of liquid vaporized will become vapor at the same velocity as the liquid [2], the momentum equations are

Vapor Momentum

$$\frac{\partial}{\partial x} (\alpha_g \rho_g v_g^2) = -\alpha_g \frac{\partial p}{\partial x} + m\, v_\ell - A_{g\ell} B_{g\ell} (v_g - v_\ell)$$
$$- A_{wg} B_{wg} v_g + \alpha_g \rho_g g \tag{5.3}$$

and

Liquid Momentum

$$\frac{\partial}{\partial x} (\alpha_\ell \rho_\ell v_\ell^2) = -\alpha_\ell \frac{\partial p}{\partial x} - m\, v_\ell - A_{g\ell} B_{g\ell} (v_\ell - v_g)$$
$$- A_{w\ell} B_{w\ell} v_\ell + \alpha_\ell \rho_\ell g \tag{5.4}$$

The mixture energy equation is

$$\frac{\partial}{\partial x} (\alpha_g \rho_g v_g H_g + \alpha_\ell \rho_\ell v_\ell H_\ell) = q'_w \tag{5.5}$$

where q'_w is heat in per unit volume.

6.   EVALUATION OF TWO-PHASE ANNULAR or SLUG FLOW

For purposes of illustration, only a detailed discussion of two-phase annular or slug flow will be presented. The remaining flow regimes can be evaluated as shown in References 2, 3, and 4.

6.1 Surface Friction

The momentum exchange between a stationary surface and phase "a" is defined as the surface friction term, F_{wa}. For two-phase flow, this friction term can be expressed as

$$F_{wa} = (\frac{\partial p}{\partial x})_{wa} = -A_{wa} \, B_{wa} \, v_a \qquad\qquad 6.1$$

where A_{wa} is the ratio of the surface area of the wall in contact with phase "a" to the total volume, and B_{wa} is the viscous drag force between the wall and phase "a" given by

$$B_{wa} = \frac{\rho_a}{8} \, |v_a| \, f_{wa} \qquad\qquad 6.2$$

The friction factor (f_{wa}) is a function of Reynolds number:

$$Re_{wa} = \frac{Dh_{wa} \, \rho_a \, |v_a|}{\mu_a} \qquad\qquad 6.3$$

where Dh_{wa} is the hydraulic diameter defined as four times the ratio of the fluid volume of phase "a" to the wetted surface area of phase "a".

For laminar flow

$$f_{wa} = \frac{64}{Re_{wa}} \qquad\qquad 6.4$$

while for turbulent flow

$$f_{wa}^{-1/2} = -0.86 \, \ln \left(\frac{e}{3.7 \, Dh_{wa}} + \frac{2.51}{Re_{wa} (f_{wa})^{\frac{1}{2}}} \right) \qquad\qquad 6.5$$

where e is the surface roughness (2).

By definition, the surface to volume ratio $(A_{w\ell})$ and hydraulic diameter $(Dh_{w\ell})$ are given by

$$A_{w\ell} = \frac{2\pi R\Delta x}{\pi R^2 \Delta x} = \frac{4}{D} \qquad\qquad 6.6$$

and

$$Dh_{w\ell} = \frac{4\alpha_\ell \pi R^2 \Delta x}{2\pi R \Delta x} = \alpha_\ell D \qquad 6.7$$

6.2 INTERPHASE FRICTION

For interphase friction, the momentum exchange between phase a and b can be expressed by the interphase friction term:

$$F_{ab} = A_{ab} \, B_{ab} \, (v_a - v_b) \qquad 6.8$$

where A_{ab} is the ratio of the surface area between the phases to the total volume, and B_{ab} is the viscous drag force between the phases given by

$$B_{ab} = \frac{\rho_c}{8} \, |v_a - v_b| \, f_{ab} \qquad 6.9$$

In this case, the density of the continuous phase (ρ_c) is equal to the vapor density (ρ_g). The interphase friction factor (f_{ab}) is a function of the interface Reynolds number:

$$Re_{ab} = \frac{Dh_{ab} \, \rho_c \, |v_a - v_b|}{\mu_c} \qquad 6.10$$

where Dh_{ab} is the hydraulic diameter defined as four times the ratio of the volume of the continuous phase to the interfacial surface area. Again by definition, the surface-to-volume ratio between the phases, $A_{g\ell}$, is given by

$$A_{g\ell} = \frac{2\pi R_i \, \Delta x}{\pi R^2 \, \Delta x} = \frac{2R_i}{R^2} \qquad 6.11$$

where R_i is the vapor stream radius. Since the vapor volume fraction (α_g) can be expressed as

$$\alpha_g = \frac{\pi R_i^2 \, \Delta x}{\pi R^2 \, \Delta x} = \frac{R_i^2}{R^2} \qquad 6.12$$

Equation (6.11) can be reduced to

$$A_{g\ell} = \frac{4 \, \alpha^{\frac{1}{2}}}{D} \qquad 6.13$$

With a similar substitution, the hydraulic diameter $(Dh_{g\ell})$ becomes

$$Dh_{g\ell} = \alpha_g^{1/2} \, D \qquad 6.14$$

At least two methods exist by which the friction factor $f_{g\ell}$ can be obtained (2). One method consists of using the same friction factor as that used for the surface friction (Equation 6.5). Another method is to use an empirical correlation developed by Wallis (2) in which experimental data can be represented by

$$f_{g\ell} = 0.02 \left(1 + 300 \left(\frac{\delta}{D}\right)\right) \qquad 6.15$$

where the liquid film thickness, δ, is obtained from

$$\frac{\delta}{R} = 1 - \alpha_g^{1/2} \qquad 6.16$$

where

$$\frac{\delta}{R} \leq 0.4 \qquad 6.17$$

6.3. ENERGY EXCHANGE

Since it is assumed that there is one mixture energy equation, the only concern need be the heat transfer from the wall to the fluid. Initially, the effects of free convection will be neglected because of the high flow rates. The effect of subcooled boiling, which is characterized by wall temperatures greater then saturation temperature of the liquid, will also be neglected. This is because at the high pressure, the saturation temperature will be very high.

According to Hughes et al. (2), the heat from the wall to a liquid surface in annular or slug flow can be expressed as

$$q_{wa} = h_{wa} \left(T_w - T_\ell\right) \qquad 6.18$$

The Colburn equation can be used to define the heat transfer to the liquid film as follows:

$$Nu = 0.023 \, Re^{0.8} \, Pr^{0.33} \left(\frac{C_{p\ell}}{C_{pf}}\right) \qquad 6.19$$

where the Nusselt number:

$$Nu = \frac{h_{w\ell} \, Dh_{w\ell}}{K_{f\ell}} \qquad 6.20$$

and the Prandtl number:

$$Pr = \left(\frac{C_{p\ell} \, \mu_\ell}{k_\ell}\right)_f \qquad 6.21$$

are evaluated at a film temperature (denoted by subscript f) equal to

$$T_f = 1/2 \left(T_w + T_\ell\right) \qquad 6.22$$

7. SOLUTION USING FINITE DIFFERENCES

The equations can be solved in finite difference form as follows. As shown in Figure 5, the tube can be divided into increments of length Δx. At the inlet to the jth increment, the conditions of the fluid are known.

7.1 Energy Balance

The total energy flowing out of the jth element is equal to the energy in at point j plus the heat in from the wall:

$$W_T \, H_{T,j+1} = W_T \, H_{T,j} + q_{w\ell} \, A_c \Delta x \qquad\qquad 7.1$$

At point j+1, the total energy is then given by

$$W_T \, H_{T,j+1} = w_{g,j+1} \, H_{g,j+1} + w_{\ell,j+1} \, H_{\ell,j+1} \qquad\qquad 7.2$$

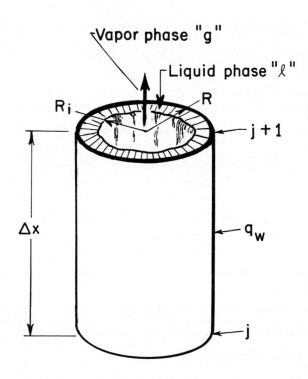

Figure 5 - Idealized flow field structure for annular or transitional flows.

By solving the equations, the temperature of the fluid at point j+1 for which the total enthalpy is equal to $H_{T,j+1}$ can be determined.

7.2 Momentum Balance

The pressure drop across the jth increment must also be equal for the gas and liquid stream. For the liquid, the momentum balance can be written as

$$\frac{(w_\ell v_\ell)_{j+1} - (w_\ell v_\ell)_j}{\Delta x} = - \alpha_\ell \frac{(p_{j+1} - p_j)}{\Delta x}$$

$$- (m_{j'} v_{\ell j'}) - (A_{w\ell} B_{w\ell} v_\ell)_{j'} - (A_{g\ell} B_{g\ell} (v_\ell - v_g))_{j'}$$

$$+ (\alpha_\ell \rho_\ell g)_{j'} \qquad\qquad 7.3$$

where j' refers to the point j+1/2 and

$$-m_{j'} = \frac{(w_{\ell,j+1} - w_{\ell,j})}{A \Delta x} \qquad\qquad 7.4$$

By assuming a value of α_g at j+1, both $v_{g,j+1}$ and $v_{\ell,j+1}$ can be found. Then, from Equation 7.3, the value for p_{j+1} can be determined.

The momentum equation for the vapor can similarly be written as

$$\frac{(w_g v_g)_{j+1} - (w_g v_g)_j}{\Delta x} = -\alpha_g \frac{(p_{j+1} - p_j)}{\Delta x} \qquad\qquad 7.5$$

$$+ m_{j'} v_{\ell j'} - (A_{g\ell} B_{g\ell} (v_g - v_\ell))_{j'} - (A_{wg} B_{wg} v_g)_{j'}$$

$$+ (\alpha_g \rho_g g)_{j'}$$

Using the same assumed value of α_g, it is possible to solve for p_{j+1} in Equation 7.5.

Since the value of p_{j+1} from Equation 7.5 should agree with the result from Equation 7.3, α_g is varied until this result is obtained. This then gives converged values of p_{j+1}; $v_{\ell,j+1}$; $v_{g,j+1}$; and $\alpha_{g,j+1}$.

7.3 Wall Temperature Calculation

To find the wall temperature, the following equation is used:

$$q_w = h_{w\ell} (T_w - T_\ell) \qquad\qquad 7.6$$

where $h_{w\ell}$ is evaluated as shown in Section 6.3 using the local properties.

8. COMPUTATIONAL RESULTS

Following the procedure outlined in Section 7 and assuming annular flow, a plot of α_g as a function of fluid temperature is shown in Figure 6 for the flow conditions described in the fluid properties example of Section 3.1. When heated in a 1.1-inch-i.d. tube at a flux of 8000 Btu/ft^{2}hr, the liquid, as expected, moves slower in upflow than in downflow. Therefore, the gas volume of α_g is smaller in upflow than in downflow. A significant rise in α_g with temperature occurs for both upflow and downflow. This is due in part to the vaporization of the liquid.

The corresponding liquid and vapor velocities are shown in Figure 7. In upflow or downflow, the vapors move faster than the liquid; however, the liquid flowing down is significantly faster than the liquid in upflow.

The expected pressure drops are shown in Figure 8. The pressure drop in the upflow is significantly larger than in downflow. By combining Equation 5.3 and 5.4, the pressure drop is given by

$$- \frac{\partial p}{\partial x} = A_{w\ell} B_{w\ell} v_\ell - (\alpha_g \rho_g + \alpha_\rho \rho_\ell) g + \frac{\partial}{\partial x} (\alpha_g \rho_g v_g^2 + \alpha_\ell \rho_\ell v_\ell^2) \qquad 8.1$$

Pressure		Frictional		Gravitational		Acceleration
	=		−		+	
Drop		Component		Component		Component

The calculations showed that the acceleration component was negligible compared to the other terms that are plotted in Figure 8. The frictional component of the pressure drop was higher in the downflow than in the upflow due to the higher liquid velocity in the downflow tube.

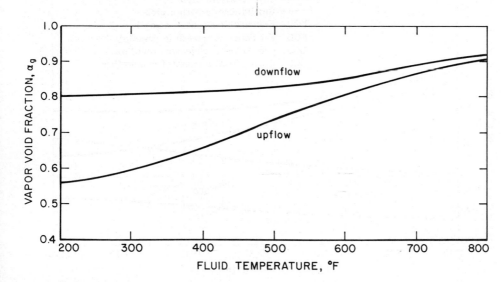

Figure 6 - Vapor void fraction versus fluid temperature.

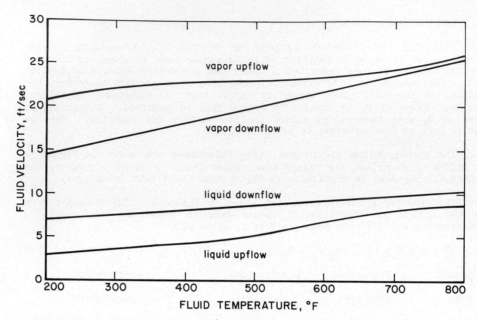

Figure 7 - Fluid velocity versus fluid temperature.

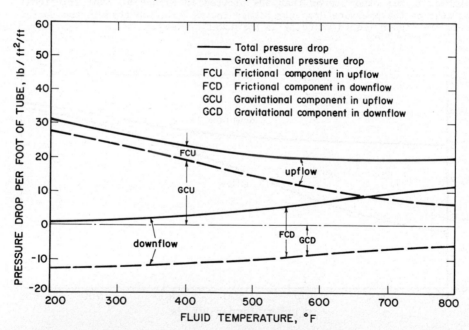

Figure 8 - Pressure drop per foot of tube versus fluid temperature.

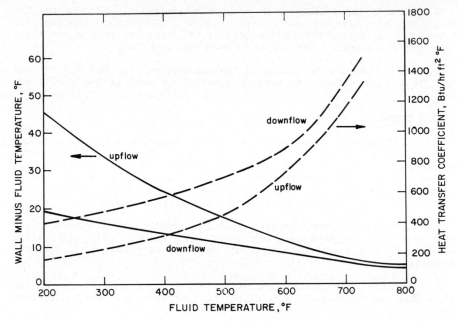

Figure 9-Wall minus fluid temperature and heat transfer coefficient versus fluid temperature.

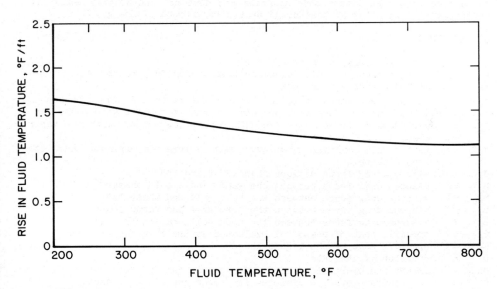

Figure 10-Rise in fluid temperature per foot of tube versus fluid temperature.

In Figure 9, the difference in temperature between the liquid and the wall, and the heat transfer coefficient is shown. Evidently in upflow, the heat transfer coefficient is smaller because the lower velocity of the fluid gives a higher temperature difference between wall and fluid.

The curve in Figure 10 relates to the temperature rise per foot of tube to the fluid temperature. No unusual behavior is seen as the fluids are heated.

9. CONCLUSIONS

In operating the experimental preheater, pressure drops through the test section will be measured as a function of flow rates. Viscosity measurements, as well as liquid samples in which thermodynamic properties can be determined, will also be obtained. These measured properties can be compared to calculated properties determined analytically. The pressure drop measurements can then be compared to the analytical results.

The experimental heat transfer results consist of measuring fluid temperatures, wall temperatures, and local heat inputs. This will allow calculation of heat transfer coefficients that can be compared to theoretical results.

The combined experimental and analytical program should improve our understanding of preheaters and allow for improved and more efficient designs.

DISCLAIMER

Reference in this report to any specific commercial product, process, or service is to facilitate understanding and does not necessarily imply its endorsement or favoring by the United States Department of Energy.

NOMENCLATURE

A = cross-sectional area of the tube
A_{ab} = surface area per unit volume between phases "a" and "b"
A_c = circumference of the tube
$A_{g\ell}$ = surface area per unit volume between the vapor and liquid phases
A_{wa} = wetted-wall surface area per unit volume in contact with phase "a"
A_{wg} = wetted-wall surface area per unit volume in contact with the vapor phase
$A_{w\ell}$ = wetted-wall surface area per unit volume in contact with the liquid phase
B_{ab} = viscous drag force between phases "a" and "b"
$B_{g\ell}$ = viscous drag force between the vapor and liquid phases
B_{wa} = viscous drag force between the tube wall and phase "a"
B_{wg} = viscous drag force between the tube wall and vapor phase
$B_{w\ell}$ = viscous drag force between the tube wall and liquid phase
C_{pf} = specific heat of phase "a" evaluated at the film temperature
$C_{p\ell}$ = specific heat of the liquid phase
D = diameter of the tube
Dh_{ab} = interphase hydraulic diameter for phases "a" and "b"
$Dh_{g\ell}$ = interphase hydraulic diameter for vapor and liquid phases
Dh_{wa} = hydraulic diameter for phase "a" in contact with the tube wall

$Dh_{w\ell}$ = hydraulic diameter for the liquid phase in contact with the tube wall

e = surface roughness factor for the tube wall

F_{ab} = friction force per unit volume exerted on phase "b" by phase "a" parallel to the flow direction

F_{wa} = friction force per unit volume exerted on phase "a" by wall parallel to the flow direction

Fr = Froude number

f_{ab} = friction factor between phases "a" and "b"

$f_{g\ell}$ = friction factor between the vapor and liquid phases

f_{wa} = friction factor between the tube wall and phase "a"

g = gravitational acceleration

H_g = enthalpy of the vapor phase

$H\ell$ = enthalpy of the liquid phase

H_T = total enthalpy of the stream

h_w = heat transfer coefficient between the wall and the fluid

h_{wa} = heat transfer coefficient between the wall and phase "a"

$h_{w\ell}$ = heat transfer coefficient between the wall and the liquid phase

J = average stream velocity

$k_{f\ell}$ = thermal conductivity of the liquid phase evaluated at the film temperature

$k\ell$ = thermal conductivity for the liquid phase

m = mass transfer rate per unit volume from the liquid to the vapor phase

Nu = Nusselt number

PR = coefficient of physical properties

Pr = Prandtl number

p = thermodynamic pressure

q_w = heat transfer rate per unit volume at the wall

$q_{w'}$ = heat transfer rate per unit volume to the stream

q_{wa} = heat transfer rate per unit volume from the wall to the liquid surface in annular or slug flow

$q_{w\ell}$ = heat transfer rate per unit volume between the wall and the liquid phase

R = radius of the tube

R_i = radius of the vapor phase flowing in the tube

Re = Reynolds number

Re_{ab} = Reynolds number at the interface between phases "a" and "b"

Re_{wa} = Reynolds number for phase "a" in contact with the wall

T_f = temperature of the film

$T\ell$ = temperature of the liquid phase

T_w = temperature of the wall-fluid interface

v_a = velocity of phase "a"

v_b = velocity of phase "b"

V_g = velocity of the vapor phase

V_ℓ = velocity of the liquid phase

W_g = mass flow rate of the vapor phase

$W\ell$ = mass flow rate of the liquid phase

W_T = total mass flow rate of the stream

X = dimensionless flow map absissa

x = spatial direction in vertical flow

Y = dimensionless flow map ordinate

Greek Letters

α_g = fraction of the total volume occupied by the vapor phase

α_{gh} = fraction of the volume occupied by the vapor phase based on equal velocity streams

$\alpha\ell$ = fraction of the total volume occupied by the liquid phase
δ = thickness of the film
μ_a = viscosity of phase "a"
μ_c = viscosity of the continuous phase
$\mu\ell$ = viscosity of the liquid phase
μ_w = viscosity of water at 68°F and 1 atm
ρ_a = density of phase "a"
ρ_c = density of the continuous phase
ρ_g = density of the vapor phase
$\rho\ell$ = density of the liquid phase
ρ_w = density of water at 68°F and 1 atm
σ = surface tension of the liquid phase
σ_w = surface tension of water and 68°F and 1 atm

Subscripts

a = phase "a"
b = phase "b"
c = continuous phase
f = property evaluated at the film temperature
g = vapor phase
j = arbitrary point in the vertical direction of flow
j' = arbitrary point at $j+\frac{1}{2}$ in the vertical direction of flow
ρ = liquid phase
T = total stream quantity
w = water at 68°F and 1 atm

REFERENCES

1. Kornosky, R.M., M. Perlmutter, W. Fuchs, and J.A. Ruether, "Apparatus Development for Measuring Heat Flux in a Direct Coal Liquefaction Preheater," DOE/PETC/TR-82/8.

2. Hughes, E.D., R.W. Lyczkowski, J.H. McFadden, and G.F. Niederauer, "An Evaluation of State-of-the-Art Two-Velocity Two-Phase Flow Models and Their Applicability to Nuclear Reactor Transient Analysis," EPRI NP-143 (1976).

3. Lyczkowski, R.W., and C.W. Solbrig, "Calculation of the Governing Equations for a Seriated Unequal Velocity, Equal Temperature Two-Phase Continuum," AIChE J., 26, 89-97, (1980).

4. Solbrig, C.W., J.H. McFadden, R.W. Lyczkowski, and E.D. Hughes, "Heat Transfer and Friction Correlations Required to Describe Steam-Heater Behavior in Nuclear Safety Studies," pp. 100-128 in Heat Transfer Research and Applications, AIChE, N.Y. (1978).

5. "Process Simulation Program Input Manual," Simulation Science, Inc., Fullerton, Calif., March 1982.

6. Oshinowo, T., and M.E. Charles, "Vertical Two-Phase Flow, Part 1, Flow Pattern Correlations." Canadian J. Chem. Eng., 52, 1974.

7. "ASPEN Plus Users Manual," Aspen Technology, Inc., Cambridge, Mass., May 1982.

8. "PROCESS Simulation Program Reference Manual," Simulation Sciences, Inc., Fullerton, Calif., 1981.

Multi-Phase Flow and Heat Transfer III. Part B: Applications
edited by T.N. Veziroğlu and A.E. Bergles
Elsevier Science Publishers B.V., Amsterdam, 1984 — Printed in The Netherlands

649

EFFECTS OF VARIABLE PROPERTIES, RADIATION AND SUSPENSION BEAD ON THE TEMPERATURE HISTORIES OF BURNING DROPLETS

R. NATARAJAN, R. VENKATRAM, K.N. SEETHARAMU AND K.N. RAO
Department of Mechanical Engineering
Indian Institute of Technology
Madras 600 036, India

ABSTRACT

In the first part of the present work, liquid-phase temperature histories of a burning drop have been calculated, taking into
account (i) the reduction in net heat transfer to the drop because
of outward mass transfer, (ii) radiative heat flux from the flame
and (iii) variable properties. Both the constant and variable
property solutions show that the maximum liquid temperatures attained at the end of drop lifetime are always less than the boiling
point at the prevailing pressure. The effects of ambient pressure
and initial drop diameter have been investigated. In the second
part of the work, the procedure of measurement of the temperature
of burning liquid drops by suspending them on thermocouple junctions has been evaluated by formulating the governing equations
for the liquid and thermocouple temperature histories. Using
typical values of parameters encountered in droplet combustion,
it is found that the thermocouple temperatures always lag behind
the liquid temperatures throughout the drop lifetime.

1.   INTRODUCTION

While in simplified quasi-steady theories of droplet combustion, the liquid-phase temperature is assumed steady and uniform
and equal to the saturation temperature at the prevailing pressure,
the temporal variation is taken into account in all unsteady analyses, and the gradients within the liquid phase are considered
for large drops in the absence of internal circulation. The temperature histories are important from the point of view of liquid-
phase pyrolysis, particularly for heavy and multicomponent fuels,
and in determining the onset of microexplosions in the case of
emulsified fuels.

2.   FORMULATION OF THE PROBLEM

In the present investigation, the temperature-time histories
of the liquid phase have been calculated by employing the thermal
energy balance for the burning droplet. It is assumed that the
liquid is well-mixed by internal circulation, and hence is at a
uniform temperature. Mass transfer shielding is explicitly taken
into account in terms of the reduction in the convective heat
transfer to the liquid as a result of the outward mass transfer.

The governing equation is:

$$m \, C_{p,1} \, \frac{dT_1}{dt} = Q_c^* - \dot{m} \, L + Q_R \tag{1}$$

where the mass transfer shielding function Q_c^*/Q_c is taken to be $(1+B)^{-1}$ and the effect of convection on heat transfer is considered through an empirical Nu-Re-B correlation [1]:

$$Nu \, (1+B) = 2 \, (1 + 0.635 \, Re^{0.5}) \tag{2}$$

The temperature histories for aniline droplets, assuming temperature-independent thermodynamic and transport properties, and neglecting radiative heat transfer, have been reported earlier [2].

In the present study the effect of variable properties such as ρ_1, C_p, $C_{p,1}$, k and L have been taken into consideration by fitting linear or polynomial equations to them in terms of T_1. This necessitates numerical solution of the governing equations. Inclusion of radiative transfer renders the governing equations non-linear.

The non-dimensionalized governing equation for the liquid temperature is:

$$d\theta_1/d\tau = E(\theta_f^4 - \theta_1^4) + F(\theta_f - \theta_1) - G \tag{3}$$

The convective heat transfer coefficient from the flame to the drop is calculated from eq.2, which takes into account the effect of outward mass transfer in reducing the convective heat transfer to the drop. The variation of the drop size with time has been assumed to follow the classical D^2-law.

In experimental investigations involving the determination of droplet liquid temperatures, the liquid drop is generally suspended from the bead of a thermocouple junction, and the liquid-phase temperature is inferred from the thermocouple output. But the presence of the thermocouple bead may have a marked effect on the liquid-phase temperature. This problem is formulated in terms of the thermal energy balance, considering the drop and the bead independently. The resulting governing equations are then solved simultaneously for obtaining the thermocouple temperature and the corresponding liquid temperature.

The non-dimensionalized governing equation for the thermocouple temperature is:

$$d\theta_t/d\tau = K(\theta_f^4 - \theta_t^4) - M(\theta_t - \theta_1) - N(\theta_t^4 - \theta_1^4) \tag{4}$$

The convective heat transfer coefficient from liquid drop to thermocouple is calculated from a Nu-Gr correlation:

$$Nu = 2 + 0.6 \, Gr^{0.25} \, Pr^{0.33} \tag{5}$$

If the radiation effects are neglected, eq.(4) reduces to:

$$d\theta_t/d\tau = M(\theta_1 - \theta_t) \tag{6}$$

The non-dimensionalized governing equation for the drop is:

$$d\theta_1/d\tau = P(\theta_f - \theta_1) + R(\theta_f^4 - \theta_1^4) + S(\theta_t - \theta_1)$$
$$+ X(\theta_t^4 - \theta_1^4) - Y \tag{7}$$

The convective heat transfer coefficient from the flame to the drop is calculated from the Nu-Re-B correlation, eq.(2). If the radiation effects are neglected, eq.(7) reduces to:

$$d\theta_1/d\tau = P(\theta_f - \theta_1) + S(\theta_t - \theta_1) - Y \tag{8}$$

3. RESULTS AND DISCUSSION

In the first part of the work, the governing equation (3) is solved numerically with the initial condition: $T_1 = 0$ at $t = 0$. The solutions for the two cases of constant properties and variable properties have been obtained for aniline, for three values of initial diameter viz. 500, 750 and 1000μm, and for four values of ambient pressure, viz. 7.05, 14.09, 28.18 and 42.28 atm., corresponding to 100, 200, 400 and 600 psig. The physico-chemical properties necessary for computation have been obtained from Ref.3. The temperature histories have been obtained with the fourth-order Runge-Kutta-Gill method.

Figure 1 shows the effect of ambient pressure on the temperature histories of a liquid droplet of 500μm initial diameter, for both cases. At all the four pressures investigated the maximum temperature attained by the liquid is less than the corresponding boiling point in both cases. The maximum temperatures attained in the constant-property case are much higher than in the variable-property case. In the constant-property solution, the maximum liquid temperature is attained just at the end of drop life-time, whereas in the variable-property case the maximum temperature is reached earlier, and the temperature decreases thereafter. This is presumably because of the balance between the different heat flux quantities which determine the net heat available to increase the liquid temperature. The different heat flux quantities depend strongly on the thermodynamic and transport properties. The temperature histories in the second case are linear upto the maximum temperature.

Figure 2 shows the effect of initial drop diameter on the temperature histories at an ambient pressure of 7.05 atm. for three drop diameters. The trends are quite similar to those in Fig.1; increasing ambient pressure and increasing initial drop size produce similar effects. Also plotted in both figs. 1 and 2 is the diameter history of a 500μm droplet.

In the second part of the work, the governing equations (4) and (7) have been solved simultaneously, with the initial condition: $T_1 = T_t = T_a$ at t=0, to yield the temperature histories for the thermocouple and the liquid drop. The numerical solutions have been obtained on an IBM 370/155 computer utilizing the fourth-order Runge-Kutta-Gill method. Five values of ambient pressure, ranging from 7.05 atm. (100 psig) to 42.28 atm. (600 psig) and three values of drop diameter, viz. 500, 750 and 1000μm, have

652

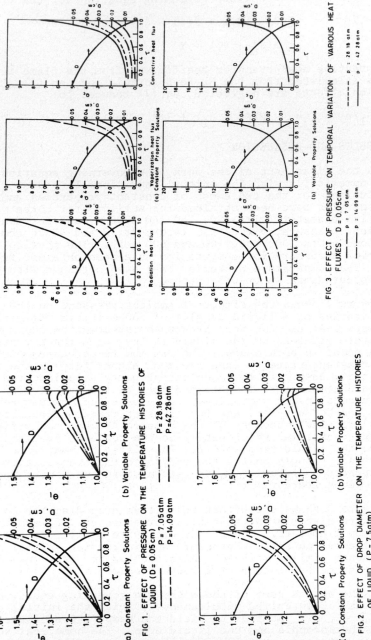

FIG.3 EFFECT OF PRESSURE ON TEMPORAL VARIATION OF VARIOUS HEAT FLUXES   D = 0.05 cm

(a) Variable Property Solutions

———— p = 7.05 atm
—·—·— p = 14.09 atm
——— p = 28.18 atm
———— p = 42.28 atm

(a) Constant Property Solutions   (b) Variable Property Solutions

FIG.1. EFFECT OF PRESSURE ON THE TEMPERATURE HISTORIES OF LIQUID (D=0.05 cm)   ———— P = 7.05 atm   ———— P = 28.18 atm   ———— P = 14.09 atm   ———— P = 42.28 atm

(a) Constant Property Solutions   (b) Variable Property Solutions

FIG 2. EFFECT OF DROP DIAMETER ON THE TEMPERATURE HISTORIES OF LIQUID (P = 7.5 atm)   ——— D = 0.05 cm   ———— D = 0.075 cm   ——— D = 0.10 cm

been considered. The diameter of the thermocouple bead is taken to be 125 µm. Only constant-property solutions have been obtained in this part of the work.

Figure 3 shows the effect of pressure on the temporal variation of the different heat fluxes contributing to the sensible heating of the liquid, with reference to the first part of the work. The radiant heat flux increases with pressure in both the constant-property and variable-property solutions, mainly through increases in T_f and ε_f. The vaporization heat flux and the convective heat flux decrease with pressure in the constant-property case, whereas they are almost independent of pressure in the variable-property case. The steep increases towards the end of drop life are related to the reduction droplet mass.

Figure 4 shows the effect of initial drop diameter on the temporal variation of radiant heat flux at an ambient pressure of 7.05 atm. In both the constant-property and variable-property cases, the radiation flux increases with drop diameter, and the temporal variations show similar trends. The convective and vaporization heat fluxes are essentially independent of initial drop diameter.

Figure 5 shows the effect of ambient pressure on the temperature histories of the liquid and the thermocouple bead for a 500µm drop, in the second part of the work. It can be seen that at all the five pressures considered, the thermocouple temperatures always lag behind the liquid temperatures, the magnitude of the difference increasing with time. It is typically of the order of 15°C at the time of consumption of the drop. The maximum temperatures attained by both drop and thermocouple decrease somewhat with increasing pressure. But since the boiling points increase with pressure, the temperature defect at the time of consumption of the drop increases with pressure. The figure also shows the variation of drop diameter with time.

Figure 6 shows the effect of initial drop diameter on the temperature histories of the liquid. At all the pressures considered, the liquid temperatures are higher for the larger drops. At the higher pressures, the temperature histories tend to become linear, particularly for the larger drops. With increasing pressure, the maximum temperature attained decreases for the smallest drops, but increases for the largest drops. The curves for the different diameter drops tend to become more widely spaced as pressure increases.

Figure 7 shows the effect of initial drop diameter on the temperature histories of the supporting thermocouple bead. The trends are quite similar to those in Fig.6. The main difference is that the liquid temperatures increase at a faster rate than the thermocouple temperatures.

4. CONCLUDING REMARKS

Extrapolation of the results of the first part of the present work to conditions close to critical point of the liquid fuel involves the taking of limits as $L \rightarrow 0$ and as $C_{p,l}$ and $B \rightarrow \infty$. It was shown in Ref.2 that in the absence of radiation the burning drop

654

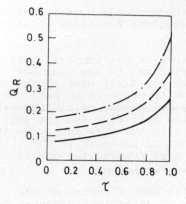
a). Constant property
solutions

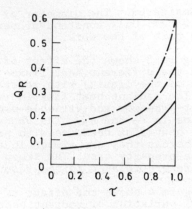
b). Variable property
solutions

FIG. 4. EFFECT OF INITIAL DROP DIAMETER ON THE TEMPORAL
VARIATION OF RADIATION HEAT FLUX (p=7.05 atm)

——— D = 0.05 cm   ——— D = 0.075 cm   —·— D = 0.10 cm

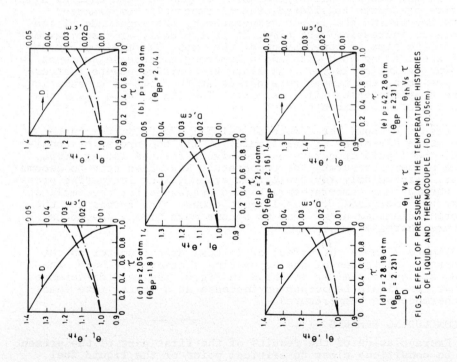

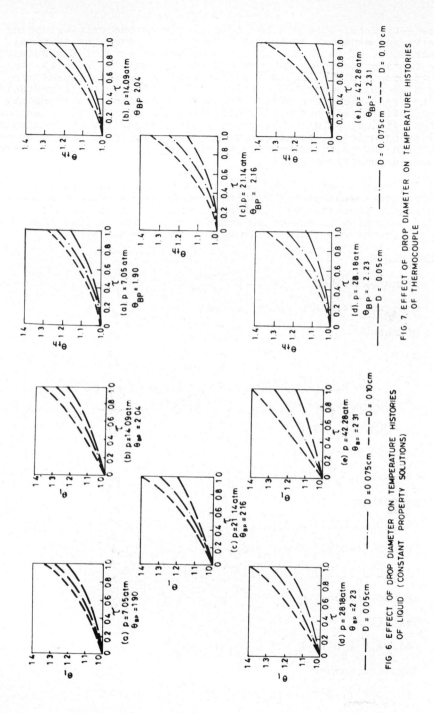

(a) p = 7.05 atm
θ_{BP} = 1.90

(b). p = 14.09 atm
θ_{BP} = 2.04

(c). p = 21.14 atm
θ_{BP} = 2.16

(d). p = 28.18 atm
θ_{BP} = 2.23
D = 0.05 cm

(e). p = 42.28 atm
θ_{BP} = 2.31
D = 0.075 cm   D = 0.10 cm

FIG. 7 EFFECT OF DROP DIAMETER ON TEMPERATURE HISTORIES
OF THERMOCOUPLE

(a) p = 7.05 atm
θ_{BP} = 1.90

(b) p = 14.09 atm
θ_{BP} = 2.04

(c) p = 21.14 atm
θ_{BP} = 2.16

(d) p = 28.18 atm
θ_{BP} = 2.23
D = 0.05 cm

(e) p = 42.28 atm
θ_{BP} = 2.31
D = 0.075 cm   D = 0.10 cm

FIG. 6. EFFECT OF DROP DIAMETER ON TEMPERATURE HISTORIES
OF LIQUID (CONSTANT PROPERTY SOLUTIONS)

is mass-transfer-shielded too effectively at the critical point to receive any sensible heat, and hence cannot reach the critical temperature. This is due to the combined effects of the outward mass transfer reducing the net convective heat transfer to the liquid drop and the liquid specific heat increasing rapidly to very high values near the critical point. Inclusion of radiation might offset the first effect somewhat, but cannot overcome the second effect, and hence will not alter the asymptotic behaviour at the critical point. These conclusions regarding the unattain-ability of critical point during droplet combustion are in agree-ment with the qualitative results obtained on the basis of thermo-dynamic arguments in Ref.4.

The second part of the work shows that for typical values of parameters employed in droplet combustion experiments, the temperature indicated by the thermocouple junction on which a liquid fuel drop is supported and burned is always less than the temperature of the liquid, throughout the drop life

NOMENCLATURE

B Transfer number

C_p Constant pressure specific heat

D Diameter

D_r D_t/D_o

E $[6\sigma\varepsilon_f\alpha_1 D_o^2 T_a^3]/[\rho_1 C_{p,1}\lambda D^2]$

F $[6h_{1f}D_o^2]/[\rho_1 C_{p,1}\lambda D]$

G $[3\ D_o^2 L]/[C_{p,1}\ D^2 T_a]$

Gr Grashof number

K $[6T_{r1}\sigma\alpha_t\varepsilon_f D_t(1-D_r^2)T_a^3]/[\rho_t C_{pt}\lambda(1-\tau)]$

L Latentheat

M $[6h_{1t}D_o^2(1-D_r^2)]/[\rho_t C_{pt}\lambda D_t]$

N $[6\sigma\varepsilon_t D_o^2(1-D_r^2)T_a^3]/[\rho_t C_{pt}\lambda D_t]$

P $[6h_{1f}D^2 D_o^2(1-D_r^2)]/[\rho_1 C_{p1}\lambda(D^3-D_t^3)]$

Pr Prandtl number

Q Heat flux

R $[6\sigma\alpha_1\varepsilon_f D^2 T_a^3 D^2(1-D_r^2)]/[\rho_1 C_{p1}\lambda(D^3-D_t^3)]$

Re Reynolds number

S $[6h_{1t}D_t^2 D_o^2(1-D_r^2)]/[\rho_1 C_{p1}\lambda(D^3-D_t^3)]$

X $\quad [6\sigma\varepsilon_t\alpha_1 T_a^3 D_o^2(1-D_r^2)]/[\rho_1 C_{p1}\lambda(D^3-D_t^3)]$

Y $\quad [1.5LDD_o^2(1-D_r^2)]/[C_{p1}T_a(D^3-D_t^3)]$

Greek Letters

α $\quad$ absorptivity

ε $\quad$ emissivity

λ $\quad$ evaporation constant

ρ $\quad$ density

σ $\quad$ Stefan-Boltzmann constant

τ $\quad$ non-dimensionalized time : $\lambda t/(D_o^2-D_t^2)$; $\lambda t/D_o^2$

θ $\quad$ non-dimensionalized temperature : T/T_a

Subscripts

a $\quad$ ambient air

c $\quad$ convective heat flux

e $\quad$ evaporation flux

f $\quad$ flame

l $\quad$ liquid

o $\quad$ initial value

r $\quad$ radius; radiation

t $\quad$ thermocouple

REFERENCES

1. Natarajan,R., 'A Note on the Correlation of Droplet Consumption Rates', Combustion Science and Technology,11, 161 (1975)

2. Natarajan,R., and Sarathy,C.P., 'Effect of Simultaneous Heat and Mass transfer on the Temperature of a Burning Droplet', II Multi-phase Flow and Heat Transfer Symposium-Workshop,Clean Energy Research Institute, Florida, April 16-18, 1979.

3. Natarajan,R., 'An experimental investigation of droplet combustion at high pressures', Ph.D. Thesis,Univ.of Waterloo, Canada, 1969.

4. Natarajan,R., 'Heat and Mass Transfer Considerations in Supercritical Bipropellant Droplet Combustion', in HEAT TRANSFER IN FLAMES,Eds: N. Afgan and J.M. Beer, Academic Press, 1974.

Multi-Phase Flow and Heat Transfer III. Part B: Applications 659
edited by T.N. Veziroğlu and A.E. Bergles
Elsevier Science Publishers B.V., Amsterdam, 1984 — Printed in The Netherlands

INFLUENCE OF AERODYNAMIC PARAMETERS ON THE BURNING CHARACTERISTICS OF
SINGLE AND TWIN-SPHERE SYSTEMS SIMULATING BURNING FUEL DROPS

R. VENKATRAM, U.S.P. SHET AND R. NATARAJAN
Department of Mechanical Engineering
Indian Institute of Technology, Madras, India

ABSTRACT

This paper describes the results of experimental investigations designed
to study the influence of aerodynamic parameters related to ambient convection, drop rotation and air turbulence on the burning characteristics such
as mass consumption rate and flame geometry of single and twin-sphere systems
simulating burning fuel drops. The porous-sphere technique has been utilized
for all the experiments described here, and shadowgraphs and schlieren photographs provide qualitative information on the combustion processes. The
interference effects under free convection in three different configurations
for a twin-sphere system can be explained on the basis of oxygen starvation
at small separation and heat loss reduction at large separation. The effect
of drop rotation is to increase the burning rates; in the rather small range
of rotational speeds investigated here the flame geometry is not altered
much. With increasing forced-convective air velocity the envelope flame
burning rate increases linearly, dropping to nearly half the value prior to
extinction, and remaining almost unchanged in the wake mode. In the presence
of grid-generated turbulence the burning rate decreases somewhat, but more
significantly the extinction velocity decreases significantly.

1. INTRODUCTION

Although spray combustion has been traditionally analyzed in terms of
the combustion of the individual drops composing the spray, it is being
increasingly recognized that the burning characteristics of a spray droplet
can be significantly affected by interference effects, which can play an
important role in determining the mode of combustion, overall burning rate and
flame geometry. The recent concern for maximizing combustion efficiency
and minimizing combustion - generated pollutants during the burning of liquid
fuels has generated a revival of interest in the interactions between burning
droplets.

Several experimental and theoretical studies of interference effects in
droplet combustion have been reported in the literature. The early experimental work of Rex et al [1], Kanevsky [2] and Fedoseeva [3] mainly dealt
with interference effects during the burning of stationary fuel droplets.
Sangiovanni and Kesten [4] have studied the effect of interference on ignition of monodispersed droplet streams. Sarasty and Natarajan [5] have investigated the effect of interference on the extinction velocities of two burning droplets. Twardus and Brzustowski [6] and Sangiovanni and Dodge [7] have
studied the flame structure of falling droplet streams. The theoretical work

of Carstens et al [8], Labowsky [9] and Williams and Carstens [10] dealt with droplet vaporization and combustion in the limiting case of negligible Stefan flow. Recently Labowsky [11] has presented a method for the calculation of the burning rates of interacting fuel droplets. The present work describes schlieren and shadowgraphic investigations of interference effects in droplet combustion utilizing the porous-sphere technique.

In several practical situations, the spray droplets injected into a combustor exhibit rotation, in combination with oscillation, spinning and swerving [12]. The effects of sphere rotation on flame geometry and burning rate have been investigated here.

The flow fields in most combustors are turbulent and ambient turbulence has a significant effect on the burning rate, flame geometry and extinction velocity. A photographic investigation of some of these effects has also been undertaken here.

2. EXPERIMENTAL TECHNIQUE

The porous-sphere technique has been utilized for all the experiments described here. Burning liquid drops have been simulated by liquid diffusion flames maintained by porous alundum spheres which were internally fed with the liquid fuel at a controlled and steady rate. The sphere was supported on a water-cooled stainless steel feed line through which the fuel was admitted from a syringe whose piston was advanced at a controlled rate by means of a motor, reduction gear and screw arrangement. In each experiment the mass consumption rate was determined as the critical fuel rate above which excess fuel started dripping from the sphere, and below which the entire sphere surface could not be maintained dry.

For the experiments on interference effects, two spheres of 8 mm diameter were supported in three configurations, viz., horizontal, vertical and inclined, and fed from two syinges separately. The inter-sphere separation was varied in steps of distance measured from center to center, and the burning rate was obtained as a function of the separation distance. In the horizontal configuration the burning rates for the two spheres were equal, while in the vertical configuration the burning rate of the top sphere was measured, with the mass burning rate of the bottom sphere remaining almost constant. The fuel employed was methanol.

For the rotating sphere experiments, a 10 mm diameter sphere was employed, with a suitable mechanism driven by a 12v d.c. motor through a set of pulleys. The variation in the rotational speed was achieved using a transistorized power supply. Four different fuels - methanol, propanol-1, n-hexane and n-heptane - were employed.

The forced convective air velocity past the burning sphere was varied from 0.2 m/s to 2 m/s, and the burning rates have been determined. The flame characteristics were determined from shadow and schlieren photographs. For studying the effects of flow turbulence on droplet combustion, wire meshes were employed in the air stream. The turbulence intensity was estimated from Baines' correlation, $Tu(\%) = 112 (x/6)^{-5/7}$, and was equal to 9.32% for the coarse mesh employed here. The sphere was located at a distance of 65 mm downstream of the grid.

3. INTERFERENCE EFFECTS UNDER FREE CONVECTION

i) Horizontal Configuration

Figure 1 contains shadow and schlieren photographs of burning drops arranged in the horizontal configuration. While at larger separation the flames are individually separated, they become merged at small separation. The schlieren photographs appear to provide a better means of flow visualization than the shadowgraphs.

Figure 4a shows that in the horizontal configuration, at spacings less than 1.3D, the burning rate is less than the value corresponding to a single sphere burning in isolation. It increases with increasing separation, reaches a maximum at 2.5D, and then decreases gradually, attaining the isolated sphere value at 4D and remaining constant thereafter. It is evident that at very small separation, the burning drops suffer from oxygen starvation, resulting in burning rates smaller than those for isolated drops. At larger separation, the burning rate is higher because of a reduction in the radiative heat losses to the surroundings, consequent on the presence of a neighboring flame surface. The burning rate is maximum at a certain separation distance, beyond which it decreases as a result of entrainment of cool ambient air between the flame surfaces, ultimately attaining the isolated sphere value. Comparison with the photographs reveals that the maximum in the burning rate is related to the limit of occurrence of merged flames.

ii) Vertical configuration

Figure 2 includes schlieren and shadowgraphs of burning drops arranged in the vertical configuration. It is seen that the flame geometry is quite independent of the separation distance; this is mainly because the individual flames are much taller than the spacings considered.

Figure 4a shows that in the vertical configuration, there is no appreciable change in the burning rate with increasing separation distance. The burning rate is slightly higher than that for an isolated drop; this is, of course, due to the enhanced free convective flow generated by the bottom sphere in which the top sphere is located, and also because of the higher ambient temperature presented to the top sphere.

iii) Inclined configuration

Figure 3 shows the interference effects in the inclined configuration. In both cases the flame geometry is affected by the separation distance, and in the second case, horizontal spacings greater than 4D result in separated flames.

Figure 4b shows that in the first case of inclined configuration, the horizontal separation is kept constant at 1D and the vertical separation is varied from 1D to 4D. The burning rate of the displaced drop increases with increasing upward displacement, due mainly to relaxation of oxygen starvation. The burning rate of the bottom sphere is constant, and slightly larger than the isolated drop value.

In the second case, the vertical separation is kept constant at 2D and the horizontal separation is varied from 1D to 4D. The burning rate of the displaced drop increases to a maximum and then decreases, as observed in the horizontal configuration. The burning rate in the inclined configuration is less than that in the horizontal configuration at the same separation, except

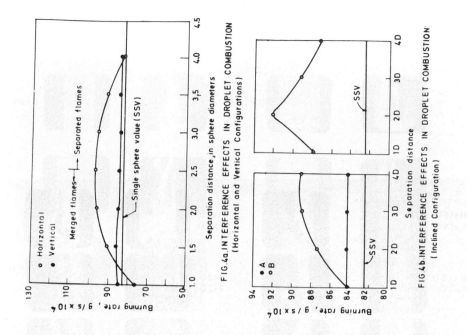

FIG.4a.INTERFERENCE EFFECTS IN DROPLET COMBUSTION
(Horizontal and Vertical Configurations)

FIG.4.b.INTERFERENCE EFFECTS IN DROPLET COMBUSTION
(Inclined Configuration)

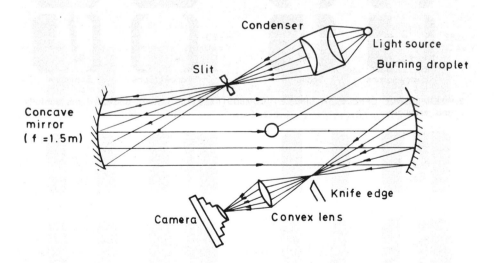

FIG.5.SCHEMATIC OF OPTICAL ARRANGEMENT

at very close separation. It is also seen that both the drops have the same burning rate.

4. EFFECTS OF DROPLET ROTATION

Figure 5 represents the schematic of optical arrangement employed in these experiments. Figures 6a and 6b show both shadowgraphs (upper row) and schlieren photographs (lower row) for four fuels, viz., propanol-1, n-heptane, methanol and n-hexane, at different rotational speeds of the sphere, viz., 0.37 and 84 rpm. Both shadowgraphs and schlieren photographs show that for the heavier fuels such as n-hexane and n-heptane a narrow vertical column of material of different density from the surroundings appears immediately above the sphere. It appears reasonable to conclude that these are pyrolysis products resulting from liquid and gas-phase cracking reactions. The flames are all quite cylindrical in the wake. While direct photographs clearly show up the near and far wake regions, the shadow and schlieren photographs do not reveal these zones. The effect of rotation in modifying flame geometry is not significant in the small range of rotational speeds investigated.

Drop rotation increases burning rates. The mass burning rates have been converted into Nusselt numbers and correlated against the effective Reynolds numbers, taking into consideration the effect of sphere rotation. Following the work of other workers in this field [13,14,15], the abscissa is chosen to be $(0.5 \, Re_R^2 + Gr)$. Additionally the effect of mass transfer shielding in reducing the heat transfer to the burning drops because of the outward mass transfer has been taken into consideration through the shielding function $(1+B)^{-1}$, which is based on laminar boundary layer theory. Figure 7 shows the correlations for the five fuels, all of which conform to the equation:

$$Nu \quad = \quad A \, (0.5 \, Re^2 + Gr) - C \qquad\qquad (1)$$

The values of A,C and the correlation coefficients are also given for each fuel. Figure 8 shows the correlations for the five fuels in conformity with the equation:

$$Nu(1+B) \quad = \quad A \, (0.5 \, Re^2 + Gr) - C \qquad\qquad (2)$$

It should be pointed out that the range of abscissa is too small to combine the five plots into one for a single correlation.

5. EFFECTS OF FORCED CONVECTION

Figure 9 shows the effect of forced convective air velocity on mass burning rates for methanol. With increasing air velocity the envelope flame burning rate increases linearly, dropping to nearly half the value prior to extinction, and remaining almost constant in the wake mode. When compared with the earlier work of Sami and Ogasawara [16],the trends in burning rate are consistent in showing a direct proportionality with sphere size, but the extinction velocity in the present case, instead of lying between their values for the 10 and 5 mm spheres, is smaller than that for their 5 mm sphere. This may be attributed to flow disturbances in the free stream in the present case.

In Figure 10, the mass consumption rates have been converted into Nusselt numbers and correlated against both approach and mean Reynolds numbers; in Fig.C the mass transfer shielding function is included in the correlation. In all three cases the parameters A and n in the standard correlations have been determined for the envelope flames; the wake flame Nusselt numbers show less

666

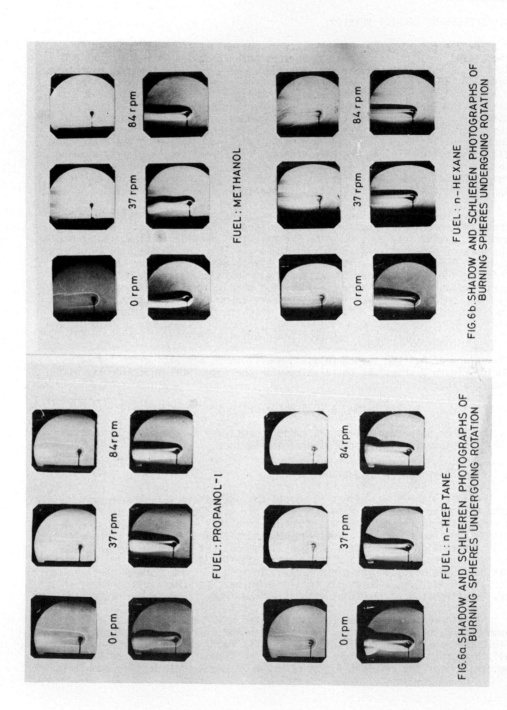

84 rpm   37 rpm   0 rpm

FUEL: METHANOL

84 rpm   37 rpm   0 rpm

FUEL: n-HEXANE

FIG.6b. SHADOW AND SCHLIEREN PHOTOGRAPHS OF BURNING SPHERES UNDERGOING ROTATION

84 rpm   37 rpm   0 rpm

FUEL: PROPANOL-1

84 rpm   37 rpm   0 rpm

FUEL: n-HEPTANE

FIG.6a. SHADOW AND SCHLIEREN PHOTOGRAPHS OF BURNING SPHERES UNDERGOING ROTATION

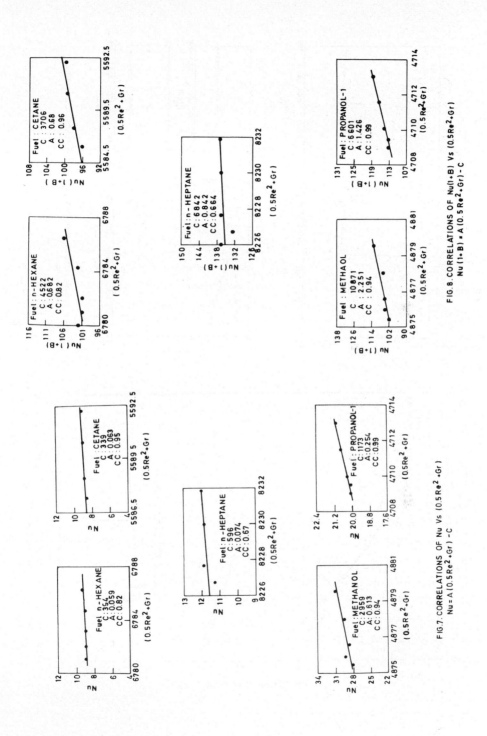

FIG.8. CORRELATIONS OF Nu(1+B) Vs (0.5Re²+Gr)
Nu(1+B) = A(0.5Re²+Gr) - C

FIG.7. CORRELATIONS OF Nu Vs (0.5Re²+Gr)
Nu = A(0.5Re²+Gr) - C

668

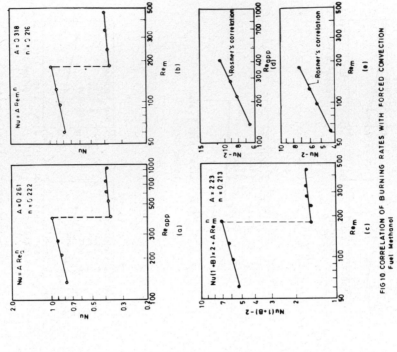

FIG 10 CORRELATION OF BURNING RATES WITH FORCED CONVECTION
Fuel Methanol

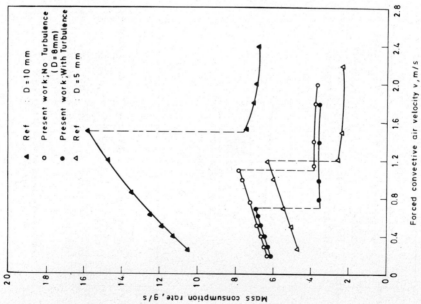

FIG. 9. EFFECT OF FORCED CONVECTIVE AIR VELOCITY ON MASS BURNING RATES
FUEL: METHANOL

V = 0 m/s

V = 0.4 m/s

V = 0.77 m/s

(a) WITHOUT TURBULENCE

V = 0.4 m/s

V = 0.63 m/s

(b) WITH TURBULENCE

FIG. 11. EFFECTS OF TURBULENCE ON ENVELOPE FLAMES
FUEL: METHANOL

V = 1.1 m/s

V = 1.8 m/s

V = 8.6 m/s

(a) WITHOUT TURBULENCE

V = 0.71 m/s

V = 1.8 m/s

V = 8.6 m/s

(b) WITH TURBULENCE

FIG. 12. EFFECTS OF TURBULENCE ON WAKE FLAMES
FUEL: METHANOL

dependence on the Reynolds numbers. Figures d and e show the mass burning rates converted into Nusselt numbers and plotted against both approach and mean Reynolds numbers according to Rosner's correlation [17]:

$$Nu = 2 + (0.555 \ Re^{0.5} \ Pr^{0.33})/(1+1.232/(Re \ Pr^{1.33}))^{0.5} \tag{3}$$

6. EFFECTS OF TURBULENCE

Figures 11 and 12 show shadowgraphs and schlieren photographs of envelope and wake flames, respectively, for methanol, with and without turbulence; the top two photographs in Fig.11 are under only free convection. Both shadowgraphs and schlieren photographs reveal the disruption of the flame zones caused by turbulence.

In the presence of grid-generated turbulence the burning rate decreased somewhat, but more significantly the extinction velocity decreased from about 1.1 m/s to 0.7 m/s.

REFERENCES

1. Rex,J.F., Fuhs,A.E. and Penner,S.S., Jet Propulsion, 26, 179 (1956).

2. Kanevsky,J., Jet Propulsion, 26, 768 (1956).

3. Fedoseeva,N.V., Advances in Aerosol Physics, 1, 21 (1971).

4. Sangiovanni,J.J. and Kesten,A., Sixteenth Symposium (Int.) on Combustion, The Combustion Institute, USA, 571 (1976).

5. Sarathy,C.P. and Natarajan,R., Symposium on Rocket Propulsion, Institute of Armament Technology, Pune, India, Dec.22-23, 1978.

6. Twardus,E.M.,and Brzustowski,T.A., Comb. Science and Tech., 17, 215 (1978).

7. Sangiovanni,J.J. and Dodge, Seventeenth Symposium (Int.) on Combustion, The Combustion Institute, USA, 455 (1979).

8. Carstens,J., Williams,A. and Zung,J., J. Atmos. Science, 28, 798 (1970).

9. Labowsky,M., Chem. Eng. Science, 31, 803 (1976).

10. Williams,A. and Carstens,J., J. Atmos. Science, 28, 1 (1970).

11. Labowsky,M., Comb. Science and Tech., 22, 217 (1980).

12.. Natarajan,R., Can. J. Chem. Eng., 52, 834 (1974).

13. Kreith,F., in Advances in Heat Transfer , Vol.5, Ed: J.P. Hartnett, Academic Press, New York, 159 (1968).

14. Nordlie, R.L. and Kreith,F., in International Developments in Heat Transfer, ASME, New York, 461 (1968).

15. Kays, W.M., and Bjorklund, Trans. ASME, 80, 70 (1958).

16. Sami,H. and Ogasawara,M., Bulletin of JSME, 13, 395 (1970).

17. Rosner,D.E., cited in Fuesh, G.M., Prog. Energy Combust. Sci., 3, 191 (1977).

Multi-Phase Flow and Heat Transfer III. Part B: Applications
edited by T.N. Veziroğlu and A.E. Bergles
Elsevier Science Publishers B.V., Amsterdam, 1984 — Printed in The Netherlands

COMBUSTION OF COAL AND COAL-WATER SUSPENSIONS IN NO. 2 DIESEL FUEL

R.S. Stagner and S.C. Kranc

College of Engineering
University of South Florida
Tampa, Florida   33620, U.S.A.

ABSTRACT

Fuel slurries have been frequently been proposed as a means of extending oil supplies and utilizing available coal. Potential applications of Diesel mixtures include gas turbines and reciprocating engines. In this investigation combustion tests of coal-oil-water mixtures were conducted in a shrouded combustion chamber at atmospheric pressure. The purpose of these tests was to study the effect of water addition to the fuel handling and combustion properties. Water may be present in the fuel as a result of transport or storage or may be added intentionally. Water is known to improve stability of coal slurries and can aid in reducing agglomeration of the coal particles during combustion.

Various fuel slurries were formulated using No. 2 Diesel, bituminous coal ground to 80 microns, and water. These mixtures were further homogenized and then pumped to a Hartmann whistle type atomizer. No surfactant was used to stabilize the slurry. Coal loads were held at a constant 20% by weight and samples were burned at excess air ratios of 150 and 200%. The exhaust enthalpy was computed from thermocouple measurements of the exhaust temperature. The thermal efficiency was then computed. The results of this investigation show that small amounts of water tend to decrease efficiency but that this trend can be reversed by increasing the water content. These conclusions are compared to the findings of other workers.

1. INTRODUCTION

Coal-oil mixtures have attracted considerable attention as alternate fuels. Not only can oil supplies be augmented but also fuel handling and burning characteristics can be adapted to suit particular needs. Significant problems remain to be solved, however, before coal-oil mixtures will be widely adopted. Borio and Hargrove (Ref. 1) emphasize the need to study stability, viscosity, atomization quality, ignition, carbon burnout, ash characteristics, and emissions as important parameters of fuel slurries.

Most workers have concentrated on suspensions formulated

676

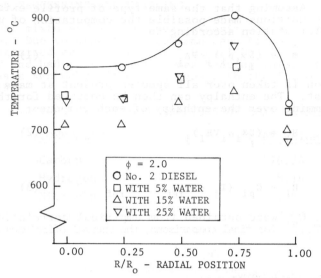

Fig. 3. Exit plane temperature profiles
for Diesel-water emulsions at
ϕ = 2.0.

where the lower heating value of the fuel is used to obtain the
thermal input, Q_{in}. In addition to incomplete combustion losses
were attributed to convection and radiation from the hot surface
of the outer shell of the combustor. These losses present no
real problem since they are nearly constant. Measurements of the
surface temperature of the shell were used to estimate these
losses at 3-4%.

6. EXPERIMENTAL OBSERVATIONS AND DISCUSSION

Most experiments were conducted at equivalence ratios ϕ of
1.5 and 2. Since the volume of the combustor was small combustion
intensities were of the order of 2.5 kilowatts/cubic meter, higher
than that for pulverized coal combustion. Tests at stoichio-
metric conditions exhibited substantial flame beyond the end of
the tube indicating the need for increased dilution and mixing to
improve combustion. Even so, when coal was added to the fuel, a
shower of luminous particles left the exit. During the experiments, few problems were encountered with the flame stability,
pumping of the fuel, or clogging of the atomizer.

Typical exit plane profiles are shown in Figures 3, 4, and
5. As would be expected, a strong gradient is observed near the
wall due to the cooling effect of secondary air admission. The
profiles are relatively uniform through the central region. Off
axis peaks observed are probably artifacts of the spray or mixing patterns developed upstream. In general, the effect of coal
addition is to lower the exhaust temperature.

After analyzing the exhaust temperature profile as outlined
above the enthalpy profile and the efficiency may be deduced.

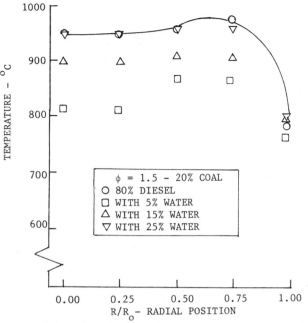

Fig. 4. Exit plane temperature profiles
for coal-Diesel-water mixtures
at φ = 1.5.

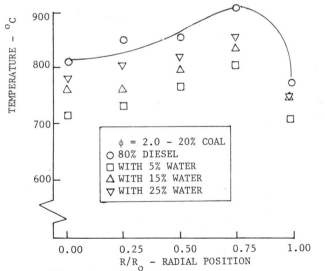

Fig. 5. Exit plane temperature profiles
for coal-Diesel-water mixtures
at φ = 2.0.

2. AVERAGE VARIABLES

2.1. Definitions and Basic Formulas

We describe the flow by using averages of certain flow properties that are locally defined either in regions occupied by gas (densities of mass, momentum and internal energy), or inside the particles (densities of mass and momentum), or on the surfaces of particles (regression distance and surface temperature). The averages are defined for a constant finite averaging volume V that has a surface S_v with a well defined normal almost everywhere, and a weight function g. (See Figure 1.) The weight function is assumed to be positive inside V, to have a continuous gradient ∇g in V and on S_v (isolated singular points are permitted), and to vanish together with ∇g on S_v. The averages are intrinsic, i. e., we average only over those parts of V where the property is locally defined. For properties defined in a volume this is achieved by a phasic function β that equals zero inside grains and has the value one elsewhere. Let V be located at x and the constant VG be the weighted average volume

$$VG = \int\limits_{V(x)} g(\xi-x) \, dV(\xi). \qquad (2.1)$$

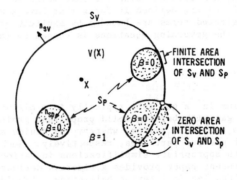

Figure 1. AVERAGING VOLUME

The gas volume fraction (porosity) $\alpha(t,x)$ we define by

$$\alpha(t,x) = \frac{1}{VG} \int\limits_{V(x)} \beta(t,\xi) \, g(\xi-x) \, dV(\xi) \quad , \qquad (2.2)$$

and the intrinsic average ϕ of a local gas property $\tilde{\phi}$ by

$$\alpha(t,x) \; \phi(t,x) \; = \frac{1}{VG} \int\limits_{V(x)} \beta(t,\xi) \; \tilde{\phi}(t,\xi) \; g(\xi-x) \; dV(\xi) \quad . \qquad (2.3)$$

The corresponding definition for the average $\overset{*}{\phi}$ of a particle property $\overset{\tilde{*}}{\phi}$ is

$$\bigl(1-\alpha(t,x)\bigr) \; \overset{*}{\phi}(t,x) \; = \frac{1}{VG} \int\limits_{V} (1-\beta) \; \overset{\tilde{*}}{\phi} \; g \; dV \quad . \qquad (2.4)$$

In order to obtain the intrinsic averages of surface properties we first define the weighted surface area SG by

$$SG(t,x) \; = \int\limits_{S_p(t,x)} g(\zeta-x) \; dS(\zeta) \quad , \qquad (2.5)$$

and then obtain the average ψ of the property $\tilde{\psi}$ that is defined on the surface of the propellant by

$$\psi(t,x) \; = \; \frac{1}{SG(t,x)} \int\limits_{S_p(t,x)} \tilde{\psi}(t,\zeta) \; g(\zeta-x) \; dS(\zeta) \quad . \qquad (2.6)$$

The integrals in (2.5) and (2.6) are taken over all particle surfaces that are inside V. Under the given assumptions about V and g, all averages are continuously differentiable with respect to t and x if the local functions are piecewise differentiable. The spacial averages are also twice continuously differentiable with respect to x if the local functions are piecewise twice differentiable.

One can derive the following formulas involving the volume averaging integral in (2.3) for gas properties

$$\int\limits_{V} \beta \; \frac{\partial \tilde{\phi}}{\partial t} \; g \; dV \; = \frac{\partial}{\partial t} \int\limits_{V} \beta \; \tilde{\phi} \; g \; dV \; - \int\limits_{S_p} \tilde{\phi} \; (\tilde{u}_{sp} \cdot n_{sp}) \; g \; dS \quad , \qquad (2.7)$$

$$\int\limits_{V} \beta \; \nabla\tilde{\phi} \; g \; dV \; = \nabla\int\limits_{V} \beta \; \tilde{\phi} \; g \; dV \; + \int\limits_{S_p} \tilde{\phi} \; n_{sp} g \; dS \quad , \qquad (2.8)$$

$$\int\limits_{V} \beta \; \nabla\nabla\tilde{\phi} \; g \; dV \; = \nabla\nabla\int\limits_{V} \beta \; \tilde{\phi} \; g \; dV \; + \int\limits_{S_p} (g \; \nabla\tilde{\phi} \; - \tilde{\phi} \; \nabla g)\cdot n_{sp} \; dS \quad . \qquad (2.9)$$

$$n \cdot \tilde{\rho} \; (\tilde{u} - \tilde{u}_{sp}) \; \tilde{u} + n \; \tilde{p} - n \cdot \tilde{\Pi} = n \cdot \overset{*}{\rho} \; \overset{*}{(\tilde{u}} - \tilde{u}_{sp}) \; \overset{*}{\tilde{u}} - n \cdot \overset{*}{\tilde{\Pi}} \; , \qquad (3.10)$$

where n is the unit normal to the interface. The local interface velocity $\tilde{u}_{sp}$ is defined in terms of the local regression rate $\dot{d}$ of the grain surface by

$$\tilde{u}_{sp} = \overset{*}{\tilde{u}} + n_{sp} \; \tilde{\dot{d}} \; , \quad \tilde{\dot{d}} > 0 \; . \qquad (3.11)$$

3.2. Derivation of the Volume Average Gas Momentum and Surface Average Regression Distance Equations

As examples, we derive the volume average gas momentum and the surface average regression distance equations in order to illustrate the technique of obtaining average equations from a set of local equations, to show how turbulence and interface quantities are defined, and to explain how errors via correlations and models are introduced and corresponding error estimates are obtained.

The volume average gas momentum equation is derived by multiplying the local momentum equation (3.2), which is defined in V(x), by the function βg, by integrating over the averaging volume V(x), and by applying formulas (2.7) and (2.8). The result is

$$\frac{\partial}{\partial t} \int_V \beta \; g \; \tilde{\rho} \; \tilde{u} \; dV + \nabla_x \cdot \int_V \beta \; g \; \tilde{\rho} \; \tilde{u} \; \tilde{u} \; dV$$

$$+ \int_{S_p} g \; \tilde{u} \; \tilde{\rho} \; n_{sp} \cdot (\tilde{u} - \tilde{u}_{sp}) \; dS + \nabla_x \int_V \beta \; g \; \tilde{p} \; dV \qquad (3.12)$$

$$= \nabla_x \cdot \int_V \beta \; g \; \tilde{\Pi} \; dV - \int_{S_p} g \; (n_{sp} \tilde{p} - n_{sp} \cdot \tilde{\Pi}) \; dS .$$

The average gas velocity u is defined as the ratio of the average momentum density [(2.3) with $\phi = \rho u$] to the average density [(2.3) with $\phi = \rho$]. This definition of u is advantageous when $\tilde{\rho}$ depends on the spatial variable and it reduces to the standard definition of an average variable (2.3) when $\tilde{\rho}$ is constant or a function of time only. The fluctuations of the values of the local density and velocity from the values of the average density and velocity are defined by

$$\tilde{\rho}' = \tilde{\rho} - \rho \quad \text{and} \quad \tilde{u}' = \tilde{u} - u \; . \qquad (3.13)$$

Substituting (3.13) into the volume integral of $\beta g \tilde{\rho} \tilde{u} \tilde{u}$ in (3.12), we obtain

$$\frac{1}{VG} \int_V \beta \ g \ \tilde{\rho} \ \tilde{u} \ \tilde{u} \ dV = \alpha \ \rho \ u \ u + \frac{1}{VG} \int_V \beta \ g \ \tilde{\rho} \ \tilde{u}'\tilde{u}' \ dV \quad . \qquad (3.14)$$

The difference between the volume average of $\widetilde{\rho u u}$ and the product of the corresponding average quantities is the volume average of the dyadic of the velocity fluctuations times the local density. We define the latter integral as the turbulent gas stress tensor and denote its model by Π_T. The volume integrals of $\beta g \tilde{p}$ and $\beta g \tilde{\Pi}$ are by definition (2.3) the gas volume fraction α times the average pressure and laminar stress tensor, respectively. These average quantities are generally modeled rather than formally averaging the local constitutive laws (3.4). The expression $\tilde{\rho} n_{sp} \cdot (\tilde{u} - \tilde{u}_{sp})$ can be replaced by $-\tilde{\rho} \overset{*}{\tilde{d}}$ via (3.9) and (3.11). A model of the surface average regression rate [(2.6) with $\tilde{\psi} = \overset{\bullet}{d}$] is denoted by $\langle \overset{\bullet}{d} \rangle$. Often $\langle \overset{\bullet}{d} \rangle$ is given as a function of the average pressure p to some power. The interphase drag is represented by the surface integral

$$\frac{1}{VG} \int_{S_p} g \ [n_{sp}(\tilde{p} - p) - n_{sp} \cdot \tilde{\Pi}] \ dS \quad , \qquad (3.15)$$

and is modeled by $\rho(1-\alpha)D$ where D is an experimentally determined correlation. This definition of drag is consistent with Ishii's[7] development but is different from Gibeling et al.[6] and Gough[3] where it is defined in terms of the surface average of the normal total gas stress tensor $n_{sp} \cdot (\Pi - \tilde{\Pi}) - n_{sp}(p - \tilde{p})$. When the average stress tensor is zero ($\Pi = 0$, the inviscid two-phase model) both definitions agree. We recognize the fact that (3.15) is a formal definition which may not be realized in an experimentally determined correlation. In such a case, the other effects due to the experiment would have to be included in (3.15). Finally, we note that

$$\nabla \alpha = -\frac{1}{VG} \int_{S_p} g \ n_{sp} \ dS \qquad (3.16)$$

from (2.8) with $\tilde{\phi} = 1$. Incorporating these changes and definitions into (3.12), using (3.9), and (3.11), and algebraically manipulating the result, we have the average momentum equation:

$$\frac{\partial}{\partial t} [\alpha \rho u] + \nabla \cdot [\alpha \rho u u] = - \alpha \ \nabla p + \nabla \cdot (\alpha \ \Pi) + \nabla \cdot (\alpha \ \Pi_T)$$

$$+ \overset{*}{\rho} \ \frac{SG}{VG} \ \overset{*}{u} \ \langle \overset{\bullet}{d} \rangle \ - \ \rho \ (1-\alpha) \ D$$

$$+ \{+ \frac{1}{VG} \int_{S_p} g \ \overset{\tilde{*}}{\rho} \ \overset{\tilde{*}}{u} \ \overset{\bullet}{\tilde{d}} \ dS - \overset{*}{\rho} \ \frac{SG}{VG} \ \overset{*}{u} \ \langle \overset{\bullet}{d} \rangle \}$$

$$+ \{\nabla \cdot [\frac{1}{VG} \int_V \beta \ g \ (\rho u u - \widetilde{\rho u u}) \ dV - \alpha \ \Pi_T]\}$$

$$\frac{\partial \overset{*}{m}}{\partial t} = - \nabla \cdot (\overset{*}{m} \overset{*}{u}) \quad , \tag{3.27}$$

$$\frac{\partial \overset{*}{d}}{\partial t} = - \overset{*}{u} \cdot \nabla \overset{*}{d} + \langle \dot{d} \rangle \quad , \tag{3.28}$$

$$\frac{\partial \overset{*}{T}}{\partial t} = - \overset{*}{u} \cdot \nabla \overset{*}{T} + \langle \dot{T} \rangle \quad , \tag{3.29}$$

where

$$B = \frac{1}{\alpha} [(1-\alpha) \nabla \cdot \overset{*}{u} - (u - \overset{*}{u}) \cdot \nabla (1-\alpha)] \quad , \tag{3.30}$$

$$H = \frac{1}{T} [\hat{e} + p/\overset{*}{\rho}) - (e + p/\rho)] \quad , \tag{3.31}$$

$$\Gamma = \frac{1}{\alpha} \frac{\rho}{\rho} \frac{\overset{*}{m}}{VG} s_p(\overset{*}{d}) \langle \overset{*}{d} \rangle \quad , \tag{3.32}$$

$$\alpha = 1 - v_p(\overset{*}{d}) \frac{\overset{*}{m}}{VG} \quad . \tag{3.33}$$

We now list the models for the correlations which close the system of differential equation. The heat dissipation function is given by

$$\Phi_2 = \frac{1}{\rho T} \Phi + \langle \Phi \rangle + \frac{1}{\rho T} \Phi_T \quad , \qquad E = 0.5 \ (\nabla u + (\nabla u)^T) \quad ,$$

$$\Phi = 2 \ \mu \ E{:}E + (\lambda - \frac{2}{3} \mu) \ (\nabla \cdot u)^2 \quad , \tag{3.34}$$

$$\Phi_T = 2 \ \mu_T \ E{:}E + (\lambda_T - \frac{2}{3} \mu_T) \ (\nabla \cdot u)^2 \quad ,$$

where the term $\langle \Phi \rangle$ is an additional correlation which is independent of turbulence. Because the value of the gradients of the average velocity will be less than the value of the average of the gradients of the local velocities, the value given by the model Φ will be an underestimate. The correlation $\langle \Phi \rangle$ provides the correction. The shear and bulk viscosity coefficients for laminar and turbulent flows (μ, λ, μ_T and λ_T) can be given by Sutherland type laws and the particular turbulence model, respectively. The heat conduction correlations are:

$$\Psi = \Psi_{gas} + \Psi_{particle} + \Psi_T \quad ,$$

$$\Psi_{gas} = \frac{1}{\alpha \rho T} \nabla \cdot (\alpha \ \kappa \ \nabla T) \quad , \qquad \Psi_T = \frac{1}{\alpha \rho T} \nabla \cdot (\alpha \ Q_T) \quad , \tag{3.35}$$

$$Q_T = \kappa_T [\nabla T - \frac{\nabla \alpha}{\alpha} (T_i - T)] \quad ,$$

where $\Psi_{particle}$ is zero except before ignition when[6]

$$\psi_{particle} = - \frac{1}{\alpha \rho T} \frac{\overset{*}{m}}{VG} s_p [h_c (T - \overset{*}{T}) + h_r (T - \overset{*}{T})] \quad ,$$

$$h_c = \frac{\kappa}{\overset{*}{D}_p /2} + 0.2 \left(\frac{\gamma}{\gamma - 1} \frac{R}{M} \frac{(\kappa \, 2 \rho)^2 |u - \overset{*}{u}|^2}{\mu \overset{*}{D}_p /2} \right)^{1/3} \quad ,$$

$$h_r = \overset{*}{\varepsilon} \sigma_{SB} (T + \overset{*}{T}) (T^2 + \overset{*}{T}^2) \quad .$$

The coefficients of thermal conductivity for laminar and turbulent flows κ and κ_T can be given by a Sutherland type law and by a particular turbulence model, respectively. The interface temperature, the effective diameter of a grain, the average grain emissivity, the Stephan-Boltzmann constant, and the combustion gas ratio of specific heats are denoted by T_i, $\overset{*}{D}_p$, ε, σ_{SB} and γ, respectively. The models for the laminar and turbulent average stress tensors are:

$$\Pi = 2 \mu E + (\lambda - \frac{2}{3} \mu) (\nabla \cdot u) I \quad , \tag{3.36}$$

$$\Pi_T = 2 \mu_T E + (\lambda_T - \frac{2}{3} \mu_T) (\nabla \cdot u) I \quad . \tag{3.37}$$

For packed bed flow[8] ($\alpha < 0.65$), the drag is

$$D = (\overset{*}{u} - u) \frac{a_p}{v_p} \frac{2}{3} \frac{1}{\alpha^2} [1.75 |u - \overset{*}{u}| + 150 (1-\alpha) \frac{\mu}{\rho \overset{*}{D}_p}] \quad , \tag{3.38}$$

and for dilute flows ($\alpha > 0.9$) the drag is

$$D = (u - \overset{*}{u}) \frac{a_p}{v_p} [0.2 |u - \overset{*}{u}| + 12 \frac{\mu}{\rho \overset{*}{D}_p}] \quad , \tag{3.39}$$

where the average frontal area of the grains is denoted by a_p. When $0.65 < \alpha < 0.9$, a linear interpolation between the two models is used.

The intergranular stress correlation is given by

$$\overset{*}{\Pi} = [\frac{-\overset{*}{\rho}}{1-\alpha} \int_\alpha^{\alpha_2} \overset{*}{a}^2 (y) \, dy] I \quad , \tag{3.40}$$

where $\overset{*}{a}(\alpha)$ is a model of the average sound speed through the particles when $\alpha < \alpha_2$ and zero elsewhere, α_2 being a model constant. The turbulence of the solid phase is assumed negligible because of the density and size of the grains, and thus Π_T^* is set to zero. The regression rate is given in terms of model constants B_o, B_1 and B_2 by

$$\langle \dot{d} \rangle = B_o + B_1 p^{B_2} \, . \tag{3.41}$$

The average energy released by burning is

$$\hat{e} = \frac{1}{\gamma - 1} \frac{R}{M} T_{flame} \, . \tag{3.42}$$

The standard Noble–Abel equation of state formulas are assumed to hold for the average variables and the average specific heats are assumed to be constant.

4. INITIAL AND BOUNDARY CONDITIONS

4.1. Initial Conditions

Typical initial conditions for interior ballistics are uniform states of gas and propellant over the entire region. Because averaging of a constant produces the same constant, the initial averages in the inner region are equal to the specified uniform states. If the initial grain or grain size distribution is not uniform, then the corresponding initial averages $\alpha(0,x)$ and $\hat{d}(0,x)$ must be computed by the averaging formulas (2.2) and (2.6), respectively, using the same V and g as for the correlations. In those parts of the inner region where the grain number m^* is zero, one has to use common sense extrapolations for u^*, T^* and d^* such that the resulting initial values are smooth functions of x.

In the boundary region, "correct" zero initial values can be specified only for variables that are uniformly zero initially. The initial values of other variables must be calculated consistently with the numerical treatment of the boundary region. Uniform non–zero initial state does not necessarily imply uniform initial values in the boundary region.

4.2. Boundary Conditions

The average governing equations were derived in Section 3 only for the inner region of the flow field. In order to solve the equations one has, therefore, to specify boundary conditions at the inner boundary. Physical boundary conditions are given only on the outer (physical) boundary and only for local gas properties and single particles. In order to derive the proper boundary conditions for the average governing equations, one therefore needs a theory that produces a complete set of conditions for average properties at the inner boundary in terms of the physical conditions at the outer boundary. It should take into account the particular V and g for which the correlations have been derived and also provide a description of the flow within the boundary region. Because such a theory has not been developed, we propose the following ad hoc treatment of the boundary region.

First we use in the boundary region the same average governing equations as in the inner region. Then one needs to specify boundary conditions only at the outer boundary. Let $\ell/2$ be the distance between the inner and outer boundaries and let ε be an estimated thickness of the gas boundary layer. Let ϕ be a function with a prescribed local boundary value $\tilde{\phi}_{wall}$ and n_i be the unit normal to the inner boundary pointing outward with respect to the inner region. We then use the following boundary value ϕ_{outerb} for gas properties:

$$\phi_{outerb}= [\frac{\ell}{2} (\phi_{innerb} + \frac{\ell}{2} \nabla\phi_{innerb} \cdot n_i) + \varepsilon \, \tilde{\phi}_{wall}] \, / \, (\frac{\ell}{2} + \varepsilon) \quad . (4.1)$$

This equation we use for the boundary values of u and T or $\partial T/\partial n$. The average gas continuity equation is used at the outer boundary to close the set of boundary conditions for the average gas properties.

Particle properties are treated as follows. First, $\overset{*}{u}$, $\overset{*}{m}$, $\overset{*}{q}$ and $\overset{*}{T}$ are computed by solving the governing equations at the wall. Then u is corrected at the wall such that loss of particles through the wall is avoided, i.e., that

$$(\overset{*}{u}_{outerb} - u_{wall}) \cdot n_{wall} \le 0$$

is satisfied if $\overset{*}{m}_{wall} > 0$. If $\overset{*}{m}$ equals zero, then the solution of the governing equations is used without restrictions at the wall.

5. CONCLUSIONS

The presented average governing equations model the transient effects of viscosity, heat conduction and turbulence in the compressible gas phase; the ignition, intergranular stress and burning in the incompressible solid phase; and corresponding interactions between the phases, e. g., drag, heat transfer and source terms. Turbulence is defined in terms of volume averages. The dependent variables and the form of the equations are chosen to facilitate the numerical solution of the system.

The following conclusions can be drawn from the theoretical exposition:

First, the proper averaging domain is a finite volume that is larger than the propellant grains. Line and surface averaging cannot be used because the corresponding averages do not have the necessary differentiability properties. Infinite volume averaging is not appropriate for interior ballistics (or other confined flows) because in such a volume the phases do not occupy complementary spaces. Time averaging is not applicable to interior ballistics because of the unsteady and rapidly changing flow conditions, including moving boundaries.

Second, the average equations are not suitable for describing flows with shocks. On the other hand, by a proper formulation of the equations, one need not explicitly follow boundaries of regions without particles.

Third, the average equations are not valid in boundary regions with thicknesses of the order of grain diameters. Lacking a proper theory, one has to use ad hoc modeling in these regions.

In summary, the derived governing equations represent a complete model of a three-dimensional two-phase interior ballistics flow. All experimental correlations and other approximations are exactly defined, allowing one to estimate their effect on the solution. The missing theory for the boundary is exposed. With this background, one can carry out numerical studies of three-dimensional two-phase flows and, using experimental data, establish the missing connections between observations and theory.

List of Symbols

$\overset{*}{d}$	regression distance
$\dot{d}$	regression rate
$\langle \dot{d} \rangle$	burning rate correlation
D	drag correlation
$\overset{*}{D}_p$	effective average grain diameter
e	specific internal energy
$\hat{e}$	e at flame temperature
g	averaging weight function
I	identity tensor of second order
$\overset{*}{m}$	weighted number of grains in V
M	molar mass
n_{sp}	unit outward normal with respect to gas on a grain
p	pressure
q	transformed pressure
R	universal gas constant
s	specific entropy
s_p	average surface area of a grain
S_p	union of all grain surfaces in V
SG	weighted area of S_p
t	time
T	temperature
$\langle \dot{T} \rangle$	correlation for rate of change of grain surface temperature
u	velocity
V	averaging volume

v_p	average volume of a grain
VG	weighted value of V
x	vector of spatial coordinates
α	gas volume fraction
β	phasic function
γ	ratio of specific heats
Γ	source term
κ	thermal conductivity coefficient
λ	bulk viscosity coefficient
μ	shear viscosity coefficient
Π	stress tensor
ρ	density
ϕ	gas property function
Φ	dissipation term
ψ	grain surface property function
Ψ	heat conduction term

Superscripts

$'$	=	fluctuations
$\sim$	=	local variable
$*$	=	particle variable
none	=	average gas variable
T	=	transpose

Subscripts

T	=	turbulent
sp	=	on the grain surface
s,q	=	partial derivative with respect to s and q, respectively

REFERENCES

1. Celmiņš, A.K.R. and Schmitt, J.A., "Volume Averaged Two-Phase (Gas-Solid) Interior Ballistics Equations," BRL Report in publication.

2. Fisher, E.B. and Trippe, A.P., "Mathematical Model of Center Core Ignition in the 175mm Gun," Calspan Report VQ-5163-D-2, 1974.

3. Gough, P.S., "The Flow of a Compressible Gas Through an Aggregate of Mobil, Reacting Particles," Ph.D. Thesis, McGill University, Montereal, 1974.

4. Krier, H., van Tassell, W.F., Rajan, S., and Vershaw, J. "Model of Flamespreading and Combustion Through Packed Beds of Propellant Grains," University of Illinois at Urbana-Champaign Report, TR-AAE-74-1, 1974.

5. Kuo, K.K., Koo, J.H., Davis, T.R., and Coates, G.R., Acta. Astron., Vol.3, pp. 574-591, 1976.

698

6. Gibeling, H.J., Buggeln, R.C., and McDonald, H., "Development of a Two-Dimensional Implicit Interior Ballistics Code," Ballistic Research Laboratory Contractor Report, ARBRL-CR-00411, 1980.

7. Ishii, M., Thermo-Fluid Dynamic Theory of Two-Phase Flow, Eyrolles, France, 1975.

8. Ergun, S., "Fluid Flow Through Packed Columns," Chem. Eng. Progr., Vol. 48, 1952, p. 89.

Multi-Phase Flow and Heat Transfer III. Part B: Applications
edited by T.N. Veziroğlu and A.E. Bergles
Elsevier Science Publishers B.V., Amsterdam, 1984 — Printed in The Netherlands

PERFORMANCE CHARACTERISTIC OF COCURRENT SWIRL TRAY

Krzysztof Szczuka and Jerzy Merwicki
Metalchem Chemical Apparatus Works
Opole, Poland

Leon Troniewski
Technical University of Opole
Opole, Poland

ABSTRACT

The paper describes the new construction of cocurrent swirl tray for mass transfer apparatus. The increased separating power of this tray is due to different flow pattern and high values of phase velocities. A study has been made of the behaviour of column with two trays which simulate conditions in an actual multi-stage column. All measurements of hydraulic parameters have been made with air-water system, whereas for experiments on mass transfer the water solutions of carbon dioxide and ammonia have been used. Because of the uselessness of published correlations for the case of swirl flow, an attempt has been made to suggest a new form of dimensionless correlations. It resulted in development of the new dimensionless geometric parameter expressing the swirl intensity. Very good accuracy of derived correlations has been achieved as a result of their new and more suitable form.

1. INTRODUCTION

A permanent development of chemical industry sets up a necessity of continuous searching for more economical column contactors. It resulted in the development of columns with cocurrent trays. First designs of such columns have appeared in literature in the beginning of the sixties. In comparison with typical packed or tray columns, the apparatus with cocurrent trays offers certain advantages such as high flow rates of gas and liquid unlimited by flooding as well as higher values of overall mass transfer coefficient. Additional increase of mass transfer rate may be achieved in two-phase swirl flow which at the same time gives facilities for more accurate separation of phases on each tray.

The principle of operation is shown in Fig. 1. Cocurrent contact device operates in upward two-phase annular or annular-mist flow which is created by high values of the gas velocity. Liquid enters the contact tube through inlet holes or sprayers and then is risen up with gas stream. After passing over contact tube, two-phase stream is separated, and then gas phase flows up into upper tray whereas liquid phase flows down to lower one. It is evident from presented orientation of phase flows that counter-current is a general flow in whole column.

Details of the research apparatus are shown on schematic
diagram in Fig. 4. The main part of this apparatus is column consisted of two cocurrent trays. All measurements were made on upper
tray of the column, where proper hydrodynamic conditions were
created by the lower tray. Water was pumped from tank through
calibrated rotameter, before passing over the contact tube. Air
coming to column was blown by fan, humidified in packed column
with cyclone separator and measured by orifice plate. The flow
rates of air and water were changed by regulating a butterfly valve
on the pipe going from fan and a valve at the outlet of the circulating pump. Liquid tank was provided with a heating source that
allowed the liquid to be maintained at any desired temperature.
The temperature of air was regulated by the the temperature of
water circulating in humidifying circuit.

3. RESULTS AND DISCUSSION

3.1. Pressure Drop

All measurements of total pressure drop during two-phase flow
in the contact tube were made with air-water system. Obtained data
were satisfactorily expressed by the formula:

$$\Delta p_T = f_{TP} \, \Delta p_G \qquad\qquad /1/$$

where Δp_G was the pressure drop occuring when only gas flowed
through the contact device. A two-phase factor f_{TP} was found to be
a function of the Reynolds numbers ratio:

$$f_{TP} = 4,761 \left(\frac{Re_L}{Re_G} \right)^{0,226} \qquad\qquad /2/$$

The main component of Δp_G is a pressure drop on vortex generator.
Thus, the following equation can be used for predicting pressure
losses during gas flow in contact tube:

$$\Delta p_G = \zeta \, \frac{w_G^2 \, \varrho_G}{2} \qquad\qquad /3/$$

For resistance coefficient ζ, the following formula was derived
from experimental data:

$$\zeta = 4,215 \, A_o^{1,209} + 1 \qquad\qquad /4/$$

In this formula, a dimensionless geometric parameter of vortex
generator A_o similar to that of Abramovich [9] is used to develop
relation between geometric dimensions, swirl intensity and resistance coefficient. A method of predicting parameter A_o is described
in detail in references [5,6,10]. Table 1. shows the values of A_o
for three different types of vortex generators used in the study.

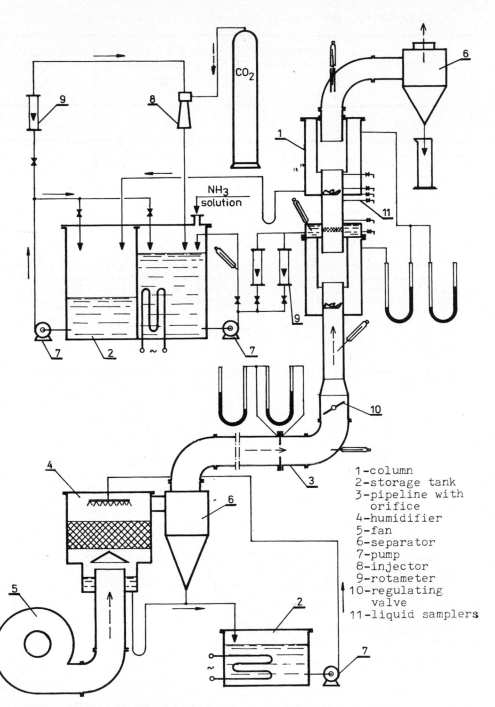

CO₂

NH₃ solution

1-column
2-storage tank
3-pipeline with orifice
4-humidifier
5-fan
6-separator
7-pump
8-injector
9-rotameter
10-regulating valve
11-liquid samplers

Fig. 4. The schematic diagram of the apparatus

Table 1.   The Main Parameters of Vortex Generators

TYPE OF VORTEX GENERATOR	NUMBER OF BLADES	ANGLE OF BLADES DEFLECTION Θ	GEOMETRIC PARAMETER A_o
I	6	30°	1,427
II	6	40°	0,883
III	8	40°	0,746

3.2.   Liquid Entrainment

The velocity of the gas phase flowing through contact tube is limited by rapid increase of liquid entrainment between the two following trays. Maximum possible value of the superficial gas velocity can be calculated from the empirical equation:

$$w_{Gmax} = 34,3 \, A_o^{-0,332} \, H_s^{0,086} \, \Gamma^{0,027} \qquad /5/$$

Fig. 5. shows the method of determining the maximum velocity of gas phase.

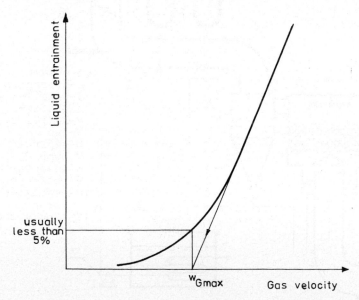

Fig. 5.   Method of determining the maximum gas velocity

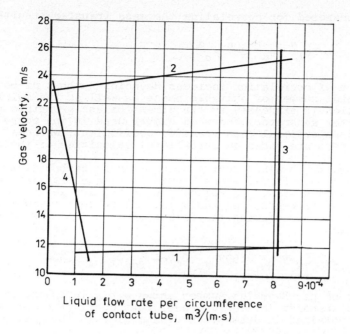

Fig. 6. Area of correct work for proposed contact device

3.3. Permissible Range of the Flow Rates

A diagram in Fig. 6. shows a permissible range of gas and liquid flow rates. The area of correct work is bounded by four lines which represent respectively:
1 - decline of cocurrent flow,
2 - rapid increase of liquid entrainment,
3 - high values of total pressure drop,
4 - irregular and pulsatory flow of the liquid.

3.4. Mass Transfer

For the experiments on mass transfer controlled by liquid phase, the desorption of carbon dioxide from water solution has been done. In the case of gas side controlled mass transfer the water solution of ammonia was used. In order to determine the extent of desorption, sampling was done by carefully withdrawing gas-free liquid samples through narrow tubes. Concentration of carbon dioxide and ammonia in liquid samples was determined by titration. Final gas composition was calculated from the material balance.

The number of past investigations into mass transfer in two-phase swirl flow is very restricted. Because of the uselessness of published correlation of mass transfer in ordinary flow for swirl flow, an attempt has been made to suggest a new form of dimensionless correlations. After detailed analysis, the following

706

formula is proposed for correlating the mass transfer results:

$$Sh = Const\ Re_L^a\ Re_G^b\ Sc^c\ A_o^d \qquad /6/$$

Proposed form of correlation includes Reynolds numbers of both phases, Schmidt number of suitable phase as well as the swirl intensity which is expressed by the dimensionless geometric parameter of vortex generator. Sherwood number used in the correlation /6/ contains superficial mass transfer coefficient, whereas Reynolds numbers are based on superficial velocities of the phases.

Some results obtained during desorption of carbon dioxide [6,11] are shown in Fig. 7 where liquid side mass transfer coefficient is plotted against gas velocity with the irrigation rate as a parameter. After regression analysis, the following equation has been achieved for predicting the values of liquid side mass transfer coefficient:

$$Sh_L = 6,546\ 10^{-3}\ Re_L^{0,597}\ Re_G^{0,784}\ Sc_L^{0,5}\ A_o^{0,304} \qquad /7/$$

The exponent on gas Reynolds number shows a big influence of gas flow on mass transfer in liquid phase. Next chart /Fig. 8/ shows the results obtained in desorption of ammonia. In the case of gas

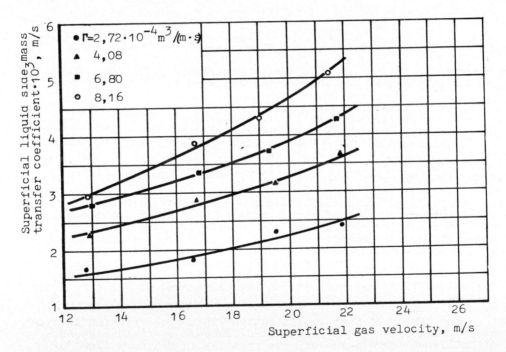

Fig. 7. Liquid side mass transfer coefficient versus gas velocity and irrigation rate for vortex generator type I

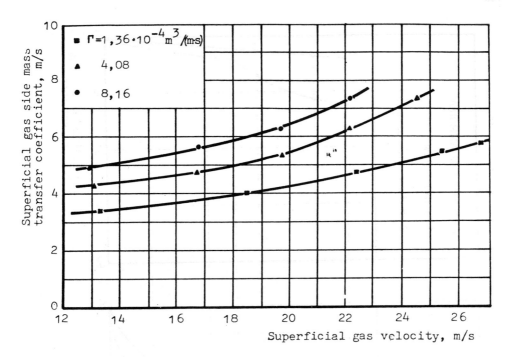

Fig. 8. Gas side mass transfer coefficient versus gas velocity
and irrigation rate for vortex generator type II

side controlled mass transfer, the following formula has been
derived from experimental data:

$$Sh_G = 2{,}847 \cdot 10^{-2} \, Re_L^{0,208} \, Re_G^{0,844} \, Sc_G^{0,44} \, A_o^{0,259} \qquad /8/$$

As can be seen from this formula, values of constant and exponent
on gas Reynolds number are very close to those of well-known
Gilliland-Sherwood correlation. The gas and liquid Schmidt numbers
were not varied much during the experiments, hence their influence
on Sherwood numbers was taken from literature data.

All equations developed in the study fit very good in the
following range of dimmensionless groups:

Re_L = 550 – 3700
Re_G = 56000 – 126000
A_o = 0,746 – 1,427

Another set of experiments was carried out to measure the
concentration changes along the length of contact tube. For these
experiments the contact tube has been equiped with six samplers.
One example of obtained results is shown in Fig. 9. As it has been
expected, the greatest changes in liquid concentration have occured

$$Sh_L = \frac{k_L a\, d}{D_L} \qquad - \text{liquid Sherwood number.}$$

REFERENCES

1. Berkovsky, M.A., Skoblo, A.I., Aleksandrov, I.A., and Sheinman, W.I., Khim.Techn.Topl.Mas., 1966, No. 5, pp. 41.

2. Nikolaev, N.A., Shavoronkov, N.M., Teor.Osn.Khim.Techn., 1970, Vol. IV, pp. 261.

3. Shavoronkov, N.M., and Malusov, W.A., Chemische Tech., 1972, Vol. 24, pp. 397.

4. Kozioł, A., Szczuka, K., and Merwicki J., Report of Inst. Chem. Engng Heat Eq., TU Wrocław, R-4/77, 1977.

5. Merwicki, J., Ph.D. Thesis, TU Wrocław, 1980.

6. Szczuka, K., Ph.D. Thesis, TU Wrocław, 1980.

7. Hoppe, K., Köhler, H., Weiner, L., and Müller, J., Chemische Tech., 1980, Vol. 32, pp. 183.

8. Merwicki, J., and Szczuka, K., Scientific Papers of TU Opole, Series: Mechanics, 1981, Vol. 15, pp. 69.

9. Abramivich, G.N., Prikladnaja gazovaja dinamika, Gostechizdat, Moscow, 1953.

10. Merwicki, J., Szczuka, K., and Troniewski, L., Inż. Chem., 1982, Vol. 3, pp. 351.

11. Szczuka, K., Merwicki, J., and Troniewski, L., Inż. Chem., 1982, Vol. 3, pp. 635.

Multi-Phase Flow and Heat Transfer III. Part B: Applications
edited by T.N. Veziroğlu and A.E. Bergles
Elsevier Science Publishers B.V., Amsterdam, 1984 — Printed in The Netherlands

THE IMPORTANCE OF HEAT TRANSFER IN A CHEMICAL HEAT PUMP UTILIZING A
GAS-SOLID SORPTION REACTION

Henrik Bjurström and Bo Carlsson
Department of Physical Chemistry
THE ROYAL INSTITUTE OF TECHNOLOGY
S-100 44 STOCKHOLM   Sweden

ABSTRACT

    The importance of heat transfer, mass transfer and sorption steps for
the performance of a chemical heat pump utilizing the adsorption of water
vapour on silica gel is analyzed. The sorption and desorption processes are
studied in a cylindrical bench-scale reactor containing a fixed annular bed
of sorbent particles. Material properties are determined with thermogravi-
metry on single particles for the sorption kinetics, calorimetry and the
Transient Hot Strip (THS) method for the thermal conductivity. The results
from dynamic and static experiments are compared and analyzed in terms of a
pseudo-homogeneous heat conduction model. The performance of the chemical
heat pump depends more on the heat transfer properties of the bed, thermal
conductivity and thermal mass, the latter including the heat turned over in
the reaction, than on the specific sorption and desorption kinetics of the
single particles themselves or on the mass transfer properties of the bed.

1. INTRODUCTION

    The first utilization of the desorption/sorption processes in a chemic-
al heat pump or sorption heat pump was for cooling purposes [1]. As the
working principles of assemblies of these processes came to be known, other
applications: heating, heat storage, conversion of heat into mechanical work
were envisaged. The ideas were not put into a large practice for various
reasons that we shall not discuss here. Low-temperature applications of
these processes, such as in waste heat recovery or solar heat utilization,
however, have attracted renewed attention.

    The suitability of a desorption/sorption process for energy conversion
purposes rest mainly on its dynamic behaviour, which is dictated by the
physical and chemical properties of the materials in the process. When a
gas-solid sorption system is used the reaction is generally performed in a
fixed-bed reactor. Knowledge not only about the properties of the single
particle, but also those of the bed, of the heat-exchanging system and of
the vapour distribution is therefore needed for the design of the reactor.

    In this work we shall discuss the use of a pseudo-homogeneous model to
describe the desorption and sorption processes in a fixed bed. Heat and
mass transfer properties are determined both on particle and bed level. Our

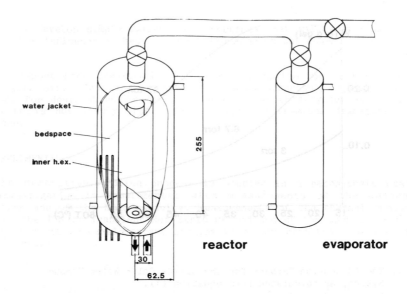

Fig. 2. Schematic Drawing of the Bench-Scale Reactor System.

Experiments. During all experiments the jacket is kept at 40°C. The inner heat exchanger is kept at 20°C during a sorption experiment and at 60°C during a desorption experiment. After completion of an experiment, the valve connecting reactor and evaporator is closed and the temperature of the inner heat exchanger is changed. When a steady state is reached in the isolated reactor, establishing thus a temperature gradient in the bed, a sudden pressure change is induced by opening the valve between reactor and evaporator. The relaxation towards a new steady state is recorded, see Figure 3 for a sorption process and Figure 4 for a desorption process. The amount of water sorbed or desorbed is determined by closingthe valve and weighing the evaporator. When an experiment is interrupted by closing off the reactor this way, the temperatures in the bed relax towards steady state through different paths which have also been recorded for the overall sorption and desorption processes shown in Figures 3 and 4, see Figures 5 and 6 for sorption and desorption, respectively. The pressure change is not a perfect step change because of the evaporator's limited capacity. The pressure has, however, attained its final value after 1 hour.

Calibration of Heat Flow Measurements. The measurements of the heat flow in the bed are calibrated by placing a paraffine with known heat conductivity (0.247 W/m,K at 22°C) in the annular bed space and applying a temperature difference between jacket and inner heat exchanger.Corrections for the parasitic heat flows can thus be made in the calculations of the true heat flow through the bed. The heat conductivity of the paraffine was measured with the transient hot strip or THS method, an improved hot wire method recently developed by researchers at the Department of Physics, Chalmers University of Technology in Gothenburg [6]. A flat metal strip is here used both as a continuous heat source and as a sensor for the temperature increase in the strip itself.

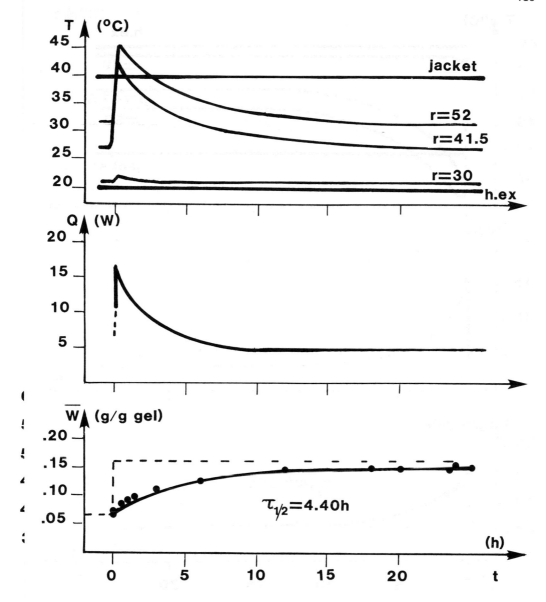

Fig. 3. Results of a Sorption Experiment. <u>Top</u>: the temperature at various points in the reactor (30 mm, 41.5 mm and 52 mm radii), in the water jacket and at the inlet to the inner heat exchanger. <u>Middle</u>: heat flow into the inner heat exchanger. <u>Bottom</u>: average water content as determined by interrupting the sorption at various times. Dashed line: initial and final equilibrium water content as calculated from equation (1).

This latter approach has become almost inevitable for modeling flow in such low permeability media as tight sands, where various factors could lead to substantial deviation from the classical Darcy's law.

2. ONE DIMENSIONAL TRANSIENT FLOW MODELS

In this study, we have used mass and momentum balances to derive equations for single and two phase flow in low-permeability porous media. The flow is assumed to be isothermal.

2.1. One Dimensional Transient Flow of Gas

The governing equations can be written as

Continuity Equation

$$\gamma \frac{\partial \rho}{\partial t} + \frac{\partial}{\partial x} (\rho \phi U) = 0 \qquad (1)$$

Momentum Equation

$$\gamma \frac{\partial (\rho U)}{\partial t} + \frac{\partial}{\partial x} (\rho \phi U^2) = -\gamma \frac{\partial P}{\partial \chi} - F_e + \gamma \rho g \qquad (2)$$

F_e is the frictional force due to flow of fluid in the porous media. The value of F_e used in this analysis was obtained from steady state pressure drop measurements in tight sand cores:

$$F_e = \frac{\mu U \phi^2}{k}$$

where k is the steady-state permeability.

2.2. Two Phase Gas-Liquid Flow Model

Two phase flow mass and momentum balances in an isothermal porous medium can be written as

Gas Phase Continuity Equation

$$\gamma \frac{\partial (\rho_g S_g)}{\partial t} + \frac{\partial}{\partial x_j} (\phi S_g \rho_g U_g) = \dot{m}_g \qquad (3)$$

Liquid Phase Continuity Equation

$$\gamma \frac{\partial (\rho_\ell S_\ell)}{\partial t} + \frac{\partial}{\partial x_j} + (\phi S_\ell \rho_\ell U_\ell) = -\dot{m}_g \qquad (4)$$

Gas Phase Momentum Equation

$$\gamma \frac{\partial (\rho_g S_g U_{ig})}{\partial t} + \frac{\partial}{\partial x_j} (\phi S_g \rho_g U_{ig} U_{jg})$$

$$= -\gamma S_g \frac{\partial P_g}{\partial x_i} + F_{eg} + F_{\ell g} + \gamma \rho_g S_g g_i \qquad (5)$$

Liquid Phase Momentum Equation

$$\gamma \frac{\partial (\rho_\ell S_\ell U_{i\ell})}{\partial t} + \frac{\partial}{\partial x_j} (\phi S_\ell \rho_\ell U_{i\ell} U_{j\ell})$$

$$= -\gamma S_\ell \frac{\partial P_\ell}{\partial x_i} - F_{g\ell} + F_{e\ell} + \gamma \rho_\ell S_\ell g_i \qquad (6)$$

Capillary Pressure Equation. In core analysis, one of the most frequently measured parameters is the capillary pressure, P_c, and it is a function of phase saturation. It is defined as –

$$P_c (S_\ell) = P_g - P_\ell \qquad (7)$$

ϕ is the total volume porosity open to flow of fluid, and γ the flow porosity, and $\dot{m}_g$ is the rate of phase change. F_{eg} and F_{e1} are gas and liquid single-phase frictional forces, respectively, between the respective phase and the pore wall. $F_{g\ell}$ is the interfacial drag force between the flowing gas and liquid phases. S_ℓ and S_g are liquid and gas saturations, respectively, in the pore volume and they are defined in such a way that

$$S_\ell + S_g = 1 \qquad (8)$$

Gas Phase Equation of State. The equation of state in terms of gas compressibility factor was used to characterize the single-component gas phase behavior:

$$\rho (P,T) = \frac{PM}{zRT} \qquad (9)$$

Liquid Phase Equation of State. Defining liquid compressibility, c, as

$$c = \frac{1}{\rho} \frac{\partial \rho}{\partial P} \qquad (10)$$

and defining some reference pressure, P_o, and corresponding density, ρ_o, we can write

$$P = P_o \exp [c (P - P_o)] \tag{11}$$

Equation 11 can be linearized to yield

$$P \simeq P_o [1 + c (P - P_o)] \tag{12}$$

3. CHARACTERISTIC CURVES

The characteristic curves for two phase flow equations were calculated numerically using the parameters of flow of gas and water in the low permeability porous media. The characteristics are real for all cases of compressible gas and water, compressible gas and incompressible water, and incompressible gas and water phases with negative slope of capillary pressure versus saturation curves, which is in agreement with the experimental data.

4. ANALYSIS OF THE MODELS

4.1. One Dimensional Transient Flow of Gas

The governing equations are Equations 1 and 2 above. For one-dimensional horizontal flow of gas through a "tight sand" core sample, we neglect the net momentum outflow term ($\frac{\partial}{\partial x} (\rho \phi U^2) \simeq 0$) and substitute Equation 9 into Equations 1 and 2. Furthermore, we assume that k, γ, z, and ϕ are constant and that $\gamma = \phi$.

With these assumptions, the governing equations become

$$\frac{\partial P}{\partial t} + \frac{\partial (PU)}{\partial x} = 0 \tag{13}$$

$$\frac{\partial (PU)}{\partial t} - \frac{zRT}{M} \frac{\partial P}{\partial x} - \frac{\mu}{k} \frac{zRT\phi}{MP} (PU) = 0 \tag{14}$$

Analytical Solution for Pressure Distribution Using Linearized Equations. Taking partial derivatives of Equation 13 with respect to t and Equation 14 with respect to x and subtracting the two resulting equations yields

$$\frac{\partial^2 P}{\partial t^2} - \frac{zRT}{M} \frac{\partial^2 P}{\partial x^2} - (\frac{\mu}{k} \frac{zRT\phi}{PM}) \frac{\partial (PU)}{\partial x} = 0 \tag{15}$$

Substituting Equation 13 into Equation 15 results in Equation 16:

$$\frac{zRT}{M} \frac{\partial^2 P}{\partial x^2} - \frac{\partial^2 P}{\partial t^2} = (\frac{\mu}{k} \frac{zRT\phi}{PM}) \frac{\partial P}{\partial t} \tag{16}$$

Linearizing the coefficients and simplifying the above equation and introducing a parameter ν we will have:

$$\frac{zRT}{M} \frac{\partial^2 P}{\partial x^2} - \frac{\partial^2 P}{\partial t^2} = (\frac{\mu}{k} \frac{zRT\phi}{MP^*}) \frac{\partial P}{\partial t} \tag{17}$$

Where ν is the parameter (in our calculation we assigned a value of 10^{-6} for ν). Equation 17 differs from the usual diffusivity equation often used to describe transient fluid flow in porous media by the term, $\partial^2 P/\partial t^2$. The initial and boundary conditions imposed on Equation 17 are the same as those of experimental analysis of a tight sand core sample. They may be written as

$$\text{at } t = 0: \quad P(0,x) = P_2 \tag{18}$$
$$\text{for all } x > 0$$

$$\text{at } t = 0: \quad \frac{\partial P}{\partial t}(0,x) = 0 \tag{19}$$

$$\text{at } x = 0: \quad P(t,0) = P_1 \tag{20}$$
$$\text{for all } t > 0$$
$$\text{at } x = L: \quad P(t,L) = P_2 \tag{21}$$

The dimensionless form of linearized Equation 17 and the associated initial and boundary conditions may be written as

$$\frac{\partial^2 \bar{P}}{\partial \bar{t}^2} - \frac{\partial^2 \bar{P}}{\partial \bar{x}^2} + 2\lambda \frac{\partial \bar{P}}{\partial t} = 0 \tag{22}$$

$$\text{at } t = 0, \ \bar{P}(0,x) = 0 \tag{23}$$
$$\text{for all } \bar{x} > 0$$
$$\text{at } \bar{t} = 0, \ \frac{\partial \bar{P}}{\partial t}(0,\bar{x}) = 0 \tag{24}$$

$$\text{at } \bar{x} = 0, \ \bar{P}(\bar{t},0) = 1 \tag{25}$$
$$\text{for all } \bar{t} > 0$$
$$\text{at } \bar{x} = 1, \ \bar{P}(\bar{t},1) = 0 \tag{26}$$

where

$$\bar{x} = \frac{x}{L}; \ \bar{P} = \frac{P - P_2}{P_1 - P_2}$$

$$\bar{t} = \frac{t}{L\sqrt{\dfrac{M}{\nu zRT}}}; \ \lambda = \frac{\mu L\phi}{2kP^*} \sqrt{\frac{zRT\nu}{M}}$$

We can write the solution $\bar{P}(\bar{t}, \bar{x})$ as a sum of two superimposed solutions $\bar{w}(\bar{t}, \bar{x})$ and $\bar{v}(\bar{x})$ without any loss of generality:

$$P(\bar{t}, \bar{x}) = \bar{w}(\bar{t}, \bar{x}) + \bar{v}(\bar{x}) \tag{27}$$

Substituting Equation 27 into the set of Equations 22-26 yields Equations 28 and 29:

$$\frac{d^2 \, \bar{v} \, (\bar{x})}{d\bar{x}^2} = 0 \tag{28}$$

$$\text{at } \bar{x} = 0: \ \bar{v} \, (0) = 1$$

$$\text{at } \bar{x} = 1: \ \bar{v} \, (1) = 0$$

Thus

$$\bar{v} \, (\bar{x}) = 1 - \bar{x} \tag{28a}$$

and

$$\frac{\partial^2 \bar{w}}{\partial \bar{t}^2} - \frac{\partial^2 \bar{w}}{\partial \bar{x}^2} + 2\lambda \, \frac{\partial \bar{w}}{\partial \bar{t}} = 0 \tag{29}$$

$$\text{at } \bar{t} = 0: \ \bar{w} \, (0,\bar{x}) = \bar{x} - 1 \text{ and } \frac{\partial \bar{w}}{\partial \bar{t}} \, (0,\bar{x}) = 0$$

$$\text{at } \bar{x} = 0: \ \bar{w} \, (\bar{t},0) = 0$$

$$\text{for all } \bar{t} > 0$$

$$\text{at } \bar{x} = 1: \ \bar{w} \, (\bar{t},1) = 0$$

The method of separation of variables was used to solve the above partial differential equation with the associated initial and boundary conditions. We can write w $(\bar{t},\bar{x})$ as

$$w \, (\bar{t},\bar{x}) = X \, (\bar{x}) \quad T \, (\bar{t}) \tag{30}$$

Substituting this into the partial differential Equation 29, yields two ordinary differential equations,

$$X'' + \alpha^2 X = 0 \tag{31}$$

$$T'' + 2\lambda \, \frac{T'}{T} + \alpha^2 T = 0 \tag{32}$$

with boundary and initial conditions of

$$X \, (0) = X(1) = 0 \tag{33}$$

and

$$T' \, (0) = 0 \tag{34}$$

Solving Equations 31–34 and putting the results back into Equation 30 yields

$$\bar{w} \, (\bar{t}, \, \bar{x}) = \sum_{n=1}^{\infty} B_n \, \sin \, (n \, \pi \, \bar{x}) \, [e^{m_1 \bar{t}} - \frac{m_1}{m_2} \, e^{m_2 \bar{t}}] \tag{35}$$

where

$$m_1 = -\lambda + \sqrt{\lambda^2 - n^2 \pi^2}$$

$$m_2 = \lambda - \sqrt{\lambda^2 - n^2 \pi^2}$$

at $\bar{t} = 0$, $\bar{w}(0,\bar{x}) = \bar{x} - 1$.

Thus

$$\bar{x} - 1 = \sum_{n=1}^{\infty} Bn \left(1 - \frac{m_1}{m_2}\right) \sin(n \pi \bar{x})$$

and

$$Bn = \frac{2m_2}{m_2 - m_1} \int_0^1 (\bar{x} - 1) \sin(n \pi \bar{x}) d\bar{x} = \frac{2m_2}{n\pi (m_1 - m_2)}$$

Hence

$$\bar{P}(\bar{t},\bar{x}) = 1 - \bar{x} + \sum_{n=1}^{\infty} \frac{2m_2}{n\pi (m_1 - m_2)} \sin(n\pi \bar{x}) \{e^{m_1 \bar{t}} - \frac{m_1}{m_2} e^{m_2 \bar{t}}\} \quad (36)$$

Analytical Expression for Velocity Distribution Using our Linearized Model. The entrance velocity $(U|_{at\ x\ =\ 0})$, was not measured experimentally. To find the velocity distribution, we need the entrance velocity as a function of time. Since such measurements have not been made, and assigning a transient entrance velocity distribution which is different for different runs as a parameter is not practical, we used Darcy's law to calculate a velocity distribution. The velocity distribution using our linearized model for pressure variation and Darcy's law was calculated. The velocity distribution as a function of time and position is

$$U(x,t) = \frac{-k}{\mu\phi} \frac{\partial P}{\partial x} = \frac{-k (P_1 - P_2)}{\phi \mu L} \frac{\partial \bar{P}}{\partial \bar{x}} \quad (37)$$

Differentiating Equation 36 with respect to $\bar{x}$ and substituting $\frac{\partial \bar{P}}{\partial \bar{x}}$ in Equation 37 results in

$$U(x,t) = \frac{k (P_1 - P_2)}{\phi \mu L} [1 - \sum_{n=1}^{\infty} \frac{2m_2}{m_1 - m_2} \cos \frac{n\pi x}{L} (e^{\frac{m_1}{L} \sqrt{\frac{ZRT\nu}{M}} t}$$

$$- \frac{m_1}{m_2} e^{\sqrt{\frac{ZRT\nu}{M}} t})] \quad (38)$$

Diffusivity Equation. Diffusivity equations have been widely used in reservoir engineering literature to describe flow in porous media. Our continuity and momentum equations can be reduced to the diffusivity equation by making further assumptions and neglecting the momentum accumulation term ($\frac{\partial(\beta U)}{\partial t} = 0$) in our momentum balance. The linearized diffusivity equation may be written as

734

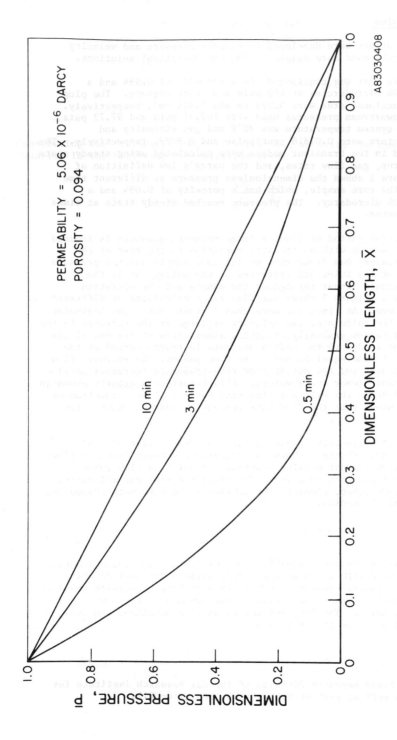

Figure 1. DIMENSIONLESS PRESSURE PROFILES CALCULATED USING LINEARIZED EQUATION

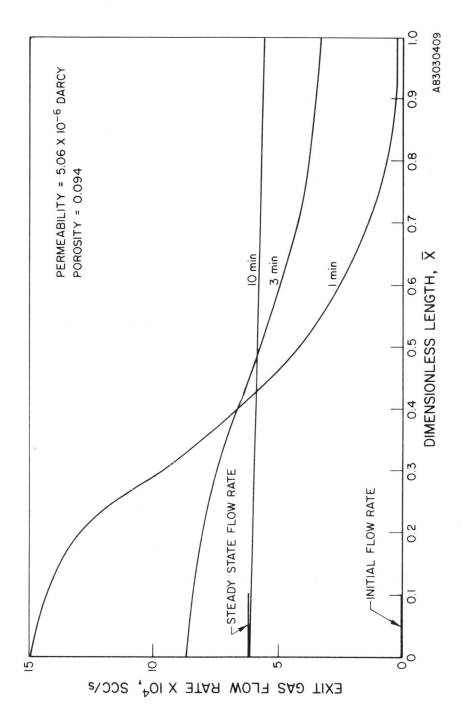

Figure 2. GAS FLOW RATE PROFILES CALCULATED USING
THE LINEARIZED EQUATION

TABLE I (Continued)

Rate of Water Evaporation

$$R_w = \frac{\partial \left(\rho_s f_w \right)}{\partial t}.$$ (8)

Combustion Front Mover

$$\vec{U}_p = \frac{0.012 \, S\rho_g \vec{v}_a}{f_c \rho_s}$$ (9)

where

$$\vec{v}_a = \left(v_x x_a - D_x' \frac{\partial x_a}{\partial x} \right) + \left(v_y x_a - D_y' \frac{\partial x_a}{\partial y} \right)$$ (9a)

Darcy's Law

$$\frac{\partial}{\partial x} \left(\frac{K}{\mu RT} \frac{\partial P^2}{\partial x} \right) + \frac{\partial}{\partial y} \left(\frac{K}{\mu RT} \frac{\partial P^2}{\partial y} \right) + G_s = 0$$ (10)

Table II summarizes the mechanisms and governing equations constituting the published oil evolution model.[5] Eq. (11) is the kerogen decomposition reaction using non-elementary stoichiometry; Equation (12) represents the degradation rate of evolved oil; Eq. (13) describes the removal rate of liquid and gaseous oil. The energy transport equation, Eq. (14), describes the conduction of energy into, transport of energy out of, and generation of heat within control volumes defining the oil shale block. This equation is currently solved by assuming one-dimensional spherical symmetry.[5] This assumption could be retained but is easily lifted using the structure and flexibility of the TRUMP code[26] which forms the basis of the two-dimensional in-situ coal gasification model describing the region outside the shale block.

The chemical reactions and rates described by Eq. (15) represent the formation of gas and liquid oil using a simulated boiling point curve function, X(T), coking reactions and production reactions. The rate constants are given by Eq. (16). The carbonate decomposition and char combustion reactions within the block have been deleted and hence these heat release rates are not included in the rate of heat generation, Eq. (17). Further details of the model are found in reference[5]. Any oil vapor produced is immediately transported to void defect regions where it is burned according to Eq. (7) if any oxygen is present, releasing heat there. The "gas" could be included with the oil vapor for purposes of heat release but its amount is small so the net effect should also be small. Carbonate decomposition has recently been added to the oil evolution code and could thus be added.[27]

TABLE II
GOVERNING EQUATIONS WITHIN THE OIL SHALE BLOCK

Oil generation

$$CH_{1.54}N_{0.028}S_{0.005}O_{0.046} \xrightarrow{k_d} 0.2 \; CH \; N_{0.066}S_{0.018} \quad \text{Carbonaceous}$$

   Raw shale   residue

$$+ \; 0.725 \; CH_{1.56}N_{0.0066}S_{0.018} \qquad \text{Oil}$$

$$+ \; 0.018 \; CO_2$$

$$\left. \begin{array}{l} 0.019 \; H_2 \\ 0.013 \; CH_4 \\ 0.014 \; C_{3.2}H_{7.5} \end{array} \right\} \quad \text{"Gas"}$$

$$+ \; 0.010 \; H_2O \qquad\qquad\qquad \text{Water} \qquad (11)$$

Oil degradation

$$CH_{1.56}N_{0.020}S_{0.002} \xrightarrow{k_c} 0.83 \; CH_{0.2}N_{0.024}S_{0.0024} \quad \text{Coke}$$

$$\left. \begin{array}{l} + \; 0.14 \; CH_4 \\ + \; 0.36 \; H_2 \end{array} \right\} \quad \text{"Gas"} \qquad (12)$$

Oil removal

$$CH_{1.56}N_{0.020}S_{0.002} \xrightarrow[k_l]{k_g} CH_{1.56}N_{0.020}S_{0.002} \qquad (13)$$

Heat Transport

$$\rho_s' C_s \frac{\partial T}{\partial t} = \nabla \cdot k_s \nabla T - v_g \rho_g c_g \nabla T + \dot{q} \qquad (14)$$

Chemical Reactions

Reaction   Rate Eqs.

$$\text{Oil precursors} \xrightarrow{k_d} \left[1 - X(T) \right] \cdot \text{Oil}^l \qquad R_{ol} = f_o \left[1 - X(T) \right] k_d \cdot \rho_{kb}$$

Chemical Reactions

$$\text{Oil precursors} \xrightarrow{k_d} X(T) \cdot \text{Oil}^g \quad R_{og} = f_o \cdot X(T) \cdot K_d \cdot \rho_{Kb}$$

$$\text{Oil}^l \xrightarrow{k_c} (f_c') \; \text{coke} \qquad\qquad R_c = f_c' \cdot k_c \cdot \rho_{lo}$$

17. E. H. Blum and R. H. Wilhelm, "A Statistical Geometric Approach to Random-Packed Beds," pp. 21-27 in Proceedings of the AICHE - I. Chem. E., Joint Meeting, London, June 1965.

18. D. E. Lamb and R. H. Wilhelm, "Effects of Packed Bed Properties on Local Concentration and Temperature Patterns," I&EC Fundamentals, 2 No. 3, pp. 173-182, 1963.

19. M. Herskowitz and J. M. Smith, "Liquid Distribution in Trickle-Bed Reactors, Parts I and II," AIChE Journal, 24, No. 3, pp. 439-454, May 1978.

20. J. N. Doggett, "Summary of Large Block Oil Shale Retorting Experiments," Lawrence Livermore Laboratory Report UCID-16699, Feb. 6, 1975.

21. R. G. Mallon and W. C. Miller, "Thermal Behavior of Oil Shale Blocks Heated in Air," Lawrence Livermore Laboratory, Preprint No. UCRL-76452, 1975.

22. R. W. Lyczkowski and Dimitri Gidaspow, "Multiphase Flow Modeling of Oil Mist and Liquid Film Formation in Oil Shale Retorting," in Multi-phase Transport Fundamentals, Reactor Safety, Applications, T. N. Veziroglu Ed., Vol. 5, pp. 2319-2346, Hemisphere Publishing Corp., Washington, D. C., 1980.

23. R. W. Lyczkowski, C. B. Thorsness and R. J. Cena, "The Use of Tracers in Laboratory and Field Tests of Underground Coal Gasification and Oil Shale Retorting," Lawrence Livermore Laboratory Preprint. No. UCRL-81252, June 16, 1978.

24. B. S. Gottfried, "Combustion of Crude Oil in a Porous Medium," Combustion and Flame, 12, No. 1, pp. 5-13, Feb. 1968.

25. B. S. Gottfried, "A Mathematical Model of Thermal Oil Recovery In Linear Systems," Soc. Pet. Eng. J., 5, No. 3, pp. 196-210, Sept. 1965.

26. A. L. Edwards, "TRUMP: A Computer Program for Transient and Steady-State Temperature Distributions in Multidimensional Systems," Lawrence Livermore Laboratory Rept. No. UCRL-14754, Rev. 3, Sept. 1, 1972.

27. J. H. Campbell, M. Gregg, and J. Taylor, "Applicability of Laboratory Data for Describing Reactions in Large Blocks," memorandum, Sept. 8, 1978.

28. T. R. Galloway and R. W. Lyczkowski, "The Behavior of Flow Channels in Rubble In-Situ Beds of Oil Shale," in Proc. of the 15th Intersociety Energy Conversion Engineering Conference, Vol. 1, pp. 268-277, American Institute of Aeronautics and Astronauts, New York, 1980.

3.2. Direct Joule Heating

The primary objective of these preliminary studies was to
find out whether transition boiling could be steadiliy maintained
along a sufficiently extended axial increment of a directly heated
cylindrical rod in forced convective in-line flow, when temperature
control of heater was established by a feedback control system as
proposed by Peterson /17/. In this feedback system, the electrical
resistance of the heater acting as a temperature sensor becomes
one arm of a comparator bridge which accomplishes a continuous
comparison between a desired heater temperature and its actual
value, and output or error signal of the bridge is then used to
control the heater power furnished by a d.c. power amplifier. Since
the controlled variable is the volume-averaged heater temperature,
a heater configuration is sought which is essentially able to inhibit the development of large temperature deviations from the
volume average that may be initiated by differently effective
boiling modes occuring in length direction of the surface due to
axially varying flow quality.

A number of experiments at atmospheric pressure have been
performed with electrically heated wires of different diameter
(0.2, 1, 2, 3 mm), heated length (10 to 100 mm), and material
(platinum, nickel) using an experimental setup including feedback
control similar to that of Peterson et al. /21/. The wires were
suspended in vertical direction within a glass tube of 15 or 25 mm
inner diameter. The water flow was in the upward direction, its
subcooling and mass flux could be varied. Except for few experiments with platinum wires of 0.2 mm diameter, a system compensation function with 4th order filtering,

$$G_c(s) = K \frac{(1 + sT_N)}{sT_N} \frac{1}{(1 + sT_1)^4} \tag{1}$$

was satisfactorily used for all operating conditions, and found
superior to that originally proposed by Peterson et al. /26/. For
example, the following values of compensation function parameters
have been found most favourable for the 3 mm wire experiments:
$K = 150$, $T_N = 0.15$ s, $T_1 = 0.001$ s.

It was the common outcome of all these experiments that, once
critical heat flux had been exceeded at a certain axial location
of heated length (usually in the downstream half), a stable film
boiling region started to develop while the greatest part of
heater length was still in nucleate boiling. The film boiling
region extended gradually as the (average) heater temperature was
increased. Only in the experiments with nickel wires of 2 and 3 mm
diameter at low mass flow (~ 68 kg/m^{2}s), the stable film boiling
region could not be seen unless the average heater temperature
had exceeded the critical heat flux temperature by more than about
30 K. That local film boiling was already present, however, could
be substantiated by additionally measuring average temperatures
of partial increments of the wire. Hence, transition boiling as
"unstable film boiling and unstable nucleate boiling alternately
existing at a given location on a heating surface" /23/ did only
occur at the boundary of stable nucleate and film boiling regions
where the largest axial temperature gradient exists.

760

These findings have further been substantiated by numerical simulation of the heat transfer system. For simplicity, only temperature variation in the axial direction and with time has been considered. Surface heat transfer has been introduced according to known boiling curve characteristics, feedback control of average (surface) temperature was represented by the same control function as used in the experiments. A sample result for the axial temperature variation of a short nickel wire as used in the experiments is shown in Fig. 1. Curve 1 (av. temp. 144°C) represents the case where critical heat flux is just reached at one location; an only slight increase of the average surface temperature by 2 K results in forming a local hot spot of about 30 K higher maximum temperature which is well in accord with our measurement. With further increase of average temperature, the hot spot becomes more and more pronounced.

Numerical experiments with different heater materials, configurations (rod, tube), and dimensions finally led to the conclusion that sufficiently flat axial temperature profiles in post-CHF region cannot be actually realized by direct heating, even when a stabilizing circuit is applied. This is primarily due to the incompatible requirements for both high thermal and low electrical conductance that are not reasonably met by any material we know. Thus, both requirements can only be satisfactorily fulfilled by different materials, i.e., an indirectly heated test section has to be applied, the design of which is strongly affected by the inherent limitions of feedback control (e.g. stability, response time). One possible design solution which yields satisfactory results is presented in the following paragraph.

3.3.   Indirect Heating

Test section assembly. Parameter studies using numerical simulation as above revealed that in a copper tube of about 10 mm wall thickness axial thermal conductance would become sufficiently effective to enforce a rather flat axial temperature distribution

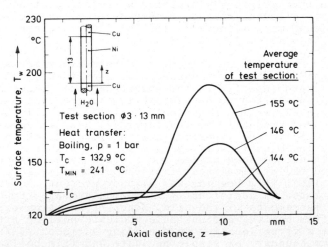

Fig. 1.   Development of local hot spot in post-CHF flow boiling

in the post-DNB regime over several centimeters of length which was considered advantageous for accurate temperature measurement and data reduction. Indirect heating of such a copper tube can most easily be established inserting a number of cylindrical heating elements coaxially on a concentric circle. Radius of this circle and number of heating elements have to be chosen such that at the surface of the inner bore, which serves as the heat transfer surface, circumferential temperature variations have become negligible. In view of sufficiently fast response time of heater power control in dependence of temperature variations of the heat transfer surface, however, distance of heater elements from inner bore as well as tube wall thickness should be selected not larger than actually required. For optimum design, thermal and control aspects have to be considered simultaneously via proper theoretical simulation.

The test section assembly finally built and used for the experiments is shown in Fig. 2. The copper tube of 5 cm length and 32 mm O.D. (10 mm I.D.) is heated by ten cylindrical heating elements of 3 mm O.D. that are located on a circle of radius 11,5 mm. Since cartridge heaters of required specific power as well as dimensions have not been found available, we use stainless steel tubes of 0.33 mm wall thickness which are electrically insulated from the copper cylinder be a high-temperature cement. The heaters have been successfully operated up to a maximum power per unit length of 350 W/cm; usually, however, heater power has been limited to about 150 W/cm which corresponds to an average heat flux at the heat transfer surface of approximately 500 W/cm^2. There certainly is some potential for a further increase of permissible heater power during longtime operation which will be exploited in the near future.

Since fast oxidation of the copper surface has been found to occur during operation in transient boiling mode, a gold coating ($\sim$ 0.5 μm thick) has been brought up on the heat transfer surface to avoid variation of surface conditions during a test run.

To measure the axial temperature distribution within the test

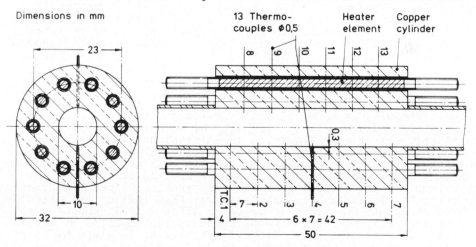

Fig. 2.   Test section for post-CHF boiling studies

section, thirteen stainless-steel sheathed Ni-CrNi thermocouples of
0.05 cm O.D. were installed close to the surface of the central
flow channel. The position of the junction of each individual
thermocouple relative to the outer tip of the sheath was determined from X-ray photographs. After calibration, the thermocouples
were press fitted into thermocouple holes (0.06 cm diameter with
a reduction to 0.05 cm in T.C. tip region) and then fixed by individual clamps. Axial location of the thermocouples is shown in
Fig. 2; junction of the T.C. was about 0.1 cm apart from flow
channel surface. The test section formed part of a low pressure
water loop with feedback control for constant fluid flow temperature at test section entrance /27/.

Temperature-Control Circuit. Design of the temperature-control
circuit has been performed simultaneously with the thermal design
of the test section according to the following criteria. First of
all, feedback control has to secure stable operation of test
section within the transition boiling region including its bounds
DNB and minimum film boiling temperature, i.e., time averaged
temperature distribution of the heat transfer surface must be in-
variably constant and amplitudes of its fluctuations with time
should be small (stability criterion). The temperature-controller
is required to adjust a stable control circuit of sufficient
damping characteristics within the given extreme values of diffe-
rential heat transfer coefficient $h(\overline{T}_w) = dq''/d\overline{T}_w$ according to
nucleate boiling and transition boiling without any changes in
controller constants (sensitivity criterion). Further, the temperature fluctuations that are inherently characteristic for the
physics of the heat transfer process should not be affected by
control. To avoid such interference, frequency region of control
action should be much lower than characteristic frequencies of
temperature fluctuations.

In principle, complete design of the controller may be
accomplished by analytical means provided these criteria are
quantified and the dynamic characteristics of all components of
the control circuit are known. Unfortunately, knowledge on dynamics
of boiling, especially transition boiling, is still completely
insufficient. Hence, fine tuning of control parameters (gain, time
constants) in accordance with actual experiments to be performed,
must be left to the experimenter. This adjustment was not found to
be at all critical.

In course of control loop design, transfer function concepts
could be applied, though the boiling process is quite nonlinear in
certain regimes of boiling curve. However, it was found that
process dynamics in terms of small increments of heater surface
temperature and power in the vicinity of any process operating
point were essentially linear. A one-dimensional multi-regional
model in the radial direction, whereby tentative extreme values
for the differential heat transfer coefficient were estimated from
previous experimental results, has been applied to evaluate the
open loop transfer function of the test section. Since the transfer
function was found to be of higher order than that of the directly
heated wires, it was feasable to add a differential component to
the controller. The corresponding compensation function of the resulting PID-controller with 4th order filtering is given by

$$G_c(s) = K \frac{1 + sT_N + s^2 T_N T_V}{T_N s (1 + T_1 s)^4} . \qquad (2)$$

Based on this function, root locus and transfer function techniques have then been applied to investigate stability and damping characteristics of the closed loop system. The resulting parameter values of eq. (2) have been essentially substantiated by proper performance of the actual experimental facility. Only small adjustments have been found necessary to optimize control characteristics of the system. The compensation function parameters finally selected for performing the present experiments are: $K = 3,900$, $T_N = 10$ s, $T_V = 0.6$ s, $T_1 = 0.106$ s (equiv. to a corner frequency of 1.5 Hz of the low-pass filter).

The compensation function was realized physically by analog computer components and a Krohn-Hite filter (No. 3323). A simplified block diagram of the final control loop is presented in Fig. 3. Temperature signals from the thermocouples within the test section are fed into a signal conditioner which evaluates a suitably weighted average signal. In a comparator, this voltage signal is compared with a control point voltage equivalent to a set point or reference temperature which establishes the process operating point. The difference of both signals is then amplified, passed through a 4th order low-pass filter, and fed forward to the PID-controller. The controller which performs amplifying and compensating functions, finally produces the input signal to the power amplifier. The output of the DC power supply unit (max. current 2,500 A at 15 V) establishes the heater power level and also introduces low frequency variations in power in accordance with the base input voltage delivered by the controller. It may be noted that the measured Bode diagram of the power supply component reveals a corner frequency of approximately 200 Hz; up to about 500 Hz, the transfer function is almost invariant to variations in power amplitude. Therefore, the power supply system could safely be considered to contain no significant dynamics in view of the low-frequency control.

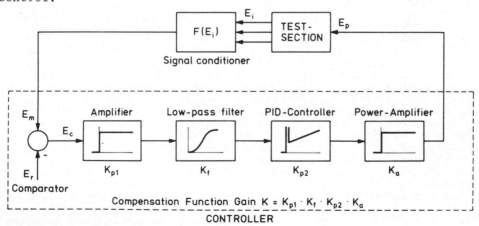

Fig. 3. Simplified block diagram of control system

It was found that stable operation in all boiling regions could be obtained using only three thermocouples as temperature sensors for control (TC 2,4, and 6). Weighting of the voltage signals in the signal conditioner is performed according to the formula

$$E_m(\overline{T}) = \sum_{i=1}^{3} \alpha_i E_i \Big/ \sum_{i=1}^{3} \alpha_i , \qquad 0,1 \leqq \alpha_i \leqq 1 . \qquad (3)$$

Until now, fixed values for the weighting constants α_i were applied throughout, though one might suspect that there always exists one choice of the individual constants which minimizes the amplitudes of power variations induced by the controller. Improvement of system performance in view of accuracy and sensitivity may also result from using more than just three, or even all wall thermocouples as temperature sensors for control. The interesting considerations involved here deserve further systematic study.

4. EXPERIMENTAL PROCEDURE

The water loop is designed to perform boiling experiments at or near atmospheric pressure at water temperatures in the range from $20\,^{\circ}C$ up to saturation temperature at test section entrance. Mass flow may be varied from 0.012 to 1.25 kg/s. Degassed, distilled water with an electric conductivity of less than 2 $\mu S/cm$ is used. Electric power of cartridge heaters which serve as water preheaters is controlled by an PI-controller such that water temperature at test section entrance varies less than $\pm$ 0.2 K during the experiment.

During each test run, pressure, flow rate, and water temperature at test section inlet were fixed, while set-point wall temperature was varied. Usually, the "boiling curve of test section" in terms of set-point wall temperature versus average wall heat flux was first traversed from nucleate to film boiling and then in reserve. Small incremental changes of set-point temperature were chosen in those regions of the boiling curve where large variations of its gradient occur. After each change, steady-state conditions were established prior to data collection. The test section power and all wall thermocouple signals were scanned in 1 second intervals for 20 seconds, stored, and then averaged by a data processing unit (Hewlett-Packard, HP 3054/85A). Steady-state conditions were considered to be established when averaged data of consecutive collections in 1 minute intervals did not reveal any changes but statistical variations around constant mean values. Flow parameters that were practically invariant with time were only collected once during each scanning interval applying 100 ms integration time per value. Water temperatures were measured by Ni-NiCr thermocouples at the turbine flowmeter used for flowrate measurement and at test section entrance and exit. Data averages of usually several scannings for each operating point were carefully processed in the on-line computer, i.e. corrected for calibration errors etc., converted to engineering units, and recorded for further evaluation. In evaluating the power transfered to the fluid via the inner heat transfer surface, heat loss through the outer insulated walls as found from heat loss calibration runs was considered.

5. DATA REDUCTION

A two-dimensional, steady-state, multi-regional heat conduction analysis of the test section has been performed to determine axial distributions of temperature and heat flux at the heat transfer surface. Assuming that circumferential variations of temperature induced by the arrangement of heating elements are negligible close to heat transfer surface, the test section has been modelled by three annular zones, where the heating elements are represented by the middle annular region of same thickness as the diameter of the heating element with a weighted average of its thermal conductivity. The specific heat generation rate in this zone was determined from the measured average power reduced by the heat losses. Thus, the outer walls could be considered adiabatic which seemed permissible in view of both the small percentage of the actual heat losses (usually less than 1 percent) and resulting simplicity.

The calculational procedure was as follows: First, the discrete boundary collocation method has been applied to develop an analytic series solution of the temperature field in radial and axial coordinates. This solution satisfies the measured temperature boundary condition at each thermocouple position in terms of specific radial and axial coordinate values. This way, the different radial locations of each thermocouple junction due to inevitable manufactering tolerances of both thermocouples and depth of thermocouple holes were properly considered. Usually, deviations of radial locations from the arithmetic mean have been within $\pm$ 0.2 mm. From this solution, the temperatures at the arithmetic mean radius r_m of all thermocouple junctions have been evaluated, i.e., this 2-D-solution is merely used to correct for the differences in radial thermocouple locations. These corrected temperatures at r_m were then fitted piecewise by cubic spline functions considering adiabatic boundary conditions at the end surfaces of the section. Thus, the discrete inner boundary condition as measured is replaced by a continuous boundary condition as required for construction of an analytic series solution with an arbitrary number of terms. Such a solution has then been applied to extrapolate the solution known at r_m to the inner heat transfer surface. All parameters of interest such as wall temperature, heat fluxes in the radial and axial directions, ordinary heat transfer coefficients were then evaluated as function of axial coordinate. The number of series terms to be retained within the solution is primarily determined by the accuracy limit set for the heat fluxes. (Only in those cases where the axial temperature variations are rather small, the boundary collocation solution with number of series terms equal to the number of wall thermocouples used may immediately be applied to evaluate the parameters at the inner heat transfer surface with satisfactory accuracy).

6. EXPERIMENTAL RESULTS

In this paper, results obtained from measurements with test section No. 4 will be reported which supplement previous results /27/. In the present experiments, water pressure and mass flow rate were fixed to 1.1 bar and 200 kg/m^{2}s, respectively. Subcooling at test section entrance was varied from 2.5 to 41 degree K.

6.1. Closed Loop Transfer Functions

To validate the one-dimensional four-zone model of the test section used for evaluation of the compensation function of the controller, frequency transfer functions have been measured in closed-loop operation of the test section. At fixed operating points, with test section either in nucleate boiling, at maximum average heat flux, or in the transition region, and fixed values of control parameters (gain, time constants), a sinusoidal power variation furnished by a function generator has been superimposed to its time-average value. The cross-correlation function of this signal and the resulting weighted temperature signal (cf. paragraph 3.3) has been recorded at different frequencies. From the cross-correlation function, the familiar Bode diagram can be constructed.

In Fig. 4, the time-averaged axial temperature distributions corresponding to the three different process operating points that have been investigated is shown. The dots in the Fig. denote the thermocouples used as temperature sensors for control. Note that large axial temperature gradients are present when the test section experiences transition boiling. For this case, which in open loop operation would be unstable, and the maximum heat flux case, the corresponding Bode diagrams are presented in Fig. 5. The diagrams reveal that, in both cases, the damping characteristics of the control loop are satisfactory. Temperature variations at frequencies less than about 0.5 Hz are well controlled, while fluctuations above 1.0 Hz are compensated by less than about 10 percent, and thus may be considered to be only slightly affected by the controller. In case of transition boiling, measurements indicate a resonance peak to occur at a frequency of roughly 1.5 Hz. Presumably, this frequency corresponds to the characteristic rewetting frequency at the given operating point of the test section, however,

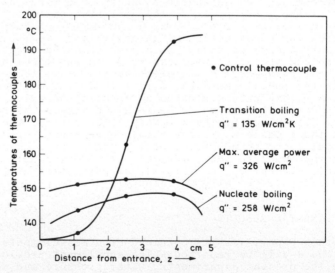

Fig. 4.   Temperatures sensed by control thermocouples
at different operating points

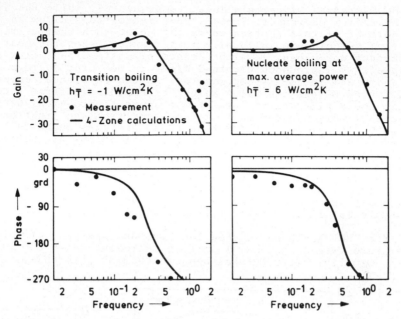

Fig. 5. Bode diagram of closed control loop at two operating points
(water at 1.1 bar, G = 200 kg/m^{2}s, ΔT_{SUB} = 11.5 K)

systematic studies using signals from single thermocouples are
required to substantiate this assumption. The theoretical curves of
the 1-D four-zone model agree reasonably well with the experimental
results, when suitable values for the incremental heat transfer
coefficient are applied. Correct determination of appropriate values
from the experiments is rather difficult in case of largely differ-
ing heat transfer mechanisms at the three thermocouple positions as
found in transition boiling situations. In the present case, better
fit of theory with experiment was obtained choosing h_T=-1.0 W/cm^{2}K instead of -1.73/W/cm^{2}K which was deduced from the experiment.

6.2. Axial Surface Temperature Distributions and Local Temperature Fluctuations

In Fig. 6, axial distributions of temperatures at the heat
transfer surface are plotted in dependence of the preset reference
temperature of the controlling thermocouples, which approximately
corresponds to the average wall temperature. The entrance subcooling was fixed at 2.5 K. The dots represent the corrected temperatures at thermocouple locations (TC 8 had failed). At an axially
averaged heat flux of 221.5 W/cm^2, the critical heat flux temperature is reached or slightly exceeded at position Z ∿ 1.8 cm, while
the rest of the surface is still in nucleate boiling. In performing
the experiments, a local excess of T_C is positively detected by an
increase of temperature or power fluctuations compared to those
experienced when the entire surface is in nucleate boiling.
A further strong verification may be obtained by the fact that any
very slight increase of reference temperature now results in sub-

stantial increases of certain local temperatures, while the average wall heat flux decreases. These strong indications allow an accurate setting of test section operating point such that critical heat flux temperature is locally just reached. When the heat transfer surface was clean, local occurance of maximum heat flux could be repeatedly reproduced within 2 W/cm^2 in measured average heat flux and 0.5 K in local temperatures. As the reference temperature is increased beyond the value which corresponds to first occurence of CHF, the amplitudes of local temperature fluctuations become appreciably more pronounced as may be seen from the right-hand side of Fig. 6. It has to be noted that these sketches ought to be considered as qualitative only, since time variations of thermocouples signals as recorded are plotted here (as well as in Figs. 7 and 8). From Fig. 7, one might deduce that the maximum amplitudes of the wall temperature fluctuations reach a peak value at few degrees above critical heat flux temperature, T_C. With further increase of wall temperature, the maximum amplitudes decrease again, until they almost die out in the stable film boiling region.

After CHF has been reached locally, any increase of reference temperature extends the axial surface increment where CHF is exceeded. Finally, after CHF has been reached at both ends, transition

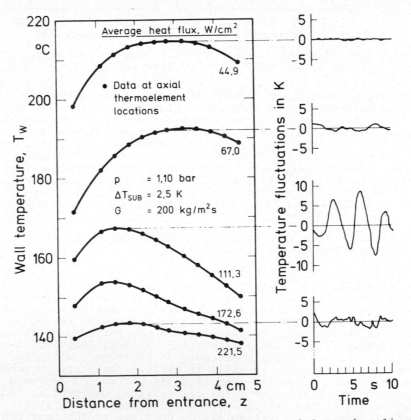

Fig. 6. Axial temperature distributions at low inlet subcooling

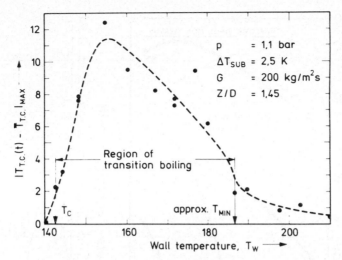

Fig. 7. Maximum fluctuation amplitudes in transition boiling

boiling prevails along the entire test section. The location of
the hot spot where CHF first occured keeps almost unchanged, until
minimum heat flux temperature is exceeded downstream of the initial
hot spot (curve of 67 W/cm^2 av. heat flux).

It may be noted that occurence of transition boiling along
the entire heated length is characteristic for low subcooling only,
where critical heat flux is also rather low. At higher subcoolings,
the axial temperature distributions are quite different as may be
seen from Fig. 8. Again CHF is reached first in the center region
of the test section, however, an only slight increase in reference
temperature forces this local hot spot to move downstream to the
end of test section, and transition boiling spreads out in the upstream direction, which results in strong axial temperature gradients. At the entrance region of the test section, subcooled nucleate boiling is still present, even though stable film boiling
may already exist at the end of test section. This behavior is
rather similar to that experienced in wire experiments. With a
further increase in average wall temperature, the steep axial gradient moves upstream. Due to corresponding axial conduction, radial
heat flux at the entrance region is increased, until CHF is finally
exceeded at the entrance, too. As a result, the shape of the axial
temperature distribution again changes drastically; it becomes
comparably flat, and transition boiling is established along the
entire test section surface as typical at very low subcooling
(cf. Fig. 6). Since, with increase of subcooling, CHF becomes a
strong function of local fluid quality, it cannot be expected that
CHF is actually exceeded at both ends at a common average wall
temperature.

With respect to control, it has to be noted that the drastic
changes of the axial temperature field from rather flat to curved
shape and vice versa are accompanied by large temperature fluctuations. One might conclude that, at this event, the set-point wall
temperature (a weighted average of three thermocouple signals at

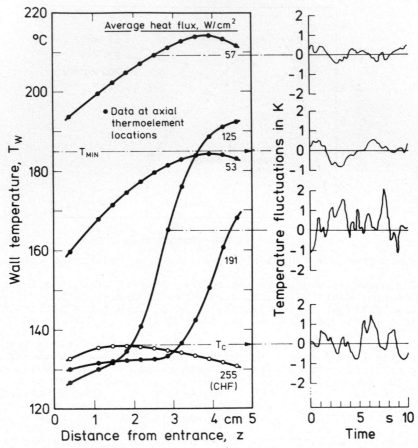

Fig. 8. Axial temperature distributions at moderate inlet subcooling (p = 1.1 bar, ΔT_{SUB} = 11.5 K, G = 200 kg/m^{2}s)

different axial locations) can equally be satisfied by both wall
temperature distributions. Changes from one distribution to the
other are primarily induced by statistical variations of the controlling parameters of the corresponding heat transfer processes.
It certainly would be an interesting task to study this stability
problem of second order in a systematic manner.

In performing the experiments, the test section was forced to
stabilize one of both situations by increasing (or decreasing) the
reference temperature by an increment of about 20 K. It was then
possible to reduce (increase) the reference temperature up to
about 10 K before fluctuations started to increase again. It seems
important to point out, that the resultant gaps of reference temperature of about 10 K are not regions where the test section is
out of control in the sense of an unmanageable instability. It is
merely a region where measurements did not seem meaningful in view
of the large power and temperature variations with time.

6.3. Boiling Curves

Having once evaluated wall heat flux and temperature as a function of axial coordinate, boiling curves may be constructed for any fixed value of relative distance from the entrance, z/D. In view of the rather short test section ($L/D = 5.0$), entrance effects are expected to play a significant role. It was found, however, that this is only the case in the regions of total film boiling and minimum film boiling temperature. In Fig. 9, sample boiling curves at fixed $z/D = 2.5$ (center of test section) are presented for different inlet subcoolings. At subcooling of 41 K, hydrodynamic instability resulted in large power and temperature fluctuations. Hence, it was impossible to obtain meaningful results, except well within the total film boiling region, without changing hydrodynamic loop characteristics. A similar experience has previously been reported by Ragheb et al. /15/.

Critical Heat Flux. CHF increases with increasing inlet subcooling, while the critical heat flux temperature decreases. At fixed inlet subcooling, CHF increases slightly with axial distance from the inlet; highest and lowest values were found to differ by less than 10 percent. For example, at ΔT_{SUB} = 21.5 K, CHF values at z/D = 2.0, 2.5, and 3.0 are 265, 269, and 281 W/cm^2, respectively. The corresponding CHF temperatures are 135.4, 135.1, and 134.0 degree C, respectively. The local equilibrium quality at CHF varied from 0.023 to 0.043 corresponding to the inlet subcoolings of 21.5 to 2.5 K. It seems worth noting that the boiling curves were found to be linear in the neighborhood of CHF in both the nucleate and the transition boiling branches.

Transition Boiling. In the range of inlet subcooling studied, its effect on the transition boiling curve does not seem to be large. That the curve for ΔT_{SUB} = 2.5 K falls in between the curves for the higher subcoolings is probably due to typical differences in upstream heat transfer at low and medium entrance subcooling as discussed in section 6.2 in terms of the associated axial temperature profiles. Note that data are also obtained in the neighborhood of CHF where other steady-state measuring methods usually fail.

Minimum Film Boiling Temperature. A distinct minimum of boiling curve can only be detected at the higher inlet subcoolings. The scatter of data in this region is essentially due to the fact that both rather flat and strongly curved axial temperature distributions may exist at a given average temperature, depending on the direction the boiling curve is traversed. Especially at locations of the test section farther downstream, transition and film boiling branches may thus even overlap in a certain wall temperature increment. In the center region and farther upstream, only a discontinuity of the boiling curve is usually apparent. Only at very low subcooling, where temperature distributions with a steep axial gradient do not occur, a continuous boiling curve is observed, (cf. Fig. 9).

Though a minimum of the boiling curve in a strict mathematical sense cannot be defined, one can nevertheless state that a minimum heat flux is measured at a certain wall temperature. In our measurements, minimum heat fluxes were obtained at wall temperatures between 185 and 210 degree C. These values are considerably lower than reported by others. One explanation for this difference

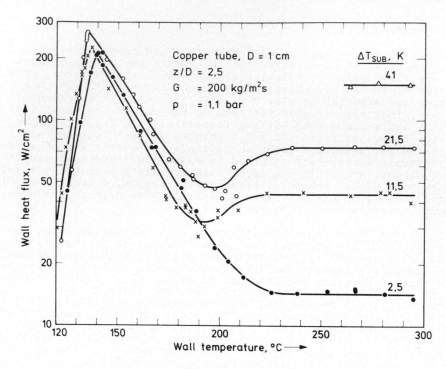

Fig. 9. Effect of inlet subcooling on
flow boiling curves of water

may be found in the lower wettability of the clean gold surface
used here. It seems further that both temperature control and high
thermal conduction of the test section strongly inhibit, that local
cold spots due to short-time rewetting can become stable and eventually spread out rather high wall temperatures. In other words,
in this test section, film boiling is stabilized against statistical
perturbations of rather large amplitude, which consequently results
in the low minimum film boiling temperatures observed.

Film Boiling. In the temperature region from about 240 to
300 degree C, film boiling heat flux is practically independent of
wall temperature regardless of inlet subcooling. Inlet subcooling
as well as axial location, however, have a strong effect on film
boiling. The present few measurements seem to indicate that film
boiling heat flux is inversely proportional to the inlet subcooling
as displayed by Fig. 10. This effect can be explained by considering
the local subcooling or quality effect /3/. As may also be seen
from Fig. 10, heat flux strongly decreases with increasing local
quality in the low void region and becomes independent of local
quality near saturation. In the low void region,the flow is in the
inverted annular regime, where heat transfer is primarily by conduction through the vapor film. With increasing local quality,
vapor film becomes thicker, thus reducing the heat transfer rate.
Near saturation, the flow regime changes with a resultant change in
the relationship between film boiling heat flux and quality.

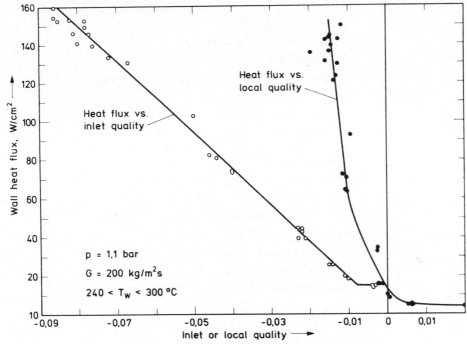

Fig. 10. Film boiling heat flux as function
of local and inlet quality

7.   FINAL REMARKS

An experimental technique has been presented to perform temperature-controlled measurements of the complete flow boiling
curve of water. The technique seems well adequate for accurate
measurement of CHF heat flux and CHF temperature of water at subcooled inlet conditions and investigating the transition boiling
region. The method may also be successfully applied to study the
low temperature film boiling region, where heat flux is essentially
independent of wall temperature. Its potential to possibly contribute to the clarification of some unresolved fundamental aspects
of flow boiling such as stability at CHF and minimum film boiling
temperature, thermal nonequilibrium in inverted annular flow etc.
has not been exploited yet at all.

ACKNOWLEDGEMENTS

The financial support of the German Research Foundation is
gratefully acknowledged. The authors would further like to express
their appreciation to M. Strangalies and H. Weihs and his co-workers
for their invaluable help and excellent performance in preparing
the experimental facilities. Thanks are also due to Mrs. M. Bauditt
for preparing the graphs and Mrs. U. Krause for typing the manuscript.

NOMENCLATURE

CHF	critical heat flux
D	tube diameter, mm or cm
E	voltage, V
E_i	output voltage of single thermocouples, V
G	mass flux, $kg/cm^2 s$
G_c	compensator transfer function
$h_{\overline{T}}$	incremental or differential heat transfer coefficient, $W/cm^2 K$
K	general gain term
L	test section length, mm or cm
p	pressure, bar
q''	wall heat flux, W/cm^2
s	Laplace variable
$\overline{T}$	time-averaged temperature, $^\circ C$
T_c	CHF temperature, $^\circ C$
$T_{T.C.}$	temperature measured by thermocouple, $^\circ C$
T_1, T_N, T_V	time constants, $1/s$
T_{min}	minimum film boiling temperature, $^\circ C$
T_W	wall temperature, $^\circ C$
ΔT_{SUB}	inlet subcooling, K
T.C.	thermocouple
x	flow quality
z	axial distance from inlet, cm
α_i	weighting parameter (cf. eq. (3))

REFERENCES

1. Groeneveld, D.C., "Prediction Methods for Post-CHF Heat Transfer and Superheated Steam Cooling Suitable for Reactor Accident Analysis", Centre d'Etude Nucléaires de Grenoble report TT/SETRE/82-4-E/DCGr, 1982.

2. Groeneveld, D.C., "A General CHF Prediction Method for Water Suitable for Reactor Accident Analysis", Centre d'Etude Nucléaires de Grenoble report DRE/STT/SETRE/82-2-E/DGr, 1982.

3. Fung, K.K., Gardiner, S.R.M., and Groeneveld, D.C., "Subcooled and Low Quality Flow Film Boiling of Water at Atmospheric Pressure", Nucl. Engng. Design 1979, Vol. 55, pp. 51 - 57.

4. Fung, K.K., "Subcooled and Low Quality Film Boiling of Water in Vertical Flow at Atmospheric Pressure", Argonne National Laboratory, Argonne, Il., Report No. ANL 81 - 78 (= NUREG/CR-2461), 1981.

5.   Stewart, J.C., and Groeneveld, D.C., "Low Quality and Sub-
     cooled Film Boiling of Water at Elevated Pressures", <u>Nucl.
     Engng. Design</u>, 1981, Vol. 67, pp. 259 - 272.

6.   Groeneveld, D.C., and Gardiner, S.R.M., "A Method for Obtain-
     ing Flow Film Boiling Data for Subcooled Water", <u>Int. J. Heat
     Mass Transf.</u>, 1978, Vol. 21, pp. 664 - 665.

7.   Groeneveld, D.C., and Fung, K.K., "Forced Convective Tran-
     sition Boiling - Review of Literature and Comparison of
     Prediction Methods", Chalk River Nuclear Laboratories, Chalk
     River, Ontario, Report AECL-5543, 1976.

8.   Poletarkin, P.G., Petrov, V.I.; Dodonov, L.D., and Aladyer,
     I.T., "A New Method for the Investigation of Heat Transfer
     in the Boiling of Liquids", <u>Dokl. Akad. Nauk S.S.S.R.</u>, 1953,
     Vol. 90, pp. 775 - 776. (In Russian).

9.   Ellion, M.E., "A Study of the Mechanism of Boiling Heat
     Transfer", California Inst. of Technology Report JRL-MEMO-
     20 - 88, 1954.

10. Mc Donough, J.B., Milich, W. and King, E.C., "An Experimental
     Study of Partial Film Boiling Region with Water at Elevated
     Pressures in a Round Vertical Tube", <u>Chem. Engng. Progr.</u>
     Symp. Ser., 1981, Vol. 57, No. 32, pp. 197 - 208.

11. Ramu, K., and Weismann, J., "Transition Flow Boiling Heat
     Transfer to Water in a Vertical Annulus", <u>Nucl. Engng. Design</u>,
     1977, Vol. 40, pp. 285 - 295.

12. Weismann, J., Kao, Y.K., and Rahrooh, G.,"Transition Boiling
     Heat Transfer in a Vertical Tube", <u>Amer. Soc. Mech. Engrs.</u>
     paper No. 79-HT-47, 1979.

13. Wang, S., Kao, Y.K., and Weismann, J., "Studies of Transition
     Boiling Heat Transfer with Saturated Water at 1 - 4 bar",
     <u>Nucl. Engng. Design</u>, 1982, Vol. 70, pp. 223 - 243.

14. Cheng, S.C., Ng, W.W.L., and Heng, K.T., "Measurements of
     Boiling Curves of Subcooled Water under Forced Convective
     Conditions", <u>Int. J. Heat Mass Transf.</u>, 1978, Vol. 21,
     pp. 1385 - 1392.

15. Ragheb, H.S., and Cheng, S.C. and Groeneveld, D.C. "Obser-
     vations in Transition Boiling of Subcooled Water under Forced
     Convective Conditions", <u>Int. J. Heat Mass Transf.</u>, 1981,
     Vol. 24, pp. 1127 - 1137.

16. Nelson, R.A., "Forced Convective Post-CHF Heat Transfer and
     Quenching", <u>J. Heat Transf.</u>, 1982, Vol. 104, pp. 48 - 54.

17. Peterson, W.C., "A Comparator Bridge Circuit", <u>IEEE Trans.
     Ind. Electron. Control Instrum.</u>, 1969, Vol. IECI - 16,
     pp. 161 - 164 .

18. Sakurai, A., "Temperature-Controlled Boiling Heat Transfer", Kyoto daigatu genshi enerugi kenkynsho iho, 1974, Vol. 46, pp. 1 - 11. (In Japanese)

19. Sakurai, A., and Shiotsu, M., "Temperature-Controlled Pool-Boiling Heat Transfer", Proc. 5th Intern. Heat Transfer Conf., Tokyo, Vol. IV, pp. 81 - 90, 1974.

20. Sakurai, A., and Shiotsu, M., "Studies on Temperature-Controlled Pool-Boiling Heat Transfer", Techn. Rep. Inst. Atom. Energy Kyoto Univ. No. 175, 1978.

21. Peterson, W.C., Aboul Fetouh, M.M., and Zaalouk, M.G., "Boiling Curve Measurements from a Controlled Forced Convection Process", Proc. Brit. Nucl. Energy Soc., Conf. on Boiler Dynamics and Control in Nuclear Power Stations, London, pp. 18.1. - 18.6., 1973.

22. Peterson, W.C., Zaalouk, M.G., and Güceri, S.I., "On the Dynamics of Transition Region Flow Boiling", Nucl. Sci. Engng., 1976, Vol. 61, pp. 250 - 257.

23. Berenson, P.J., "Experiments on Pool-Boiling Heat Transfer", Int. J. Heat Mass Transf., 1962, Vol. 5, pp. 985 - 999.

24. Cheng, S.C.; and Ng, W., "Transition Boiling Heat Transfer in Forced Vertical Flow via a High Thermal Capacity Heating Process", Lett. Heat Mass. Transf., Vol. 3, 1976, pp. 333 - 342.

25. Ng, W., and Cheng, S.C., "Steady State Flow Boiling Curve Measurements via Temperature Controllers", Lett. Heat Mass Transf., 1979, Vol. 6, pp. 77 - 81.

26. Peterson, W.C., Thacker, A., and Avery, W.L., "A Feedback System for Control of an Unstable Process", IEEE Trans. Ind. Electron. Control Instrum., 1969, Vol. IECI - 16, pp. 165 - 171.

27. Kleen, U., and Johannsen, K., "Measurement of Transition Boiling Data of Subcooled Water at Forced Convective Conditions and Atmospheric Pressure via a Steady-State Process", Jahrestagung Kerntechnik '82 (Dt. Atomforum, ed.), Mannheim, 1982, pp. 105 - 108.

TEMPERATURE CONTROLLED MEASUREMENT OF BOILING CURVES
FOR LOW QUALITY UPWARD FORCED FLOW IN TUBES

Hein Auracher and Hartmut Albrodt
Institut für Technische Thermodynamik
und Thermische Verfahrenstechnik
Universität Stuttgart
Pfaffenwaldring 9, 7000 Stuttgart 80, F.R.G.

ABSTRACT

A technique has been developed providing for steady state
temperature controlled experiments in the transition region by
an electrically heated system. The test facilities are designed
for experiments with refrigerants in a wide range of reduced
pressures, mass fluxes and qualities. The experiments were carried
out in a 0.364 m long test section downstream of a vertical
nickel-tube of about 2.5 m total length, 14 mm inside diameter
and 1 mm wall thickness. Along the main-evaporator of the test
section a cylindrical copper block of 35 mm outside diameter is
soldered onto the nickel-tube. This copper cylinder is devided
into six sections of different lengths between 26 and 64 mm.
The smallest section serves as main heater, whereas the adjacent
ones are used as guard heaters. Each copper section can be heated
separately and the different heat inputs are also separately
measured. Furthermore the temperatures of the copper sections can
be controlled independently. A number of radially and axially
distributed thermocouples are embedded in the test section to
determine the wall temperature and the temperature distribution
along the tube.

Boiling curves for Freon 114 are presented for different
mass fluxes ($1000 < \dot{m} < 4000$ kg/m^{2}s) at constant pressure (p= 3 bar)
and quality (x=0,03). It is shown that significant temperature
changes along the main heating section can be avoided by means
of guard heaters. The temperature oscillations in the transition
region were measured by thermocouples 0.3 mm beyond the heating
surface and were found to be less than $\pm$ 0.7 K. The controlling
system thus permits reliable measurements of heat transfer coefficients in the transition region.

1. INTRODUCTION

In heat transfer practice it is important for different
reasons to know the total shape of boiling curves [$\dot{q}$ = f (T_{wall} -
T_{fluid})] including the transition region. For example, accurate
critical heat flux data are required to design heat exchangers
as economically as possible and to observe on the other hand a
safety margin from burnout. In reactor safety analysis

transition boiling data and minimum film boiling temperatures
are required to calculate the temperature history of fuel rods
during hypothetical accidents as LOCA etc..

Because of the negative slope of $\dot{q},\Delta T$-curves steady-state
experiments in heat flux controlled systems are not possible.
The few available data result therefore mostly from transient
tests in quenching experiments. In fast transient tests, however,
the heat transfer rates are probably different from such found
by steady-state experiments at the same wall temperature.
Furthermore the results could be inaccurate due to boundary
effects at the inlet and outlet region of the test section as
shown by Newbold and co-workers [1] and Cheng et al.[2]. Few
attempts for steady-state measurements have been made. Stephan
and Hoffmann [3] obtained boiling curves for low quality flow
of Freon 114 by averaging heat transfer rates along a 0.31 m
long horizontal tube, heated from outside by temperature-
stabilizing water. Yilmaz and Westwater [4] used an inside steam
heated horizontal tube arranged in vertically upward forced flow
of Freon 113 to measure boiling curves for different flow
velocities near the atmospheric pressure. Ng and Cheng [5] used
the same test section as in [2] - an electrically heated
cylindrical copper block of 95.3 mm outside diameter and a
center bore of 12.96 mm without guard heaters - to also measure
boiling curves under steady-state conditions for subcooled
water in vertical forced flow at atmospheric conditions and low
mass fluxes ($\dot{m} < 203$ kg/m^{2}s).

Systematic experiments on boiling curves for a wide range
of the governing parameters are missing. Particularly in the low
quality post-critical heat flux region - i.e. the transition
boiling region and the inverted annular film boiling region
included the minimum film boiling temperature - a significant
gap in data and prediction methods [6] exists.

2. EXPERIMENTS

2.1 Test Loop

Fig. 1 shows a schematic drawing of the experimental system.
The liquid flows from a centrifugal pump (1) via a mass flow
metering device (2) and a heater (3) into the vertical test
section (A). The mass flow is measured by parallel arranged
turbine flow meters (4) having different measuring ranges. Small
pressure fluctuations coming from the pump are eliminated by a
pressure stabilizer (5). The fluid flow is manually controlled
by a valve (7) and a by-pass line (13), respectively. In (3) the
fluid is heated by temperature controlled oil (6). By means of
a platinum resistance thermometer (T) at the test section inlet
(14) this temperature is measured and controlled to be constant
at each test run. After the test section liquid and vapor are
separated in (8) and the vapor flows into the condenser (9) where
it is liquefied by cooling water (15). By controlling the water
temperature and its mass flow the system pressure is held
constant during each test run. The liquefied fluid flows back to
the bottom of the separator (8) and, together with the unevap-
orated mass flow fraction, via a cooler (10) to the suction

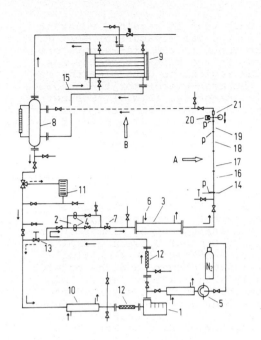

A. Test Section, Vertical
B. Test Section, Horizontal
1. Centrifugal Pump
2. Mass Flow Measurement
3. Heater
4. Turbine Flow Meter
5. Pressure Stabilizer
6. Heating Oil
7. Control Valve
8. Separator
9. Condenser
10. Cooler
11. Filter
12. Compensating Tube
13. By-Pass Line
14. Test Section Inlet
15. Cooling Water
16,17,18. Pre-evaporator
19. Main-evaporator
20. γ-Ray Gauge
21. Capacitance Meter

Fig. 1. Schematic Drawing of the Experimental System

side of the pump. Alternatively the liquid can divert into a by-pass line with a filter (11) before flowing to the pump. By means of the cooler (10) cavitation inside the pump can be avoided. Several thermocouples and resistance thermometers are installed in the loop to check and control the system conditions. The whole apparatus is insulated with aluminum - coated glass wool.

2.2 Test Section

The entire test section (A) consisted of a three stage pre-evaporator (16-18), 0.622 m per stage, the main-evaporator (19; 0.364 m), a γ-ray attenuation gauge (20) and a capacitance meter (21).

The γ-ray gauge was located downstream of the main-evaporator during the experiments and was used only for a quali-tative determination of the flow pattern. Systematic experiments about the interaction between heat transfer and flow pattern were not carried out in the present investigations. Such experiments will be reported later. The capacitance meter was used to deter-mine the void fraction. A detailed description of this device and calibration results are given elsewhere [7].

Boiling takes place in a nickel-tube of 14 mm inside dia-meter and 1 mm wall thickness. In the pre-evaporator sections, the tube is electrically heated by cartridge heaters, rolled into a copper shell around the nickel tube. During the measure-

ments the quality was controlled such that it remained constant
in the center plane of section D in the main-evaporator (see Fig. 2).
This required a tuning of the heat input to the pre-evaporators
according as the heat transfer mode in the main-evaporator was
in the high or in the low heat flux region. To determine the relevant saturation temperature in the main-evaporator the absolute
pressure is measured by a precision manometer (Fig.1). Furthermore the pressure difference relative to the main-evaporator
outlet and the test section inlet is determined. The size of the
test laboratory also allows a horizontal installation of the
entire test section to study the influence of gravity.

2.3 Main-evaporator

Fig. 2 shows the main-evaporator (see 19 in Fig.1). A cylindrical copper block of 294 mm total length and 35 mm outside
diameter is soldered onto the nickel-tube. A gap clearance of
0.035 mm between outside nickel- and inside copper diameter was
found to be useful for a satisfactory contact between nickel and
copper during the soldering process. A solder with a melting
point of 270°C was used. Both the inside copper- and the outside
nickel-surface were first covered by a thin solder layer. Then
the heaters of the copper sections (see Fig.2) were turned on
and controlled in a way, that the copper block could be slided
softly onto the nickel-tube. It was found in preliminary tests
that this procedure guarantees a uniform contact between both
tubes without voids.

The copper block is divided into six sections of different
lengths between 26 and 64 mm by small slots of 0.45 mm width.
Only a copper bridge of 1 mm thickness is left between the
sections. Each of this procedure copper sections can be heated separately
by a sheathed resistance wire, rolled and brazed into a coiled
channel on the copper surface. The different heat inputs can also
be measured separately. Section D serves as the main heater. All
data and parameters of the presented boiling curves, i.e. heat
flux, wall temperature, temperature of the boiling liquid and
quality are given with respect to section D. The adjacent sections
E and C are used as guard heaters. Optionally the sections A, B
and F can also be heated. The whole equipment is designed for
maximum heat fluxes of $5 \cdot 10^5$ W/m^2 on the boiling surface.

Two axially installed thermocouples (D8, D9) are used to
measure the wall temperature in section D. They are soldered into
an axial slot of the nickel-tube 0.3 beyond the inner surface.
Both thermocouples - and all others located in the test section -
are metal sheathed with a 0.5 mm outside diameter and consist of
NiCr-Ni-wires of 0.09 mm thickness each. The tips of both thermocouples are located in the middle of section D (see Fig.2). To
determine the real inside surface temperature the measured
temperature is corrected by means of Fouriers one dimensional
heat conduction equation introducing the actual heat input to
section D.

Several thermocouples (type 2) are installed radially along
the entire test section, as shown in Fig.2. They are soldered
into holes of 0.8 mm diameter and their tips are 0.3 mm beyond
the inside surface. The thermocouples D1 to D7 yield the temper-

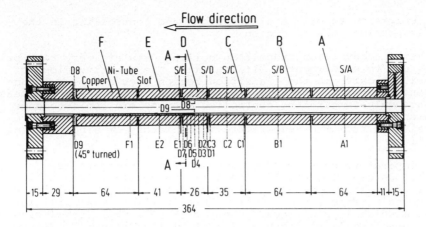

Cross Section A ÷ A:

Section D of the Main-
evaporator:

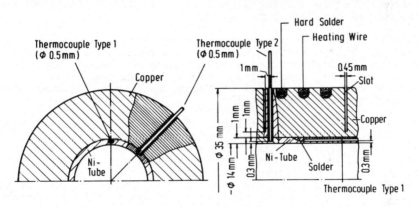

Fig. 2. Main-evaporator

ature distribution along the main heater and those on both sides
of the slots, C3/D1 and D7/E1 respectively, are used to ensure
the temperature equalization between main- and guard-heaters. Due
to the non-isothermal installation of all type 2-thermocouples
the measured temperature may deviate from the correct absolute
value. This, however, is not relevant, since the type 2-thermo-
couples are only used to adjust and measure the temperature distribution along the sections, whereas the type 1-thermocouples
(D8/D9) yield the correct temperatures for the heat transfer determination. In each of the sections A to E one of the radially
located thermocouples is taken as a sensor to give the actual
temperature for the controlling system (S/A to S/E, Fig.2), which
permits to control each copper section temperature independently.
This, particularly, allows to equalize carefully the temperatures
at the boundaries of the main heater D.

2.4 Remarks on experimental technique and controlling in the transition boiling region

Forced convection transition boiling data are very rare primarily due to experimental difficulties. The negative slope of the boiling curve in the transition region does not permit conventional heat flux controlled experiments. In steady-state measurements only temperature controlled systems are applicable or transient tests have to be carried out where the boiling curve is evaluated from the temperature time history of a body during quenching.

Temperature controlled steady-state measurements are for instance possible if a stabilizing fluid is applied. Such a fluid may be used both for heating and stabilizing the boiling fluid at a certain temperature [3] or primarily for only stabilizing a system which is electrically heated [8; 9]. This stabilizing fluid technique is applicable if a stability criterion can be fulfilled that permits stable operation in the transition region [10; 11; 12]. If fluids with low normal boiling points are tested, pressurized water can be used as a stabilizing fluid. If, however, water itself is the boiling liquid to be tested, a suitable stabilizing fluid can hardly be found.

A second approach to carry out temperature controlled measurements in the transition region is the application of high thermal inertia test sections, mostly made of copper. Newbold et al. [1], Fung [13], Cheng et al [14; 2] and Ragheb et al. [15] used this technique in transient tests, Ng et al. [5] made also some steady-state experiments.

Both methods, the stabilizing fluid technique and the high thermal inertia technique - either transient or steady-state - usually suffer from at least one serious drawback. In transition boiling a liquid front exists somewhere in the inlet region of the heated section reducing the temperature due to high heat transfer coefficients in this area. At the outlet region, on the contrary, the fluid quality is higher, a vapor film has established and thus the wall temperature increases due to a relatively low heat transfer coefficient. This is true in two-phase flows with higher quality where no continuous liquid layer exists after the dryout front. In the low quality and subcooled boiling regions, the vapor film collapses after the heated section, resulting in a temperature drop near the outlet. In both cases axial temperature gradients and therefore variations in the axial heat flux distribution are present. If e.g. nucleate boiling exists at the inlet region and film boiling at the outlet the heat flux may vary by more than a factor of 10 along the test section. Consequently a careful evaluation of the axial surface heat flux distribution by means of the two-dimensional heat conduction equation is required to obtain correct boiling curves. This, however, can often not reliably be done due to the limited number of installed thermocouples. As a result, different boiling curves for different axial positions may be obtained, or if a mean heat flux and a mean temperature difference is introduced, the resulting boiling curve is levelled off compared with the real one, especially in the area of the minimum film boiling temperature.

If transient tests are carried out with high thermal inertia test sections another drawback must be considered. During the quenching process, the vapor production changes and consequently the flow pattern also. Furthermore, due to a pressure drop change, the mass flux may vary too. It is, in addition, difficult to evaluate reliable boiling curves from the temperature-time readings, due to uncertainties in determining the real heated surface temperature and the actual heat flux.

A steady-state method is therefore preferable. However, in each case, special attention must be payed to the temperature variation along the test section. Guard heaters at both ends are a useful method to reduce the temperature differences as shown by Newbold et al. [1] in transient experiments with a copper block test section. However, if steady-state experiments are to be carried out, this requires separate controlling systems for both the temperatures of the test section and those of the guard heaters. The stabilizing fluid technique would, in this case, demand extensive installations connected with serious design problems and experimental difficulties. Therefore an electrically heated system seems to be the most promising and elegant one, provided a suitable controlling system for both the main heater and the guard heaters is available.

The experiments presented here were carried out with a controlling system that was especially designed and adapted to the test section shown in Fig.2. In each section one thermocouple of type 2 (S/A to S/E, see Fig.2) serves as a sensor to give the deviation between set-point and actual temperature at the thermocouple tip beyond the inside surface. A modified ON-OFF-controller is used to stabilize the desired temperatures in the different sections. However, in the transition region the local heat transfer mode is characterized by a rapid change between vapor- and liquid-contact-intervals resulting in a fast oscillation of the local wall temperature. Therefore, due to the thermal inertia of the heating system, the controller does not work satisfactorily in the entire transition region. A modulation of the thermocouple deviation signal is required in these cases. This modulation function is determined by means of an electrical analogue model of the heating system taking into account a model for the inside heat transfer mode. For each section to be controlled a modulator unit as well as the coupled power supply was designed. The different heat fluxes are computed from the power inputs to the heating wires. Heat losses through the insulation were calibrated as a function of temperature level in preliminary tests and are subtracted from the power inputs. A detailed description of the whole heating and controlling system is in preparation and will be presented later.

Data processing is carried out by a Hewlett/Packard 9835A-data transfer unit consisting of a 3455A-digital voltmeter, a 60-channel scanner, a plotter and a printer. Furthermore a 6-channel transient recorder is available that permits a simultaneous recording of 6 analog signals.

784

3. RESULTS

The main purpose of the present investigation was to develop
an accurate technique for the measurement of boiling curves,
especially with respect to the transition region. In the first
step, one boiling curve (Fig.3) was carefully measured and analyzed. Emphasis was layed on the study of temperature distributions and oscillations in the main heater in different boiling
regions and on the reproducibility of the data.

A linear-linear plot as used in Fig. 3 seems to be the most
suitable one to show in particular the change in heat transfer
between critical heat flux and minimum film boiling temperature.
The presented curve was measured during different runs with both
step by step increasing and decreasing wall temperature. The reproducibility was found to be satisfactory.

Axial wall temperature distributions. Some typical axial
wall temperature distributions along section D are shown in Fig.4.
The corresponding heat transfer data are marked in Fig. 3 (a ÷ m).
Noticeable temperature changes were only found in the transition
region. The maximum temperature change along section D however
did not exceed 1 K. Judging this it has to be considered, that the
different temperatures in section D (T_{D1} ÷ T_{D7}) were not recorded
simultaneously but within a time interval of approximately 0.15
seconds between the temperature readings. Regarding the temperature oscillation plots (Fig.6) it can be concluded that the detected temperature changes along section D can be a result of
local temperature oscillations during the recording procedure.
Partly the temperatures at the boundaries were not completely
equalized with those in the adjacent sections (e.g. d and e).
In these cases the axial heat losses were approximated by a calibration function and included in the heat flow balance for
section D.

Wall temperature oscillations. Figs. 5 to 7 show some plots
of temperature oscillations occuring in the different boiling
regions (A ÷ F, Fig.3). Type 2-thermocouple D4 was used as sensor.
Its DC-signal was recorded via a 10 Hz low-pass filter by the
transient recorder. Clearly, due to the thermal inertia involved

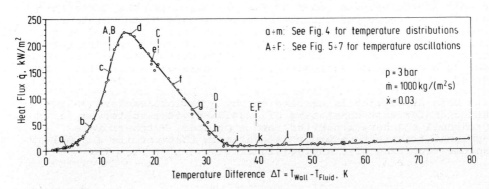

Fig. 3. Boiling Curve for Freon 114

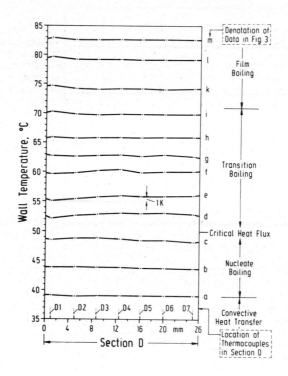

Fig. 4. Axial Wall Temperature Distributions in the
Main Heater in Different Boiling Regions

in the temperature measuring method, the detected temperature
oscillations are not primarily the result of local dryout or
wetting of the wall surface beyond the thermocouple. This effect
could only be measured by fast-response surface thermocouples [16].
The purpose of the presented oscillation measurements was to show
overall temperature fluctuations of the heated wall.

In adiabatic flow and also in nucleate boiling the temperature oscillations are insignificant (Fig.5). No difference was
observed between the measured $\dot{q}, \Delta T$-data whether a controller was
used or not. This is also true for the film boiling region (Fig.7).
There, a long wave oscillation occurs, if the system is controlled. Its impact on the experimental accuracy is however negligible. In the transition region (Fig.6) the temperature amplitudes
are higher but they do not exceed a margin of $\pm$ 0.7 K in the
region near the critical heat flux (C, Fig.3) and of $\pm$ 0.3 K near
the minimum film boiling temperature (D, Fig.3). Regarding the
extent and the slope of the transition boiling curve it can be
concluded that neither the axial wall temperature distributions
nor the overall temperature oscillations are of significant effect

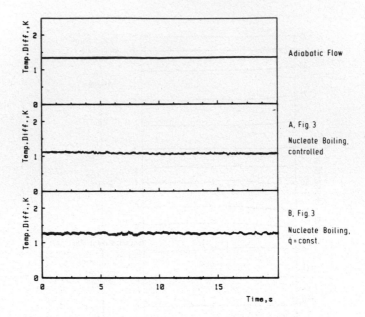

Fig. 5. Wall Temperature Oscillations in Adiabatic Flow
and in the Nucleate Boiling Region ($\Delta T = 11.8$ K;
A, B in Fig. 3)

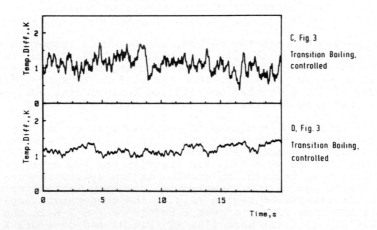

Fig. 6. Wall Temperature Oscillations in the Transition Region,
$\Delta T = 20.9$ K, C in Fig. 3; $\Delta T = 31.9$ K, D in Fig. 3.

on the measured heat transfer.

Variation of mass flux. Fig. 8 presents boiling curves in
log-log-coordinates for mass fluxes of 1000, 2000 and 4000 kg/m²s
at constant pressure (3 bar) and quality ($\dot{x}$ = 0,03). Experiments
with smaller mass fluxes were not carried out, since in this case
the liquid would have to be subcooled at the main-evaporator inlet
to establish the desired quality of 3 % in section D, because the
guard heater technique requires a heating of section C and partly
also of section B. Due to the short heating distance probably no
thermodynamic equilibrium would have been reached in section D
and the real quality would have been somewhat higher than the
calculated one along with a corresponding subcooling of the liquid
[17]. The experimental results would not be comparable with those
in Fig. 8 for saturated boiling. A simultaneous void fraction
measurement with the γ-ray gauge is required in this case and
will be carried out later.

Convective boiling, nucleate boiling and critical heat flux.
The presented curves in Fig. 8 clearly show the different boiling
regions. It seems expedient to compare these results with some
heat transfer prediction methods from literature. This is however
only useful for the better understood boiling regions between
convective boiling and critical heat flux. It should be noted that
no systematic comparison of data and prediction methods shall be
carried out here. Only a test of the experimental system with
respect to the measurement of entire boiling curves is intended.

Fig. 9., e.g., shows the measured heat transfer coefficients
vs. the heat flux for convective and nucleate boiling. The typical
shape of such plots is obtained. In the low heat flux region, the
heat transfer mode is determined by convection. The heat transfer
coefficients (α) can therefore be calculated based on relationships for one-phase flow such as the Dittus-Boelter equation,

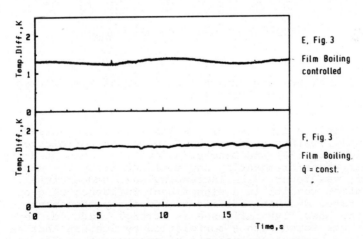

Fig. 7. Wall Temperature Oscillations
in the Film Boiling Region,
ΔT = 39.5 K; E,F in Fig. 3.

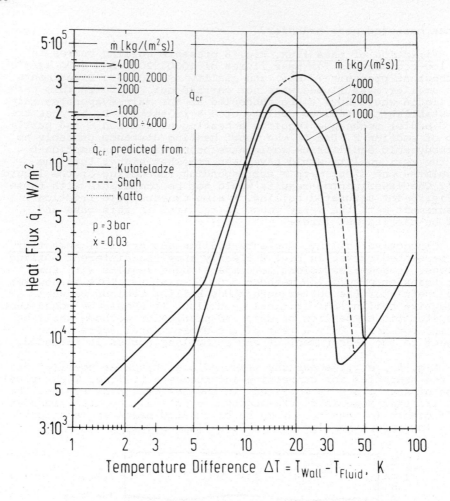

Fig. 8. Boiling Curves for Different Mass Fluxes
Fluid: Freon 114

where α is only a function of the mass flux at constant pressure. Consequently, rather good agreement with the correlations of Guerrieri/Talty [18] and Dengler/Addoms [19], who used the single-phase heat transfer equation and modified it by the Martinelli parameter, was found. With increasing wall temperature the onset of nucleation results in a significant influence of $\dot{q}$ on the heat transfer rate. Nucleate boiling is now the predominant mechanism whereas the mass flux influence is strongly reduced. Very good agreement was found with a correlation by Stephan/Abdelsalam [20] derived by a regression analysis of numerous data for pool boiling of refrigerants. Stephan/Auracher [21] proposed a modification of this correlation to allow also for forced convection nucleate boiling calculations. This relationship however is only valid for

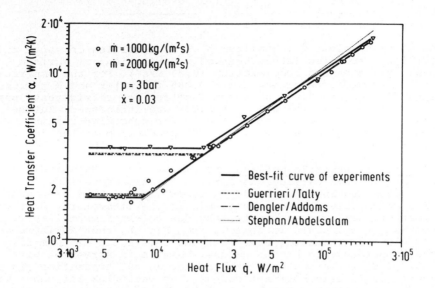

Fig. 9. Heat Transfer Data in the Convective and
Nucleate Boiling Region for Freon 114

lower mass fluxes and higher qualities and overpredicts the heat
transfer coefficient in the present parameter range.

Correlations for both the convective and the nucleate boiling
region have been proposed e.g. by Pujol/Stenning [22] where the
boiling number, to characterize nucleation, and the Martinelli
parameter, to characterize convection, were introduced, and by
Rohsenow [23] and Chen [24] where the contributions of nucleation
and convection are calculated separately and then added. Such
correlations yield a more or less significant deviation, either
positive or negative, from the experimental data. Nevertheless,
it can be concluded that the present results fit into the generally accepted characteristics of heat transfer in this part of
the boiling curve.

This is also true for the critical heat flux (CHF) data.
A comparison was made with the graphical methods of Shah [25] and
Katto [26]. Shah's procedure underpredicts the measured data and
shows no considerable influence of the mass flux (see Fig. 8).
On the contrary a moderate overprediction is obtained by Katto's
method. At $\dot{m}$ = 4000 kg/m²s a higher CHF-value is determined than
for the two smaller mass fluxes. As a matter of fact, one has to
consider that both methods are developed for uniformly heated
tubes which is not exactly true in the present case. The best
fit, especially with respect to the mass flux influence, was found
by an equation of Kutateladze [27], which is probably not generally known and therefore presented here.

$$\dot{q}_{cr} = K \cdot \rho_v \cdot \Delta h_v \cdot \left(\left(\frac{\rho_1 - \rho_v}{g \cdot \sigma} \right)^{1/4} \cdot \frac{(1-\dot{x}) \cdot \dot{m}}{\rho_1} \right)^{1/2} \cdot \left(\frac{g \cdot \sigma \cdot (\rho_1 - \rho_v)}{\rho_v^2} \right)^{1/4} \cdot \qquad (1)$$

Where: $\dot{m}$: mass flux; $\dot{x}$: quality; ρ_v, ρ_1 : densities of vapor and liquid; Δh_v : specific latent heat of vaporization; σ: surface tension; g : standard acceleration (9,81 m/s^2). For the empirical constant K Kutateladze proposed K = 0.085 for water and K = 0.045 for liquefied gases. Since no recommendation for refrigerants has been made, a best-fit value of K = 0.035 was introduced. Kutateladze's equation was proposed primarily for subcooled boiling but obviously it can also be used for a CHF-prediction in the low quality region.

Transition and film boiling. No comparison of data and prediction methods was carried out for the transition boiling results presented in Figs. 3 and 8, respectively. This is because very few comparable data are available in the literature. Furthermore most of them are not trustworthy due to experimental shortcomings (see chap. 2.4). It follows from Fig. 8, that an increase in mass flux increases the heat flux and shifts the minimum film boiling temperature to higher values. This is in agreement with findings of different authors [28] as far as the transition region is concerned. In film boiling, however, no mass flux influence was found. The transition region of boiling curves presented in the literature is mostly not as steep as determined by the present measurements. It is believed that such a levelling off in the transition region of boiling curves often results from boundary effects at the test sections, as discussed in the previous chapter.

In the literature it is sometimes proposed [6] to derive a prediction method for transition boiling heat transfer by introducing the CHF-value as upper limit and the heat flux at the minimum film boiling temperature as lower limit and to connect both values by a straight line in a log-log plot. This is probably the best approach to obtain a reasonable prediction method as long as no better understanding of the governing physical mechanisms in forced convection transition boiling is achieved. However, the linear-linear plot in Fig. 3 suggests that a linear relationship between $\dot{q}$ and ΔT rather than an exponential function should be introduced. In each case a disadvantage of this method is, that the minimum film boiling temperature and the corresponding heat flux are still subject to a large degree of uncertainty. In the present experiments, too, a higher scattering of the data than in other boiling regions was observed (see Fig. 10). Therefore the curve for $\dot{m}$ = 2000 kg/m^{2}s in Fig. 8 is partly dotted since the data-scattering did not permit an exact determination of the minimum film boiling temperature. A great amount of additional data for different mass fluxes and also for different pressures and qualities are required to improve the information on heat transfer in this region. It is planned to carry out systematic experiments with the presented technique in the near future.

To give a better insight into the experimental results obtained for the film boiling region a linear-linear $\dot{q}, \Delta T$-plot containing all measured data is shown in Fig. 10. No systematic influence of the mass flux was found. This is in agreement with Freon 113-experiments by Rankin [29] and Dougall/Rohsenow [30],

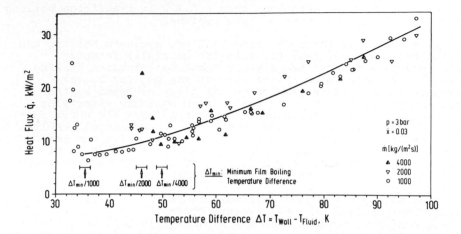

Fig. 10. Heat Transfer Data in the Film Boiling Region
for Freon 114

but not with some others as discussed for instance by Groeneveld/
Gardiner [31].

By the capacitance meter (21 in Fig. 1) void fractions of
55 % on the average were measured. Thus the flow pattern is most
likely of the churn- or wispy-annular type [32]. The film boiling
heat transfer connected with such a flow regime has not been well
studied up to now and reliable prediction methods are not available. Therefore an important task for future work is to measure
simultaneously heat transfer and flow pattern - which is intended
to be carried out by means of the γ-ray gauge (20, Fig. 1) - and
to determine in addition the surface wetting e.g. by fast response
thermocouples [16]. Then the necessary prerequisites are available
for a better derivation of prediction methods.

This is also true particularly for the minimum film boiling
temperature difference (ΔT_{min}). As mentioned above, a significant
data-scattering was observed in this region. Nevertheless, from
the plot in Fig. 10 the increase of $\dot{q}$-values which marks the
location of ΔT_{min} can be determined within a margin of about ± 2 K.
This increase takes place at significant higher wall temperatures
at a mass flux of 4000 kg/m²s compared with that of 1000 kg/m²s.
The result for 2000 kg/m²s does not fit well into this concept
and more experiments have to be carried out to clear up these
findings.

4. CONCLUSIONS

1. The technique of a temperature controlled electrically
heated system with guard heaters at both ends of the test section
proved to be an effective tool for the steady-state measurement
of boiling curves. No significant axial wall temperature change

along the test section was observed. The temperature oscillations of the heating system were less than $\pm$ O.7 K in transition boiling and negligible in the other boiling regions.

2. The results for convective boiling, nucleate boiling and for the critical heat flux fit into the generally accepted characteristics of heat transfer in these well-studied boiling modes. In the transition region most of the boiling curves presented in the literature are not as steep as found by the present experiments. This may result from boundary effects at test sections without guard heaters. In this case the curve is levelled off in the transition region. The experimental results suggest that a linear relationship between heat flux and temperature difference in the transition region is likely to be realistic at least under low quality conditions. More experimental data are however required for a further prove.

A higher scattering of data was observed in the film boiling region. Nevertheless, the temperature controlled measuring technique permits to determine the minimum film boiling temperature for steady-state conditions within a margin of approximately $\pm$ 2 K.

3. Further experiments are required and planned for a wide range of pressures, mass fluxes and qualities to prove the results and to extend the knowledge to different system conditions. Since heat transfer is closely connected with the existing flow regime, simultaneous measurements of heat transfer and flow pattern are important. For this purpose a γ-ray attenuation technique and a capacitance meter are available.

ACKNOWLEDGEMENTS

The authors express their gratitude to the German Ministry of Research and Technology (BMFT) for providing the financial support. This project is part of the BMFT-Reactor Safety Research Program managed by the Gesellschaft für Reaktorsicherheit (GRS).

Furthermore the authors wish to thank Mr. J. Daubert for his work during the design of the test facilities and for his outstanding ideas in the development of different measuring techniques.

REFERENCES

1. Newbold, F.J., Ralph, J.C., and Ward, J.A., "Post-Dryout Heat Transfer under Low Flow and Low Quality Conditions", AERE-R 8390, Harwell (1976).

2. Cheng, S.C., and Ragheb, H., "Transition Boiling Data of Water on Inconel Surface under Forced Convective Conditions", Int. J. Multiphase Flow, 5, 281-291 (1979).

3. Stephan, K., and Hoffmann, E.G., "Transition and Flow Boiling Heat Transfer inside a Horizontal Tube", Int. J. Heat Mass Transfer, 20, 1381-1387 (1977).

4. Yilmaz, S., and Westwater, J.W., "Effect of Velocity on Heat Transfer to Boiling Freon-113", J. Heat Transfer, 102, 26-31 (1980).

5. Ng, W.W.L., and Cheng, S.C., "Steady State Flow Boiling Curve Measurements via Temperature Controllers", Lett. Heat Mass Transfer, 6, 77-81 (1979).

6. Groeneveld, D.C., and Rousseau, J.C., "CHF and Post-CHF Heat Transfer: An Assessment of Prediction Methods and Recommendations for Reactor Safety Codes", Adv. in Two-Phase Flow and Heat Transfer, NATO Adv. Res. Workshop, Spitzingsee, F.R.G. (1982).

7. Auracher, H., and Daubert, J., "A Capacitance Method for Void Fraction Measurements", Proc. 3rd. Conf. IMEKO Techn. Com. on Flow Measurement - TC9, Budapest/Hungary, Sept., 1983 (To be published).

8. Ellion, M.E., "Study of Mechanism of Boiling Heat Transfer", Jet Prop. Lab. Memo., 20-88, (1955).

9. Hesse, G., "Heat Transfer in Nucleate Boiling, Maximum Heat Flux and Transition Boiling", Int. J. Heat Mass Transfer, 16, 1611-1627 (1973).

10. Stephan, K., "Stabilität beim Sieden", Brennst.-Wärme-Kraft, 17, 571-578 (1965).

11. Stephan, K., "Übertragung hoher Wärmestromdichten an siedende Flüssigkeiten", Chem.-Ing.-Techn., 38, 112-117 (1966).

12. Stephan, K., "On Methods of Studying Heat Transfer in Transition Boiling", Int. J. Heat Mass Transfer, 11, 1735-1736 (1968).

13. Fung, K.K., "Forced Convective Transition Boiling", M.Sc. Thesis, Dep. Mech. Eng., University of Toronto, 1976.

14. Cheng, S.C., Ng , W.W.L., and Heng, K.T., "Measurements of Boiling Curves of Subcooled Water under Forced Convective Conditions", Int. J. Heat Mass Transfer, 21, 1385-1392 (1978).

15. Ragheb, H.S., Cheng, S.C., and Groeneveld, D.C., "Observations in Transition Boiling of Subcooled Water under Forced Convective Conditions", Int. J. Heat Mass Transfer, 24, 1127-1137 (1981).

16. Lee, L., Chen, J.C., and Nelson, R.A., "A Surface Probe for Measurement of Liquid Contact in Film and Transition Boiling on High Temperature Surfaces", Rev. of Scient. Instruments, in press.

17. Rouhani, S.Z., "Void Measurements in the Region of Subcooled and Low Quality Boiling", Part 2, AE-239, 1966.

18. Guerrieri, S.A., and Talty, R.D., "A Study of Heat Transfer to Organic Liquids in Single-tube, Natural-circulation, Vertical-tube Boilers", Chem. Eng. Prog. Symp. Ser., 52, 69-77 (1956).

19. Dengler, C.E., and Addoms, J.N., "Heat Transfer Mechanism for Vaporization of Water in a Vertical Tube", Chem. Eng. Prog. Symp. Ser., 52, 95-103 (1956).

20. Stephan, K., and Abdelsalam, M., "Heat Transfer Correlations for Natural Convection Boiling", Int. J. Heat Mass Transfer, 23, 73-87 (1980).

21. Stephan, K., and Auracher, H., "Correlations for Nucleate Boiling Heat Transfer in Forced Convection", Int. J. Heat Mass Transfer, 24, 99-107 (1980).

22. Pujol, L., and Stenning, A.H., "Effect of Flow Direction on the Boiling Heat Transfer Coefficient in Vertical Tubes", Proc. Int. Symp. Concurrent Gas-Liquid Flow, Univers. Waterloo, Canada, 401-453 (1968).

23. Rohsenow, W.M., "Heat Transfer with Boiling", Modern Developm. in Heat Transfer, Acad. Press, New York, 85-158 (1963).

24. Chen, J.C., "Correlation for Boiling Heat Transfer to Saturated Fluids in Convective Flow", Int. Eng. Chem., Proc. Design Dev., 5, 322-329 (1966).

25. Shah, M.M., "A Generalized Graphical Method for Predicting CHF in Uniformly Heated Vertical Tubes", Int. J. Heat Mass Transfer, 22, 557-568 (1979).

26. Katto, Y., "General Features of CHF of Forced Convection Boiling in Uniformly Heated Vertical Tubes with Zero Inlet Subcooling", Int. J. Heat Mass Transfer, 23, 493-504 (1980).

27. Kutateladze, S.S., "Critical Heat Flux to Flowing, Wetting Subcooled Fluids", Energetica, 2, 1959.

28. Groeneveld, D.C., and Fung, K.K., "Forced Convective Transition Boiling, Review of Literature and Comparison of Prediction Methods", AECL 5543, Chalk River, Canada, 1976.

29. Rankin, S., "Forced Convection Film Boiling Inside Vertical Pipes", Ph.D. Thesis, Iniv. Of Delaware, 1961.

30. Dougall, R.S., and Rohsenow, W.M., "Film Boiling on the Inside of Vertical Tubes with Upward Flow of the Fluid at Low Qualities", MIT Report No. 9079-26, 1963.

31. Groeneveld, D.C., and Gardiner, S.R.M., "Post-CHF Heat Transfer Under Forced Convective Conditions", in: Light Water Reactors, Vol. 1, ASME, 43-73 (1977).

32. Collier, J.G., Convective Boiling and Condensation, Mc Graw-Hill, 2nd Ed. 1981.

Multi-Phase Flow and Heat Transfer III. Part B: Applications
edited by T.N. Veziroğlu and A.E. Bergles
Elsevier Science Publishers B.V., Amsterdam, 1984 — Printed in The Netherlands

NEW TRENDS IN NEUTRON DIAGNOSIS OF MULTI-PHASE FLOWS

Esam Hussein
Department of Engineering Physics, McMaster University
Hamilton, Ontario, Canada   L8S 4M1, and
Nuclear Studies and Safety Department, Ontario Hydro
Toronto, Ontario, Canada   M5G 1X6

ABSTRACT

Neutron methods are very attractive for diagnosis of multi-phase systems. In addition to being non-intrusive, they are suitable for measurements of flows in thick metallic (high pressure) pipes, in packed beds and through rod or tube bundles. A classification of neutron methods is first presented, and then existing techniques are critically reviewed. Two new methods that widen the scope of the neutron probe are introduced. The first method extends the application of neutron diagnosis to multi-phase flow systems by showing a technique for determining the phase content in a steam-water-oil flow. The technique is based on utilizing selective neutron backward scattering to determine the carbon and oxygen content in the system, while using forward scattering to determine the hydrogen content. The second method utilizes the contrast between macroscopic absorption cross section of water and a reference material (heavy-water or graphite) in the neighbourhood of the flow. By measuring the depression of the neutron fluence in the reference material, caused by introducing the flow, the void fraction can be deduced. This method also liberates neutron diagnosis from the conventional use of neutron beams and, consequently, facilitates its use in field applications.

1. INTRODUCTION

The study of multi-phase flows in many industrial processes requires the measurement of flow parameters, such as phase volume fraction, phase distribution and flow rates. The measurements may be made by a variety of techniques as discussed by Hewitt[1] and Banerjee and Lahey[2]. Radiation-based techniques are particularly attractive because they are non-intrusive. Gamma-rays, x-rays and neutrons have been utilized in the diagnosis of two-phase flow systems. Gamma and x-ray techniques depend on the interaction of photons with the electrons of the flow material, consequently providing an indication of the electron density of the system materials. This electron density is then utilized to estimate flow parameters related to the phase content. When metallic walls are present in the flow system and the flow material is hydrogenous (low atomic number), gamma and x-ray techniques lose, to a great extent, their capability of diagnosing the flow. This is because of the large contrast

in the electron density between metals and hydrogenous materials, which
makes radiation much more sensitive to the system walls than the the flow
material. Neutrons, on the other hand, interact with the material nuclei
and are more affected by low mass number materials (such as hydrogenous
fluids), than by large mass number materials (such as metallic walls).
Therefore, neutrons are more suitable for use in flows through thick
metallic (high pressure) pipes, in packed-beds and through rod or tube
bundles.

In this paper, the neutron techniques that are used, or being
considered for use, for diagnosis of multi-phase flow systems are
critically reviewed. The techniques are classified, as shown in Table 1,
according to the nature of the neutronic process utilized in the
technique. The new trends reported in this paper appear in Sections 5
and 6 dealing with the single scattering techniques and fluence
depression techniques, respectively. It is shown how the single
scattering process can be used to estimate the oil content in a
steam-water-oil flow. The fluence depression method is proposed here as
a potential alternative to beam-dependent techniques, since no neutron
beam is required for the depression measurements.

2. TRANSMISSION TECHNIQUES

Transmission techniques are schematically presented by detector 1 of
Figure 1. The transmitted neutrons consist mainly of uncollided
neutrons. However, a number of collided neutrons succeed in reaching the
detector. The number of uncollided neutrons increases exponentially as
the density of the flow in the test section decreases and, therefore, it
can be used to measure the void fraction in two-phase flows. Since the
number of collided neutrons increases as the flow density increases,
their contribution to the neutron detector has to be either minimized or
accounted for.

The number of collided neutrons reaching the detector depends on the
energy of the neutron beam, the size and material of the test section,
the detector aperture, the distance between the test section and the
detector, and the energy of the detected neutrons. The collision
probability can be decreased by increasing the energy of the neutron beam
or by reducing the size of test section and amount of materials used. By
decreasing the detector aperture, the number of collided neutrons seen by
the detector decreases, but this is at the expense of reducing the
detector angle of view and; consequently, only the flow density within
this angle can be estimated. Increasing the distance between the
detector and the test section is probably the easiest way to reduce the
contribution of collided neutrons; this is usually at the expense of the
count rate. Another method of reducing the contribution of collided
neutrons to the detector response is to discard all neutrons of energy
less than that of the neutron beam, since collided neutrons lose part of
their energy during the scattering process.

Thermal neutron transmission techniques have been intensively studied
by Harms and co-workers(3-8). Their studies indicate that an adequate
estimate of the cross-section averaged flow density requires either an a
priori knowledge of the phase distribution, or a series of chordal
measurements to provide the phase profile. Sets of chordal measurements

taken at different orientations can be used to obtain the spatial phase distribution, as shown in Reference 7. Thermal neutron transmission was used for steam-water pipes of no more than 25 mm in diameter. For larger pipe sizes, the contribution of collided neutrons becomes excessive, as compared to the contribution of uncollided neutron which sharply drops.

A study at Grenoble, France(9) has compared the use of beryllium cooled (10^{-4} - 10^{-3} eV), thermal (0.025 eV) and epicadmium (0.5 eV) neutrons, extracted from a swimming pool type research reactor, for void fraction measurements in steam-water pipe of 12 mm in diameter. The study shows that the transmission of either cold neutrons or thermal

TABLE 1

Classification of Neutron Techniques

Technique	Source Energy	Detected Energy	Measured Parameter	Reference
Transmission	Cold	Cold	Void Fraction	9
	Thermal	Thermal	Void Fraction	3-9, 11
	Thermal	Thermal	Void Fraction Distribution	7
	Epi-cadmium	Epi-cadmium + Thermal	Void Fraction	9
	Epi-cadmium	Epi-cadmium	Flow Regime Profile	10
Dispersion	Cold	Cold	Void Fraction	9
	Thermal	Thermal	Void Fraction	9, 12
Multiple Scattering	Epithermal	Thermal	Void Fraction	10, 13-17
Single Scattering	14 MeV	3-12 MeV	Void Fraction Distribution	18-20
	14 MeV	0-12 MeV	Flow Composition	Present Paper
Depression	Thermal	Thermal	Void Fraction	Present Paper
Activation	14 MeV	Gamma Rays	Flow Rate + Flow Density	25-26

neutrons can be used to measure the void fraction, provided that the radial void distribution is known. On the other hand, epicadmium neutron measurements showed an almost flat response to the change in void fraction, and the count rate measured was very small. This is because the amount of water in the 12 mm diameter pipe is not enough to absorb a significant number of fast neutrons. Moreover, the detector used in this study, a boron-lined proportional counter, is not very efficient for fast neutrons. For pipes of larger diameters and using a more efficient detector, such as high pressure Helium-3 detectors, the transmission of epicadmium neutrons can be used for void fraction measurements and flow regime identification, as shown by Banerjee et al(10).

In both the work of Harms and the Grenoble group, neutron beams extracted from reactors were employed. These beams are usually well thermalized, consequently enabling a straightforward thermal neutron transmission measurement. For neutron beams extracted from an isotopic source surrounded by appropriate moderating materials, one has to make sure that the beam does not contain a significant epithermal component and is reasonably well collimated. This is because the epithermal neutrons contribute to the collided detector component when they are thermalized inside the test sections. When well collimated neutrons collide inside the test section, they tend to scatter away from the detector. If the extracted beam is not well collimated more collided neutrons tend to reach the detector. Generally, thermal neutron transmission techniques tend to give the best resolution if the neutron beam used is well thermalized and well collimated. A device based on the thermalization of a Californium -252 source is reported by Frazzoli and Magrini(11).

In addition to their strong dependence on the flow regime, transmission techniques are not very sensitive at low void fractions, because of the exponential nature of their response function. Transmission techniques are successful at high void fractions, provided that the contribution of collided neutrons to the detector response is minimum.

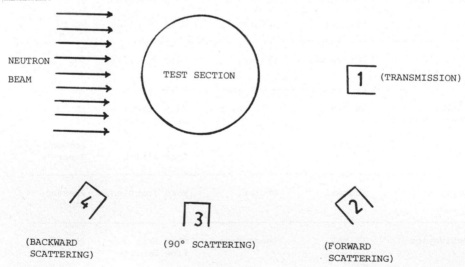

Fig. 1 Schematic Representation of Neutron Methods

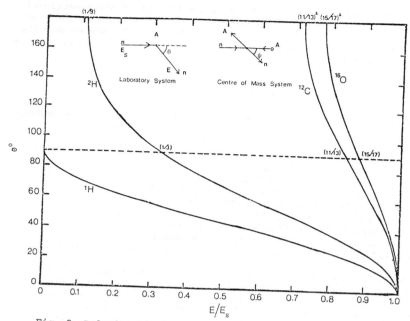

Fig. 2 Relationship between Energy and Angle of Scattering

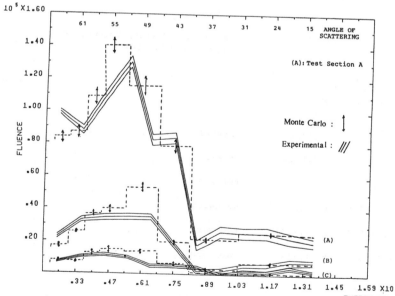

Fig. 3 Comparison between Measured and Calculated
Energy Spectra of Scattered Neutrons

806

material, one can observe a large fluence depression. Since the fluence depression increases with increasing the pipe diameter, the proposed technique can be used effectively for large pipes, particularly with large void fractions. The technique is also effective at high pressures, since Σ_{ao} is proportional to the flow density and f_α decreases with increasing Σ_{ao}, thereby resulting in an increase in the fluence depression.

A schematic diagram of a self-shielded assembly that can be used for measuring the fluence depression near a pipe is shown in Figure 6. Note that one can introduce more than one pipe into the assembly for simultaneous void fraction measurements. The amount of regular moderating material (H_2O in Figure 6), can be increased to accommodate large pipe sizes. The dimensions in Figure 6 are approximate and are used only to give the reader an idea of the size of the proposed device.

One last note regarding the proposed fluence depression technique: the analysis provided here is very simple and by no means accurate. It is only used to introduce the idea of the technique. More accurate

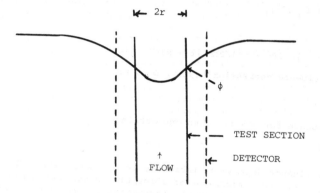

Fig. 4. Fluence Depression caused by Flow

analysis is being undertaken, and the results will be reported in the near future. The effect of the flow regime, pipe wall and detector perturbation will be considered.

7. ACTIVATION TECHNIQUES

In all the aforementioned techniques, neutron diagnosis is based on neutron measurements. However, activation techniques depend on gamma ray measurements, with the latter being produced as a result of the neutron activation of one of the fluid isotopes. Neutron activation is used for flow rate measurements in sodium and water fluids by activating the $^{23}Na(n,\alpha)^{20}F$ and $^{16}O(n,p)^{16}N$ reactions, respectively[24-28]. The first reaction produces 1.63 MeV photons by the decay of ^{20}F (11.2 s half life), while the second reaction results in 6.13 MeV photons by the decay of ^{16}N (7.13 s half-life). In order for these reactions to occur, a high energy neutron source is required. Pulsed 14 MeV neutron sources are used for flow rate measurements.

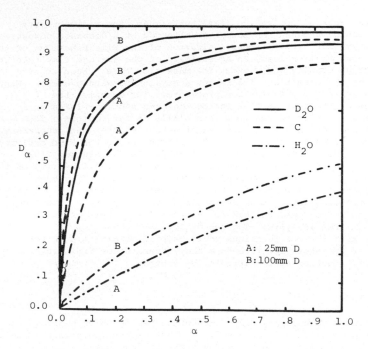

Fig. 5 Fluence Depression D_α vs. Void Fraction α

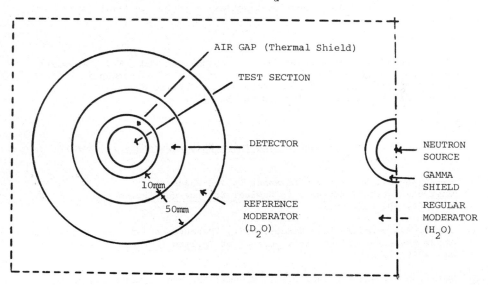

Fig. 6 A Schematic Design of an Assembly for Fluence Depression
Measurement for Void Fraction Determination

808

In Pulsed Neutron Activation (PNA) techniques, the radioactivity induced by the pulsed neutron source is observed at a distance downstream from the activation site. Proper evaluation of the time behaviour of the count rate distribution leads to an estimate of the average flow velocity and the average density and, hence, the mass flow rate. However, the evaluation of the mass flow rate is not a straightforward process. Many factors that affect the evaluation procedure. For example, the distribution of the neutron flux inside the test sections leads to a non-uniform activation of the flow, especially in large pipes. The photons are also attenuated before reaching the detector and, therefore, the count rate depends on how close the detector is to the high activity regions of the flow. The flow regime also affects the neutron flux distribution and the gamma attenuation. The molecular diffusivity of the radionuclides also affects the measurements, particularly for high temperature and low flow rate fluids in long pipes. Also, the velocity profile of the flow affects the behaviour of the count rate with time. Because of these reasons, one has to be careful in interpreting the results of the PNA analysis. The use of the SENT technique in conjunction with the PNA technique could help in obtaining a better estimation of the mass flow rate from the latter, since SENT can provide the phase distribution at the irradiation section.

8. CONCLUSIONS

The paper presented many different approaches by which neutrons can be used for diagnosing multi-phase flows. The discussion presented shows that neutron diagnosis is in its early state of development. There is still room for improvement and innovation. More work is needed to further explore the field of neutron diagnosis and provide fresh ideas and concepts.

ACKNOWLEDGEMENTS

The discussions with D.L. Bot of Bot Engineering Ltd., M.Z. Farooqui of Ontario Hydro and P. Yuen of WNRE have led to the new concepts introduced in this paper. Mrs. L. Ferguson of Ontario Hydro and co-workers are to be thanked for typing this paper.

REFERENCES

1. Hewitt, G.F., Measurement of Two-Phase Flow Parameters, Academic Press, London, 1978.

2. Banerjee, S., and Lahey, "Advances in Two-Phase Flow Instrumentation", Advances in Nuclear Science and Engineering, Vol. 60, pp. 851, 1981.

3. Harms, A.A., Lo, S., and Hancox, W.T., "Measurement of Time-Averaged Voids by Neutron Diagnosis", Journal of Applied Physics, Vol. 42, pp. 4080, 1971.

4. Harms, A.A., and Forrest, C.F., "Dynamic Effects in Radiation Diagnosis of Fluctuating Voids", Nuclear Science and Engineering, Vol. 46, pp. 408, 1971.

5. Hancox, W.T., Forrest, C.F., and Harms, A.A., "Void Determination in Two-Phase Systems Employing Neutron Transmission", Proc. AIChE-ASME,

National Heat Transfer Conference, Paper No. 72-HT-2, Denver, Colorado, 1972.

6. Younis, M.H., Harms, A.A., and Hoffmann, T.W., "Source Fluctuation Effects in Radiation Diagnosis of Voided Fluidic Systems, Nuclear Engineering and Design, Vol. 24, pp. 145, 1973.

7. Zakaib, G.D., Harms, A.A., and Vlachopoulos, J., "Two-Dimensional Void Reconstruction by Neutron Transmission", Nuclear Science and Engineering, Vol. 65, p. 145, 1978.

8. Younis, M.H., Hoffmann, T.W., and Harms, A.A., "Neutron Diagnosis of Two-Phase Flow", Nuclear Instrumentation and Methods, Vol. 187, pp. 489, 1981.

9. Delhaye, J.M., "The Latest Grenoble Advances in Two-Phase Flow Instrumentation", ANS/ASME/NRC International Topical Meeting on Nuclear Reactor Thermal Hydraulics, Saratoga Springs, New York, 1980.

10. Banerjee, S., Chan, A.M.C., Ramanathan, N., and Yuen, P.S.L., "Fast Neutron Scattering and Attenuation Techniques for Measurement of Void Fractions and Phase Distribution in Transient Flow Boiling", Proceedings 6th International Heat Transfer Conference, Vol. 1, pp. 351, 1978.

11. Frazzoli, F.V., and Magrini, A., "Neutron Gauge for Measurement of High Void Fraction in Water-Steam System", Nuclear Technology, Vol. 45, pp. 177, 1979.

12. Archer, P., and Harms, A.A., Void-Distribution Effects in Void Measurements Using Scattered/Uncollided Neutrons", Trans. ANS, Vol. 23, pp. 983, 1979.

13. Banerjee, S., Hussein, E., and Meneley, D.A., "Simulation of a Neutron Scattering Method for Measuring Void Fraction in Two-Phase Flow", Nuclear Engineering and Design, Vol. 53, pp. 393, 1979.

14. Banerjee, S., Yuen, P., and Vandenbroek, M.A., "Calibration of a Fast Neutron Technique for Measurement of Void Fraction in Rod Bundles", Journal of Heat Transfer, Vol. 101, No. 2, pp. 295, 1979.

15. Rousseau, J.C., Czerny, J., and Riegel, B., "Void Fraction Measurement During Blowdown by Neutron Absorption in Scattering Methods, presented at OECD/NEA Specialists Meeting on Transient Two-Phase Flow, Toronto, August 1976.

16. Frazzoli, F.V., Magrini, A., and Mancini, C., "Void Fraction Measurement in Water-Steam Mixture by Means of a Californium-252 Source, Journal of Applied Radiation and Isotopes, Vol. 29, pp. 314, 1978.

17. Yuen, P.S.L., Ph.D. Thesis, McMaster University, under preparation.

18. Hussein, E., Banerjee, S., and Meneley, D.A., "A New Fast Neutron Scattering Technique for Local Void Fraction Measurement in Two-Phase Flow", Proceedings of 2nd International Topical Meeting on Nuclear Reactor Thermal Hydraulics, Santa Barbara, Calf., pp. 1431, 1983.

19. Hussein, E., Bot, D.L., Banerjee, S., and Meneley, D.A., "Design
    Aspects of a Fast Neutron Scattering Technique for Phase Distribution
    Measurement in Two-Phase Flow", submitted for presentation at IUTAM
    Symposium on Measuring Techniques in Gas-Liquid Two-Phase Flows,
    Nancy, France, 1983.

20. Hussein, E., "Fast Neutron Scattering Method for Local Void Fraction
    Distribution Measurement in Two-Phase Flow", Ph.D. Thesis, McMaster
    University, to be issued, 1983.

21. Miller, W., and Meyer, W., "Standardization of Neutron Spectrum
    Unfolding Codes", Nuclear Instrumentation and Methods, Vol. 205,
    pp. 185, 1983.

22. Bot Engineering Ltd., 7393 Twiss Road, Campbellville, Ontario, Canada.

23. Weinberg, A.M., and Wigner, E.P., The Physical Theory of Neutron
    Chain Reactors , University of Chicago Press, 1958.

24. Kehler, P., "Pulsed Neutron Activation Techniques for the
    Measurements of Two-Phase Flow", Proc. USNRC Review Group Conf.
    Advanced Instrumentation for Reactor Safety Research, NUREG/CP-0007,
    US Nuclear Regulatory Commission, 1979.

25. Perez-Griffo, M.L., Block, R.C., and Lahey, R.T., "Measurement of
    Flow in Large Pipes by the Pulsed Neutron Activation Method", Nuclear
    Science and Engineering, Vol. 82, pp. 13, 1982.

26. Larson, H.A., Price, C.C., Curran, R.M., and Sachett, J.I., "Flow
    Measurement in Sodium and Water Using Pulsed-Neutron Activation",
    Nuclear Technology, Vol. 57, pp. 57, 1982.

Multi-Phase Flow and Heat Transfer III. Part B: Applications   811
edited by T.N. Veziroğlu and A.E. Bergles
Elsevier Science Publishers B.V., Amsterdam, 1984 — Printed in The Netherlands

OPTIMIZING THE COMPTON PROFILE APPROACH TO TWO-PHASE
DISTRIBUTION MEASUREMENTS

Samim Anghaie, Kenneth D. Finlon, Alan M. Jacobs
Department of Nuclear Engineering Sciences
University of Florida
Gainesville, Florida 32611, U.S.A.

ABSTRACT

A new non-intrusive technique, which we have termed "Compton Profile
densitometry" offers a unique possibility of accurate measurement of void
fraction distribution as well as real-time numerical visualization of flow
patterns in optically opaque two-phase systems. This approach is based on
high-resolution energy discrimination of scattered, emerging, detected radiation. Relative positioning of illuminated chord and detector can essentially eliminate the contribution of multiple scattering to the detected
spectrum over portions of the relevant spectrum interval. For a fixed
channel geometry, decreasing the detector view angle of once-scattered
photons from the illuminated diameter of the pipe results in a decrease
in the multiple-scattering factor. The optimum choice of the detector position which would yield the highest quality data and lowest multiple scattering,
seems to be in the most forward scattering angle direction. However, sensing the most backward scattered radiation is a very efficient arrangement
for measurement of the density distribution in the half diameter of the
pipe, ranging from the beam entry point to the pipe centerline axis.

1. INTRODUCTION

Utilizing gamma-rays in designing two-phase flow sensing techniques has
been widely accepted. Non-intrusive measurement of a variety of two-phase
flow parameters by employing a system of external gamma-ray sources and detectors in a variety of configurations has been suggested and tested. The source,
two-phase flow system and detector geometrical arrangement which have been
used by a large group of investigators, for measurements of two-phase flow
parameters include:

1. Collimated beam transmission technique for the chordal average void
    fraction measurements [1-5].

2. Collimated, few-beam transmission technique for chordal average void
    fraction and discrimination between some of the more important
    regimes [6-11].

3. Transmission for reconstructive tomography for the measurement of
    density distribution in pipes [12-14].

4. Collimated source and detector using the scattered field for local
    void fraction measurements [15-20].

An extensive review of these techniques, including their mathematical descriptions and experimental implementation, is given in Reference [21].

Figure 1 depicts the schematic set-up for the conventional gamma-ray techniques. The geometric arrangements in these techniques are such that either: (1) unscattered radiation transmission is sensed and thereby only illuminated section average density parameters are deduced, or (2) singly-scattered radiation is sensed through a highly-delineating geometric collimator thereby allowing measurements of a local parameter in a single internal location. The employment of very restrictive collimation systems in these techniques introduces additional systematic or statistical errors as well as having the drawbacks of high cost and poor time response. Often, these restrictions combined with inadequate interpretation of radiation transport in the fluid field have caused gamma-ray methods to be criticized as being more a qualitative indicator of two-phase flow parameters rather than a quantitative measurement device.

Limitations of conventional gamma-ray techniques have motivated the development of Compton scattered spectrum profile techniques which seem to yield significantly more information. In this technique detector selectivity of photon energy and photon kinematic relations replaces conventional field-of-view collimator density field location discrimination.

The multiply-scattered, detected radiation can be successfully removed by calculational techniques for unraveling the two-phase density distribution. However, the unavoidable presence of the multiply-scattered radiation in the acquired spectrum may cause "blurring" of the numerically visualized flow re-

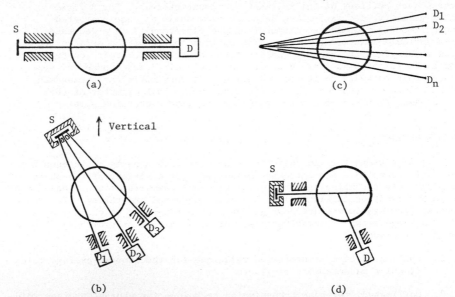

Figure 1. Conventional Sources/Detectors Configurations for Gamma-Ray Densitometry; (a) Single Collimated Beam and Collimated Detector Technique, (b) Collimated Few-Beam and Detectors Technique, (c) Single-Source, Multiple-Detectors for Reconstructive Tomography, and (d) Collimated Source and Detector Using the Scattered Field.

gimes. Moreover, if the multiple-scattered photon field is dominant, the method becomes intractable. A detailed theoretical and experimental treatment of this technique, including a full analysis of a particular experiment, is given in Reference [21] and [22]. This paper is devoted to optimization of the once-scattered-to-multiple-scattered detected intensity ratio as well as the spectral nature of the multiply-scattered field.

2. THEORY

A particular variant of the densitometry technique is illustrated in Figure 2. A radioactive source, S, emits gamma-rays which are collimated to a narrow beam before illuminating a chord of a cylindrical object representing a distribution of liquid and vapor. An energy sensitive gamma detector, D, is illustrated with a collimator of sufficiently wide angle to view the entire illuminated chord.

For the source-detector configuration shown in Figure 2, presuming that only singly-scattered photons are detected, there is a one-to-one correspondence between locations, Z, along the illuminated diameter of the pipe and energies, E, of detected gamma rays, i.e., the Compton scattering kinematics relations

$$E(\theta) = E_o/[1 + a\ (1 - \cos\theta)] \tag{1}$$

where E_o = gamma-ray source energy (monoenergetic source)

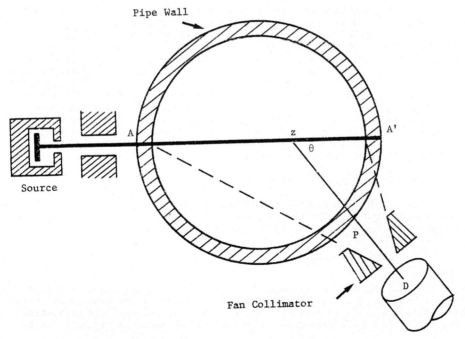

Figure 2. A Typical Geometrical Arrangement for Compton Profile Densitometry.

a = gamma-ray source energy in the units of the electron rest mass

energy (i.e., $m_o c^2$ units) = E_o(keV)/511

θ = polar scattering angle = $\tan^{-1}[h/(Z_D - Z)]$

Z_D = detector coordinate on z-axis

h = normal distance from the detector to z-axis

Photons proceed from source point S to scattering volume centered at Z via radiation beam path SAZ. Those photons which scatter with polar angle θ (and requisite azimuthal angle) proceed to detector point D via radiation beam path ZPD. The locations A and P represent the respective points of rays entering and leaving the system of the two-phase flow. As the interrogation point Z proceeds from point A to point A', the angle of scattering θ monotonically increases. Thereby, the scattered photon energy from point Z monotonically decreases.

In this method of delineating transit path by energy history, the equivalent of direction selectivity of a geometric restricting collimator is the energy resolution of the detector. Using equation (1) and the geometry of Figure 2, the spatial resolution element, δ, is given by

$$\delta = [h/a \sin^3\theta][1 + a(1 - \cos\theta)] \, \epsilon \qquad (2)$$

where ϵ is the fractional energy resolution of the detector, h is the normal distance from D to Z-axis, and θ is the photon scattering angle which essentially identifies the location of point Z. In order to demonstrate the magnitude involved, consider the numerical result for a configuration where h is approximately the radial dimension of the pipe, R, θ is approximately 90°, and E_o is about 500 keV. For this case, the fractional spatial resolution of the measured void fraction distribution, i.e., δ/R, is about 2ϵ. Clearly, NaI(Tℓ) crystal detectors, with typical ϵ = 0.1, yield insufficient spatial resolution for consideration in most application of interest. However, solid state detector-cyrostat systems, with characteristic ϵ ranging from .01 to .001, provide ample spatial resolution. In fact, high-resolution solid state detectors apparently yield far better resolution in spatial location than any conventional geometric restricting collimator system for photons with energy above 500 keV.

2.1 Two-Phase Distribution by Compton Profile

Denoting the density distribution along the illuminated z-axis by $\rho(Z)$ and the unconvoluted detector response due to scattered photons of energy E by D(E), there is a first order straight-forward relation

$$\rho(Z) = \frac{\Lambda_e}{F} \frac{D(E)}{B(E)} \qquad (3)$$

where F includes the source strength, beam collimation description, detector efficiency, and Klein-Nishina differential Compton scattering cross section for a polar angle resulting in detector entrance; B(E) is the multiple-scattering factor; and Λ_e for the source-detector configuration shown in Figure 1 can be written as

$$\Lambda_e = \Lambda_o(Z) + \Lambda_s(Z)$$

$$= \exp[\mu(E_o) \int_0^Z \rho(z)ds + \mu(E) \int_Z^\rho \rho(s)ds] \qquad (4)$$

where $\mu(E_o)$ and $\mu(E)$ are the total attenuation coefficient of primary and scattered photons, respectively. Employing the Klein-Nishina formula for free-electron Compton scattering cross section, the factor F may be expressed as[1]

$$F = \omega(E) \; S \; A_D \; n \; (Z_e N_o/M)(h \; \sin\theta)^{-1}(r_o^2/2)$$

$$x \; [1 + a(1-\cos\theta)]^{-2} \; [1 + \cos^2\theta + \frac{a^2(1 - \cos\theta)^2}{1 + a\;(1-\cos\theta)}] \qquad (5)$$

where $\omega(E)$ is the detector efficiency for photons of energy E, S is the total source strength (photons/sec), A_D is the planar detector effective area, n is the cosine of the detector reception angle, Z_e is the atomic number, N_o is Avogadro's number, M is the atomic mass number, h is the normal distance from the detector to z-axis, θ is the polar scattering angle, r_o is the classical electron radius (2.818×10^{-15}m) and a is the primary photon energy in units of the electron rest mass energy.

The actual measurement results of a particular approach to the Compton profile technique for several two-phase distribution simulations are shown in Figures 3 through 7 [23,24]. In this particular experiment, various steady-

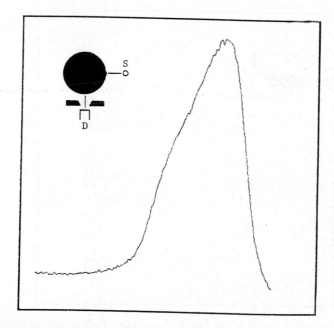

Figure 3. Scattering Profile for a Single-Phase Flow Simulation

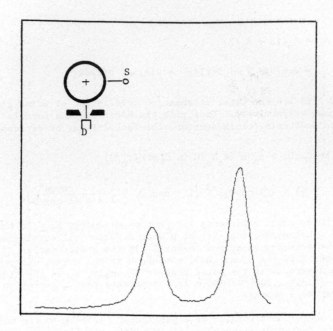

Figure 4.   Scattering Profile for an Annular Flow Simulation.

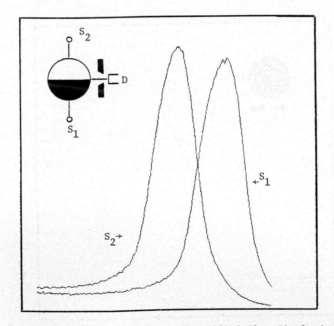

Figure 5.   Scattering Profile for a Stratified Flow Simulation When
the Source is Positioned Above, and Then Below the Pipe.

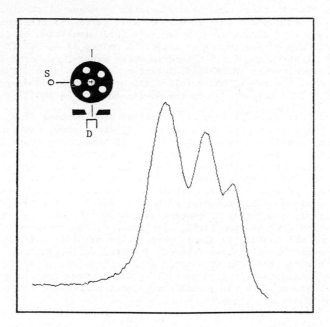

Figure 6.   Scattering Profile for Pentagonal Void-Array.

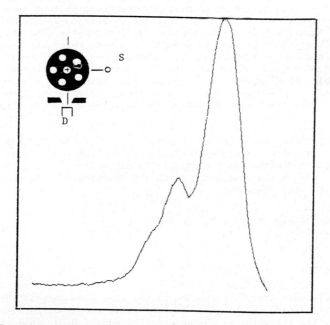

Figure 7.   Scattering Profile for Pentagonal Void-Array.

two-phase flow distributions have been simulated by means of machined Lexan pieces and air. Lexan has a mass density of 1.13 gm/cc and electron population density of 3.59×10^{23} electrons/cc which is a good simulation of gamma-ray attenuation and Compton scattering by water. A source of Am-241 (E_o = 59.54 Kev) with a total strength of 1.5 Ci and an Ortec Hyper-Pure Germanium Low-Energy photon spectrometer (HPGe-LEPS) detector have been used. The HPGe-LEPS detector and amplifier system has a full-width-at-half-maximum energy resolution of $\Delta E = 0.35$ Kev for photons in the Am-241 energy region ($\Delta E/E \simeq 0.007$).

Figure 3 is the spectrum from the simulated single-phase flow; Figure 4 that from the simulated annular flow regime. Figure 5 shows the spectra from a simulated stratified flow when the source is positioned above, and then below the pipe. Figures 6 and 7 are the spectra obtained from a pentagonal void-array phantom. The voids are circular and approximately two-tenths of the phantom diameter is dimension. Figure 6 is for a void near the source; Figure 7 far from the source. These spectra demonstrate the ability of the Compton profile measurement technique while providing a clear indication of the magnitude and effects of multiple scattering of photons on the quality and resolution of the delineated density distribution field. In Figure 7, the presence of a void, located far from the source, is almost obscured by multiple scattering radiation. Even though the multiply-scattered, detected radiation blurring of the density distribution field can be removed by calculational techniques [21,22], optimization of the geometric arrangement, incident gamma-ray energy and illumination beam shape may provide a real time, and accurate numerical visualization of the two-phase distribution.

2.2 Multiple Scattering

The Compton scattering profile measurement of a two-phase distribution system depends primarily on those photons which collide only once before reaching the detector. The detection of multiply-scattered photons is accounted for and can be removed by introduction of multiple scattering factor B(E) in two-phase density equation. To solve equation (3) for density distribution $\rho(Z)$, the explicit description of B(E) is needed. For many relevant cases, B(E) is not a strong function of the details of the density distribution in the scattering field; therefore, it can be accurately approximated in terms of the average density of fluid. For cases of a two-phase flow regime with distinctive liquid-vapor phases (i.e., fully developed annular flow or stratified flow), the multiple-scattering factor can be calculated in terms of the density of each phase.

The Fano-Spencer-Berger method of orders-of-scattering [25] can be applied to the problem of photon multiple scattering in the Compton profile method. The orders-of-scattering method is essentially an expansion of the photon flux, $\Phi(\underline{r}, E, \underline{\Omega})$, in terms of the components which are unscattered, once-scattered, twice-scattered, etc. That is,

$$\Phi(\underline{r}, E, \underline{\Omega}) = \sum_{J=0}^{\infty} \Phi^{(j)}(\underline{r}, E, \underline{\Omega})$$

$$= \Phi^{(0)} + \Phi^{(1)} + \Phi^{(2)} + \ldots$$

(6)

where $\Phi^{(j)}$ represents that contribution to the flux from photons which have suffered j scatterings. $\underline{r}$, E and $\underline{\Omega}$, respectively, represent location, photon energy, and photon trajectory. The detector response, D(E), is a simple function of $\Phi(E)$ involving $\underline{r}$ and $\underline{\Omega}$ integrations over the detector exposed location

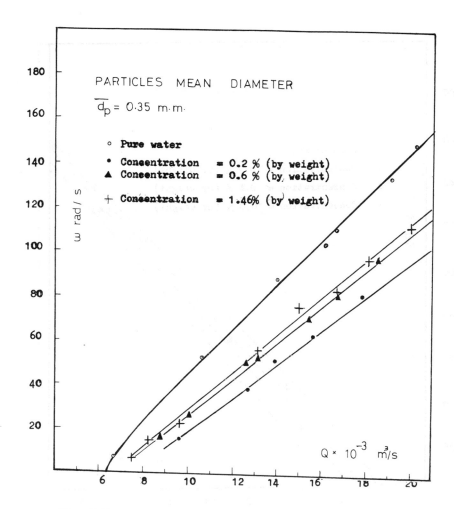

Fig. 3 The effect of concentrations on the performance of
Turbine Flowmeter(Particles diameter =0.35mm)

blades and particles is increased. It could be said that the drag reduction is encountered by the losses due to the shock between particles and blades.

The concentration in this investigation is kept as low as possible. The upper limit of which is 1.5% by weight (0.5% by volume), usually this is the maximum concentration tolerated in the meter operation. Which is important to note is that the maximum drop is encountered with the lowest concentration. At relatively low flow rate this can alter the reading of meter by 15%.

The effect of grain size on the meter is shown in figure(5). The results of figure (5) are expected from the experiments in the field of two phase flow [4] . At low concentration with mean particles diameter lower than the microscale of length of the turbulent, the particles do not affect the mechanism of flow or its influence is small. The influence is increased with increasing the diameter.

5. CONCLUSION

Based on this investigation the following conclusions and recommendations are offered:

1. The performance of the turbine flowmeter is affected by the presence of solid materials in very low concentration. The maximum deviation is observed at the lowest concentration of(0.2% by weight). Increasing concentration reduces the deviation.

2. The effect of the presence of solid materials in general is dependent on the grain size. Increasing the particles diamater increases the deviation. This is confirmed by experiments in the field of multi phase flow.

REFERENCES

1. Hochreiter, H.M., "Dimensionless Correlation of Coefficients of Turbine Type Flowmeters", Trans. ASME. PP. 1363-1368. October 1958.

2. Comolet, R., "Theorie de la Turbine Debimetrique", Romanian Journal of Techanical Sciences, Applied Mechanics, Vol. 15, No. 5, PP. 1175-1188, Bucarest, 1970.

3. Rayan, M.A., and Mansour, H., "Performance of Turbine Flowmeter Handling Water and Water-Solid Mixture" Proceedings of 16th Southeastern Seminar on Thermal Sciences, Flo., U.S.A. April 1982.

4. Bouvard, M., Pethovie, "Modification des Characteristiques d'une Turbulence Sous l'influence de Particules Solids en Suspension" La Houille Blanche No. 1 (1973).

Multi-Phase Flow and Heat Transfer III. Part B: Applications   851
edited by T.N. Veziroğlu and A.E. Bergles
Elsevier Science Publishers B.V., Amsterdam, 1984 — Printed in The Netherlands

THERMAL ANALYSIS OF THE FUEL BEARING PAD IN THE CANDU REACTOR -
PREDICTION OF NUCLEATE BOILING

M.H. Attia and N. D'Silva
Tribology and Mechanical Processes Unit
Mechanical Research Department
Ontario Hydro, Ontario, Canada

ABSTRACT

A solution approach has been developed to determine the
thermal response of a CANDU fuel bearing pad and the extent of
the nucleate boiling region. This approach recognizes the fact
that local boiling is both controlling and being controlled by
the conditions of heat transfer at the boundaries. The finite
difference model accounts for the volumetric effect of the
thermal contact resistance at the bearing pad/pressure tube
interface. Information on the distribution of the coefficient
of heat transfer over water-cooled surfaces, which is not
available in the literature, has been generated. Analysis of
the results indicated the importance of considering the nonlinear
behaviour of the system in predicting its state of equilibrium.

1. INTRODUCTION

Due to the repeated movement of the fuel bundle in the CANDU
reactor, and in order to space the bundle from the pressure tube,
bearing pads are attached to the outer elements. A proper design
of the bearing pad requires a prior knowledge of the conditions
that lead to nucleate boiling in the crevice region between the
bearing pad and the pressure tube (Figure 1). Therefore, the
present work is devoted to developing a heat transfer analysis
model which allows us to predict the occurrence of nucleate
boiling in the crevice region formed by any bearing pad configuration and the pressure tube. The proposed model is based on a
two-dimensional, steady state, finite difference idealization.

Due to the fact that there is no data available in the
literature pertaining to the distribution of the coefficient of
heat transfer h along the boundary of the bearing pad, a wide
range of h-values have been assumed in the previous works cited
in [1] and shown on Figure 4 (cases A to C). Analysis of these
results indicated that compatibility between the assumptions,
on which the analysis is based, and the results obtained is not
satisfied. This manifests itself in the following observations:

  i - A condition of single phase was assumed in the
      crevice region; nevertheless, nucleate boiling
      was predicted within gaps smaller than $\sim$ 100 µm.

ii - A condition of fully developed nucleate boiling
was imposed on the boundaries of the subchannel.
Results, however, indicated that partially
developed boiling (or rather a single phase regime)
is expected.

In the present study, more consideration is given to the
modelling of the problem, and the solution approach, to surmount
these limitations. Here, the conditions of heat transfer at the
water-cooled surfaces are neither assumed nor imposed on the
system. On the contrary, they are treated as unknowns and are
considered as an integral part of the sought-for solution.
This concept will be expanded upon in section 4.

2.   STATEMENT OF THE PROBLEM

For a proper idealization of the problem at hand, three
different regions should be considered to express the thermal
boundary conditions;

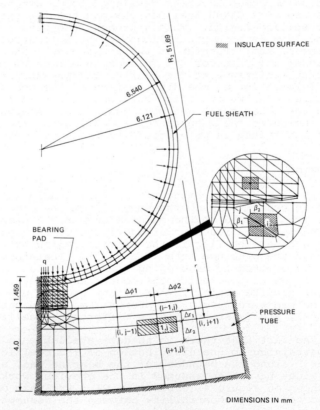

FIGURE 1

NODAL REPRESENTATION OF THE SHEATH - BEARING PAD — PRESSURE TUBE DOMAIN
FINITE DIFFERENCE APPROXIMATION OF TWO - DIMENSIONAL HEAT CONDUCTION

- the contact region at the bearing pad/pressure tube interface,

- the crevice region, formed by the radiused surface of the bearing pad and the pressure tube, and

- the outer subchannel, bounded by the sides of the bearing pad, the sheath and the pressure tube.

The problem is solved, therefore, in two stages. First, estimating the micro- and macro-contact configuration to define the various regions, and the distribution of the thermal contact resistance. Second, determining the temperature field and the distribution of the heat flux along the water-cooled surfaces.

The two-dimensional, steady state temperature field within the domain of the bearing pad and the pressure tube has been determined using the finite difference method, whose nodal network representation is shown in Figure 1. Due to symmetry, only a half of the bearing pad and the sheath, and a sector of the pressure tube are considered. By definition, these lines of symmetry are adiabatic.

The analysis is based on the following assumptions:

1. The apparent contact area A_a is determined by Hertzian theory which assumes that contacting solids are perfectly elastic and their principal curvatures are large relative to the characteristic dimensions of A_a.

2. The microscopic contact configuration, which defines the thermal contact resistance, is based on the assumption that the heights of the surface asperities are Gaussian and are randomly distributed over the apparent contact area.

3. The mass flow in the narrow crevice is governed by Reynold's theory; the cooling water is Newtonian, incompressible and of constant viscosity. It also assumes that there are no surface tension, gravitational or inertia effects.

4. The heat input from the fuel is uniformly distributed over the sheath I.D., at a rate of 1100 kW/m^2.

5. The usual correlations for nucleate boiling are applicable to narrow gaps. It is also assumed that a wide range of nucleate sites exist on the surfaces forming the crevice. Consequently, the onset of nucleate boiling is independent of the surface conditions.

6. The change in the saturation temperature, due to the change in the concentration of LiOH, resulting from boiling is neglected.

3. FORMULATION OF THE PROBLEM

3.1 The Temperature Field in the Fuel Sheath - Bearing Pad - Pressure Tube Domain

As shown in Figure 1, the two-dimensional heat flow field is divided into a grid, whose size is varied in relation to the expected pattern of temperature gradient. Apart from the disturbed contact zone, two different types of elements are used to conform to the boundaries of the domain and to accommodate elements of different sizes; the triangular and the annual segment elements.

The energy balance on the cylindrical elemental volume surrounding each nodal point (i, j), shown on Figure 1, results in a set of finite difference equations of the form [2]:

$$C_1 \cdot t_{i-1,j} + C_2 \cdot t_{i+1,j} + C_3 \cdot t_{i,j-1} + C_4 \cdot t_{i,j+1} + C_n \cdot t_{i,j} + D = o \quad (1)$$

where, $C_1 = \dfrac{1}{2\Delta r_1} (r - \dfrac{\Delta r_1}{2}) (\Delta\phi_1 + \Delta\phi_2)$

$C_2 = \dfrac{1}{2\Delta r_2} (r + \dfrac{\Delta r_2}{2}) (\Delta\phi_1 + \Delta\phi_2)$

$C_3 = \dfrac{\Delta r_1 + \Delta r_2}{2 \cdot r \cdot \Delta\phi_2}$, $C_n = \sum\limits_{i=1}^{4} C_i$

$D = \dfrac{q \cdot a_{i,j}}{k_s}$

where, q, $a_{i,j}$ = the input heat flux, and the elemental area surrounding the nodal point, respectively.

k_s = the thermal conductivity of the solid.

As for the triangular element, the energy balance for each nodal point i (see the insert of Figure 1), results in the following relation:

$$\sum\limits_{j=1}^{n} C_{ij} t_j + C_n t_i + D = o \quad (2)$$

where, j represents all neighbouring nodes with which the node i exchanges heat

$C_{ij} = \dfrac{\cot \beta_1 + \cot \beta_2}{2}$

$C_n = \sum\limits_{j=1}^{n} C_{ij}$

$D = \dfrac{q \cdot a_i}{k_s}$

where, a_i = the area of the polygon surrounding node i.

The set of equations obtained by applying Eqs. 1 and 2 for various nodal points are then solved by the Gauss-Seidel iteration method. Convergence has been accepted when the absolute difference, for each temperature, between two successive iterations is <0.0005; i.e. till converging to three significant digits.

The computer code has been tested for both the bearing pad and the pressure tube domains. By imposing isothermal conditions on the pressure tube surfaces, the results were compared to the known analytical solution [3], and the relative error was found to be <0.0025.

3.2 The Boundary Conditions

The contact region. The thermal contact resistance is a volumetric effect, and not a surface effect as it is usually treated. For an accurate prediction of the thermal response of the bearing pad, especially when the physical dimensions are relatively small and the temperature gradient is steep, the contact zone should be properly modeled to maintain the physical characteristics of the thermal constriction resistance phenomenon. It has been observed that the temperature distribution in a plane perpendicular to the interface exhibits a 'pseudo' temperature drop Δt_c between extrapolated values on either side of the interface. In an idealized situation, the so called 'distributed zone' can be replaced by the unidimensional heat conduction element. Within this volume, the thermal contact resistance R_c is contained,

$$R_c = \frac{1}{h_c} = \frac{\Delta t_c}{q} \tag{3}$$

Following Fenech and Rohsenow's theory [4], the width of the disturbed zone can well be approximated as two- to three-fold the average radius of contact r_c [1].

Knowing the local contact pressure, the standard deviation of surface asperity heights σ and the mean surface facets slope $\overline{\tan \theta}$, the micro-contact configuration can be defined in terms of the average radius of contact r_c, the ratio of the actual-to-apparent contact areas ε and the equivalent gap thickness between surface asperities in contact Δ. These parameters that are estimated from the relations developed by R.K. Dukiewicz [5], T. Tsukizoe, et al [6], Sanokawa [7], and T.N. Veziroglu [8], can be expressed in a simplified functional form as follows:

$$r_c = f \{p_c^+, p_m^-, \overline{\tan \theta}^-, \sigma^+\} \tag{4}$$

$$\varepsilon = f \{p_c^+, p_m^-\} \tag{5}$$

$$\Delta = f \{p_c^-, p_m^+, \sigma^+\} \tag{6}$$

where p_m denotes the flow pressure of the metal, which depends on the slope of the asperities, the coefficient of friction and the work hardened layer. Eq. 4 shows that the local contact pressure p_c; for example, has a positive effect on r_c. In other words, the increase in p_c leads to an increase in r_c. This sign convention will be used throughout this paper. The theory developed by K. Sanokawa [7,9] is used to determine the distribution of the thermal contact resistance R_c. In a simplified functional form, the local value of R_c is related to the micro-contact configuration and the material properties as follows:

$$R_c = f \{ (\Delta^+, r_c^-, \varepsilon^-), (k_s^-, k_f^-) \} \tag{7}$$

where k_s and k_f symbolize the thermal conductivities of the solid and the interstitial fluid, respectively. For different $\bar{k}=k_f/k_s$ ratios, the dependence of the thermal contact resistance on $\bar{\Delta}$, r_c and ε is presented on Figure 2, in terms of the equivalent additional length ℓ of a thermal resistor made of the same material as the parent solid bodies.

To test this theory, the experimental results reported by Celli [10] for Zr/Zr contact in air; ie, for $\bar{k} = 0.0026$, indicated that under a contact pressure of ~0.25 kg/mm², the equivalent length ℓ equals ~3.5 mm of Zr. This value is compared to 5.5 mm of Zr predicted by Sanokawa's theory. The ratio between these two values, that are in a good agreement, is taken as a correction factor for the prediction of R_c in water. From Figure 2, the equivalent length ℓ for the case of Zr/Zr in water, where $\bar{k} = 0.03$, is approximately 0.46 mm. This value is in very good agreement with Vezrioglu's empirical correlation [8] which predicts ℓ ~0.44 mm.

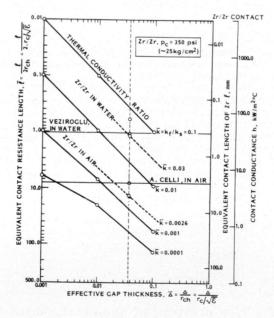

FIGURE 2

THERMAL CONTACT RESISTANCE AS A FUNCTION OF
THE MICRO-CONTACT CONFIGURATION AND THE
THERMAL CONDUCTIVITY RATIO $\bar{k}$

The crevice region. The flow of a fluid film in a narrow
passage can be described by Reynold's theory. In its reduced
form, the distribution of the flow velocity v is related to the
pressure gradient along the crevice dp/dy, by the following
equation which satisfies the equilibrium of the pressure and
viscous forces:

$$\frac{d^2v}{dx^2} = \frac{1}{\mu} \cdot \frac{dp}{dy} \tag{8}$$

For thin film layers, the ratio of the heat transferred by convection to that transmitted by conduction across the film can be
estimated by calculating the weighted Peclet Number Pè [11]:

$$P\grave{e} = \frac{\bar{v} \cdot H^2}{\alpha \cdot L} \tag{9}$$

where $\bar{v}$ = the average velocity at a given film thickness H,
obtained by virtue of Eq. 8

α = the fluid thermal diffusivity

L = the length of contact.

For the given design conditions, and for crevice heights H in
the range of up to 65 μm, Pè is estimated to be approximately
zero, implying that unless nucleate boiling is taking place,
conduction is the dominant mode of heat transfer across the
crevice.

The outer subchannel region. Within the outer subchannel
region, the coefficient of heat transfer $h_{s/c}$ depends on whether
the flow is single- or two-phase. The available correlations
for single phase liquid flow inside tubes, h_t, can be used to
estimate $h_{s/c}$, knowing the two ratios $h_{s/c}/h_b$ and h_b/h_t, where
h_b stands for the average coefficient of heat transfer of the
bundle. For the given design data, these ratios have been found
to be approximately unity [12,13]. Therefore,

$$h_{s/c} \sim h_b \sim h_t = f \{\bar{Re}, \bar{Pr}\} \tag{10}$$

where Re and Pr are the Reynolds and Prandtl number, respectively.
The average of the values of the coefficient of heat transfer
inside tubes h_t determined by Colburn's [14] and Miller's [15]
correlations has been calculated and rounded off to 60 kW/m² °C
(∿10,000 BTU/ft²hr°F).

Sub-cooled nucleate boiling occurs when the surface temperature t_w exceeds the saturation point t_{sat} and the coolant
temperature t_c at a certain location:

$$t_w > t_{sat} > t_c$$

The nucleate boiling is a local surface phenomenon and depends
neither on the liquid velocity nor on the surface geometry.
Available correlations can, therefore, be applied to CANDU type
rod bundles. To predict whether nucleate boiling is taking

place in the crevice region, the local surface temperature and heat flux should be compared to the onset-, and fully developed-boiling curves. These curves are expressed in terms of the amount of superheat of the surface $[\Delta t_{sat}]_{ONB}$ and $[\Delta t_{sat}]_{FDB}$ that are required to start and to establish a fully developed nucleate boiling regime, respectively. Bergles and Rohsenow's correlation [16] expresses $[\Delta t_{sat}]_{ONB}$ as a function of the surface heat flux q_{ONB} and the pressure p:

$$[\Delta t_{sat}]_{ONB} = f\ \{q^+_{ONB},\ p^-,\ \psi\} \tag{12}$$

where ψ is a parameter containing the physical properties and depends on the surface/fluid combination. For the given design pressure p, Thom's correlation [17] is recommended to predict $[\Delta t_{sat}]_{FDB}$:

$$[\Delta t_{sat}]_{FDB} = f\ \{q^+_{F.D.},\ p^-\} \tag{13}$$

A partially developed subcooled nucleate boiling exists between the subsaturation- and the fully developed subcooling boiling-regimes. The geometric interpolation method proposed by Bergles and Rohsenow [16] is followed. The heat flux during partial boiling q_{PB} is therefore determined in terms of the heat flux for single phase forced convection q_{SPL}, the heat flux for fully developed regime q_{FD} and the surface heat flux for fully developed nucleate boiling at a surface temperature corresponding to the onset of boiling $q_{FD/ONB}$:

$$q_{PB} = f\ \{q_{SPL},\ q_{FD},\ q_{FD/ONB}\} \tag{14}$$

4. SOLUTION OF THE PROBLEM

The proposed solution method recognizes the fact that the thermal response of the bearing pad, as defined by the distribution of the surface temperature and the surface heat flux along the boundary, is both controlling and being controlled by the distribution of CHT along the wetted surface, which is unknown in advance. This thermal coupling results in a closed-loop interaction, as shown on Figure 3. As the heat flows into the bearing pad, both the temperature field and the heat flux adjust themselves according to the distribution of CHT. This may result in different regions of single phase liquid and nucleate boiling. Thus, the distribution of CHT will be altered; being enhanced at some location by the boiling process. Consequently, the temperature field and the heat flux will be redistributed and a new boiling region is established. This cycle is repeated as many times as required till the steady state of thermal equilibrium is reached. As shown on Figure 3, the block which represents the region of nucleate boiling is interrelated to the other two basic elements of the loop by the boiling curves and the coefficient of heat transfer-surface temperature relationship.

A simplified scheme of solution is developed to take into account this nonlinear behaviour. As an initial guess, the CHT is assumed to be uniformly distributed along the water-cooled

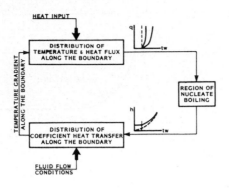

FIGURE 3
SCHEMATIC REPRESENTATION OF
THE THERMAL CLOSED-LOOP INTERACTION

surfaces and is equal to the single phase value. Under these
boundary conditions, and for the heat flux input to the sheath
I.D., the temperature field and the distribution of the heat
flux along the boundary are calculated, allowing us to predict
whether local nucleate boiling is taking place and, if so,
whether it falls inside the partially-, or fully-developed
regimes. As a result the values of the local CHT can be updated.
The interaction scheme proceeds until a sufficiently small
difference between two successive CHT distribution profiles is
reached.

5. RESULTS AND DISCUSSION

To evaluate the significance of the proposed solution method,
the problem has been solved at two different positions along the
core. These positions were chosen for comparison with some
reported results, cited in [1]. Moreover, at position II, the
difference between sheath O.D. temperature and the coolant
saturation temperature is maximum, and thus the nucleate
boiling tendency reaches its highest level.

5.1 The Distribution of the Coefficient of Heat
    Transfer CHT Along Water-Cooled Surfaces

As mentioned earlier, the cornerstone of the present
approach is the recognition of the effect of the distribution of
the CHT on the thermal behaviour of the bearing pad, and the
need for its determination as a necessary and integral part of
the solution. The distributions of the CHT obtained in the
present study along with those assumed in previous studies are
given in Figure 4. Figure 4 indicates that the distribution
of CHT is highly nonuniform and depends noticeably on the
location of the bearing pad along the core of the reactor. It
also shows the significant difference between the assumed and
calculated distributions of CHT. This difference, which is a
measure of the deviation of the results of previous studies
from the correct solution, is vividly reflected on the temperature predictions as will be seen in section 5.2.

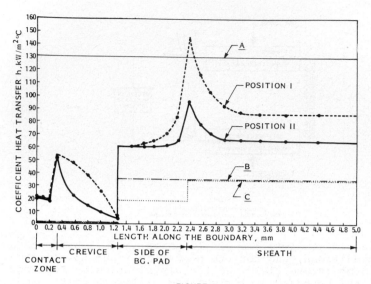

FIGURE 4

COMPARISON OF THE CALCULATED DISTRIBUTION OF COEFFICIENT OF
HEAT TRANSFER WITH THOSE ASSUMED IN PREVIOUS WORK (A TO C)
(Positions I and II are at 4.25 and 4.85, respectively, from the inlet side)

5.2 Temperature and Nucleate Boiling Predictions

The temperature rise θ along the sheath I.D. above the
coolant temperature t_C has been found to vary between 41 and 72°C
and from 37 to 68.2°C at positions I and II respectively. At
the sheath O.D., the temperature rise above t_C has been found to
vary in the range of 15.7 to 19.7°C and 11.5 to 14°C at positions
I and II respectively. These ranges are in a reasonable agreement with the design specifications and predictions [1]. The
comparison of the results that correspond to cases A and C with
those obtained in the present study at the same position I
(Figure 5), reveals the significant errors that would be associated with such predictions, had the nonlinear behaviour of the
system been neglected. The relative error for the surface temperature rise could, in fact, be as large as 100%.

The results shown on Figures 5 and 6 allow us to predict
whether nucleate boiling is expected to occur, at positions I
and II respectively. From these graphs, it can be concluded
that nucleate boiling is confined within a crevice gap of
approximately up to 45 μm at position I while it is extended
over the entire crevice region at position II.

At position I, parts of the crevice and the sides of the
bearing pad fall within the single phase regime, but as one
moves toward point D (Figure 5) along the side of the pad or
the sheath O.D., the boiling process changes from a partially-
to a fully-developed nucleate boiling regime. At position II
Figure 6 shows that the outer subchannel experiences fully

Substituting in eq. (20), and rearringing the terms,

$$V = \frac{1.527 \times 1.149 \times 1.6425}{1.72}$$

$$= 1.676 \ m^3$$

Selecting on appropriate area to maintain a velocity of approximately 1.8 m/s throughout the air washer, the cross-sectional area

$$A = \frac{1.527}{1.8} = 0.85 \ m^2$$

The selected dimension of the cross-sectional area are taken: 0.95 m x 0.95 m, and the length of the air washer is taken 2.0 m. The general Layout of the apparatus is shown in Fig. 8.

5.   APPLICATION

As an illustration of the use of the apparatus, the following example is considered:
Calculate the required number of compound evaporative cooling systems of the same type given in this paper to keep the temperature inside a poultry house at 36°C using the following data:

Size of poultry house,   50 m x 10 m x 2.8 m
Number of birds,   5000
Average weight of bird,   1.2 kg
Outside dry bulb temperature,   40°C
Outside wet bulb temperature,   24°C

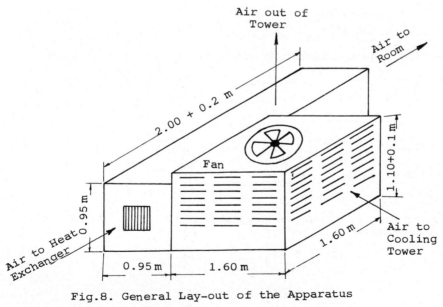

Fig.8. General Lay-out of the Apparatus

The ASHRAE 1981 handbook of fundamentals, Chapter (9), gives the following equation for estimating the heat production from poultry in watts,

$$Q = 8.72 \ M^{0.75}$$

where M is the average weight of the bird in kg. Thus, the total heat generated by the poultry is,

$$Q_T = \frac{5 \times 10^3}{10^3} \ (8.72 \times (1.2)^{0.75})$$

$$= 50 \ kW$$

For poultry, the ratio of sensible to total heat produced is assumed an average of 0.67, thus,

Sensible heat Load = 33.5 kW
Lantent heat load = 16.5 kW

Assuming an additional 15% of the sensible heat is transmitted through the walls of the poultry house, thus the total sensible heat to be removed is 38.5 kW, and the sensible heat ratio for the poultry house is 0.71.

Referring to fig. 2.a., $t_1 = 40^\circ C$, $t_2 = 35^\circ C$, $t_3 = 25^\circ C$; to determine point (4) a line of SHR = 0.71 is drawn from point (3) to intersect the dry bulb temperature line 36ºC at point (4). Then, the enthalpy difference $(h_4 - h_3)$ can be determined;

$$h_4 - h_3 = 83 - 66.5$$

$$= 16.5 \ kJ/kg_a$$

Also, $\vartheta_3 = 0.866 \ m^3/kg_a$

Therefore, the required number of apparatuses is,

$$n = \frac{Q_T}{\frac{V}{\vartheta_3} (h_4 - h_3)}$$

$$= \frac{50}{(\frac{1.527}{0.866}) \ (83-66.5)}$$

$$= 1.72$$

$$= 2$$

The following Fig. 9 shows the suggested positioning of the two compound evaporative coolers.

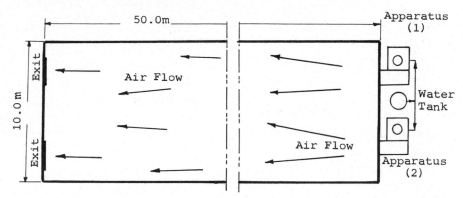

Fig.9. Location of Apparatuses Outside Poultry
House.

6.   CONCLUSION

This paper gives the design procedure and calculations necessary to construct from simply available materials a compound
evaporative cooling system that could be used in several air
cooling applications. The dimensions of the apparatus are reasonable, and its overall cost does not exceed US ∉ 1500. This
low capital investment together with its substantial reduction
in energy consumption as compared to conventional cooling systems
makes it specially suitable for application in developing countries to improve the environment of works, farm animals held in
animal houses, and plants held in plant growth chambers. The
apparatus is particularly suitable for application in standard
poultry houses built in Egypt, for which an example of application was given. This work constitutes the first step in the
analysis of the characteristics of the apparatus at different
inlet air conditions. For this aim, a prototype of this system
is being built and tested in the laboratories of the faculty of
engineering, Zagazig University.

REFERENCES

1.   Kays, W.M. and London, A.L. Heat transfer and flow friction
      characteristics of some compact heat exchanger surfaces,
      Trans. ASME, vol. 1950 p. 1075.

2.   Threlkeld, J.L. Thermal Environmental Engineering, Prentice
      Hall, Inc. 1970.

3.   Kern, D.Q. Process Heat Transfer, Mc Graw-Hill, 1950.

4.   Kusuda, T. Calculation of the temperature of a flat-plate
      wet surface under adiabatic conditions with respect to Lewis
      relation, Humidity and Moisture, volume 1, Reinhold publish-
      ing corporation, 1965, P. 29.

5.   Carey, W.F. and Williamson, G.J. Gas Cooling and Humidifi-
      cation: Design of Packed Towers from Small Scale Tests,
      Proceedings of the Institution of Mechanical Engineers,
      Vol. 163, 1950, P. 49.

Multi-Phase Flow and Heat Transfer III. Part B: Applications
edited by T.N. Veziroğlu and A.E. Bergles
Elsevier Science Publishers B.V., Amsterdam, 1984 — Printed in The Netherlands

883

IDENTIFICATION OF TWO-PHASE HYDRODYNAMIC PROCESSES FROM
NORMAL OPERATING DATA WITH APPLICATION TO A POWER BOILER

T.M. Romberg
CSIRO Division of Mineral Physics
PMB 7, Sutherland
New South Wales, 2232
Australia

ABSTRACT

Measurements of the coal feed, feedwater flow, drum level, steam flow, TSV pressure error and generator power were monitored on a 500 MW(e) drum boiler during full power operation under closed loop control. The fluctuations in these measurements (inherent noise) are analysed using multivariate time series techniques and a hydrodynamic model to identify the fundamental mode dynamics of the drum boiler.

The paper presents an overview of multivariate systems analysis theory, and shows how selected parameters of the hydrodynamic model are estimated using the measured correlation functions as reference descriptors. The sensitivity of the model results to changes in the subcooled boiling boundary, two-phase slip, friction multiplier and heat transfer correlation coefficients, is also discussed.

1. INTRODUCTION

Boiling water reactors, fossil-fired boilers and other industrial two-phase heat transfer processes are normally operated under closed loop control to satisfy load demands imposed by the external system. In this mode of operation the process variables (coolant pressures, temperatures, flow rates, heat fluxes, etc.) fluctuate about the steady state level, particularly if the plant is used for peak load rather than base load operation. Attempts in the early sixties to utilise this "inherent noise" were encouraging (e.g. the Halden Reactor studies [1]), but more stringent control and safety regulations, coupled with significant developments in process instrumentation and computer technology in recent years, has led to a rapid growth in the number of investigations attempting to monitor and analyse noise signals for process performance [2], diagnostic analysis and surveillance [3], as well as the evaluation of mathematical models [4,5] for process design and control studies. The inherent noise method is particularly relevant to the analysis of complex large scale commercial plant, where traditional methods for superimposing test signals are impractical during normal operation.

In contrast to most modelling studies which use inherent noise techniques and autoregressive (AR) or autoregressive moving average (ARMA) parametric models to identify the process (e.g.[3]), the present study uses a distributed hydrodynamic model derived from fundamental mass, energy and momentum conservation equations. Previous papers [8,9] have shown how the model and multivariate spectral analysis methods can be used to assess the two-phase flow stability characteristics of a boiling channel from measurements of the

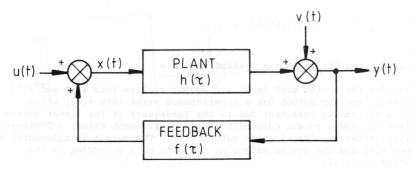

Fig. 2. Block Diagram of Closed Loop System

measurements are x(t) and y(t), and u(t) and v(t) are extraneous noise sources superimposed on the input and output respectively. The convolution equations relating the variables are

Feedforward : $y(t) = E\left[x(t-\tau)h(\tau)\right] + v(t)$,

Feedback : $x(t) = E\left[y(t-\tau)f(\tau)\right] + u(t)$.

(6)

The corresponding crosscorrelation functions for the feedforward and feedback loops are respectively

$$R_{xy}(\tau) = E\left[(R_{xx}(\tau-\lambda)\ \tilde{h}(\lambda)\right] \qquad \tau>0$$

$$= R_{xv}(\tau) \qquad\qquad , \quad \tau<0 \qquad\qquad (7)$$

and $$R_{yx}(\tau) = E\left[R_{yy}(\tau-\lambda)\ \tilde{f}(\lambda)\right] \qquad \tau>0$$

$$= R_{yu}(\tau) \qquad\qquad . \quad \tau<0 \qquad\qquad (8)$$

In principle, therefore, the feedforward (process) weighting function $\tilde{h}(\tau)$ can be identified from the crosscorrelation estimates in the positive lag domain, and the feedback (controller) weighting function $\tilde{f}(\tau)$ can be identified from the crosscorrelation estimates in the negative lag domain. In practice, however, these results only apply to non-deterministic broadband variables; narrowband filtering increases the correlation between neighboring estimates of the crosscorrelation function [10], and may cause overlap at the zero lag axis.

2.4 Multivariate System Analysis

The concepts discussed in the previous sections are readily extended to the analysis of multivariate linear systems. However, it should be noted that the most general multiple input multiple output (MIMO) system model cannot be identified in the minimum least squares sense unless certain constraints are placed on the residual noise matrix which effectively reduce the MIMO system to a set of independent multiple input single output (MISO) subsystems [9].

The output $x_n(t) = y(t)$ of a MISO system is a linear sum of the inputs $x_j(t)$ (j=1,m;n=m+1) as defined by the convolution equation

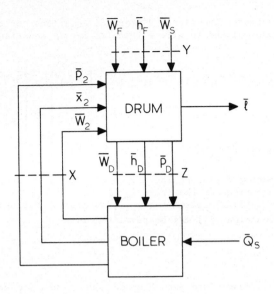

Fig. 3. Block Diagram of Drum Boiler Model

$$x_n(t) = E\Big[\sum_{j=1}^{m} x_j(t-\lambda) \; h_{jn}(\lambda)\Big] + v(t) \quad , \tag{9}$$

where $h_{jn}(\lambda)$ are the corresponding input-output weighting functions. The crosscorrelation between the ith input variable and the output is given by

$$R_{in}(\tau) = E\big[x_i(t) \; x_n(t+\tau)\big]$$

$$\tag{10}$$

i.e. $\quad R_{in}(\tau) = \sum_{j=1}^{m} E\big[R_{ij}(\tau-\lambda) \; \tilde{h}_{jn}(\lambda)\big] \quad , \quad (i=1,m)$

which is the multivariate form of equation (5).

3.   HYDRODYNAMIC MODEL

The hydrodynamic model represents the drum and boiler dynamics by the transfer matrix "blocks" depicted schematically in Figure 3.

The boiler model is derived from the one-dimensional partial differential equations for the conservation of mass, energy and momentum, in conjunction with empirical correlations to predict the onset of subcooling boiling, and model the flow characteristics in the single phase (subcooled liquid) and two-phase (subcooled, saturated boiling) regions of the evaporator [8,12]. The equations are perturbed in time, linearised, and Laplace transformed to yield nonlinear ordinary differential equations which are integrated spatially using finite difference techniques.

In recent modifications, the finite difference equations for the jth space node have been manipulated to give a transfer matrix equation of the form

$$
\begin{bmatrix} \bar{w}_j \\ \bar{h}_j \\ \bar{p}_j \end{bmatrix} = \begin{bmatrix} \bar{T}(w_j, w_{j-1}) & \bar{T}(w_j, h_{j-1}) & \bar{T}(w_j, p_{j-1}) \\ \bar{T}(h_j, w_{j-1}) & \bar{T}(h_j, h_{j-1}) & \bar{T}(j_j, p_{j-1}) \\ \bar{T}(p_j, w_{j-1}) & \bar{T}(p_j, h_{j-1}) & \bar{T}(p_j, p_{j-1}) \end{bmatrix} \begin{bmatrix} \bar{w}_{j-1} \\ \bar{h}_{j-1} \\ \bar{p}_{j-1} \end{bmatrix} + \begin{bmatrix} \bar{Q}_{1j} \\ \bar{Q}_{2j} \\ \bar{Q}_{3j} \end{bmatrix} \bar{Q}_s, \quad (11)
$$

which relates the frequency dependent mass flow ($\bar{w}$), specific enthalpy ($\bar{h}$), and pressure ($\bar{p}$) perturbations at the inlet (j-1) and exit (j) by a (3x3) transfer matrix T_j, and a (3x1) column vector Q_j to account for heat source ($\bar{Q}_s$) perturbations. Equation (11) may be rewritten as

$$
X_j = T_j X_{j-1} + Q_j \bar{Q}_s . \quad (12)
$$

If the boiler circuit is subdivided into N space nodes, then the transfer matrix blocks are cascaded N times to yield the overall input-output transfer matrix equation

$$
X = HZ + P\bar{Q}_s , \quad (13)
$$

where
$$
H = H_N = \prod_{j=1}^{N} T_j = T_N H_{N-1} , \quad (14)
$$

$$
P = P_N = Q_N + T_N P_{N-1} , \quad (15)
$$

and Z is a (3x1) column vector of the mass flow, enthalpy and pressure perturbations at the downcomer inlet. The continuity in flow conditions at modal boundaries with local area changes/restrictions and at the boiling boundary are conserved by coupling equations.

The drum model is derived from the macroscopic equations for the conservation of mass, energy and volume in the separate steam and water phases, and the total system [16]. The equations are perturbed, linearised and Laplace transformed to yield a matrix equation of the form

$$
Z = BX + CY , \quad (16)
$$

where
Z' = vector of downcomer inlet perturbations = $[\bar{w}_d, \bar{h}_d, \bar{p}_d]$,

X' = vector of evaporator exit perturbations = $[\bar{w}_2, \bar{x}_2, \bar{p}_2]$,

Y' = vector of feedwater/steam perturbations = $[\bar{w}_F, \bar{h}_F, \bar{w}_S]$,

and B,C are (3x3) complex matrices similar in from to the T_j matrix in equation (11).

Insertion of equation (16) in (13) yields

$$X = G^{-1}[HCY + P\bar{Q}_s] \quad , \tag{17}$$

where the (3x3) system characteristic matrix

$$G = [I-HB] \quad . \tag{18}$$

Note that the drum boiler system stabilty is given by the locus of the determinant

$$\Delta = |G| \gg 0 \tag{19}$$

in the complex plane (Argand diagram), that is, the determinant must not encircle the origin.

The drum level perturbation is calculated from the vector equation

$$\bar{\ell} = EX + FY \quad , \tag{20}$$

where E and F are (1x3) row vectors. Inserting equation (17) in (20) yields

$$\bar{\ell} = [EG^{-1}HC + F]Y + EG^{-1}P\bar{Q}_s = U'W \quad , \tag{21}$$

where $U' =$ row vector of inputs $= [\bar{w}_f, \bar{h}_f, \bar{w}_s, \bar{Q}_s]$, and W is a (4x1) column vector of input–output transfer functions. Inverse Laplace transformation of equation (21) yields (cf. equation (9)),

$$\ell(t) = u_5(t) = E\left[\sum_{j=1}^{4} u_j(t-\lambda)w_{j5}(\lambda)\right] \quad . \tag{22}$$

The crosscorrelation between the ith input $u_i(t)$ and the output is given by

$$R_{i5}(\tau) = E[u_i(t)u_5(t+\tau)] \qquad (i=1,4)$$

$$= \sum_{j=1}^{4} E[R_{ij}(\tau-\lambda)\tilde{w}_{j5}(\lambda)] \quad . \tag{23}$$

4. BOILER PLANT AND DATA ACQUISITION

4.1 Description of Plant

The drum boiler investigated in this study is a 500 MW(e) unit situated at Wallerawang in New South Wales, Australia. A simplified schematic of the boiler and its associated turbine–generator units is shown in Figure 4. The primary circuit consists of five interacting sub-systems : a steam drum, downcomers, circulating pumps, evaporator (water walls), and risers. External

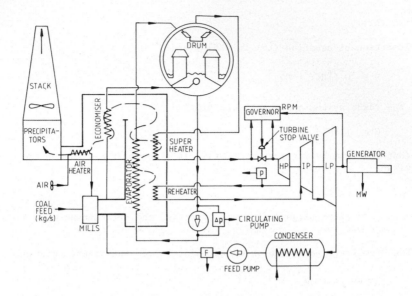

Fig. 4. Simplified Schematic of Wallerawang Power Station

connections to and from the turbines via the spray superheaters and economiser have not been included in the present model, and are being incorporated in further modifications.

Subcooled water flows from the drum through vertical downcomers to four circulating pumps, which pump it to lower headers at the base of the furnace. The lower headers distribute the flow through about 1200 vertical pipes spaced around the outer and central walls of the evaporator. These pipes (or channels) are connected to upper headers which, in turn, are connected to the steam drum by riser pipes. The two-phase (saturated water/steam) mixture enters the upper part of the drum, and is directed to turbo separators by a cylindrical shell mounted inside the drum. The steam leaving the turbo separators passes through a series of plate dryers before entering the spray superheaters. The saturated water mixes with the subcooled feedwater entering from the economiser through horizontal holes in the sparge tubes above each of the eight downcomers.

4.2 Data Acquisition

Measurements of the variations in coal feed to the pulverising mills, feedwater flow, steam flow to the turbine, pressure error at the turbine stop valve (TSV) and drum level were recorded on the boiler during full power operation in a boiler-follow control mode (coal-to-mills is controlled to minimise the error between the upstream TSV pressure and its setpoint of 15.8 MPa). The analog signals obtained from the station control panels were recorded on magnetic tape using an AMPEX FR1300 tape recorder. The tape records were subsequently digitised at 1.6 second intervals to give 3600 data points/record, which were stored on the disc units of an IBM3033S mainframe computer. The station computer was also programmed to print a snapshot log at one minute intervals of the drum pressure and level, feedwater flow,

Table 1 : Steady State Data from Computer Log

Drum pressure	16.45 MPa
" level	48.1 cm (53.3 cm datum)
Economiser inlet temp.	187°C
" outlet temp.	268°C
Feedwater flow	389 kg/s
Downcomer inlet temp.	349°C
Circulating pump Δp	228 kPa
Coal feed to mills	47.3 kg/s
Generator power	480 MW

economiser inlet and outlet temperatures, coal flow, pressure drop across the circulating pumps, generator power and downcomer inlet temperatures. A five channel pen recorder was also used to monitor in parallel selected signals recorded on tape.

4.3 Steady State

The data from the snapshot log was averaged over the recording period (2 hours) to obtain the steady state information given in Table 1.

Energy balances based on this data yielded economiser, evaporator and superheater powers of 145, 543 and 325 MW respectively. The recirculation flow calculated from the pump characteristic was 2243 kg/s (4 pumps).

This basic data and the primary circuit dimensions supply the input data for the hydrodynamics computer code. The primary circuit steady state conditions were matched by adjustment of the loss factor for the gags (orifices) inserted at the inlet of the water wall tubes. The computations yielded an evaporator exit quality of 26.5%, which corresponds to a steam flow of 594 kg/s. However, the measured steam flow to the superheater was 389 kg/s, which indicates that 205 kg/s of saturated steam was condensed in raising the feedwater temperature to almost saturation, as confirmed by the downcomer inlet temperatures.

5.   BOILER IDENTIFICATION

5.1 Data Analysis and Results

The correlation functions for the MISO models discussed in sections 2 and 3 were computed from the digitised data using a CSIRO time series analysis program SISOD. The power spectral density estimates for the five records (feedwater flow, steam flow, coal feed, TSV pressure error, drum level) were calculated after broadband digital filtering to remove the d.c. and frequencies above the Nyquist frequency (0.3125 Hz). The spectral density estimates had dominant amplitudes in the 0.001-0.1 Hz frequency range, and the digital filter settings were adjusted to bandpass information in this range.

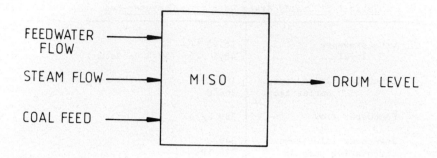

Fig. 5. Multivariate Model of Drum Boiler

The correlation functions for the MISO model relating feedwater flow, steam flow, coal feed (inputs) to drum level (output), Figure 5, were computed as outlined in section 2. The maxima/minima for the various combinations of crosscorrelation functions and their associated lag in seconds are summarised in Table 2.

The coal feed and TSV pressure error are highly correlated (-0.92) due to feedforward control action. The TSV pressure error and the steam flow, as inferred from the pressure after the HP turbine stage (Figure 4), are also highly correlated (+0.93) and thus by process and control mechanisms, the steam flow is highly correlated with the coal feed (-0.84). The lag of only 8 seconds suggests the dominance of the latter mechanism. Hence the steam flow is not an independent input, and is eliminated from the MISO model in the boiler identification. The feedwater flow is predominantly correlated with the drum level via the control loop, as indicated by the lead (negative lag) of 19.2 seconds.

5.2 Identification Procedure

The aim of the identification procedure is to determine the model weighting functions, $\tilde{w}(\tau)$ in equation (23), which are compatible with the drum boiler weighting functions $\tilde{h}(\tau)$, equation (10), by minimisation of the root mean square (RMS) errors between equivalent crosscorrelation functions.

The hydrodynamic model result at the same steady state operating level (as logged) is computed from

$$R_{in}(\tau) = \sum_{j=1}^{m} E\left[\hat{R}_{ij}(\tau-\lambda)\tilde{w}_{jn}(\lambda) \right] \quad , \tag{24}$$

Table 2 : Normalised Crosscorrelation Maxima/Minima

Crosscorrelation	Maxima/Minima	Lag(s)
Coal feed ⟶ Feedwater flow	+0.21	+ 4.8
" ⟶ Drum level	+0.49	+20.8
" ⟶ TSV pressure error	−0.92	+ 1.6
" ⟶ Steam flow	−0.84	+ 8.0
Feedwater flow ⟶ Drum level	−0.51	−19.2
" ⟶ TSV pressure error	+0.35	−36.8
" ⟶ Steam flow	+0.28	−32.0
Drum level ⟶ TSV pressure error	−0.50	−19.2

where $i=1,m$ and $\hat{R}_{ij}(\tau)$ denotes the correlation functions computed from the measurements which are used to perturb the hydrodynamic model. The RMS error between the predicted input-output crosscorrelation $R_{in}(\tau)$ and the corresponding measured crosscorrelation $\hat{R}_{in}(\tau)$ is given by

$$\varepsilon_{RMS} = E\left[R_{in}(\tau) - R_{in}(\tau)\right]^2 \tag{25}$$

Substitution of equations (10) and (23) in (25) yields

$$\varepsilon_{RMS} = \left\{ E\left[E\left[\hat{R}_{ij}(\tau-\lambda)(\tilde{h}_{jn}(\lambda) - \tilde{w}_{jn}(\lambda))\right]\right]^2 \right\}^{\frac{1}{2}} . \tag{26}$$

Thus the RMS error tends to zero as $\tilde{w}_{jn}(\tau) \longrightarrow \tilde{h}_{jn}(\tau)$.

5.3 Coal Feed - Drum Level Analysis

The hydrodynamic model coal feed - drum level crosscorrelation at the same steady state operating level is computed from

$$R_{13}(\tau) = E\left[\hat{R}_{11}(\tau-\lambda)w_{13}(\lambda) + \hat{R}_{12}(\tau-\lambda)w_{23}(\lambda)\right] , \tag{27}$$

where the subscripts 1,2,3 denote the coal feed, feedwater flow (inputs) and drum level (output) respectively, and $-\tau_m \leqslant \tau \leqslant \tau_m$ (τ_m = maximum lag).

In utilising equation (27), the model requires power perturbations in the evaporator water walls as an input, and so some assumptions are necessary regarding the mill/furnace transfer function. The internal grinding and combustion processes are extremely complex, and in this analysis it is assumed that the coal feed and water wall power can be represented by a simple transfer function model of the form

894

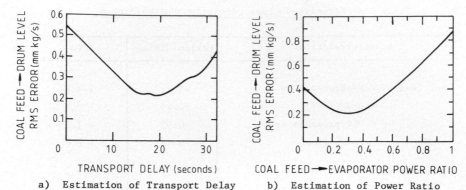

a) Estimation of Transport Delay   b) Estimation of Power Ratio

Fig. 6. Estimation of Mill/Furnace Transport Delay and Power Ratio

$$H_{MF}(s) = K_{MF} e^{-s\tau_{MF}} , \tag{28}$$

where τ_{MF} is the mill/furnace (MF) transport delay, s is the Laplace operator, and K_{MF} is the static gain. The static gain K_{MF} denotes a coal feed (CF) - evaporator power (EP) ratio, which can be approximated by

$$\frac{\sigma_{EP}}{M_{EP}} = K_{MF} \frac{\sigma_{CF}}{M_{CF}} , \tag{29}$$

where M_{EP}, M_{CF} and σ_{EP}, σ_{CF} are, respectively, the evaporator power and coal feed mean values and standard deviations.

In order to avoid a non-minimum phase identification problem, the mill/furnace transport delay τ_{MF} must be estimated first. The change in coal feed-drum level RMS error, equation (25), as a function of τ_{MF} is given in Figure 6a, and has a minimum at 19.2 seconds. The reference model for these calculations assumed :

a)   subcooled boiling,
b)   homogeneous flow,
c)   no slip between the steam/water phases, and
d)   no thermal storage in the water wall tubes.

Similarly, estimation of K_{MF} by minimisation of the RMS error yields an "optimum" value of 0.3 (Figure 6b).

Relaxation of the above thermal-hydraulic assumptions by varying the coefficients of the subcooled boiling boundary, two-phase friction, slip and heat transfer correlations [8] over physically realistic ranges, resulted in only marginal changes (<4%) in the minimum RMS error of 0.214, calculated in the range $-\tau_m < \tau < \tau_m$. The two-phase distribution parameter (C_o) and the weighted mean drift velocity (V) in the Zuber and Findlay slip correlation [8], had the most influence on the RMS error. Increasing the distribution parameter from its homogeneous value (C_o=1), increased the RMS error. However, decreasing the weighted mean drift velocity resulted in a minimum RMS error at V=-1 m/s. These values suggest that the two-phase flow in the water walls was probably annular, and the steam phase underwent some deceleration in the upper headers and horizontal riser tubes. Some improvement in the sensitivity of the RMS error could be achieved by invoking the physical

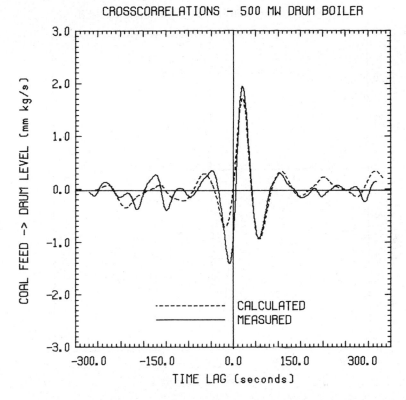

Fig. 7. Comparison of Coal Feed—Drum Level Crosscorrelations

realizability condition, equation (4), and confining the range to $0 \leqslant \tau \leqslant \tau_m$.

The comparison between the calculated and measured crosscorrelations obtained by this identification procedure is shown in Figure 7. The process (positive lag) results are in good agreement overall, although further refinements may produce marginally better agreement in the region of the peak response. The negative lag (feedback) results agree less favourably, which is consistent with the open loop structure of the present model.

5.4 Feedwater Flow — Drum Level Analysis

The hydrodynamic model feedwater flow-drum level crosscorrelation is computed from

$$R_{23}(\tau) = E\left[\hat{R}_{21}(\tau-\lambda)w_{13}(\lambda) + \hat{R}_{22}(\tau-\lambda)\tilde{w}_{23}(\lambda)\right] , \qquad (30)$$

where the subscripts denote the same variables as in equation (27), and $-\tau_m \leqslant \tau \leqslant \tau_m$ as before. The result obtained using the same steady state and parameters estimates as computed for the coal feed-drum level

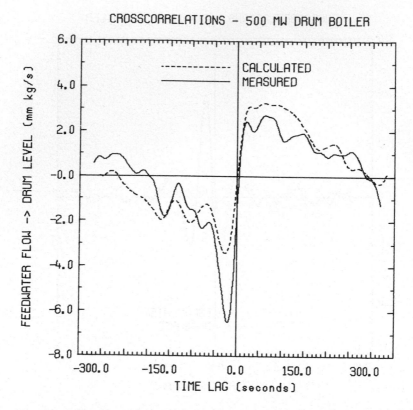

Fig. 8. Comparison of Feedwater Flow—Drum Level Crosscorrelations

crosscorrelation, Figure 6, is given in Figure 8. The good agreement, particularly the results in the positive lag domain, shows that the fundamental mode hydrodynamic characteristics of the drum boiler are adequately described by the model. As before, the negative lag (feedback) results are influenced by the action of the drum level control loop in the boiler operation.

6.   CONCLUSIONS

The paper has demonstrated how multivariate noise analysis techniques may be used to assess the hydrodynamic performance of a power boiler from data logged during normal operation. Crosscorrelations computed from measurements of the variations in coal feed, feedwater flow, drum level, TSV pressure error, steam flow and generator power recorded on a 500 MW(e) coal-fired drum boiler operating in a boiler-follow control mode, showed that the drum level variations were predominantly caused by coal feed and feedwater flow variations. The computed correlation functions were also used to evaluate a hydrodynamics model of the drum boiler. Coal feed-drum level and feedwater flow-drum level crosscorrelations computed from the measurements were in good agreement with those calculated by the hydrodynamic model after unknown mill/furnace parameters had been estimated. Changes in the subcooled boiling boundary, two-phase slip, friction multiplier and heat transfer correlation coefficients over physically realistic ranges had only a marginal affect (<4%) on the RMS error between the measured and calculated crosscorrelations. The two-phase slip correlation coefficients (Zuber and Findlay distribution parameter and weighted mean drift velocity) had the most influence on the RMS error. The optimised values suggest that the two-phase flow in the water walls was probably annular, and the steam phase decelerated in the upper headers and horizontal riser tubes. These aspects will be investigated further when the hydrodynamic model has been modified to include the economiser, spray superheaters and boiler controllers.

ACKNOWLEDGMENTS

Helpful discussions with Professor N.W. Rees, Head, School of Electrical Engineering and Computer Science, University of New South Wales, and Mr P.T. Nicholson, Electricity Commission of New South Wales, who also assisted with the data logging, and the practical assistance provided by the staff at Wallerawang Power Station, are gratefully acknowledged. Mr G.W. Herfurth and Mr D. McColm, CSIRO technical staff, also contributed to this investigation.

REFERENCES

1. Eurola, T., "Reactor Noise Experiments on Halden Boiling Water Reactor", Noise Analysis in Nuclear Systems, USAEC Rep. TID-7679, 1963, pp. 449-468.

2. Fukunishi, K., "Coherence Analysis of Boiling Water Reactor Noise", J. Nuc. Sci. & Technol., 1977, Vol.14, No.5, pp. 351-358.

3. Upadhyaya, B.R., and Kitamura, M., "Stability Monitoring of Boiling Water Reactors by Time Series Analysis of Neutron Noise", Nuc. Sci. & Eng., 1981, Vol.77, pp. 480-492.

4. Gall, C.J., "Assessment of the System Response Characteristics from 'Noise' Measurements at a Single Point with Particular Application to Flow Stability", J. Br. Nucl. Energy Soc., Vol.12, pp. 175-181.

5. Matsubara, K., Oguma, R., and Kitamura, M., "A Multivariable Autoregressive Model of the Dynamics of a Boiling Water Reactor", Nucl. Sci. & Eng., 1978, Vol.65, pp. 1-16.

6. Kerlin, T.W., Zwingelstein, G.C., and Upadhyaya, B.R., "Identification of Nuclear Systems", Nuclear Technology, 1977, Vol.36, pp. 7-38.

7. Upadhyaya, B.R., Kitamura, M., and Kerlin, T.W., "Multivariate Signal Analysis Algorithms for Process Monitoring and Parameter Estimation in Nuclear Reactors", Annals of Nuclear Energy, 1980, Vol.7, pp. 1-11.

8. Romberg, T.M., and Rees, N.W., "Multivariate Hydrodynamic Analysis of Boiling Channel Flow Stability from Inherent Noise Measurements", Int. J. Multiphase Flow, 1980, Vol.6, No.6, pp. 523-551.

9. Romberg, T.M., and Rees, N.W., "On the Modelling of Boiling Channel Dynamics Using Spectral Methods", Int. J. of Control, 1981, Vol.34, No.2, pp. 259-284.

10. Jenkins, G.M., and Watts, D.G., Spectral Analysis and its Applications , Holden-Day, San Francisco, 1968.

11. Otnes, R.K., and Enochson, L., Applied Time Series Analysis , John Wiley, New York, 1978.

12. Romberg, T.M., "Noise Analysis of Coolant Dynamics in Boiling Two-Phase Flow Systems", Ph.D. Thesis, University of NSW, 1978.

13 Gustavsson, I., Ljung, L., and Soderstrom, T., "Survey Paper : Identification of Processes in Closed Loop - Identifiability and Accuracy Aspects", Automatica, 1977, Vol.13, pp. 59-75.

14. Akaike, H., "Some Problems in the Application of the Cross-spectral Method", B. Harris (Ed.), Spectral Analysis of Time Series, John Wiley, New York, 1967.

15. Priestley, M.B. "Estimation of Transfer Functions in Closed Loop", Automatica, 1969, Vol.5, pp. 623-632.

16. Brown, D.H., "Transient Thermodynamics of Reactors and Process Apparatus", Advances in Nuclear Engineering, Pergamon Press, New York, 1957, Vol.II, pp. 526-534.

Multi-Phase Flow and Heat Transfer III. Part B: Applications
edited by T.N. Veziroğlu and A.E. Bergles
Elsevier Science Publishers B.V., Amsterdam, 1984 — Printed in The Netherlands

MULTI-PHASE FLOW CONSIDERATIONS IN SIZING EMERGENCY RELIEF SYSTEMS FOR RUNAWAY CHEMICAL REACTIONS

Hans K. Fauske, Michael A. Grolmes and Joseph C. Leung
Fauske & Associates, Inc.
16W070 West 83rd Street
Burr Ridge, Illinois 60521, U.S.A.

ABSTRACT

Based upon available literature data recommendations are made as to the choice of two-phase flow models to assure an adequate emergency relief system (ERS) design. For large process vessels and in the absence of flow regime characterization data under runaway conditions, a safe ERS design requires consideration of homogeneous vessel behavior and vent line flow characteristics based upon homogeneous equilibrium flow. For the majority of cases this approach may in fact represent a best estimate assessment.

1. INTRODUCTION

Emergency relief systems for pressure vessels are being used extensively in the chemical industry to minimize the potential for chemical explosions and extensive spreading of hazardous material. In addition to uncertainties related to chemical kinetics data in general, the possibility of flashing flow occurring in the relief device and its potential effect on the vent size is of particular interest* [1].

The question of vapor venting versus liquid-vapor venting depends upon the prevailing flow regime, such as bubbly, churn turbulent or droplet flow, which is generally not known during runaway conditions. In addition, the nature of the venting process is likely to be strongly influenced by the general problem of "foaming" which is aggravated by the presence of certain additives or emulsifiers and is known to be highly dependent on the particular system properties and minute quantities of impurities. In view of this highly variable picture, it is not surprising that a number of calculational methods have been published concerning sizing of emergency relief systems for runaway chemical reactions [2]. These methods include all-vapor venting, all-liquid venting [3], vapor-liquid venting with no vapor disengagement, i.e. homogeneous vessel behavior [4, 5, 6] and vapor-liquid venting with vapor disengagement, i.e. churn turbulent vessel behavior [7].

A similar status exists for describing the flow through the relief device [2]. These methods include all-vapor flow, all-liquid flow without flashing [3], homogeneous equilibrium flashing flow [1, 4, 5], frozen flow [6], non-

---

*Generally speaking, a smaller vent is needed to handle all-vapor flow than a two-phase mixture.

equilibrium flashing flow [7], and slip equilibrium flashing flow [4, 6]. It follows that a significant variation in the vent size can be obtained depending upon the choice of modeling the vessel behavior, (i.e the degree of vapor disengagement) and the vent line flow dynamics, (i.e. the two-phase critical flow). The variation in overpressure for a given vent size as a function of flow model assumed is illustrated in Fig. 1. Particularly noteworthy is the

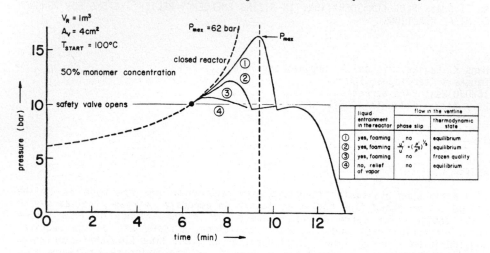

Fig. 1 Typical pressure-time curves calculated for a runaway polymerization reaction, including all-vapor and homogeneous (foaming) venting assumption. Vent flow model assumptions include homogeneous equilibrium, slip equilibrium and non-equilibrium (frozen quality) conditions. (Taken from Ref. [6].)

sizable reduction in overpressure resulting from assuming either all vapor venting or non-equilibrium flow (frozen quality) relative to assuming equilibrium flashing flow. These aspects are discussed further below with the objective of providing definitive recommendations as to the choice of models to assure a safe but not overly conservative emergency relief design.

2. VAPOR DISENGAGEMENT

The liquid and vapor motion inside a reaction vessel during pressure relief is a complex hydrodynamic problem. Fauske, Grolmes and Henry [7] have recently presented a first order method to predict liquid swell and partial disengagement which has been verified for one-component, non-reacting systems such as water, (i.e. non-viscous and non-foaming). Figure 2 illustrates typical capability of integral analysis [8] incorporating vapor disengagement based upon churn turbulent behavior [7] and non-equilibrium flashing critical flow [9] to predict Freon-12 depressurization experiments reported in Ref. [10]. Significant deviations from both homogeneous vessel behavior as well as equilibrium flashing vent flow* are required in order to predict such data.

However, significant vapor disengagement would appear to be absent with

---

*Significant deviation from equilibrium flashing flow in this case is closely related to the aperture geometry in question, (i.e. short nozzle) and is discussed in further detail in Section 3 of this paper.

901

many chemical systems because of the inherent "bubbly" and/or "foaminess" as well as high viscosity, giving rise to homogeneous two-phase flow like behavior throughout most of the venting sequence. For example, Howard has stated (see Ref. [4]) that an industrial vessel containing a 10 cp monomer discharges once or twice a year through an adequately designed relief system and only 25-30% of the liquid mass remains in the vessel at the end of blowdown. Reference [3] also reports actual case histories of polymerization kettles containing up to 4,000 gal. of reacting monomers and water which were relieved of their contents (completely emptied) through the emergency relief line. Finally, incidents reported by Burchett [11] involving runaway reactions of chloroprene in large process vessels which are interpreted in the concluding portion of this paper suggest a similar behavior, i.e. a nearly homogeneous-like venting behavior. Even small test vessels in some instances indicate little or no vapor disengagement. Harmon and Martin [12] experimented with various polymerization reactions at high monomer concentrations in a 5-gal. vessel. For those runs with an adequate relief system they reported only 0-20% liquid retention at the end of blowdown. Boyle's experience working with a polystyrene solution in ethylbenzene (1-quart volume) lead him to suggest the all-liquid venting model [3]. Huff [13] examined visually the liquid-vapor interactions in gallon-size polystyrene/ethylbenzene solutions and concluded that a zero vapor disengagement assumption "is quite realistic over much of the course of the discharge from polymerization reactors". It follows that a safe emergency relief design approach must consider a homogeneous liquid-vapor mixture entering the vent line, unless flow regime characterization data are available for a given system under prototypic runaway relief conditions which demonstrate significant vapor

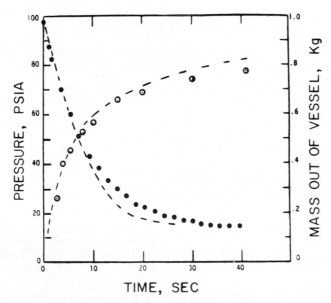

Fig. 2 Comparison between integral analysis and depressurization data with Freon-12 using top venting (nozzle size 4.6 mm).

disengagement*.

3. TWO-PHASE CRITICAL FLOW

Model selection for predicting critical flow of flashing two-phase mixtures requires consideration of non-equilibrium effects [9]. Required relaxation lengths to approach equilibrium flow conditions have been demonstrated in
a number of experiments reported in the open literature. In 1964 Fauske [14]
illustrated the effect of geometry upon the critical flow rate for saturated
water and a very wide range in the stagnation pressure, (see Fig. 3). The

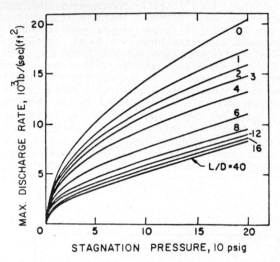

Fig. 3 Maximum discharge rates of saturated
water for 0.25-in. I.D. tube [14].

rapid decay in flows prior to reaching the asymptotic values [length-to-
diameter (L/D) ratio of approximately 16] were attributed to increasing fluid
residence times. For an L/D = 0 (sharp edged orifice), the residence time is
zero resulting in no flashing and the flow rate can be predicted by the standard incompressible single-phase flow equation [14]. On the other hand, at L/D
approximately 16 sufficient time is available to allow the flashing process to
approach equilibrium. The relatively small decreases in the flow rates noted
for larger L/D's were attributed mostly to frictional effects. For the given
tube diameter (D = 6.35 mm), these experiments suggest an essentially constant
relaxation length of the order of 100 mm over a wide range in the stagnation
pressure.

A similar trend in the critical flow behavior starting from saturated or
inlet quality conditions in terms of flow geometry dependency have been noted
by Sozzi and Sutherland [15], Flinta [16], Uchida and Nariai [17] and Fletcher
[18]. These experiments are summarized in Table 1 in terms of L/D ratios and

---

*Significant vapor disengagement implies that a sizable fraction of the con-
tents in a large process vessel would remain in the vessel following blowdown.
It is noted that top venting in a small test vessel can be quite misleading in
this regard, since the entrainment velocity is directly proportional to vessel
height [7].

relaxation lengths corresponding to a change to equilibrium critical flow behavior. Table 1 clearly shows that the L/D ratio does not correlate the

Table 1

ILLUSTRATION OF RELAXATION LENGTH, L OBSERVED
IN DIFFERENT CRITICAL FLOW EXPERIMENTS

Source	D, mm	L/D	L, mm
Fauske (water)	6.35	∿ 16	∿ 100
Sozzi and Sutherland (water)	12.7	∿ 10	∿ 127
Flinta (water)	35	∿ 3	∿ 100
Uchida and Nariai (water)	4	∿ 25	∿ 100
Fletcher (Freon-11)	3.2	∿ 33	∿ 105
Marviken Data (water)	500	< 0.33	< 166

relaxation process, while a simple length criterion of the order of 100 mm appears to characterize the residence time requirement for both tubes and nozzles covering wide variations in diameter and stagnation pressure including different fluid properties such as water and Freon-11.

Further support for the simple criterion is provided by the recent large scale Marviken data with inlet quality conditions [19]. For a nozzle diameter of 500 mm, relatively little change is observed in the critical flow rate when the L/D ratio is varied from 0.33 to 3.2 (see Table 2). It is particularly

Table 2

MARVIKEN DATA - SATURATED FLOW
$D = 500$ mm, $P_o \sim$ MPa

Critical Flow Rate kg/m^2-s x 10^{-3}	L/D
∿ 24.5	∿ 0.3
∿ 23.3	∿ 1.5
∿ 22.0	∿ 3
HEM → 21.7	-

noteworthy that the predicted homogeneous equilibrium model (HEM) critical flow rate* is only about 10% lower than the observed flow rate at L/D = 0.33, suggesting that the relaxation length for the 500 mm nozzle also is of the order of 100 mm.

In fact, excellent agreement between the HEM predictions based upon stagnation properties and the experimental data is noted for reduced critical pressure (P/Pc where Pc is the thermodynamic critical pressure) of the order of 0.1 and above which is the range of interest for most chemical systems (see Figs. 4 and 5).

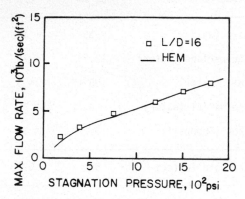

Fig. 4 Comparison between measured [14] critical flow rates (L ∿ 100 mm) and the homogeneous equilibrium model (HEM) evaluated for stagnation conditions.

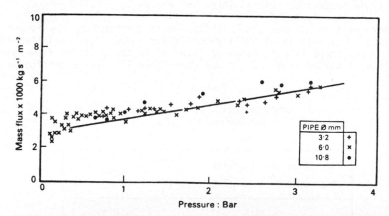

Fig. 5 Effect of pipe diameter on mass flux through a pipe of length 120 mm [18] and comparison with the HEM predictions (solid line).

---

*This model can be described by the same equations as an equivalent single-phase flow. The two phases are everywhere in equilibrium with equal velocities and temperatures. At low qualities the HEM critical flow rate can be estimated within 10 to 15% from $[h_{fg}\rho_g(cT)^{-1/2}]$ where h_{fg} is the latent heat of vaporization, ρ_g is the vapor density, c is the specific heat and T is the temperature, all properties evaluated at the stagnation condition. For further details see Ref. [20].

Since sizing of emergency relief systems for runaway chemical reactions generally involve inlet quality conditions and relatively large flow devices, the above criterion suggests that a safe as well as a best estimate prediction of the critical flow rate should be based on the HEM.

4. CONCLUDING REMARKS

For large process vessels and in the absence of flow regime characterization data under runaway conditions, a safe emergency relief system design requires consideration of homogeneous vessel behavior and vent line flow characteristics based upon homogeneous equilibrium flow. For the majority of cases this approach may in fact reflect a best estimate assessment.

An example at hand is the two incidents of runaway reaction of chloroprene in large scale vessels reported by Burchett [11]. The first incident involved a 3000-gallon reaction vessel with a 30 psig, 4 in. diameter safety disc and an equivalent sized tailpipe. The vessel vented safely, and following the incident the vessel was found to be essentially empty. The second incident involved a 2000-gallon vessel again with a 4-in. diameter safety disc (and an equivalent sized tailpipe) but set at 75 psig. In this case the runaway reaction resulted in vessel rupture. Using Burchett's measured energy release rates for the chloroprene system [11], the predicted pressure behavior for the two vessels is illustrated in Fig. 6. Assuming the 4-in. relief device is

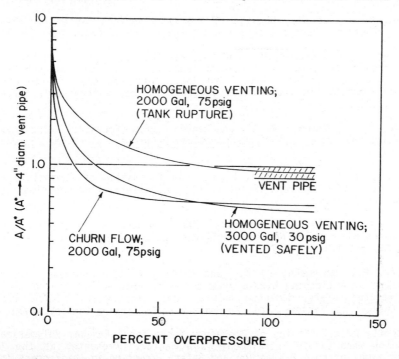

Fig. 6 Predicted pressure behavior of two reported incidents involving runaway reaction of chloroprene in large scale vessels. Shaded band represents the estimated uncertainty in the actual discharge coefficient for the incident vent line. Calculations performed with a discharge coefficient equal to 1.0.

fully available for venting, (i.e. ruling out any polymer deposits on the disc and/or partial plugging of the relief line), homogeneous venting (homogeneous vessel behavior and homogeneous equilibrium vent flow) is clearly suggested by the observed behavior. The 3000-gallon reaction vessel is predicted to vent safely while an explosive pressure runaway condition cannot be ruled out for the 2000-gallon reaction vessel. On the other hand, assuming churn turbulent vessel behavior [7], (i.e. significant vapor disengagement) clearly suggests that the 2000-gallon vessel also should have vented safely.

REFERENCES

1. Fauske, H. K., et al., "Emergency Pressure Relief Systems Associated with Flashing Liquids," Swiss Chem., Nr 7/8, pp. 73-78, 1980.

2. Duxbury, H. A., "Relief System Sizing for Polymerization Reactions," The Chemical Engineer, pp. 31-27, 1980.

3. Boyle, W. J., "Sizing Relief Area for Polymerization Reactors," Chem. Eng. Prog., 63, No. 8, pp. 61-66, 1967.

4. Huff, J. E., "A General Approach to the Sizing of Emergency Pressure Relief Systems," Paper presented at the DECHEMA's 2nd Intl. Symp. on Loss Prevention and Safety Promotion in the Process Industries, September 1977.

5. Booth, A. D., et al., "Design of Emergency Venting System for Phenolic Resin Reactors - Part 1 and Part 2," Trans. J. Chem. Eng., Vol. 58, 1980.

6. Gartner, D., Giesbrecht, H., and Leuckel, W., "Effects of Two-Phase Flow Upon the Emergency Relief of Reactor Tanks and Upon the Subsequent Atmospheric Jet Dispersion," Proc. 3rd Intl. Symp. on Loss Prevention and Safety Promotion in the Process Industry, Baske, Switzerland, September 15-19, 1980.

7. Fauske, H. K., Grolmes, M. A., and Henry, R. E., "Emergency Relief Systems - Sizing and Scale-Up," Plant/Operations Progress, Vol. 2, No. 1, January 1983.

8. Grolmes, M. A., and Fauske, H. K., "An Evaluation of Incomplete Vapor Phase Separation in Freon-12 Top Venting Depressurization Experiments, see these proceedings.

9. Henry, R. E., and Fauske, H. K., "The Two-Phase Critical Flow of One Component Mixtures in Nozzles, Orifices and Short Tubes," J. Heat Transfer, 93, pp. 179-189, 1971.

10. Gubler, M., Hannemann, R. J., and Sallet, D. W., "Unsteady Two-Phase Blowdown of a Slashing Liquid from a Finite Reservoir," Two-Phase Flows and Reactor Safety, Momentum Heat and Mass Transfer in Chemical Process and Energy Eng. Systems, Vol. 2, Hemisphere Publishing Company, 1980.

11. Burchett, D. K., "Sizing of Emergency Vents for a Runaway Polymerization Reaction when Liquid Entrainment is a Factor," Presented at the Third Intl. Symp. of Loss Prevention and Safety Promotion in the Process Industries, Basel, Switzerland, September 1980.

12. Harmon, G. W., and Martin, H. A., "Sizing Rupture Discs for Vessels Containing Monomers," Preprint No. 58a, 67th Natl. Mtg. AIChE, February 1970.

13. Huff, J. E., "Computer Simulation of Polymerizer Pressure Relief," _AIChE Loss Prevention_, Vol. 7, 1973.

14. Fauske, H. K., "The Discharge of Saturated Water Through Tubes," _Chem. Eng. Prog. Symp. Series_, Vol. 61, No. 59, 1965.

15. Sozzi, G. L., and Sutherland, W. A., "Critical Flow of Saturated and Subcooled Water at High Pressure," ASME Symposium on Non-Equilibrium Two-Phase Flow, Winter Annual Mtg., Houston, TX, November-December 1975.

16. Flinta, J., Hernborg, G., and Akesson, H., "Results from the Blowdown Test for the Exercise European Two-Phase Flow Group Meeting," Denmark, June 1971.

17. Uchida, H., and Nariai, H., "Discharge of Saturated Water Through Pipes and Orifices," _Proc. Third Intl. Heat Transfer Conf._, Chicago, IL, Vol. 5, 1966.

18. Fletcher, B., "Flashing Flow Through Orifices and Pipes," Paper presented AIChE 17th Loss Prevention Symposium, Denver, CO, August 28-31, 1983.

19. "Critical Flow Data Review and Analysis," EPRI/NP-2192, Report prepared by S. Levy, Inc., 1982.

20. Grolmes, M. A., and Leung, J. C., "Scaling Considerations for Two-Phase Critical Flow," see these proceedings.

Mul
edit
Else

LAR

T.
Law
Live

ABST

heat
rapi
prod
equi
test
Nati
were
(NWC
was
rapi
larg
faci
the
be r
Chin
thos
scal
RPT
the

1.

team
(DOE
eval
(198
prin
LNG
a va
when
expl
the
much
haza
sion
they

912

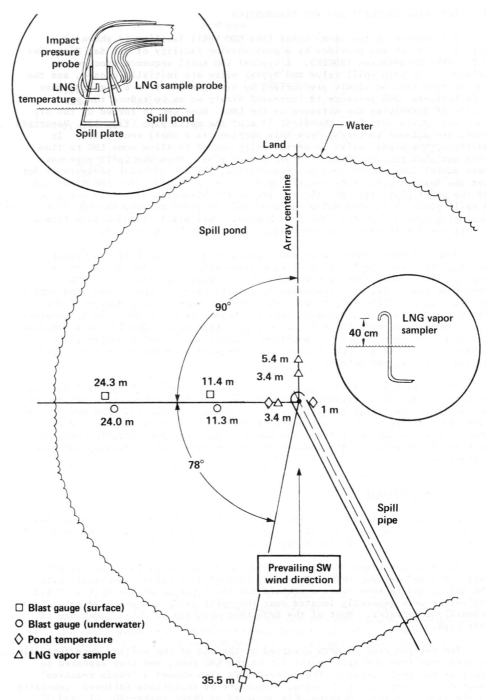

Figure 2. Coyote Series RPT Diagnostics

TABLE 1A. China Lake Burro series summary.

Test	Estimated Tank Composition (% Vol) CH_4, C_2H_6, C_3H_8	Spill Rate (m^3/min.)	Spill Volume (m^3)	Spill Plate Depth (cm)	Impact Pressure Max/Average (psia)	Pond Temperature (°C)	RPT Explosions	Max Point Source Yield (kg TNT)
B2	91.3, 7.2, 1.5	11.9	34.3	5			–	
B3	92.5, 6.2, 1.3	12.2	34.0	5			–	
B4	93.8, 5.1, 1.1	12.1	35.3	Below water	Not	Greater	–	
B5	93.6, 5.3, 1.1	11.3	35.8	At water level	Measured	than	–	
B6	92.8, 5.8, 1.4	12.8	27.5	?		17°C	Large delayed	??
B7	87, 10.4, 2.6	13.6	39.4	Above water			–	
B8	87.4, 10.3, 2.3	16.0	28.4	Above water			–	
B9	83.1, 13.9, 3.0	18.4	24.2	5 (initially)			Large early	3.5

TABLE 1B. China Lake Coyote series summary.

Test	Estimated Tank Composition (% Vol) CH$_4$, C$_2$H$_6$, C$_3$H$_8$	Spill Rate (m³/min.)	Spill Volume (m³)	Spill Plate Depth (cm)	Impact Pressure Max/Average (psia)	Pond Temperature (°C)	RPT Explosions	Max Point Source Yield (kg TNT)
C1	81.7, 14.5, 3.8	6	14	30	0.8/0.2	30	Small early / Large delayed	??
C2	70.0, 23.4, 6.6	16	8	2.5	5/5	27.6	Small early	0.23
C3	79.4, 16.4, 4.2	13.5	14.6	2.5	10/6	22.8	–	
C4a	78.8, 17.3, 3.9	6.8	3.8	25	2.4/0.4	22.4	Small early	0.001
C4b	78.8, 17.3, 3.9	12.1	6.0	25	5/3	20.6	–	
C4c	78.8, 17.3, 3.9	18.5	5.2	25	10/5	20.2	Large early	1.5
C5	74.9, 20.5, 4.6	17.1	28	6	13/8	17.2	Large delayed	3.0
C6	81.8, 14.6, 3.6	16.6	22.8	5	13/8	15	–	
C7	99.5, 0.5, 0	14.0	26	33	15/6	13.6	–	
C8a	99.7, 0.3, 0	7.5	3.7	33	2/0.6	12.8	–	
C8b	99.7, 0.3, 0	14.2	5.4	33	10/4	12.7	–	
C8c	99.7, 0.3, 0	19.4	9.7	33	14/11	12.3	–	
C9a	LN$_2$	7.2	3.6	36	2/0.2	14.1	–	
C9b	LN$_2$	9.9	3.3	36	8/3	14.8	–	
C9c	LN$_2$	13.3	8.2	36	15/10	15.8	–	
C10a	70.2, 17.2, 12.6	13.8	4.6	36	8/5	10.6	–	
C10b	70.2, 17.2, 12.6	19.3	4.5	36	14/10	10.6	–	
C10c	70.2, 17.2, 12.6	18.8	5.0	Removed	12/9	11.6	Small early	0.005

The air-blast TNT equivalents of the observed RPTs, presented in Table 1 and later in this report, were calculated from peak overpressures measured at known distances from the spill point. These yields represent only that portion of the explosion which produced the air shock, as no underwater data on over-pressure were obtained due to equipment malfunction. The calculations assume that the explosion is a point source, similar to TNT in energy release times, and that the surface shock wave reflection produces an overestimate of the explosive energy by a factor of 1.8. In earlier reports of the RPT yields [2,3], the surface reflection reduction was not imposed. The tank composi-tions in Table 1 were calculated using assays provided by SDG&E and a tank boiloff rate of 1 m^3 of LCH_4 per day. A linear interpolation of the SDG&E assay data was used to estimate the composition loaded on the tanker truck, with an additional 1 m^3 of LCH_4 assumed to have evaporated in the transfer process. Assays were taken by NWC personnel prior to each spill; however, the sampling technique was not accurate enough to be useful. The LNG exit liquid composition data was also unusable due to a malfunction of the gas analyzer system. The spill rate is an average value determined by dividing the total volume spilled by the spill duration.

Before beginning the analysis of the data of Table 1, there are a few more general comments about the facility which should be made, and comments about some tests in particular. First of all, there is the possibility of further boiloff enrichment of the contents of the first 53 m of the spill pipe during the cooldown period. Estimates by Lind [4] (the NWC facility director) indi-cate that it requires 42,800 kcal to cool this section of pipe from 20°C to −161°C and an additional 4700 kcal/hr to keep it at this lower temperature. Due to the small size of the spill pipe vent valve, Lind estimates that the volume of the 53-m section of pipe is at most only half full (~ 1.1 m^3). If one uses the richest tank composition of Table 1 (test C2) and assumes that the cooldown requirements previously mentioned are due to the evaporation of CH_4 only, the composition of the 1.1 m^3 of LNG in the 53-m section of the spill pipe prior to the spill is found to be 54% CH_4, 34% C_2H_6, and 12% C_3H_8. This would be the first and most enriched LNG to spill onto the pond. Furthermore, at the maximum spill rate it would take about 20 sec after the spill valve begins to open for the LNG from the tank itself to reach the pond. Because of these circumstances it would be desirable to know the outlet composition of the LNG during the spill. An attempt was made to make this measurement during the Coyote Series tests, but the results are of questionable accuracy, hence the actual composition of the initial portions of each spill are not known.

The B9 spill was unique in several ways. At some time after the beginning of this spill but before the largest RPT at 35 sec, the spill plate was blown loose, greatly changing the dynamics of the LNG/water interaction. The largest RPT observed during all of the China Lake tests occurred during test B9, when there was no spill plate and the spill rate was near maximum. A list of the largest of these explosions is given in Table 2. As can be seen, the RPTs began immediately and persisted for over a minute. Only those RPTs above the dashed line in Table 2 occurred early enough that they could have been affected by the cooldown-enriched LNG; all below the dashed line occurred while LNG of storage-tank composition was spilling out of the pipe. The B9 spill was unique in that it was the only spill conducted without a spill plate, at the maximum spill rate, and with the pond temperature greater that 17°C. As will be dis-cussed later, these three parameters (spill plate depth, spill rate, and pond temperature) are the important ones which define the occurrence of RPTs during large-scale spills.

Several of the spills during the Coyote series were conducted solely for the purpose of studying RPTs (C4, C8, C9, C10). Each of these typically con-sisted of three short spills at low, medium, and high spill rates. The other

TABLE 2. Occurrence times and magnitudes of major Burro 9 RPT explosions. Dashed line marks time at which all cooldown-enriched LNG has been spilled.

Time[a] (s)	Side-on Pressure[b] (psi)	TNT equivalent[c] (g)
6.5	0.12	36
7.1	0.15	64
9.2	0.27	295
21.4	0.57	1890
35.1	0.72	3500
43.2	0.10	23
46.0	0.12	36
54.1	0.12	36
54.9	0.13	45
66.9	0.19	120
72.7	0.12	36

[a] t = 0 is start of spill-valve opening.
[b] Measured at distance of 30 m.
[c] Equivalent free-air point-source explosion of TNT.

parameters (LNG composition, spill plate depth, water temperature, etc.) were essentially constant during the three consecutive spills. Spill tests with liquid methane (C8) and liquid nitrogen (C9) were performed to examine the effect of composition on the occurrence of RPTs. Unfortunately, the results of these two tests were clouded by the pond water temperature effect, which is discussed in the next section. The C1 spill was to have been a three-spill sequence; however, the spill valve jammed in the open position during the first spill allowing the entire contents of the tank to slowly spill out over a period of about six minutes. Most of the LNG was spilled during the first two minutes, but a small quantity continued to dribble out for at least four more minutes as the tank slowly depressurized. At about five minutes, six to eight large RPTs occurred. These were located at the edges of the LNG pool and occurred in rapid succession as if the initial explosions triggered the rest. Although no overpressure data were obtained during this spill, observers compared the RPTs to those of test C4c, equivalent to about 1 kg of TNT.

4. RPT DATA ANALYSIS

Correlations between the various recorded parameters and the frequency and yield (magnitude) of the RPTs can be determined from the data of Table 1. First, the effect of water temperature and spill plate depth are examined with regard to the occurrence of early RPTs. Correlations of RPT yield with impact pressure (Section 4.2), spill rate (Section 4.3), and LNG composition (Section 4.4) are discussed next. Some constraints have been imposed on the spills used in these last three comparisons. First, the emphasis here is primarily on the early type of RPTs. The delayed RPTs are believed to be a result of boiloff enrichment and consequently are not as repeatable as the early RPTs. Second, only those spills for which the water temperature was greater than 17°C, and the spill plate beneath the surface of the water are considered. The last section (4.5) deals with the energetics of the RPT explosions as determined by the overpressure measurements and point source calculations.

4.1. Water Temperature and Depth

Examination of the pond temperature column of Table 1 shows that, with one exception (test C10c), all of the RPTs occurred for water temperatures greater than 17°C. In fact, the first two spills of the C10 sequence repeated the spill conditions of the earlier spills (tests C1-C5) except for the cold water temperature (10.6°C). These two spills (C10a, C10b) confirmed our suspicions that the colder water was inhibiting the occurrence of RPTs. Several RPTs occurred during the C10c spill, although the water temperature was less than 17°C, indicating that removal of the spill plate and allowing the LNG to interact with the pond bottom also effects the occurrence of RPTs.

Numerous laboratory-scale experiments [1,5,6] involving simple spills of hydrocarbon cryogens on water indicate that the ratio of the water temperature (T_W) to the cryogen superheat limit temperature (T_{SL}) must be in the range of 1.0-1.1 in order for RPTs to occur. But the results of the larger-scale Burro and Coyote spills are contrary to these laboratory-scale results. For the most enriched spill (test C2), the ratio of the superheat limit of the mixture (222 K) to the measured pond temperature variations (290-303 K) is 1.31-1.36. This includes the effect of the cooldown enrichment (54% CH_4, 34% C_2H_6, 12% C_3H_8). The ratios would be even larger if the tank-mixture superheat limit temperature was used. According to the laboratory-scale results, the colder water temperatures should have <u>increased</u> the occurrence of RPTs, and not inhibited them as the data indicates. The pond temperatures necessary for the China Lake data to conform to the laboratory-scale results (T_W/T_{SL} = 1.0-1.1) are well below the freezing point of water.

Our experiences during the Burro and Coyote series also led us to believe that the depth of water above the spill plate had an effect on the occurrence of RPTs. No early RPTs occurred when the spill plate was located at or above the water surface, while the largest RPTs occurred when the spill plate was absent (B9). To further verify this hypothesis, the last spill (C10c) was conducted without a spill plate. RPTs did occur on this spill, even for a pond temperature of 11.6°C. It thus appears that both the depth of penetration into the water and the water temperature are important factors affecting the occurrence of RPTs for large-scale tests.

4.2. Impact Pressure

The concept of dynamically collapsing the vapor barrier in a liquid/liquid film boiling system to achieve superheating has been pursued by several investigators [7,8]. The most extensive study of this impact/collapse concept was done by Jazayeri and Reed at MIT [8] and a summary of these results is included in Fig. 3. In these experiments, the water was actually impacted upon the cryogens (methane, ethane, propane, nitrogen), and the results were reported in terms of peak overpressures recorded at a distance of 10 cm and impact velocities determined from high-speed photography. In Fig. 3 the overpressure data have been converted to source yield using the free-air point source equivalent. The impact velocities have been converted to impact pressures using $p = 1/2\ \rho V^2$, where ρ is the density of the water and V is the impact velocity. The MIT results show a gradual increase in RPT strength with impact pressure. Furthermore, the experiments also showed that cryogen/water systems which did not produce RPTs in simple spills could be made to do so with sufficient impact pressures.

The impact pressure and explosive yield data from the China Lake experiments are also shown in Fig. 3. They indicate no real trend of RPT yield with impact pressure for the large-scale spills. There is, however, a considerable

918

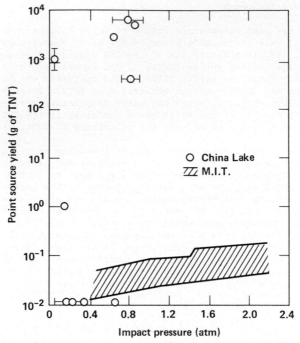

Figure 3. Comparison of China Lake and MIT Impact
Pressure Data with RPT Explosive Yield.

increase in the magnitude of the RPTs going from the small to the large spills;
approximately five orders of magnitude at an impact pressure of 0.8 atm. The
observation of a fairly large RPT at an impact pressure of less than 0.1 atm
(C1) and no RPT at an impact pressure of 0.7 atm (C3) indicates that impact
pressure alone may not be a dominant parameter responsible for RPTs in large-
scale spills.

4.3. Spill Rate

When RPT yield is compared with LNG spill rate a definite trend is apparent and is shown in Fig. 4. Of all the parameters examined for correlation
with RPT yield, spill rate produced the best agreement. The data indicate an
apparent threshold or abrupt increase in the RPT explosive yield at a spill
rate of about 15 m^3/min. Unfortunately, this rapid increase of RPT yield
occurs close to the spill rate limit of the China Lake facility. Consequently,
the scaling of the RPT yield to higher spill rates is uncertain at this time
and needs to be investigated further. There was no correlation between RPT
yield and the total volume of LNG spilled. This lack of correlation with spill
volume, but good correlation with spill rate, seems to offer two possible
mechanisms which may affect the occurrence and magnitude of large-scale LNG
RPTs. One would be the increased mixing of the LNG and water associated with
the higher spill rates. The other is the increased mass of LNG interacting
with the water at any instant, also associated with the higher spill rates.
Further experiments, with extensive diagnostics, will be required to resolve
the mechanisms responsible for the RPT/spill rate correlation. Ultimately the

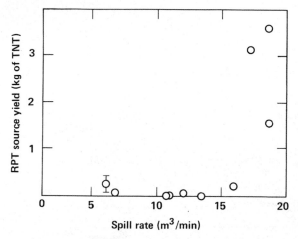

Figure 4.   The Effect of Spill rate on RPT Yield for Large-Scale Spills.

question of the possible magnitude of the LNG RPT explosions for spill rates greater than 20 m^3/min must be addressed, as current dockside loading of LNG tankers involves flow rates of about 50 to 100 m^3/min. The lack of data at high spill rates and uncertainties about RPT mechanisms make it impossible to scale current results to these larger spill rates. Experiments to study the RPT yield variation with LNG composition, water temperature and depth of penetration for spill rates of at least 50 m^3/min are necessary to address the possible hazards of the loading arm accident.

4.4. LNG Composition

In the past ten years, a considerable effort has been spent on the correlation of LNG composition and the possibility of RPT explosions [1,6,7]. In the early 1970s, Enger and Hartman [1] of the Shell Pipeline Corp. conducted one of the most thorough and extensive studies of LNG/water RPTs to date. Even though the amount of LNG spilled was small ($\sim$ 0.1 m^3), the number of spills was large, lending considerable credibility to the results. A summary of these results for the methane, ethane, and propane mixtures is shown in Fig. 5. Enger and Hartman concluded that only LNG with compositions within the envelope shown in the figure, or LNG allowed to "age" into this envelope, could produce RPT explosions. Small amounts of n-butane and water temperature variations tend to enlarge the envelope of Fig. 5; however, Enger and Hartman state that no early type RPTs ever occurred for LNG mixtures containing more than 40% methane. The large-scale China Lake RPT results for the early RPTs, also shown in Fig. 5, are clearly well out of the Enger-Hartman envelope. Furthermore, there is a lack of consistency for the large-scale spills, i.e., LNG of similar composition may or may not produce an RPT. Apparently there are other mechanisms, besides the prespill composition, which dominate the occurrence of RPTs for these large spills. The asterisk in Fig. 5 is an estimate of the maximum amount of cooldown enrichment for the most enriched LNG of the Coyote series (C2). This calculation shows that even the first cubic meter of LNG during that spill was still well out of the Enger-Hartman envelope.

This does not mean that enrichment has nothing to do with RPT explosions for LNG spilled on water. Certainly the delayed type of RPT must be a result

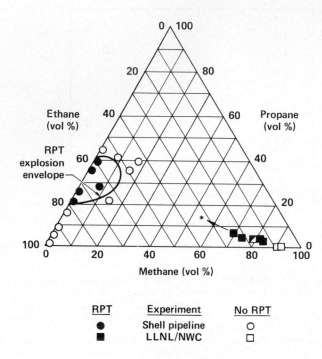

Figure 5. Large-Scale RPTs and the Enger and Hartman RPT Explosion Envelope.

of enrichment or "aging" on the pond. It is even quite possible that some form
of local or transient enrichment may play a part in the early type of RPTs.
Nonetheless, for large-scale spills the results indicate that large RPTS can
occur even for methane prespill compositions as high as 90%.

Since this is an extremely important conclusion with regard to LNG spills
on water, much time has been spent on the verification of the tank assays given
in Table 1. As mentioned previously, the data obtained by the LNG exit sample
probe is not reliable enough to be used as an accurate measure of the composition of the LNG that actually reached the pond. One other measurement, the
LNG exit temperature, does offer some information concerning the spilled LNG
composition, and seems to substantiate the tank composition calculations of
Table 1. The temperature of an equilibrium mixture of methane and ethane is
given in Fig. 6 [9]. Obviously, measurement of the mixture temperature is not
a very sensitive method for determining compositions for methane concentrations
greater than about 50% since the mixture temperature only changes 8°C over this
range (50-100%). However, one would expect to see LNG exit temperatures warmer
than -150°C if the mixture composition were to approach the Enger RPT compositional envelope of Fig. 5. The LNG exit temperature for a typical spill involving an early RPT (C2) is shown in Fig. 7. The accuracy of these temperature
measurements is ± 1°C in the range of -170 to -150°C (103-123°K), as was verified in situ during the pure CH_4 (C7 and C8) and LN_2 (C9) spills. The
average exit temperature of the C2 spill (-154°C) corresponds to a methane concentration of about 64% (Fig. 6), which is quite close to the 70% calculated
using the SDG&E data (Table 1). Analysis of the LNG exit temperatures for all
of the Coyote series spills substantiates the estimated tank assays of Table 1.

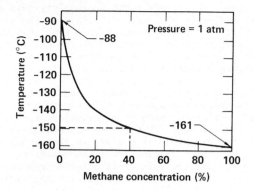

Figure 6. Equilibrium Temperature Versus Composition
for a Liquid-Methane/Ethane Mixture.

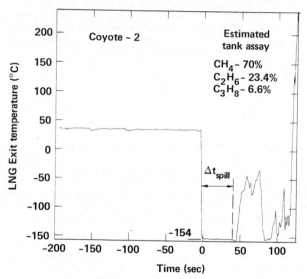

Figure 7. LNG Exit Temperature.

4.5. RPT Energetics

A very important issue concerning LNG/water RPTs has to do with the energetics of the explosion itself. We are concerned with several important
questions including: How large can RPTs get? Is the blast wave from an RPT
explosion equivalent to that from a chemical explosion? In order to gain some
preliminary insight into the energetics of LNG/water RPT explosions, blast-wave
overpressure measurements at various distances (both above and below the water
surface) were attempted during the Coyote test series by Lind, the NWC test
director. Unfortunately the underwater blast gauges did not operate reliably,
hence a complete analysis of the energetics of the China Lake LNG/water RPT
explosions is not possible. Nevertheless, the existing surface-blast pressures
will be discussed and analyzed as to accuracy and any other possible implications they may contain which may be of help to future experiments.

As mentioned previously, the air-blast TNT equivalents used in this report were calculated from peak reflected overpressure data measured at known distances from the spill point. The calculations assume that the explosion is essentially a point source at the pond surface, that the energy release is on the same time scale as that of TNT, and that the blast wave has reached the pressure gauge by propagating unobstructed through air at a uniform temperature and pressure. None of these assumptions are rigorously correct. It is not known how much LNG is involved in these RPT explosions, nor are the kinetics of the energy release completely understood at this time. However, any pressure disturbance of finite amplitude which is produced in times less than 0.1 sec will produce a shock wave within a distance of approximately 10 source diameters. Consequently, if the magnitude of the RPT is large, and if pressure measurements are made at the proper distance from the center of the exploding mass of LNG, the point-source TNT equivalents calculated from the overpressure data should give a good relative estimate of the total energy release.

The more critical violations of the free-air point source assumptions for the TNT equivalent calculations involve the blast wave propagation to the point of the measurement, and the attenuation of the air shock due to the submergence of the explosion beneath the water surface. For the RPT explosions at China Lake the blast wave does not propagate through a homogeneous medium. It originates and initially travels through a very cold (-161°C), foggy, methane/air mixture, which within a few meters changes to air at typically 30°C. The analysis is further complicated since the origin and spatial extent of the explosion is not known precisely and the cold foggy vapor cloud is not symmetrical. This means that the effect of the inhomogeneous propagation medium on the blast wave structure is directional. It is not surprising then that overpressure data from the same RPT explosion, but recorded at different distances and orientations, do not produce the same point-source TNT equivalents.

The TNT equivalents as calculated for two RPT explosions which occurred during the C4c and C5 spills are shown in Table 3. The values shown here differ somewhat from those of Table 2 because of a difference in the way the pressure data were processed. The overpressure data used for the calculations of Table 3 were processed with a system response of 20 kHz at LLNL, whereas the TNT equivalents of Tables 1 and 2 were calculated from the data processed at only a 1 kHz response at the NWC. The yields for each explosion have been calculated using both the peak reflected overpressure measurements and the pressure impulse, and are based on the empirical blast parameter computations of Baker [10]. The yields, as determined from the peak pressure measurements, for the C4c RPT at 12.8 sec vary by as much as a factor of 12 (0.1-1.2 kg TNT), while those of the C5 RPT at 101 sec vary by as much as a factor of 4 (1.1-4.4 kg TNT). The correlation of the calculated yields for the C5 RPT using the 11.4-m and 35.5-m peak overpressures is most likely fortuitous, since at their respective locations (Fig. 2) one would expect the greatest differences in blast wave propagation characteristics due both to asymmetry conditions and to interactions with the spill pipe and its support structures. An additional and typical correction for an explosion on a flat surface is to reduce the yields as calculated by the free air point source calculation by 56%. This correction is a result of the reflection of the spherical blast wave from the pond surface.

Since an RPT blast wave is likely to be different than its equivalent TNT counterpart at the point of initiation, calculations of free-air point source yields based on pressure impulse (the area under the pressure-vs-time data) should produce more consistent results than those using peak overpressures. The TNT equivalents calculated using pressure impulses are also listed in Table 3; as expected, the variation in the yields is reduced considerably. Estimates of the surface yields of LNG/water RPTs should be made using pressure impulse data rather than peak overpressure data.

TABLE 3. Explosive yield variations for two RPT explosions.

(a) C4c RPT @ t = 12.8 sec		
Gauge location (m)	Yield from peak pressure (kg of TNT)	Yield from impulse (kg of TNT)
11.4	0.6	3.2
24.3	1.2	3.1
35.5	0.1	1.1

(b) C5 RPT @ t = 101 sec		
Gauge location (m)	Yield from peak pressure (kg of TNT)	Yield from impulse (kg of TNT)
11.4	1.1	2.0
24.3	4.4	5.5
35.5	1.1	3.2

The pressure-vs-time data for the two RPTs of Table 3 are shown in Figs. 8 and 9. Similar time spans for the pressure data in each figure enable comparisons of the structure of the shock wave at three different locations. The general shape and durations of the overpressures are typical of those from chemical explosions of equivalent yields. The initial pressure spike seen in all but the C4c data at 11.4 m is of too short a duration to be an air shock. If the spike is due to the RPT explosion, then that portion of the blast wave pressure has been transmitted via the blast gauge pipe support system. Although this spike is not completely understood at this time, it is not of primary interest here. The airblast component of the pressure pulse is our main interest. Consequently, these spikes are not included in the TNT equivalent calculations of Table 3. However, to a certain extent they may have influenced the calculated TNT equivalents of Table 1 and earlier reports [2,3]. As mentioned earlier, these pressure data were processed by the NWC using a recorder with a maximum frequency response of 1 kHz, whereas those reported in Table 3 and Figs. 8 and 9 have been processed at 20 kHz. The degree to which this spike was included in the overpressures and corresponding RPT yield estimates of Tables 1 and 2 is unknown. If the error was consistent throughout the test series, then the yields are still useful for comparing the air blast of one spill with that of another. Luckily, the slower response data processing actually reduced the erroneous pressure spike contribution to such a degree as to agree with the more accurate and appropriate pressure impulse results for TNT equivalents (see Tables 1 and 3). However, while accounting for surface shock reflections and using the pressure impulse rather than a peak overpressure will certainly improve the accuracy of the surface component of the RPT yield estimates, one must have both surface and underwater blast data to accurately determine the energetics of LNG/water RPT explosions.

924

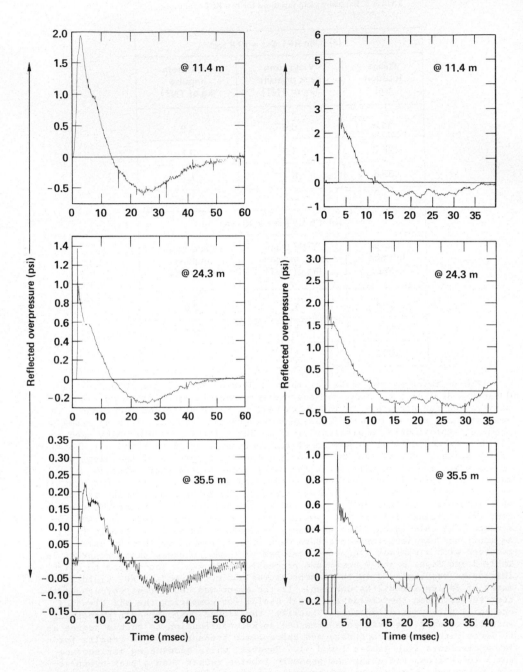

Fig. 8. Overpressures at Three
Locations for a C4c RPT Explosion.

Fig. 9. Overpressures at Three
Locations for a C5 RPT Explosion.

TABLE 4. Water depth effect on explosive mass calculations.

	Calculated Mass		
Actual mass	2.5 cm deep	30 cm deep	Calculated mass/ actual mass @ 30 cm
61 g	126 g	0.065 g	1.1×10^{-3}
61 g	46 g	0.065 g	1.1×10^{-3}
101 g	230 g	0.13 g	1.3×10^{-3}

The accuracy of calculated free air yields is further in doubt for explosions which occur underwater. In their study of LNG/water RPT explosions, Enger and Hartman [1] realized this and stressed the importance of both surface and underwater blast measurements. In order to estimate the effect of the submergence, they set off numerous explosions of known yields at two different depths underwater. Although both surface and underwater overpressure measurements were made, there was considerable scatter in the underwater data due to shock reflections inside the small 1-m-deep tank. A comparison of actual and calculated yields for those cases involving the same amount of explosive, but at two different depths (2.5 cm and 30 cm) is shown in Table 4. The explosive mass calculations of Table 4 were made using the air-blast overpressure data as reported by Enger and Hartman [1]. As can be seen, the calculations are very sensitive to the depth of the explosive under the water. In fact, using air-blast data alone, one can easily underestimate the yield of an explosion 30 cm underwater by a factor of 1000. The point-source yield calculations appear to be quite accurate for the 2.5-cm-deep explosions if one uses the 1.8 (56%) surface reflection correction [10].

4.6. Delayed RPTs

The primary emphasis of the data analysis up to this point has been on early RPTs. This type of RPT, until recently [2,3], was not expected to occur for the LNG compositions used in the Coyote experiments. Delayed RPTs have occurred in other experiments [1], and, being attributed to boiloff enrichment of the heavier hydrocarbons, are possible for practically any LNG composition spilled on water. Boiloff enrichment was observed in all three China Lake test series (Avocet, Burro, Coyote). Whereas measurement of the liquid composition during a spill proved to be most difficult, the LNG vapor sampling diagnostic (Fig. 2) appeared to function quite well. This system was designed to determine the time of occurrence of liquid pool enrichment by monitoring the increase in the ethane and propane vapor concentrations. Valencia-Chàvez [9] conducted numerous laboratory-scale experiments at MIT in which he monitored the composition of the LNG vapors as a function of the mass evaporated for different initial mixtures of methane, ethane, and propane. These experiments were performed by quickly spilling from 0.15 to 1 liter of the LNG mixture into a well-insulated vessel containing approximately 1.7 liters of distilled water. The results of one of these experiments are shown in Fig. 10. Similar

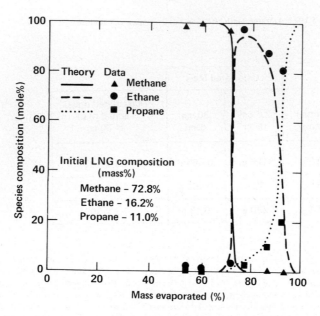

Figure 10. LNG Vapor Composition as a Function of Mass Evaporated.

results were observed during the Avocet and Burro test series at China Lake.
The theoretical curves of Fig. 10, are the result of calculations assuming
thermodynamic equilibrium between the liquid and vapor phases, and using a
Redlich-Kwong equation of state with binary interaction parameters selected
from a best fit to existing experimental data. These results show that essentially no ethane or propane vapors appear until practically all of the liquid
methane has evaporated. Consequently the vapor composition above the spill
pond may be used as an indicator of the pond's LNG enrichment. Therefore the
China Lake pond vapor sample data, if it were representative of the LNG responsible for the explosion, should indicate the degree of enrichment present when
the delayed RPTs occurred. For purposes of comparison with the China Lake tank
compositions, the mole percentage (% volume) equivalents of the initial LNG
mass percentage of Fig. 10 are 85.2% methane, 10.1% ethane, and 4.7% propane.

Only two spills produced delayed RPTs during the Coyote test series (C1
and C5). The results of the pond vapor sample measurements for the C5 spill
are shown in Fig. 11. Due to the spill valve malfunction of test C1, the pond
sample concentrations were too low to be of value for this analysis. The enrichment ratio is the ratio of the number of moles of the heavier hydrocarbons
(ethane + propane) divided by the total number of moles of the mixture (methane
+ ethane + propane). The figure shows the temporal variation of the enrichment
ratio of the LNG vapors just above the pond surface and the times of occurrence
of RPT explosions. According to Fig. 10, an increasing enrichment ratio should
occur at the end of the spill when the methane component is essentially gone
and the mixture should be within the Enger and Hartman RPT compositional envelope (Fig. 5). The delayed RPTs of the C5 spill appear to fit the expected
pattern; they begin to occur just as the enrichment ratio starts to increase at
about 110 sec. We conclude that the delayed RPTs which occurred at China Lake
are a result of boiloff enrichment (aging), which drives the composition of the
spilled LNG into the RPT explosion envelope (Fig. 5).

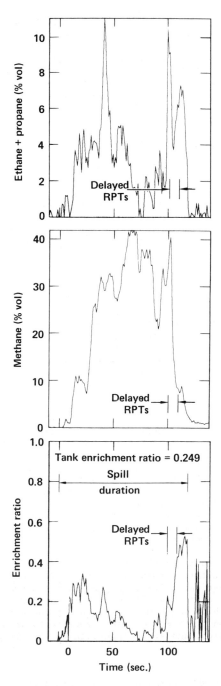

Figure 11. Comparison of Delayed RPT Occurrence and Vapor Composition for the C5 Spill.

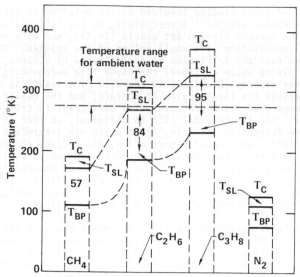

Figure 12. Relationship of Ambient Water Temperature and
Cryogen Critical, Superheat Limit, and Boiling-Point Temperatures.

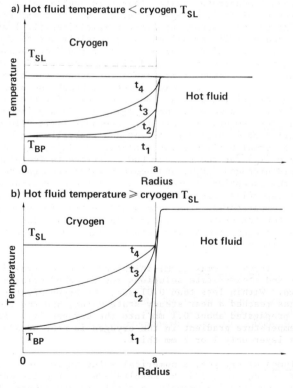

Figure 13. Transient Heat Conduction to a Small Cryogen Sphere of Radius a.

transfer is considered in this example. (The presence of convective cells within the small droplet would tend to increase the rate of heat transfer.) Two different cases are depicted in Fig. 13: (a) one where the hot fluid (water) temperature is less than the cryogen T_{SL}, and (b) one where the hot fluid temperature is greater than the cryogen T_{SL}. The development of the temperature gradient in the cryogen is shown at four different times for each case (t_1, t_2, t_3, t_4). In Fig. 13(a) the initial development of the temperature gradient is much the same as for the semi-infinite case (t_1), that is until it reaches the center of the sphere (t_2). From this point on, the interface temperature begins to increase (t_3) and bulk heating of the drop begins. After a sufficient period of time (t_4), the entire drop will equilibrate to the water temperature. There will be no vaporization in this case because of a lack of nucleation sites; in fact, it is this same heat transfer process which occurs in the bubble column experiments [16,17,18] used to experimentally measure the superheat limits of various cryogens. For methane, ethane, and propane drops with diameters less than 1 mm, surrounded by another fluid at a higher temperature, this equilibration of temperatures is accomplished within a few sec.

We now consider the example in Fig. 13(b) where both the hot fluid temperature and the initial interface temperature are greater than the cryogen T_{SL}. This example involves two critical assumptions. The first is that in order for stable film boiling to be initiated in the liquid/liquid heat transfer regime, the cryogen interface temperature must be at or very near to T_{SL}. While there are sound qualitative arguments to support this assumption, no experimental data exists with which it can be substantiated. The second assumption is that the flattening of the temperature gradient in the interior of the drop (bulk heating) is much faster than the reduction of the drop radius due to evaporation. If this were not true, the drop would evaporate before bulk heating to T_{SL} could occur. Once again the validity of this second assumption is difficult to assess, and will most likely require a series of clever experiments and/or sophisticated calculations.

Under the two previously discussed conditions, we now examine the heat transfer for the droplet of Fig. 13(b). As the temperature gradient begins to form (t_1), the interface temperature approaches T_{SL} with no vapor generation. However, when the interface temperature reaches T_{SL} it can go no higher, and the cryogen begins to evaporate in the stable film boiling mode (t_2). At this point the droplet interface temperature would most likely begin to decrease, as most of the heat conduction to the cryogen would be lost to the vaporization process. However, should the droplet interface temperature remain at T_{SL}, then we would have a case of maximum heat transfer to the cryogen and the times required for bulk heating under these conditions would be a minimum. This scenario is depicted in Fig. 13(b), i.e., the cryogen interface temperature remains at T_{SL} and heat transfer continues within the drop interior (t_3) until the entire volume reaches T_{SL} (t_4) and explodes.

Numerous transient heat-transfer calculations for methane drops of various radii have been made with the assumptions that the drop interface temperature is constant at T_{SL} (169 K) and that the drop radius does not change with time. Since the interface is held constant at T_{SL}, the results give an indication of the minimum time required for bulk heating to occur for small cryogen drops. The calculations were made using a two-dimensional implicit finite-element heat transfer code [19], and the results for methane drops of 1-mm and 2-mm radii are shown in Fig. 14. The physical properties used were those of liquid methane at its normal boiling point. In the case of the 1 mm radius drop (Fig. 14a), bulk heating is accomplished in 5 sec. For the China Lake spill conditions, it is not expected that small buoyant LNG drops would remain submerged for periods longer than about 5 sec. A 2-mm radius drop requires in excess of 15 sec for complete bulk heating to T_{SL}. These calculations indicate that in order for a

5.2. Propagation Processes

There are currently four propagation models which address the theory of cryogen/water RPT explosions, those of Anderson [14], Fauske [15], Fowles [21], and Harlow [22]. The models of Anderson and Fauske include initiation processes (discussed in the previous section), while the models of Fowles and Harlow concern themselves only with the propagation process itself.

<u>Anderson</u>. This model [14] is composed of three phases: a preliminary mixing phase, a trigger and growth phase, and a shock-wave propagation phase. The model was conceived as a result of numerous experiments involving the interaction of 1-3 liters of Freon 22 with water. In some experiments the Freon was injected unconstrained into the water, while in others the Freon was constrained in a plastic balloon until it was beneath the surface of the water. Peak underwater overpressures in excess of 40 atm. were recorded during these experiments.

The preliminary mixing phase of the Anderson model requires only that the majority of the Freon 22 be under the water surface and in one coherent mass. This requirement is based primarily on the experimental observations that RPTs did not generally occur until this configuration was achieved. It is not clear whether the mixing phase of this model is necessary for the initiation or the propagation phases. The second phase of the model (trigger and growth) includes the generation of a finite pressure pulse produced when random turbulent fluctuations collapse the vapor film layer causing liquid/liquid contact and violent nucleate boiling. This pressure disturbance must be of sufficient magnitude to cause further collapsing of the vapor film and to initiate a shock wave which then propagates along the vapor film surface.

Whereas Anderson admits that the first two phases of his model are hypotheses, he does offer some calculations to support the third phase--the shock propagation mechanism. In this phase, as the initiation shock wave passes over the Freon/water interface, which is in the stable film boiling mode, it accelerates the two liquids toward each other. This produces a high velocity impact which finely fragments and intermixes the two liquids. The high-speed movies show that this shock wave actually travels at subsonic speeds with respect to the speed of sound of either Freon or water. While this is an established fact for shock propagation in two-phase media, these finite overpressures must travel at, or faster than, the speed of sound in the bulk Freon and water surrounding the intermixed film region. Anderson offers few details of the shock wave collapse of the vapor film; however, a process similar to this is also the main thrust of the Harlow model [22] to be discussed later.

Anderson calculates the magnitude of the impact velocity by using the measured overpressure data and a simple one-dimensional momentum balance calculation. Assuming that the shock pressure effects are the same in the Freon as in the water, that the pressure history as recorded at the wall of the water tank represents the interface pressure variations, and that the vapor film thickness is 2.5 cm, the impact velocity is found to be 61 msec. The particular example used in these calculations was not a very severe RPT, producing a peak overpressure of slightly less than 20 atm. It is of questionable accuracy to assume that the pressure history from a transducer mounted in the wall of a 30-cm-diameter water tank is representative of the actual pressure fluctuations which occur in the vapor film. The magnitudes and rise times of these overpressures (20 atm, 1 msec) clearly indicate a shock wave structure; however, it is well known that any finite pressure disturbance abruptly generated in a liquid will steepen into a shock wave in a very short distance. Analysis of underwater blast pressures in small containers is a most difficult task and may not be reliable for accurate estimates of vapor film pressure variations or RPT impulse energetics. Nevertheless, with an impact velocity of 61 m/sec, Anderson's model calculations show sufficient energy to fragment and mix the two liquids into particles on the order of 0.2 mm in diameter. The model assumes that the time of the fragmentation and mixing is of the same order as the film collapse time ($\sim$ 1 m sec), that the pressure of the impacting liquids is sufficient to suppress any vaporization during this mixing process, and that the energy released upon vaporization of the completely fragmented and mixed liquids is of sufficient magnitude and speed to maintain the shock wave propagation. The model calculations involve assumptions for the magnitude of many of the key parameters. The first two phases of the model (mixing, triggering) are based on experimental observations, although more experiments and calculations will be required to actually quantify these effects. The model is weakest in the description of the propagation phase.

Fauske. The model proposed by Fauske [15] also contains three phases: premixing, triggering and propagation. In contrast to the model of Anderson, Fauske's model concentrates primarily on the premixing phase, i.e., if the premixing conditions are fulfilled, the trigger and propagation phases are assumed to follow. With this approach, the claim is made that this model is able to calculate the maximum possible RPT explosion for LNG spills into water. The model proposes that when LNG is injected into water at ambient temperatures, it will fragment into millions of tiny droplets which will be in the stable film boiling mode. These tiny droplets mix with water until an appropriate LNG/water/vapor mixture is formed, at which time a trigger event causes rapid vaporization of all of the droplets, producing the energy necessary for the explosion.

The model uses the simple equation

$$L \simeq 3d \sqrt{\frac{\rho}{\rho_c}}$$

to approximate the penetration length L to total breakup for a jet of LNG of density ρ_c and diameter d when injected into water of density ρ. For the spill conditions of China Lake, the calculation gives L $\simeq$ 1.3 m, indicating that the LNG jet would be almost totally fragmented. Fauske argues that the presence of the spill plate does not alter this effect [15]. Some calculations of LNG penetration depth were performed at LLNL in the early design phases of the China Lake spill facility. The model developed was not designed to predict jet breakup, but rather to estimate the maximum depth of penetration of the coherent LNG jet as a function of spill rate. The model incorporates momentum dissipation due to turbulent entrainment in both the shear layer and the fully developed jet, plus LNG/water buoyancy effects, jet spreading, and buoyancy contributions due to vapor generation. The results from the LLNL calculations show a steady-state penetration depth of 2.8 m for an initial LNG jet velocity of 18.6 m/sec. In the opinion of the author, the penetration length as calculated by the Fauske model is more applicable to the transient injection of a finite LNG mass into water, and therefore would not accurately describe any fragmentation which might have occurred for the steady-state flow conditions of the China Lake experiments.

The Fauske model estimates the average droplet diameter to be about 5 mm, assuming capillary breakup of the LNG jet. This means about 10^7 droplets for each cubic meter of LNG spilled. Using an LNG/water/vapor mixture ratio of 0.3/0.3/0.4 involving 100 kg of LNG and "maximum thermodynamic effiency" of energy release, the model predicts the maximum yield possible for the China Lake experiments to be approximately 10 kg TNT. The maximum thermodynamic energy used in these calculations (443 J/g LNG) was obtained from the calculations of Briscoe and Vaughan [23], and is about twice as large as the maximum available energy (209 J/g) according to the superheat limit concept [1,6]. It is doubtful that <u>all</u> of this energy would be released at the same time. Nevertheless, the maximum yield estimate of this model is very close to the maximum calculated air-blast energy at the China Lake tests. However, as mentioned in section 4.5, calculations of RPT TNT equivalents using air-blast data alone could seriously underestimate the yields if the explosions occurred underwater. This would mean that the maximum yield estimate of this model (10 kg) may be too low. Furthermore, the model indicates that the magnitude of LNG RPTs should be independent of spill rate, which is in direct conflict with the experimental results (Fig. 4).

Fowles. The two remaining RPT theories to be examined are considerably more complex than those of Anderson or Fauske but involve only the propagation mechanism of the explosion; i.e., they assume the presence of a shock wave to initiate the process. The RPT model of Fowles [21,24] is by far the most sophisticated of those in existance to date and has been recently addressed specifically to the cryogen/water system [25]. The model is fundamentally the same as the one currently used for chemically driven detonations. For the case of RPT detonations it is applied to a superheated fluid in which the energy stored in the metastable thermodynamic state supports the detonation wave. However, unlike the standard Chapman-Jouguet (CJ) theory, Fowle's model allows for both strong and weak detonations depending on the choice of phase-change reaction rate and fluid viscosity. This flexibility may be a big plus since very little is understood to date about the thermophysics of the superheated fluids.

The one-dimensional model assumes that the initiating pressure pulse produces a certain degree of phase change at the appropriate two-phase equilibrium condition. The two-phase fluid is then allowed to expand isotropically from this compressed state back to the original ambient pressure. Therefore, by assuming a degree of phase transformation (ratio of gas and liquid), and specifying the initial superheat state of a fluid, the model is able to predict the detonation velocity, peak detonation pressure, and energy used to sustain the detonation wave.

TABLE 5. Calculated detonation wave properties for methane and propane

	Initial temperature (K)	Degree of superheat (K)	Detonation energy (J/g)	Detonation velocity (m/sec)	Detonation pressure (atm)
CH_4	172 ($\sim T_{SL}$)	60	82.9	460	32.2
C_3H_8	253	22	22.6	1053	2.1
	263	32	33.8	987	3.1
	290	59	65.2	796	7.5
	300	69	77.3	725	9.9
	326 ($\sim T_{SL}$)	95	112.3	526	19.2

The weak points of the Fowles model are that it is currently only one-dimensional, it requires that the phase-change reaction rate be assumed, and it requires bulk superheating of large volumes of the cryogen. The accuracy of the calculation also depends heavily on the accuracy of the equation of state used. Unfortunately, while many of the existing equations of state are extremely accurate in the pure liquid and pure gas phases, their accuracy within the coexistence dome (binodal curve on a phase diagram) is undetermined as yet.

Recent calculations by Hixson [25] invoke the CJ assumptions that the phase change reaction goes to completion and that the detonation wave speed is a minimum. The calculations for methane used a modified (12-term) Benedict-Webb-Rubin equation of state and were done for one initial degree of superheat only. Calculations were done for propane at several initial conditions using the nine-term Benedict-Webb-Rubin equation of state. The initial pressure was one atmosphere in all cases. The results of these calculations are shown in Table 5. At the minimum degree of superheat (T_{SL}), the detonation energy is about 40% of the available superheat energy for methane, and about 47% of that for propane. This model (if validated) is capable of making estimates of the maximum RPT possible for a particular spill scenerio.

Hixson also attempted some experiments to validate the RPT detonation model of Fowles, but was unable to obtain significant degrees of superheat with his apparatus. Although the experimental results were marginal, there was an indication of acceleration of the initial shock front for the metastable propane tests. There was also an indication that the degree of phase transformation was not complete behind the accelerating shock wave. This conclusion (if correct) underlines the importance of knowing the phase change reaction rates for these cryogens as functions of both temperature and pressure.

There are several interesting aspects of the Fowles RPT model in regard to the China Lake tests. Because of the close analogy of this model with the chemically sustained detonation theory, there is an inherent critical size or "failure diameter" criterion. In order for material of a certain size to explode, the chemical reation rate must first be fast enough to drive the detonation wave through the material before the effects of side rarefactions become significant. If the detonation velocity is low, then the volume of the material must be large, and in fact, will have some minimum or threshold size. This may be a possible explanation of the abrupt increase in the measured RPT yield with spill rate for the China Lake tests (see Fig. 4). This effect may also place a lower limit on the size of future experiments designed to investigate this phenomenon.

Another interesting result has to do with the magnitude of the detonation energies calculated by the Fowles model for methane and propane (Table 5). If one assumes that the efficiency of the ethane detonation is similar to those of methane (40%) and propane (47%), then the detonation energy for the maximum degree of superheat for ethane would be about 87 J/g. Summation of the total air-shock RPT yield energy for all of the B9 spill explosions (Table 2) gives a value of about 6 kg TNT or 25 MJ of energy. As mentioned previously, 25 MJ is a minimum value; and the total yield could have been much larger. For the detonation energies as calculated by Hixson, the individual volumes of methane, ethane, and propane required to produce the total B9 air-shock energies are 0.72 m^3, 0.53 m^3, and 0.4 m^3 respectively. This assumes that these volumes achieve maximum superheat. If this is not the case, more material would be required. The total amounts of these three constituents spilled during the B9 test were 20.1 m^3, 3.4 m^3, and 0.7 m^3 respectively. Clearly there was more than enough methane to have produced 25 MJ of explosive energy, whereas a large portion of the entire spilled volumes of ethane and propane would be required. This implies that it is highly unlikely that the B9 RPTs were produced by the

ethane and propane alone, especially since 25 MJ is a lower limit to the actual
energy release of the explosions.

Harlow and Ruppél. The final RPT model to be discussed is relatively new,
and was developed by Harlow and Ruppél [22]. This model deals only with the
propagation mechanism of the detonation wave, specifically with propagation
along the interface between two fluids in an initially stable film boiling
mode. The equations of motion are solved using a particle-in-cell method
adapted for multiphase flow conditions. The model handles both the transient
and steady-state regimes of the gas/liquid dynamics, provided some appropriate
initial conditions are supplied. The code has been shown to produce reasonable
qualitative results [22] when "physically reasonable values" of the momentum
coupling between the liquid and the vapor are assumed. There is also mention
of an "unresolved" interface stability which is responsible for the mixing of
the liquid and vapor after the film is collapsed. Whereas the results of the
calculations look realistic, the properties of the materials are not. For instance, the velocity and pressure distributions displayed indicate the formation of a strong pressure wave (340-3000 atm) propagating at Mach 1.06 in the
liquid, which is a physical impossibility for any of the cryogens responsible
for RPT explosions. Further analysis of this model will require that it use
realistic cryogenic liquid and vapor properties and dimensions. More details
and justifications for the liquid/vapor mixing process are also needed.

This concludes the section on the analysis of the current cryogen/water
RPT models and how they relate to the China Lake experiments. None of the
models examined are completely rigorous in their description of the RPT explosion process. Some require assumptions about the magnitudes of parameters
critical to their success. Some appear to be in conflict with the China Lake
RPT results. Others require thermodynamic data which are not currently available. The fact that there is lack of understanding of the RPT process is
further emphasized by the number of different models that have been offered to
explain this phenomenon. More fundamental and well-documented experiments
would help to narrow the field, and provide direction for the improvement and
validation of RPT models.

6. CONCLUSIONS AND UNANSWERED QUESTIONS

A detailed analysis of the data obtained during 26 large-scale spills of LNG
on water has been presented. These experiments involved large volumes (30 m^3)
and high spill rates (19 m^3/min), and are the most recent contributions to
almost 20 years of effort toward the understanding of cryogen/water RPT explosions. The following conclusions may be drawn from the China Lake experiments:

1) Large-scale RPT explosions appear to fall into two different regimes:

   • Early RPTs—close to the spill point, primarily underwater.

   • Delayed RPTs—near edge of LNG pool, at surface.

2) The occurrence of RPTs appears to be inhibited by colder water temperatures.

3) The probability of early RPTs appears to be greater as the depth of
   penetration of the LNG into the water increases.

4) Large-scale spills produce early RPTs at LNG compositions well outside the
   Enger-Hartman explosive envelope as determined for small-scale spills.

5) The magnitude of early RPTs increases dramatically at higher spill rates.

6) Accurate determination of RPT yields require both surface and underwater pressure impulse measurements.

An analysis of current cryogen/water RPT models leads to the general conclusion that no single model accurately predicts all of the observed phenomena. It is clear that the RPT explosion is not fully understood at present.

Many questions still remain concerning the mechanisms responsible for both the initiation and propagation of these explosions. The same is true when it comes to understanding the energetics of RPT explosions. Unfortunately, the China Lake results may have actually added to the number of unanswered questions rather than reducing them. The following questions must be answered in order to understand and mitigate the hazards associated with these RPT explosions:

1) What do the heat transfer curves for liquid/liquid boiling look like? To date practically all the theories about LNG/water RPTs assume that the liquid/liquid heat transfer regimes of conduction, convection, nucleate boiling, and film boiling are the same as those derived from solid-to-boiling-liquid heat transfer data. Some fundamental studies on liquid/liquid heat transfer need to be conducted if the LNG/water RPT phenomenon is to be understood.

2) Is existing knowledge of the thermodynamic properties within the binodal curve adequate? While current equations of state may be adequate in the liquid and gaseous regions, their accuracy within the coexistence dome has not been demonstrated. Accurate measurements of liquid-to-gas phase transformation rates are required both for input to, and evaluation of, the current RPT models.

3) Are existing models adequate to explain large-scale RPT explosions? Some of the China Lake results appear to violate existing LNG/water RPT models. Existing models need to be reexamined in more detail in light of these new results. It is possible that new and more complex models need to be developed, or need to evolve from the current models.

4) How do RPT occurrence and yield depend on water temperature, water depth, and LNG composition? The China Lake results indicate that these parameters are important in determining the magnitude and probability of LNG RPTs. Perhaps some of these variations can be resolved by laboratory-scale experiments somewhat larger than previous ones. However, due to the apparently strong size dependence of these RPTs, a series of systematic, well-instrumented, China Lake-scale experiments will most likely be required.

5) How large can RPTs get? LNG spills on water at rates up to 100 m^3/min are of concern to the LNG industry. Unfortunately, the China Lake results are not adequate enough to allow accurate scaling of RPT yields to these higher spill rates. Model validation against current RPT data and additional experiments at higher spill rates will be necessary to establish these scaling relationships.

ACKNOWLEDGEMENTS

The analysis presented in this paper was made possible by the efforts of many others who are not listed as authors. While we cannot mention them all, we would like to extend our special thanks to Doug Lind and his team at NWC for the operation of the facility, the the LLNL scientists, engineers, and technicians for many months of dedicated effort in obtaining the data, and to Bill Hogan, our LLNL Liquefied Gaseous Fuels Program Leader.

Work performed under the auspices of the U.S. Department of Energy by the Lawrence Livermore National Laboratory under contract number W-7405-ENG-48.

REFERENCES

1. T. Enger and D.E. Hartman, "LNG Spillage on Water," Shell Pipeline Corp., Feb. 1972, Technical Progress Report 1-72.

2. R.P. Koopman, R.T. Cederwall, D.L. Ermak, H.C. Goldwire, Jr., W.J. Hogan, J.W. McClure, T.G. McRae, D.L. Morgan, H.C. Rodean, and J.H. Shinn, "Analysis of Burro Series of 40-m^3 LNG Spill Experiments," J. Hazardous Materials, 6, (1982) 43-83.

3. T.G. McRae, "Preliminary Analysis of RPT Explosions Observed in the LLNL/ NWC LNG Spill Tests," LLNL, UCRL-87564, 1982, presented at the Gas Research Institute LNG Safety Research Workshop, Massachusetts Institute of Technology (MIT), Cambridge, Mass., March 22, 1982.

4. C.D. Lind, Naval Weapons Center, China Lake, Ca., personal communication.

5. D.L. Katz, and D.M. Sliepcevich, "LNG/Water Explosions: Cause and Effect," Hydrogen Process., 50 (1971) 240-244.

6. W.M. Porteous, and R.C. Reid, "Light Hydrocarbon Vapor Explosions," Chem. Eng. Progr., 72, (1976) 83-89.

7. W.M. Porteous, "Superheating and Cryogenic Vapor Explosions," Ph.D. thesis, MIT, Cambridge, Mass., 1975.

8. B. Jazayeri, "Impact Cryogenic Vapor Explosions," M.S. thesis, MIT, Cambridge, Mass., 1975.

9. J.A. Valencia-Chàvez, "The Effect of Composition on the Boiling Rates of Liquefied Natural Gas for Confined Spills on Water," Int. J. Heat Mass Transfer, 22, (1979) 831-838.

10. W.E. Baker, "Explosions in Air," Univ. of Texas Press, Austin Tx., 1973.

11. R.C. Reid, "Superheated Liquids: A Laboratory Curiosity and, Possibly, an Industrial Curse," Chem. Eng. Educ., XII (1978) (1977 Award Lecture, three-part series), (2) 60, (3) 108, (4) 194

12. L.S. Nelson, "Steam Explosions in the Molten Iron Oxide/Liquid Water System," High Temp. Sci., 13, (1980) 235-256.

13. M.L. Corradini, D.E. Mitchell, and L.S. Nelson, "Recent Experiments and Analysis Regarding Steam Explosions with Simulant Molten Reactor Fuels," presented at Winter Annual ASME Meeting, Washington, D.C., Nov. 15-20, 1981.

14. R.P. Anderson, and D.R. Armstrong, "R22 Vapor Explosions," presented at Winter Annual ASME Meeting on Nuclear Reactor Safety Heat Transfer, Atlanta, Ga, Nov. 27, 1977.

15. H.K. Fauske, "Scale Considerations and Vapor Explosions (Rapid Phase Transitions)," presented at the Gas Research Institute LNG Safety Research Workshop, MIT, Cambridge, Mass., March 22-24, 1982.

16. E. Nakanishi, and R.C. Reid, "Liquid Natural Gas-Water Reactions," Chem. Eng. Progr., 67, (1971) 36.

17. M. Blander, D. Hengstenberg, and J.L. Katz, "Bubble Nucleation in n-Pentane, n-Hexane, n-Pentane + Hexadecane Mixtures, and Water," J. Phys. Chem., 75, (1971) 3613-3619.

18. W.M. Porteous, and M. Blander, "Limits of Superheat and Explosive Boiling of Light Hydrocarbons and Hydrocarbon Mixtures," AIChE J., 21, (1975) 560-566.

19. W.E. Mason, "TACO--A Finite Element Heat Transfer Code," LLNL, UCID-17980, Feb. 1980.

20. S.J.D. Van Stralen, C.J.J. Joosen, and W.M. Sluyter, "Film Boiling of Water and an Aqueous Binary Mixture," Int. J. Heat Mass Transfer, 15, (1972) 2427-2445.

21. G.R. Fowles, "Vapor Phase Explosions: Elementary Detonations?," Science, 204, (1979) 168-169.

22. F.H. Harlow and H.M. Ruppèl, "Propagation of a Liquid-Liquid Explosion," Los Alamos National Laboratory, Los Alamos N. Mex, LA-8971-MS, Aug. 1981.

23. F. Briscoe, and G.J. Vaughan, "LNG/Water Vapour Explosions - Estimates of Pressures and Yields," presented at Gastech 78 LNG/LPG Exhibition and Conference, Monte Carlo, Monaco, Nov. 7-10, 1978.

24. R.L. Rabie and G.R. Fowles, "The Polymorphic Detonation," Phys. Fluids, 22, (1979) 422-435.

25. R.S. Hixson, "Vapor Phase Detonations in Light Hydrocarbons," Ph.D. thesis, Washington State University, Pullman, Wash., 1980.

AUTHOR INDEX